The scientific enterprise, today and tomorrow

The scientific enterprise, today and tomorrow

Adriano Buzzati-Traverso

Published in 1977 by the
United Nations Educational, Scientific
and Cultural Organization
7 Place de Fontenoy, 75700 Paris

Published in Italian under the title
La sfida della scienza, by
Edizioni Scientifiche e Tecniche Mondadori, Milan

Published in Spanish under the title
La empresa científica hoy y mañana, by
Editorial Santillana, Madrid

Composed by Imprimerie des Presses
Universitaires de France, Vendôme, France
Layout and page make-up by
CRYRE, Creaciones y Realizaciones, SL, Madrid, Spain
Illustrations by Mondadori, Milan, Italy
Printed by Imprimerie Oberthur, Rennes, France

ISBN 92-3-101268-1

Preface

The book to which the following paragraphs are but a brief preamble is a study different from its congeners, and different in several respects. It was born as an idea expressed during the sixteenth session of the General Conference of the United Nations Educational, Scientific and Cultural Organization in 1968. The conception of the form and tone of the study were subsequently reinforced as the result of two international meetings of scientists; they sought to define the scope and depth of the material to be included in an undertaking of this magnitude. 'Books', propounded Francis Bacon, 'must follow sciences, and not sciences books.'[1] One of the respected scientists taking part in those deliberations, almost ten years ago, was the French physicist Pierre Auger; he had been responsible for an earlier undertaking on the part of Unesco to review the state of the scientific arts, a publication we called *Current Trends in Scientific Research* (1961).

For readers unfamiliar with Professor Auger's work, it was a reflection of the cumulative scientific and technical temper of the times: a stocktaking of the scientific and engineering disciplines in a world for the most part recovering psychologically as well as economically from a brutal and devastating war. We realize now that it was this war which brought applied, experimental and even theoretical science to the innermost cores of our social institutions—whether household, school, church, museum, laboratory, office, factory or government agency—and changed perhaps forever the roles of science and technology in their interactions with the structures and values of our culture as a whole.

The Second World War had another major impact, too, besides its obvious military and political results: it triggered the release of many peoples from subservience to foreign masters, and new nation-States appeared in many parts of the world. These States had little scientific capability and needed it desperately if they were to advance their capacity to enjoy the fruits of scientific progress. There was required, then, not only the rebuilding of scientific skills in the industrialized world but the building of them in the newer and poorer nations.

The present volume could not be, therefore, merely another edition or a catalogue of the recent evolution and perfection of man's scientific and technical knowledge. It could not because, put quite simply, the world changed too much during the 1960s—the decade inaugurated by the appearance of *Current Trends in Scientific Research*. Man and his culture found themselves suddenly confronted with a host

1. *Proposition touching Amendment of Laws.*

of elusive problems which seemed to many to be directly or indirectly traceable to the emergence of science as a major force in human society, a process which had been gathering momentum for three centuries. Even more frightening was its progeny, technology, now running rampant after less than two centuries since the beginning of the industrial revolution. Everywhere people's lives were being affected increasingly by the technological revolution.

The danger inherent in succumbing to the lure of increased technology without due reflection on what we expect from it to improve our quality of life led Unesco to attempt to find a new way to examine, analyse and try to put forth rational solutions to some of the problems plaguing mankind today. Professor Adriano A. Buzzati-Traverso, the main author of this work and a former Assistant Director-General for Sciences at Unesco, has thus taken a hard look at the development of science and technology in the present day, how they are organized and used and what are their effects on the cultures they are so rapidly modifying. The impact of science and technology is such that the gradual evolution of cultures has been transformed, indeed, into a revolution. Professor Buzzati-Traverso has eschewed, except for brief references, examination of the history of science and techniques; libraries already abound with such volumes and many of them are referred to in the course of his analysis.

This brings us to another characteristic which makes the present book unusual in comparison with the publications commonly associated with Unesco, and its programmes approved by each biennial General Conference. Stated tersely, Professor Buzzati-Traverso's views do not always coincide with those of Unesco (nor with mine, for that matter). Yet it will be clear to the reader why some of the reasoning and conclusions have been proffered by the principal author. He asks, and attempts to answer, many provocative questions: What have been the true benefits of science to man and his institutions? Is it possible for 'developing' lands to absorb the shock of technology transfer on their ancient, tradition-dominated cultures without destroying these? Do we really need science and technology?

Dare we even continue to engage in laboratory research and experiments?

In order to arrive at coherent and useful replies to such questions, Professor Buzzati-Traverso leads us through the remarkably complex but surprisingly productive organization of science, round the world. It was my own countryman, Voltaire, who counselled a young scientist:

If you want to apply yourself seriously to the study of nature, allow me to tell you that you must begin by not constructing a system.[1]

Scientists may not have constructed a system of the organization of their disciplines since Voltaire's words were written more than two centuries ago, but Professor Buzzati-Traverso's guided tour through the meadows, forests and lakes of scientific activity throughout the world will leave the reader with little doubt that a humanly organized pattern is quite evident. How the pattern is applied to the betterment or the detriment of life is one of the more vivid impressions this book will leave on the reader's memory.

So the present book is an examination of man's condition during the period many have already called the Golden Age of Science. It is up to each one of us, as individuals, to decide if the right questions have been asked and if the pages which follow contain feasible solutions to some of the many problems with which all of us are preoccupied. What is clear, however, is that science and technology are no longer separate entities; they form an integral part of our cultures and societies, and their needs make ever increasing demands on our educational systems. This volume, considered as a companion to those other volumes published by Unesco that review the broader problems of culture and education, notably *Main Trends of Research in Social and Human Sciences* and *Learning to Be*, clearly shows the intimate linkage between education, science and culture.

René MAHEU[2]

1. Besterman's definitive edition of Moland, paper D2463, of 15 April 1741.
2. This preface was written shortly before its author's untimely death in December 1975.

Acknowledgements

I should like to acknowledge two types of debts. The first is to those who have helped me directly in the preparation of this book and the second is to those whose help has been indirect.

My primary gratitude goes to Robin Clarke, who, over a period of two years (while I was Assistant Director-General for Sciences at Unesco) helped me immensely in discussing the layout of the volume, in collecting pertinent literature, in analysing it, and in editing the text of Part II. Second, I wish to thank the editorial team of the *Enciclopedia della Scienza e della Tecnica Mondadori* of Milan, who have prepared the text of Part II. Third, I want to express my appreciation to David Dickson who has helped me in preparing the preliminary draft of Part III.

My principal indirect debt is to the following eminent scientists who generously gave of their time to review the first draft of Parts I, III and IV, and whose invaluable suggestions I have taken into serious account in the preparation of the final text: Professor M. P. Kassas, Professor Alex Keynan, Sir Peter Medawar, Professor Anton Mjolk, Dr Ashok Parthasarathi, Dr Aurelio Peccei, Professor Abdus Salam, Dr Jean-Jacques Salomon, Professor D. de Solla Price, Academician Vladimir Toutchkevitch and Dr Carlos M. Varsavsky who have been asked by Unesco's Secretariat to review the first draft of Parts I and III. Their advice has been exceedingly useful in the preparation of the final text.

Also, I wish to thank F. W. G. (Mike) Baker, who in his capacity of Secretary of the International Council of Scientific Unions has submitted the text of Part II to specialists in various fields for comment and criticism. Many others have helped me in many ways and particularly in suggesting additions and corrections to the list of major scientific and technological breakthroughs (Part I, Chapter 3). To these, too numerous to be mentioned by name, I wish to express my thanks.

I am also indebted to a number of publishers and individuals for permission to quote from their work. These are listed as appropriate.

Lastly, but above all, I wish to acknowledge my gratitude to René Maheu, former Director-General of Unesco, who did me honour when he offered me the opportunity to be the author of this volume, and to many other colleagues within Unesco's Secretariat who, with unfailing friendliness, have offered their knowledge and experience to make my effort easier, while I was working with them and afterward.

Just a few more words to relieve the friends mentioned from all responsibility for any erroneous statement in this book's contents or for sharing in the views expressed.

A. A. B.-T.

Contents

Introduction

The subject-matter of this book—the scientific enterprise as it looks at the dawn of the last quarter of the twentieth century—is presented in four parts, corresponding to the four themes defined in the course of Unesco's preparatory work: 'Science in the face of human needs', 'Science in the making and its frontier', 'Organization of scientific research' and 'Science and society'. Minor changes have been made to the final titles used in the four parts of the present volume. The general structure of this work had been set, therefore, by the time I was asked to take the responsibility of preparing and writing the volume, which I accepted.

In Part I, I have tried to present the most significant traits of the scientific enterprise as seen from three main points of view: man's urge to satisfy his curiosity about the world, and the ensuing thrill of discovery; the triumphs of pure and applied science during this century and particularly in the last two decades, and the resulting progress in medicine, agriculture and engineering: the involvement of the scientist in the development of weapons, the resulting museum of horrors of the atomic, chemical, biological and computerized arsenals, and the related problem of the responsibilities of scientists. Then follows a preliminary discussion of the modern predicament: technological progress has enriched immensely human life but at the same time has created unexpected and complex problems for modern societies.

Part II is an extensive survey of the present status of scientific research in a large number of disciplines; these have been grouped into three main sections: Universe, Matter, Life. An attempt has been made to summarize the major advances since 1960, so that this part could be considered as an updating of Professor Pierre Auger's 1961 volume, *Current Trends in Scientific Research*. I wish, however, to draw attention to one major difference between that volume and this: in the former, the survey of advances in technological (including medical) and industrial fields was more extensive than the one on fundamental research (106 pages against 70 pages); in the latter, the presentation is limited to the fundamental sciences, which are discussed in greater detail. The difference is justified by the fact that the present volume is the outcome of a resolution of the Unesco General Conference (1968), and because Unesco's mission in science is aimed primarily at basic research (Auger's *Current Trends in Scientific Research* was the outcome, rather, of a resolution of the General Assembly of the United Nations, 1260/XIII/1958). Although the responsibilities of both the United Nations and Unesco extend to fundamental and applied research, there have been no official consultations with the United Nations and its Specialized Agencies in the preparation of the present volume. Problems related to the development of technologies and their impact on society, however, are extensively discussed in Parts I, III and IV of this volume. Another, minor, difference with respect to the work by Auger should be mentioned: no attempt has been made to cover advances in mathematics. This is because of my own conviction that it is practically impossible to survey recent advances in mathematics in terms understandable to the non-specialized reader, to whom this volume is addressed.

While the general outline of Part II has been my responsibility, the actual drafting of the text was done by a group of specialized science writers and reviewed by a group of referees whom the International Council of Scientific Unions was kind enough to propose (see 'Acknowledgements').

Part III aims at illustrating how the scientific enterprise is organized at the national as well as at the international level, I have attempted to proffer answers to questions like the following: What is the magnitude of the scientific effort, measured in terms of manpower and of invested sums, in the world, within regions and in individual countries? What is the function of government, universities, private industry, scientific academies and associations, and of individual research workers in the present stage of scientific, technological and industrial development? What are the problems that politicians and decision-makers face when dealing with major scientific and technological projects? How significant is the action of international organizations in fostering scientific progress, particularly in the developing countries? What are the likely future developments within the national and the world-wide research systems?

After reviewing what science and technology have accomplished especially during the last few decades, where the frontiers of scientific advance lie today and how the scientific enterprise is organized, Part IV attempts to summarize the great number of interrelated problems of modern societies that have been caused, directly or indirectly, by scientific and technological progress: Should science be considered today as the main hope or the greatest threat for the further advance of mankind? What is the position of the scientist in modern society? How are science and technology viewed by the sociologist and the philosopher? Can science still be considered an objective pursuit of truth? How justified are the criticisms raised by the movements of counter-culture? What are the alternative solutions available to man for ensuring his own survival? These kinds of problems are analysed critically, and I have taken personal stands on some of the major issues. Part IV ends with a proposal to attempt to attribute a new meaning to the word *progress*, and with an appeal to the scientific community to get under way a 'new Enlightenment' that should re-evaluate the purpose of human action in the light of today's knowledge.

Part I

Science and the needs of man

*Intueri Naturam. Quo
munere? Curiosum esse.*

Lucretius

1   The prime mover

Science is a game: it can be exhilarating, it can be useful, it can be frightfully dangerous. It is a play prompted by man's irrepressible curiosity to discover the universe and himself, and to increase his awareness of the world in which he lives and operates. Throughout history, men have tried to increase their knowledge and understanding of nature and have often questioned the validity of their cultural heritage handed down by previous generations. Their deep-seated propulsion towards inquiry and discovery was and is aimed at scrutinizing the various aspects of this heritage, explaining its foundations, discovering its consequences; in other words, man has enjoyed playing a game which brought about the reward of an increased knowledge and critical awareness. This critical approach can thus be equated with philosophical and scientific thinking. At all times, in fact, from the remote past to today, the philosopher (and the scientist) has provided a dissenting voice, capable of formulating objections, of casting doubts and of expounding subtle arguments aimed at undermining widespread erroneous beliefs and accepted ideologies; looking with hindsight at past events, it often appears that the scientist is a genuine expression of the time he lives in, but yet his stand is detached and forward-looking. For such characteristics the philosopher and the scientist have often been looked at askance by the possessors of power and by the defenders of the status quo; sometimes they have been ignored and sometimes, both yesterday and today, persecuted. But in spite of this incompatibility between critical curiosity and authority, the philosopher and the scientist have succeeded in playing a determining role in the evolution of civilization.

And it is this role, as we shall see, which today is being questioned in some quarters. The very pride of science, its objectivity, is criticized and accused of being the primary source of the alienation of man.

The science under discussion—that of the 1970s—is historically directly connected with two major intellectual events, similar but not identical, which occurred about twenty-five centuries ago and 500 years ago: the flowering of the school of Ionic philosophers, followed by a period of diffusion of the new ideas, that had a remarkable social impact, due to the Stoics; and the rise of modern science in Western Europe, followed by the Enlightenment.

Critical awareness

The flourishing of an extensive critical awareness at the beginning of the sixth century B.C. in Ionia, the coastal cities and islands of Asia Minor, marks the birth of Western science and philosophy. It is true that a good share of the quantitative mathematical spirit that permeates modern science and gave it its greatest successes derives from Babylonian astronomy and mathematics. It is also certain that the splendid civilizations of China and India had made many fundamental scientific discoveries and inventions which were later transmitted to Europe. But in the outlook of the Ionians there was something new which determined their admirable creativity; they did not create their ideas from nothing, but succeeded in building a new system of analytical tools and concepts on the fertile soil of the cultural heritage, particularly its technological and scientific content, bestowed on them by earlier Mediterranean civilizations. Indeed, we are more closely linked to the thoughts and deeds of the Greeks than to those of any other civilization.

More so than at any other time, no clear line can be drawn between the philosophical and the scientific thinking of the Greeks in the sixth and fifth centuries B.C. Scientific and philosophical thinking are certainly not mutually antithetic, but are rather the two sides of only one rationality, which through advances and setbacks, has characterized the history of man. The variety of *Weltanschauungen* brought forth by philosophers and scientists has never been entirely satisfactory, never final. But this very lack of absolute precision warrants the rationality of the system, for criticism prompted by curiosity refuses to accept absolute truths or dogmatic doctrines, today as well as yesterday. And it may well be that the equal pace of advance in the two parallel domains, philosophy and science, ensured a harmonious development of that civilization at the prime of Greek thought. The most general trends of ideas—humanitarianism, critiques of the gods, interest in politics as a new science—developed by the Sophists between 540 and 400 B.C., are directly derived from Protagoras' principle, which considers man as the 'measure of all things'. And such a relativistic approach led to the enunciation of the need for mutual respect and tolerance, as expressed by Pericles:

As we live free in public life, similarly, in the daily mutual watching deriving from everyday actions, we do not feel hurt if somebody behaves as he pleases, nor do we inflict on him through our anger the slightest nuisance, for, if this is not a true punishment it is still something unpleasant.[1]

The heroic age of Greek science was followed by a brilliant succession of great thinkers, as exemplified by Plato, Aristotle, the Hellenistic School of Alexandria and the later Roman-Hellenistic period. But then a decline set in, interrupted briefly by the bright but short-lived meteor of Arabic science, particularly significant for mathematics and astronomy, chemistry and pharmacy and, to some extent, medicine. With this exception together with that of the relatively lively mediaeval mechanics and astronomy, the voice of science was very subdued, while Christian thinkers debated the philosophical implications of revealed truth.

Astronomy, experimental science and the Enlightenment

With the downfall of scholasticism, the appearance of Occam's empiricism, but primarily thanks to the Renaissance and the birth of modern astronomy and experimental science, the modern era appeared and man regained faith in himself.

The scientific movement that emerged in the seventeenth century brought a promise of progress and of the realization of man's deepest hopes. It proposed not only to liberate men from ignorance but also to free them from superstition, religious hatred, irksome toil and war. For science was then not the pursuit of technicians; it was 'the new philosophy' as Francis Bacon called it, the 'active science', the first genuine alternative that had been contrived to dogma, myth and taboo. The world of medieval gloom that denigrated the powers of man gave way to an optimistic faith in his capacities.[2]

The new season of human thinking in the seventeenth century led to an impressive array of basic discoveries in astronomy (Copernicus, Galileo, Kepler, Tycho Brahe); in physics and mathematics (Galileo, Newton, Descartes, Pascal, Gilbert, Torricelli, Huygens); in bi-

1. Thucydides, *History of the Peloponnesian War*.
2. L. S. Feuer, *The Scientific Intellectual*, New York, Basic Books, 1963.

ology (Vesalius, Malpighi, Harvey . . .); and to the appearance of the philosophers of the scientific revolution who became the spokesmen for the new hedonistic moral philosophy. The utilitarian aim of science was expounded particularly by Francis Bacon: 'Now the true and lawful goal of the sciences is none other than this: that human life be endowed with new discoveries and powers.' The powers of curiosity, of scientific research for improving the material condition of man were further described and glorified by the Enlightenment and *les philosophes* of the *Encyclopédie*. With prophetic spirit d'Alembert, the famous mathematician, in the 'Discours préliminaire' of the *Encyclopédie* wrote: 'Because when once the barriers are broken, the human spirit often goes faster than it intends.'

The ideal of indefinite material progress and of the perfectibility of man, which was to become engrained in today's industrialized societies, was born. There was a good reason for such drastic change in the history and the hopes of men—so the *philosophes* believed: the scientists of the previous century hit upon a method of discovery, a method which would guarantee future progress. Bacon, Galileo, Leibnitz, Newton insist on the significance of the new method of inquiry. The original title proposed by Descartes for his *Discours de la Méthode* was: 'The Prospect of Universal Science which can Elevate our Nature to its Highest Perfection'. The new 'scientific method' did not consist of a set of rules to be followed in order to discover truth, but was rather the composite convergence of measurement, of mathematical analyses and hypotheses, the adoption of experiments to ascertain the validity of assertions, the geometrization of space and the acceptance of a mechanical interpretation of reality. As Victor Weisskopf recently put it:[1]

Science developed only when men refrained from asking general questions such as: What is matter made of? How was the universe created? What is the essence of life? Instead they asked limited questions such as: How does an object fall? How does water flow in a tube? Thus, in place of asking general questions and receiving limited answers, they asked limited questions and found general answers. It remains a great miracle, that this pro-

cess succeeded, and that the answerable questions became gradually more and more universal.

As Einstein said: 'The most incomprehensible fact is that nature is comprehensible.'

The new approach, together with progress in instrumentation and diffusion of printing processes led to scientific revolution. For quite some time technological and industrial developments seemed not to be directly affected by scientific advance. Looking back, we may say that this was not surprising for major technological breakthroughs had occurred in the past, in China, in India, Mesopotamia, Greece and Rome without the impulsion of science, as we conceive of it today. And, indeed, the early stages of the industrial revolution were hardly influenced by discoveries in physics and chemistry. But later, particularly during the latter part of the nineteenth century and more so during recent decades, the institutionalization of science determined its self-propelling character. This we shall discuss in later chapters.

Science as the prime mover of modern societies

Prompted by curiosity, science became the prime mover of the modern world. Over a period of about two centuries, slowly at the beginning, but gradually faster and faster, scientific knowledge and technologies exploded. In less than 0.002 per cent of human history—300 years out of 2 million—we progressed from believing that the Sun went round the Earth to a detailed knowledge of astronomical objects so far away that the light they emit takes thousands of millions of years to reach us. Science has revealed the organization of the planetary system, of the galaxy, of galactic clusters and finally of the Universe itself (Part II, Chapter 12). In doing so its knowledge has been accumulated not just from astronomical instruments but from investigations of the nature of the smallest elementary particles. The laws of nature thus revealed are held to be valid throughout the universe

1. V. F. Weisskopf, 'The Significance of Science', *Science*, No. 176, 1972, p. 138–46.

and—even more important—both for living and inanimate matter. The nature of life interpreted by modern biology would have been nearly incomprehensible to someone living even fifty years ago, so speedy has been the advance in our knowledge (Part II, Chapter 25). Other examples given in Chapter 3 offer an idea of the pace of progress in the last two hundred years.

The applications of science, as we shall see (Table 1), turned out to be an exceedingly effective method for raising the level of techniques and ensuring an abundance for most of the needs of human societies. Now that more and more of us are enjoying such abundance, we begin to wonder whether such riches might not be the source of many problems that the human race is facing today. In the advanced countries signs of disruption brought about by technology are clear. In the less-advanced countries the hope still prevails that science and technology are the major sources to be tapped to obtain the good life.

Is science the 'measure of man'?

The very success of science and technology has given rise to the idea that all problems can be 'solved' if treated by objective analysis. An apparent consequence of this attitude is that if problems cannot be tackled scientifically, they can either be ignored or defined out of existence. Vital problems such as the nature and the needs of man, expressed in terms of factors which determine his happiness, remain unstudied. According to Sir Peter Medawar,[1] this stems not from a question of choice of research field but from the very nature of science:

No scientist is admired for failing in the attempt to solve problems that lie beyond his competence. The most we can hope for is the kindly contempt earned by the Utopian politicians. If politics is the art of the possible, research is surely the art of the soluble.

If this is true, it may have grave implications and prompts the harsh criticism of such representatives of the 'counter-culture' as Theodore Roszak,[2] who argues that our scientific world suffers from the delusion that:

If a problem does not have a technical solution, it must not be a *real* problem. It is but an illusion . . . a figment born of some regressive cultural tendency.

According to one school of thought the scientific-technical approach to life has never served man's best interests but has enslaved him in a system in which he is now powerless to act. Jacques Ellul has put the point the most forcibly:[3]

The human being is no longer in any sense the agent of choice. Let no-one say that man is the agent of technical progress . . . and that it is he who chooses among many possibilities. In reality he neither is nor does anything of the sort. . . . He can decide only in favour of the technique that determines what gives the maximum efficiency. But this is not choice. A machine could effect the same operation.

As we shall discuss later, many problems of the world of the seventies seem to ensue from the belief that science is the best guide, the infallible one, of man; or, in other words, that man could optimize his own realization as a rational being through the rational use of knowledge. Perhaps the increasing tempo of scientific discoveries, and especially of technological developments arrived at haphazardly, has changed the nature of scientific endeavour; perhaps Protagoras' principle, according to which man is the measure of all things, is no longer true. The time has come to assess the limits of what science can do as a guide to human action. What are the limits of our ability to assess the consequences of our decisions? To within what limits is science a liberating force for man? What sources, other than the genuine scientific one, uphold the humanity of man, and up to what point is scientific knowledge compatible with those other sources? Such are the questions about science which science is now called upon to answer.

1. P. B. Medawar, *The Art of the Soluble*, London, Methuen, 1967.
2. T. Roszak, *The Making of a Counter Culture*, Garden City, New York, Doubleday & Co., 1969.
3. J. Ellul, *The Technological Society*, New York, Knopf, 1964.

2 Exhilarating science

The atmosphere prevailing in a laboratory actively engaged in research is one of excitement. To the scientist, working is the most fascinating game: his free imagination can set the rules and change them, so that his unlimited curiosity may find pause and satisfaction. The scientist is a lucky player, for the toy he is playing with is constantly changing and has no boundaries: the toy is the Universe. The writings of creative scientists of all times bear witness to such joyful spirit of play. More than three centuries ago Kepler wrote: 'As God the Creator played, so he also taught nature, as His image, to play the very game which He played before her.' When Einstein was trying to describe the functioning of his mind, he said it was a sort of 'combinatory play', an 'associative play' with images. And a psychologist of science, Abraham H. Maslow, recently wrote:[1]

It is possible for healthy scientists to enjoy not only the beauties of precision but also the pleasures of sloppiness, casualness and ambiguity. They are able to enjoy rationality and logic but are also able to be pleasantly crazy, wild, emotional. They are not afraid of hunches, intuitions, improbable ideas. It is pleasant to be sensible, but it is also pleasant to ignore common-sense occasionally. It is fun to discover lawfulness, and a neat set of experiments that solve a problem can and do produce peak experiences. But puzzling, guessing and making fantastic and playful surmises is also part of the scientific game and part of the fun of the chase. Contemplating an elegant line of reasoning or mathematical demonstrations can produce great esthetic and sacral experiences, but so also can the contemplation of the unfathomable. All this is exemplified in the greater versatility of the great scientist, of the creative, courageous and bold scientists. This ability to be either controlled and/or uncontrolled, tight and/or loose, sensible and/or crazy, sober and/or playful seems to be characteristic not only of psychological health but also of scientific creativeness.

A case history: the discovery of the structure of DNA

A striking illustration of such intense human involvement of the scientist in his work is provided by the personal account of the discovery of deoxyribonucleic acid (DNA) (see Part II, page 183) written by James D. Watson in his book *The Double Helix*.[2]

At the age of 22, after receiving his Ph.D., under the supervision of the virologist,

1. A. H. Maslow, *The Psychology of Science*, New York, Harper & Row, 1966. Copyright by Abraham H. Maslow, reprinted by permission of Harper & Row, Publishers, Inc.
2. J. D. Watson, *The Double Helix*, New York, Athaeneum, 1968.

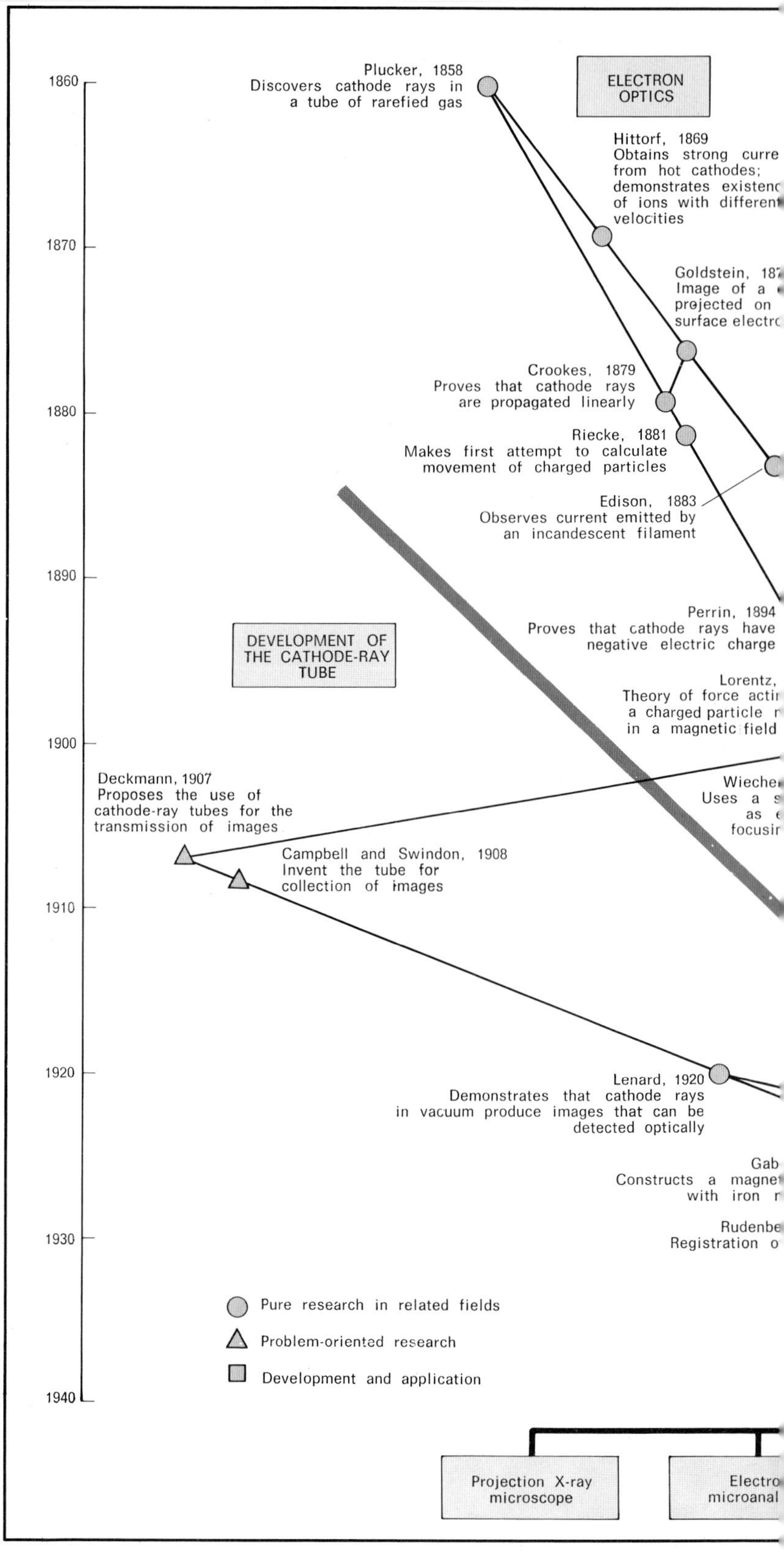

Fig. 1. The history of the electron microscope dates back to the middle of the last century. Scientists of all countries working in different disciplines have contributed to its development. Around the turn of the century, efforts began to concentrate on using electron beams to produce images. The 'birth certificate' of this instrument (a sheet of paper with a sketch by Ernst Ruska of the first complete model) bears the date 9 March 1931 and was produced at the Technische Universität in Berlin. On 7 April of the same year, Ruska took the first photograph showing the image of a metal grid traversed by an electron beam. Today there are thousands of electron microscopes in use all over the world, in the service of physicians and biologists, physicists, geologists and metallurgists. (Reproduced from *Traces*, Vol. 1, 1968, Illinois Institute of Technology Research Institute, Chicago.)

Plucker, 1858
Discovers cathode rays in a tube of rarefied gas

ELECTRON OPTICS

Hittorf, 1869
Obtains strong curre from hot cathodes; demonstrates existenc of ions with different velocities

Goldstein, 187
Image of a projected on surface electro

Crookes, 1879
Proves that cathode rays are propagated linearly

Riecke, 1881
Makes first attempt to calculate movement of charged particles

Edison, 1883
Observes current emitted by an incandescent filament

DEVELOPMENT OF THE CATHODE-RAY TUBE

Perrin, 1894
Proves that cathode rays have negative electric charge

Lorentz,
Theory of force actin a charged particle n in a magnetic field

Deckmann, 1907
Proposes the use of cathode-ray tubes for the transmission of images

Wieche
Uses a s as focusir

Campbell and Swindon, 1908
Invent the tube for collection of images

Lenard, 1920
Demonstrates that cathode rays in vacuum produce images that can be detected optically

Gab
Constructs a magnet with iron r

Rudenbe
Registration o

○ Pure research in related fields

△ Problem-oriented research

□ Development and application

Projection X-ray microscope

Electro microanal

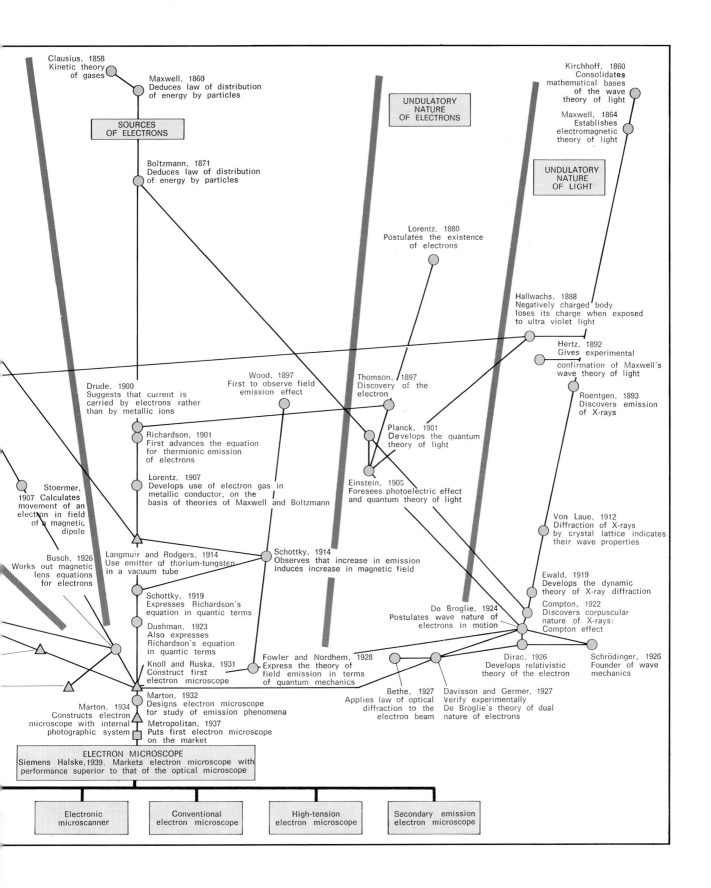

Clausius, 1858
Kinetic theory
of gases

Maxwell, 1860
Deduces law of distribution
of energy by particles

SOURCES
OF ELECTRONS

Boltzmann, 1871
Deduces law of distribution
of energy by particles

Kirchhoff, 1860
Consolidates
mathematical bases
of the wave
theory of light

Maxwell, 1864
Establishes
electromagnetic
theory of light

UNDULATORY
NATURE
OF ELECTRONS

UNDULATORY
NATURE
OF LIGHT

Lorentz, 1880
Postulates the existence
of electrons

Hallwachs, 1888
Negatively charged body
loses its charge when exposed
to ultra violet light

Hertz, 1892
Gives experimental
confirmation of Maxwell's
wave theory of light

Wood, 1897
First to observe field
emission effect

Thomson, 1897
Discovery of the
electron

Drude, 1900
Suggests that current is
carried by electrons rather
than by metallic ions

Roentgen, 1893
Discovers emission
of X-rays

Richardson, 1901
First advances the equation
for thermionic emission
of electrons

Planck, 1901
Develops the quantum
theory of light

Lorentz, 1907
Develops use of electron gas in
metallic conductor, on the
basis of theories of Maxwell and Boltzmann

Einstein, 1905
Foresees photoelectric effect
and quantum theory of light

Stoermer,
1907 Calculates
movement of an
electron in field
of a magnetic
dipole

Von Laue, 1912
Diffraction of X-rays
by crystal lattice indicates
their wave properties

Busch, 1926
Works out magnetic
lens equations
for electrons

Langmuir and Rodgers, 1914
Use emitter of thorium-tungsten
in a vacuum tube

Schottky, 1914
Observes that increase in emission
induces increase in magnetic field

Ewald, 1919
Develops the dynamic
theory of X-ray diffraction

Schottky, 1919
Expresses Richardson's
equation in quantic terms

De Broglie, 1924
Postulates wave nature of
electrons in motion

Compton, 1922
Discovers corpuscular
nature of X-rays:
Compton effect

Dushman, 1923
Also expresses
Richardson's equation
in quantic terms

Knoll and Ruska, 1931
Construct first
electron microscope

Fowler and Nordhem, 1928
Express the theory of
field emission in terms
of quantum mechanics

Dirac, 1926
Develops relativistic
theory of the electron

Schrödinger, 1926
Founder of wave
mechanics

Marton, 1932
Designs electron microscope
for study of emission phenomena

Marton, 1934
Constructs electron
microscope with internal
photographic system

Metropolitan, 1937
Puts first electron microscope
on the market

Bethe, 1927
Applies law of optical
diffraction to the
electron beam

Davisson and Germer, 1927
Verify experimentally
De Broglie's theory of dual
nature of electrons

ELECTRON MICROSCOPE
Siemens Halske, 1939. Markets electron microscope with
performance superior to that of the optical microscope

Electronic
microscanner

Conventional
electron microscope

High-tension
electron microscope

Secondary emission
electron microscope

S. Luria, at the University of Indiana, Jim Watson came to Europe to carry out biochemical research on nucleic acids. Since he was a senior in college, he had the desire to learn what the gene was: by that time it was already clear that the hereditary material of all organisms was formed by genes, and that these were in control of the activities of living cells. There was already available evidence to show that DNA played an essential role in the hereditary process of bacteria, plants and animals. The significance of DNA for life, however, was far from being undisputed: before Watson arrived in Cambridge in 1951 the molecular physicist, Francis H. C. Crick—with whom Watson was to discover, two years later, the structure of DNA and share the Nobel Prize in 1962—had thought, only occasionally, about DNA and its role in heredity. But Jim Watson had already taken his decision; the gene was made of DNA and he was going to solve the riddle of heredity.

Unfortunately for Watson, DNA was a complex chemical compound and he had to study chemistry, a subject he spontaneously disliked. Working under the guidance of a very distinguished chemist in Copenhagen did not stimulate Watson at all: he remained totally indifferent to nucleic-acid chemistry. But the almost haphazard encounter with Maurice H. F. Wilkins, during a small scientific meeting in Naples, opened new expectations: Wilkins was an X-ray crystallographer and he excited Watson's interest about X-ray work on DNA. 'Suddenly I was excited about chemistry', he writes. 'Before Maurice's talk I had worried about the possibility that the gene might be fantastically irregular. Now, however, I knew that genes could crystallize; hence they must have a regular structure that could be solved in a straightforward fashion.' Watson attempted to talk to Wilkins to explore the idea of whether he could go to London and work with him, but he had disappeared. 'I proceeded to forget Maurice, but not his DNA photograph. A potential key to the secret of life was impossible to push out of my mind.' Jim Watson moved to Cambridge where he had heard that Max Perutz and John Kendrew were interested in the structure of large biological molecules and were using X-ray diffraction

techniques. There, he immediately discovered the 'fun of talking to Francis Crick', and together they started to construct a set of molecular models and 'begin to play'.

Months followed, filled by infinite discussions, unsuccessful attempts, slight disappointments. After a lecture by the astronomer, Tommy Gold, on the 'perfect cosmological principle', Francis Crick started elaborating on the possibility of a 'perfect biological principle'. Then the tempo of discovery began to accelerate. 'The idea of the genes being immortal smelled right', writes Watson, 'and so on the wall above my desk I taped up a paper sheet saying DNA→RNA→protein. The arrows did not signify chemical transformations, but instead expressed the transfer of genetic information from the sequences of nucleotides in DNA molecules to the sequences of amino acids in proteins'. The 'central dogma' of molecular biology was born.

Shortly thereafter a period of frantic work followed, in the attempt to develop a convincing model of DNA's structure, compatible with its X-ray diffraction pictures. After a succession of lucky hunches and despite some setbacks Watson and Crick, haunted by the risk that the famous chemist Linus Pauling might succeed in solving the problem before them, finally built the now celebrated model of the double helix. Thus was made the discovery. The exhilarating news was divulged and the great experts came to admire the model of the double helix.

The pitch of his [Francis'] excitement was rising each day. . . . Pauling's reaction was one of genuine thrill. . . . In almost any other situation Pauling would have fought for the good points of his idea. The overwhelming biological merits of a self-complementary DNA molecule made him effectively concede the race.

A new era of biological research had begun. As J. D. Watson himself wrote in the preface to his book:

. . . science seldom proceeds in the straightforward logical manner imagined by outsiders. Instead, its steps forward (and sometimes backward) are often very human events in which personalities and cultural traditions play major roles.

The thrill of discovery

The exhilaration, which comes when a break-through and elegant intellectual processes converge, has been experienced by a multitude of scientists, whose discoveries have laid mile-stones in the history of science, but it is re-peatedly renewed by the much wider audience of students when reading the works of the masters. And, even if with lesser intensity, it represents the subjective satisfaction that every research worker encounters when he suc-ceeds in solving the problem on which he has been working, a fulfilment derived either by the results of well-designed experiments, or by the creation of novel theoretical interpretation.

As in other fields of intellectual endeavour, different persons, according to their psycho-logical requirements, are attracted by the multifarious aspects of scientific work. A complex typology of scientific talents could be worked out, and it would certainly include the theorizer, often indulging in speculation, which may occasionally lead to formulate novel interpretation of known facts, and to outline critical experiments to test the validity of the theory; the gadgeteer, whose mechan-ical ability makes possible the development of complex analytical tools, which allow un-expected breakthroughs; the collector, who carries out systematic surveys of vast arrays of phenomena, leading to generalizations that sometimes make possible the emergence of unifying principles; the extremely specialized and narrow-minded researcher, who perseveres over the years in gathering increasingly pre-cise data on seemingly uninteresting subjects, which, however, may become turning points in the development of a discipline; the passionate encyclopaedic, who is capable of retaining a bewildering amount of information and is led by his aesthetic leanings to propose vast gen-eral theories, as exemplified by Charles Darwin and his evolutionary theory.

One could continue. The point I want to make is that the sources of exhilaration can be different for different temperaments, each of which approaches scientific research from dif-ferent angles.

And, when discussing this issue, one should not forget the so-called Ortega hypothesis:[1]

For it is necessary to insist upon this extraordinary but undeniable fact: experimental science has pro-gressed in great part thanks to the work of men astoundingly mediocre, and even less than me-diocre. That is to say, modern science, the root and symbol of our actual civilization, finds a place for the intellectually commonplace man and allows him to work therein with success. In this way, the majority of scientists help the general advance of science while shut up in the narrow cell of their laboratory, like the bee in the cell of its hive, or the turnspit in its wheel.

And, judging from its dances, it may well be that the bee is exhilarated in the cell of its hive.

In this connexion, it may be useful to add that a recent study of papers published by American physicists shows that, contrary to the Ortega hypothesis, a relative small number of scientists produce work that becomes the base for future discoveries; in other words, this analysis suggests that, in the case of phys-ics at least, only a few scientists contribute to scientific progress.

The stimulation of participating in a universal enterprise

Under ideal conditions, scientific endeavour has a universal character. A discovery, in order to be valid, should satisfy the criterion of being capable of verification by anyone, any-where in the world, with no ideological or national limitations. In its early days, science was a matter for decision of private individuals and co-operation between scientists belonging to different countries, in terms of correspon-dence, exchange of publications and visits, was completely free, and sometimes respected by governments even in times of war. Thus, the tradition of science as a universal enterprise was established, and still represents an ideal for the working scientists of today, even if in recent times the increasingly closed links be-tween political power and scientific develop-ments have brought about a progressive decay

1. J. R. Cole and S. Cole, 'The Ortega Hypothesis', *Science*, No. 178, 1972, p. 68–375.

of such an ideal state (as we shall examine in Part III, Chapter 30).

As J. D. Bernal put it:[1]

The internationalism of science was also, from the start, a positive one. Those men we recognise as the earliest scientists, the philosophers of Ionian Greece, had acquired their science by mixing with other people, either by travelling themselves or by talking with travellers. The very birth of the scientific spirit in Greece was largely due to the internationalism in ideas and techniques that Greek traders introduced into the ancient world. The poet was recognised among people who could understand his language: the priest or the prophet in the regions where common beliefs held: but the scientist, the man of natural learning, could go everywhere, and everywhere be well received.

Even in time of war, the individual scientist feels the joy of doing a kind of work that transcends the hatred and enmity of the times, for he knows that when hostilities are over, he will be able to re-establish contacts with colleagues whom he knows to be sincere citizens of the world even though they are subject, like himself, to the folly of the times. And this feeling of belonging to a universal community, pursuing common aims for the advancement of knowledge certainly exerts a great appeal to the scientist, who considers himself, perhaps with a certain lack of humility, *au-dessus de la mêlée.*

The development of the scientific enterprise bears witness to the extranationality of the scientific enterprise: innumerable examples can be found in every discipline. Particularly for those who have not had the opportunity of indulging in the study of the history of science, we shall consider only one example, the electron microscope. The history of its development has been recently analysed in a report[2] in which systematic studies were made of the role of research in the overall process, which eventually leads to technological innovation, the use of which has been essential for the advancement of our knowledge in the fields of biology, chemistry, metallurgy, electronics and other sciences.

The development of this instrument can be traced back to the second half of the nineteenth century, when fundamental discoveries were made on the nature of light of the electron, on electron emission and on electron optics. Here the names of Kirchhoff (in Germany), Maxwell (in England), Lorents (in Denmark), Boltzmann (in Germany), Plucker (in Germany), Crookes (in England), Edison (in the United States) are outstanding. It then follows a period of theoretical elaboration of experimental results, where one finds a Frenchman (Perrin), a Dutchman (Lorentz), two Britons (Thompson and Wood) and several Germans (Hertz, Roentgen, Hallwachs, Planck, Drude and Weichert). The understanding of the basic processes of the physics of light, electrons and 'cathode rays' is further developed by Von Laue (in Munich), Einstein (in Switzerland), Lorentz (in Leiden), Stoermer (in Norway). Then follow the first attempts to take advantage of the recent discoveries for technological innovations, and we find Langmuir and Rogers in the United States, Deckmann in Germany and Campbell and Swinton in England. Further fundamental research, essential for the final development, is carried out by Compton (in the United States), de Broglie (in France), Schroedinger (in Switzerland), Dirac (in England), Davisson and Germer (in the United States), Bethe, Schottky, Lenard and Busch (in Germany), Fowler and Norheim (in England), Dushman (in Canada) and Marton (in Belgium). Finally, research oriented towards the aim of guiding and controlling beams of electrons takes over and, once again, contributions are coming from scientists and technologists of many countries: Gabor (an Hungarian working in Berlin), Knoll and Ruska (in Germany), Rudenberg (in Germany) and Marton (in Belgium).

Political tensions had reached unprecedented levels during the period of less than a century covered by this analysis, but those scientists, together with their closest co-workers, knew and felt that in spite of hate and violence their work had a value which would extend much beyond the vagaries of nationalistic emotions.

1. J. D. Bernal, *The Social Function of Science*, London, Routledge, 1939.
2. *Technology Retrospect and Critical Events in Science (TRACES)*, Vol. 1. Prepared by the Illinois Institute of Technology Research Institute for the National Science Foundation, Washington, D.C., 1968.

The excitement of the public

Thus far, we have examined aspects of scientific inquiry, which bring about a sense of elation to the practising scientist, but a similar feeling is occasionally shared by the public at large, by the informed laymen, if results can be made understandable and if such understanding pervades the collective unconscious and then changes their social philosophical outlook.

In the sixteenth century, Church and State battled for supreme power over the people: however contradictory their views, there was no higher power to which to turn. The limits of knowledge, of personal freedom, even of private thoughts were dictated by these powers. But the advent of the new cosmology of Copernicus and Kepler, and the even more spectacular impact of Galileo's telescope—deeper indeed in its effect than any book—disclosed unexpected horizons for man, and his position in the Universe became drastically different. The subsequent scientific developments culminating in Newton's principles promptly made understandable to a wide non-specialized public, led to the publication of the *Encyclopédie ou Dictionnaire Raisonné des Sciences des Arts et des Métiers* and to the Enlightenment. For the first time man had discovered within himself a higher court of appeal based on reason, and the pursuit of truth was no longer tarnished by religious or parochial overtones. If science's only contribution was the liberation of the human spirit that ensued, it would have had to be counted a triumph.

Two centuries later, biology was called upon to contribute one further fundamental change in human perspective: Darwin's theory of evolution through natural selection pointed to a continuous lineage from primitive, and relatively simple organic forms to the magnificent variety of living forms of today and to man himself. Again the emotion of the discoverer rebounded in the halls of Victorian England, where Thomas Huxley was preaching the new gospel.

Later in this century, Einstein's theories, together with astronomical discoveries, changed once more the limits of the Universe: the Earth is but one of a milliard planets situated not in the middle but on the edge of one of a milliard galaxies. Just as in the last decades, space exploration gave man the possibility of observing his solitude on a fragile planet.

And the biologists repeat the same message. The human animal works in much the same way as the lowliest lemur and even shares with the smallest bacterium a precisely similar code for specifying the way in which hereditary information is passed from one generation to the next. Even the ecologists are showing that man like any other plant or animal has his own ecological niche into which, within limits, he must fit. And finally, science has revealed no master plan or predetermined role for man to play. So, concludes Jacques Monod:[1]

Man knows at last that he is alone in the indifferent immensity of the universe. His duty, like his fate, is written nowhere. It is for him to choose between the kingdom and the darkness.

These are just four of the most striking examples of the sharing of the exhilaration of science by the public: of course, innumerable ones could be found in the history of science, particularly in the last few decades, when science writers in the daily press have attempted, and often succeeded, in conveying to the layman the feeling of excitement prevailing in the laboratories. Public exhilaration tends to come a long time after laboratory exhilaration and becomes more evident when the discovery is bound to change man's view of himself, as in the previous examples, or when technological developments ensuing from scientific advances change his daily life. As we shall see in the following chapter, useful science has changed profoundly the way of living of individuals and societies, particularly as a result of progress in medicine and the resulting decrease in death rates, which brought about extended life expectancies. For this reason, I think, immediate reaction to, and wonder for scientific achievements, are likely to occur more often when the layman is made

1. J. Monod, *Le Hasard et la Nécessité*, Paris, Le Seuil, 1970.

aware of the personal benefits he may accrue: typical examples have been the 'miracle drugs', the 'pill', heart transplantation and many similar ones. And indeed advances in the physical sciences, superficially less immediately related to man's welfare, can be made more easily understood and appreciated if the writer presents them in terms clearly related to the satisfaction of human needs.

Early last century, the famous scientist Sir John Herschel could write:

The advantages conferred by the augmentation of our physical resources through the medium of increased Knowledge and improved Art have this peculiar and remarkable property—that they are in their nature diffusive, and cannot be enjoyed in an exclusive manner by few. . . . To produce a state of things in which the physical advantages of civilised life can exist in a high degree, the stimulus of increasing comforts and conveniences and constantly elevated desires must have been felt by millions.

In such pervasive quality of physical advantages, derived from scientific research, lies the difficulty of explaining, and therefore teaching, to the millions the nature of scientific research, its appeal, and the ensuing exhilaration. Understanding means have a 'feel for the Tactics and Strategy of Science', as the physicist J. B. Conant put it. Only if one succeeds in this difficult task will he be able to convey to the layman the thrill of scientific discovery. Otherwise, as the poet Wystan H. Auden broods *After Reading a Child's Guide to Modern Physics*:

> The passion of our kind
> For the process of finding out
> Is a feat one can hardly doubt,
> But I would rejoice in it more
> If I knew more clearly what
> We wanted the knowledge for—
> Felt certain still that the mind
> Is free to know or not.

Is discovery different from other types of creation?

Because of the difficulty of conveying to the educated layman a genuine understanding of the nature, motivations, procedures and aims

of science (referred to in the preceding pages), writers, teachers, lawyers, politicians, civil servants, businessmen and artists often consider the scientist and his work in esoteric terms. Already in the past the scientist and the inventor were often looked at as magicians and thus were born the myths of Prometheus, of the Golem, of Faust and others. But in recent times the extraordinary success of the applications of science, as exemplified by atomic energy, electronic computers and 'Man on the Moon', have reinforced the feeling of the man in the street that the scientist is a monster, a sort of court magician, a different kind of animal from the rest. 'Is it not because we have failed to assimilate science into our western culture that so many feel spiritually lost in the modern world?'—James B. Conant is asking.[1] And he replies:

So it seems to me. Once an object has been assimilated it is no longer alien; once an idea has been absorbed and incorporated into an integrated complex of ideas, the erstwhile foreign intruder becomes an element of strength. And in this process of assimilation, labels may well disappear. When what we now roughly designate as science has been fully assimilated into our cultural stream, we shall perhaps no longer use the word as we do today. When that time arrives, and I have no doubt that it will, [understanding science] will be fused into the age-old problem of understanding man and his works: in short, secular education.

But if we read the accounts given by scientists of their own experiences in the course of the process of discovery, we find that the act of scientific creation is probably similar to that of the artist. The discovery of the structure of DNA, presented earlier in this chapter, is a case in point.

There appears to be a convergence of views today on the nature of scientific discovery. Thus Gunther S. Stent,[2] a scientist, analysing the fantastically rapid progress of molecular genetics in the past twenty-five years, reaches

1. J. B. Conant, *On Understanding Science—An Historical Approach*, New Haven, Yale University Press, 1947.
2. G. S. Stent, 'Prematurity and Uniqueness in Scientific Discovery', *Scientific American*, December 1972, p. 84–93.

the conclusion that 'every creative act in the arts and sciences is both commonplace and unique'. Thus Abraham H. Maslow, a psychologist, wrote:

Creative persons have often reported their reliance on hunches, dreams, institutions, blind guessing, and gambling in the early stages of the creative process. Indeed, we could almost define the creative scientist in this way—as the creative mathematician is already defined—i.e., as one who reaches the truth without knowing why or how. He just feels something to be correct and then proceeds *post hoc* to check his feelings by careful research. The choice of hypothesis to test, the choice of this rather than that problem to invest oneself in, is proved correct or incorrect only *after* the fact. We may judge him correct because of the facts that he has gathered, but he himself did not have these facts to base his confidence on. Indeed the facts are the consequence of his unfounded self-confidence, not because of it. We call a scientist talented for just this reason, that he is often right *in spite of* insufficient evidence. The lay picture of the scientist of one who keeps his mouth shut until he is sure of his facts is quite incorrect, at least for talented, breakthrough scientists.

Analysing the matter further, Maslow adds:

In a word, scientists are not a different species. They share with others the characteristic of curiosity, the desire and even the need to understand, to prefer seeing to being blind, and to prefer more reliable to less reliable knowledge. The specialized abilities of the professional scientist are intensifications of these general human qualities. Every normal person, even every child, is a simple undeveloped, amateur scientist who can in principle be taught to be more sophisticated, more skilled, more advanced. A humanistic view of science and of scientists would certainly suggest such a domestication and democratization of the empirical attitude.

On this view, then, there exists no essential difference between the act of creation of the poet, the painter, the sculptor, on the one hand, and that of the scientists, on the other. But even if we accept such a point of view, we may still raise the question: is scientific discovery different from the work of art? And if so, where lies the main difference? For even assuming a basic similarity of mental structures and manual abilities in all men, i.e. that the means through which we express ourselves and operate are common to us all, the end products of our action can be exceedingly different.

Similarities between the works of art and scientific discoveries have been identified in their both being attempts to bring to light and communicate truths about the world. But while the artist concentrates primarily his attention on the inner subjective world of the emotions and passions, the scientist, on the contrary, is interested in the external, physical world. The former expresses feelings, relations between personal impressions and conveys aesthetic impressions; the latter establishes relationships between events, and their regularities, that can be exactly repeated, verified and, to use Professor Ziman's expression, become 'public knowledge'. This very characteristic of scientific discovery, of dealing with tangible objects and events and of being interpersonal knowledge, warrants its main characteristic and its main difference from the products of the artist. The scientist feels that there is but one Universe to discover, and this feeling is reinforced by the claim of sociologists of science (such as R. K. Merton) that no single scientist is crucial for scientific advance. The innumerable examples of discoveries made independently, by two or more scientists approximately at the same time, seem to indicate that indeed the object on which scientists concentrate their study is one and only one entity.

But there is one more, and a significant, difference between artistic and scientific activity, to which we will return later: the products of scientific effort open the way to concrete, practical and useful as well as dangerous applications. It may well be that such a difference is at one and the same time science's pride and curse.

15

3   Useful science

According to the *Oxford Dictionary of the English Language*, 'useful' is defined: 'having the qualities to bring about good or advantage; helpful in effecting the purpose'. It is hardly worth stating that science and its applications have been useful to satisfy, to an increasingly high degree, the intellectual and material needs of man. While in the previous chapters we have discussed some aspects of the fulfilment the scientist and the public derive from the scientific enterprise, we will now concentrate on the good resulting from the application of scientific discoveries. But we will postpone to Part IV a more detailed discussion of the nature of the needs of man and on how far modern science has been successful in meeting them.

More than thirty years ago J. D. Bernal in his classic volume *The Social Function of Science* stated:[1]

Most scientists as well as laymen are content with the official myth that that part of the work of the pure scientists which may have human utility is immediately taken up by enterprising inventors and business men, and thus in the cheapest and most commodious way possible put at the disposal of the public. There has always been a close interaction between the development of science and that of material techniques. Neither would be possible without the other, for without the advance of science techniques would fossilize into traditional crafts, and without the stimulus of techniques science would return again to pure pedantry. It does not follow, however, that this association is either conscious or efficient; in fact, the application of science to practical life has always been faced with the greatest difficulties and is even now, when its value is beginning to be recognised, carried on in the most haphazard and ineffective way.

The interrelationships between science, technology and economic development have been and are today very complex, and have undergone frequent change in the course of history.

Meeting the needs of man

Varied, changing and complex as his needs are, man, throughout his existence has endeavoured to use his intelligence and skills for improving his lot. From the primitive stages of early man, throughout the long period characterized by the discovery of agriculture, the establishment of urban settlements, the rise and flourishing of the great civilizations of the East and West and until the end of the Middle Ages in Europe, the contacts between science and technology have been sporadic and, by and large, not highly significant. At

1. J. D. Bernal, *The Social Function of Science*, London, Routledge, 1939.

17

the time when the simplest sciences of mathematics, mechanics and astronomy appeared in the relatively late stage of civilized city life, man was already master of a vast array of sophisticated technologies. Archaeology and history bear witness to the great technological achievements of man, before the birth of modern science, such as the sophisticated use of metals, constructions that compare well with the feats of modern engineering, refined potteries and textiles, and effective weapons. Throughout the classic period and a good part of the Middle Ages the technological superiority of the Orient over the West was clear. The remote civilization of China was technologically much more advanced than those of the Mediterranean basin. Obviously, it is quite difficult to compare complex phenomena, such as the influence of technology on society, occurring in widely separated countries having very distinct cultures. But it is generally agreed among historians of science that towards the end of mediaeval times there occurred a shift in the trend, and, slowly at first, but rapidly in later centuries, Western Europe became technologically superior.

One of the most significant events in the history of the world was the birth of modern science and its repercussions on human activities. Before the advent of the 'new philosophy', the description of natural phenomena was the task of the philosopher; but their utilization for practical purposes was left to the artisan. And the latter knew hardly anything beyond the nature of his craft, which he was performing according to traditional methods transmitted from father to son; he had no theory to explain his action and machines. Only with the Renaissance and the seventeenth century it became clear, even if only to few, that natural phenomena were important both for science and the crafts, and that these could help each other. The understanding of nature was essential for putting it under man's power and control. Bacon, Galileo and Descartes expressed the dreams of the liberation of man from the daily toil through science. But the dreams were very slow in coming true. Indeed, also in previous times science and technology did interact, as, for example, in the evolution of the clock during the Middle Ages. Such loose interaction continued long after the birth and development of modern science. Even during the industrial revolution, which began in England in the second half of the eighteenth century, relatively few were the examples where the impact of science on industrial production were felt. The chemical industry, strictly related to the development of textile manufacture, benefited from some of the scientific concepts of the time; the evolution of the steam engine was greatly influenced by the theory of latent heat of Professor Joseph Black, of the University of Oxford, with whom James Watt had been closely associated, but, as it has been said: thermodynamics owes much more to the steam engine than the steam engine to thermodynamics. The production of optical lenses for special uses was influenced by scientific investigations such as those of Michael Faraday at the Royal Institution; and the development of the miner's safety lamp, which was a major advancement in the exploitation of coal mines, was the outcome of laboratory research, under the guidance of Sir Humphrey Davy and W. R. Clanny.

Only in the course of the nineteenth century, and more clearly in the latter half, the advancement in scientific knowledge exerted a major influence on the development of technologies. Entirely new fields of production appeared, such as electric power and chemical, especially organic, synthesis. Nevertheless, the dependence of industry upon scientific research remained to a large extent fortuitous. Towards the end of the century the technological advancements of each nation, and hence its ability to compete with the others in industrial production, became more and more dependent upon the number of scientists and engineers, as well as on the level of education of the population. Within the continuous flow of technological innovations, the speed of advance became the more rapid the closer the connexions and interactions between the laboratory of the fundamental scientist, the specialized research team of the engineer and the production line of industry and agriculture.

In the twentieth century science is for the first time on its own. Much more scientific research has been carried out and a far greater number of technological breakthroughs have occurred during the last fifty years than in the whole previous history of modern science. Our understanding of the Universe, of matter and of life has gone far beyond the wildest dreams of our predecessors. And, at the same time, the impact of technological change has become felt not only in the limited number of countries where research activities were promoted but also, literally, by every citizen of the world. Infant mortality dropped markedly from the levels that had prevailed over the millenniums, life expectancy doubled in the course of a few decades, modern communication, transportation and the extended use of new materials have abruptly changed the face of the planet and the life of almost every single human being. The benefits accrued are certainly immense, but, as we shall examine later, the problems created by the sudden rise in population, production, pollution and other factors have hardly vindicated the dreams of Francis Bacon and of the French Encyclopaedists.

The table which follows is a list of some major scientific and technological landmarks from 1900 to 1970. The liberation of nations, the wealth of the world, the health of the people, the quantity and quality of their nourishment, their standard of living, their time for leisure and, perhaps, their happiness, have progressed under the direct or indirect impact of science and technology. But it is during this period that the politicization of science has occurred; scientists became nationalistic, partisan, and in many other ways very 'impure' philosophers, as we shall discuss further in Parts III and IV.

TABLE 1. Major scientific and technical landmarks from 1900 to 1975

Date	Universe	Matter	Life	Technology
1900		Radium	Mendel laws rediscovered	Turbopropeller
		Quantum theory	Serum therapy against diphtheria	Nernst lamp
				Rigid airship
			Freud's interpretation of dreams	Working model of magnetic recorder
1901		Crignard reactions	Adrenaline synthesized	First application of rotary drilling
			Malaria cycle explained	Synthetic resins
			Conditioned reflexes explained	
			Establishment of blood groups	
1902	Ionosphere proposed	Sex chromosomes		Mercury arc rectifier
1903		Adsorption chromoatography		First powered flight
				Safety glass
1904				Diode
				Radio echoes foreseen
				TNT shells
				Rayon invented
				Liquefaction of helium
1905	Special relativity		More hormones discovered—name given	Gaede pump
1906	Inversions of Earth's magnetic field proposed		First co-enzyme	Triode
			Specific test for syphilis	Crystal radio detector
				Pure ductile titanium metal produced
1907			Homo heidelbergensis	Bakelite
				Gramophone
1908			Hardy-Weinberg law	100-inch telescope
			Hybrid maize	
1909		Geiger detector		

Date	Universe	Matter	Life	Technology
1910		Isotopes	First gene localized in chromosome	Gyro-compass
1911	Cosmic rays discovered	Super-conductivity Cloud chamber Rutherford's atom. model	Tumour-inducing virus discovered	Electric motor starter Air conditioning
1912	Continental drift proposed	X-ray diffraction and cristallography	Vitamins discovered	Cellophane introduced
1913	Spectral classification	Bohr atom model Concept of atomic number	First chromosome map Freud's *Totem and Taboo*	Ford T-model Industrial synthesis of nitrogen compounds Tungsten filament light
1914				Stainless steel Tungsten carbide (VIDLA) first produced
1915				
1916	General relativity		Non-disjunction as proof of chromosome theory of heredity	
1917		Theory of stimulated emission (laser)		Industrial production of acetone
1918	Polar front theory of cyclones			Acetate fibre
1919		Rheology Splitting of N-atom	Structure of cellulose	First regular radio programmes
1920				
1921	Predominance of red shift			
1922	Friedmann universe	Explanation of periodic system with quantum theory of atom	Insulin isolated Lysozyme discovered	Tetraethyl lead developed to cure 'knock'
1923	White white stars		Gene theory	Bulldozer
1924			Oparin's theory on the origin of life	Electronic logging techniques invented First motor freeway First insecticide used
1925	Ionosphere detected	Exclusion principle Wave mechanics New quantum theory	Nerve conduction explained	Polystyrene discovered
1926			First enzyme crystallized Vitamin B₁ Publication of *Conditioned Reflexes*	Television demonstrated Firing of first liquid fuel rocket
1927	Theory of expanding universe	Relativistic wave equation of electron	Mutations produced by X-rays	Trans-Atlantic solo flight
	Concept of galaxies	Uncertainty principle Electron diffraction discovered	Theory of the gene	
1928		Neutrino predicted Raman spectroscopy Anti-particles predicted	Penicillin discovered Ascorbic acid isolated— vitamin C	
1929	Hubble's law	Cyclotron	Chemical transmission of nerve impulses discovered	Pre-stressed concrete

Date	Universe	Matter	Life	Technology
1930	Pluto discovered	Electrophoresis	Genetic theory of natural selection	Catalytic cracking Efficient rocket flight 66 polyamide fibre (nylon) BUNA synthetic rubber
1931	First radiotelescope	Prediction of existence of neutrino		First television emitter
1932		Discovery of neutron Deuterium discovered Positron discovered Disintegration of nucleus	Ramapithecus discovered Prostaglandins Sulfa drugs	Sulphonamides Neoprene Kroll process for titanium metal Polarizing glass
1933	Neutron stars proposed	Theory of beta radio-activity	Proconsul discovered	Modern Cellophane
1934		Nuclear forces between nucleons Meson predicted Neutron bombardment of uranium	First virus crystallized	Communication satellites predicted
1935		Pion predicted	Cell membrane structure proposed Cortisone isolated	Radar
1936	Hubble's constant	Compound nucleus theory of nuclear reactions		
1937	Interstellar magnetic field proposed Solar fusion explained	Moon discovered	Electron microscope Vitamin A synthesized Krebs cycle	Jet engine Xerography
1938		Meson observed Fission reaction Superfluidity discovered	Glycolysis elucidated	Acrylic fibres produced 6 polyamide fibre (perlon) nylon introduced
1939			Theory of nerve cells Oxidative phosphorylation discovered	First jet flight Possibilities of nuclear energy outlined Penicillin introduced DDT's insecticidal properties discovered
1940		Trans-uranic elements	Rh factor	Cybernetics Polyester fibres (terylene) developed Cavity magnetron Betatron
1941		Plutonium discovered		Shell moulding
1942				First chain reaction First electronic computers Silicones
1943				V.2 rocket
1944			DNA discovered to be hereditary material Structure of penicillin Streptomycin	Reinforced glass fibre Isotopes as tracers in chemical processes

Date	Universe	Matter	Life	Technology
1945		Nuclear magnetic resonance		Full-scale gaseous diffusion plant A-bomb Ball-point pen Chlorinated hydrocarbons for insect control Minimum tillage Foliar feeding
1946		ENIAC electronic computer	Genetics of bacteria	Cloud seeding first used Phototype-setting
1947		Pion discovered Baryons mesons discovered Holography proposed	Australopithecus prometheus	Sound barrier broken First off-shore drilling Chloromycin Organo-chloro plant hormone insecticides Direct application of anhydrous ammonia
1948	Polarization of starlight discovered Solar radio waves	Partition chromatography	Polio virus grown in tissue cultures	Transistor Long-playing (LP) records 200-inch telescope
1949	Radio waves detected from other galaxies First bathyscaph	k-meson discovered	Theory of immunology	Terramycin discovered
1950				Orlon
1951		Shell model of nucleus	Cortisone synthesized	Commercial computer Power steering Terylene introduced Chemical weed control
1952		Quantum excitation of nucleus Bubble chamber	Infectious heredity One gene, one enzyme theory	H-bomb Fuel cell Oxygen steel making
1953		Hyperons discovered Concept of 'strangeness' in particle physics	DNA structure First polypeptide hormone synthesized	250 kW nuclear reactor Field effect Inertial guidance Dacron Systemic biocides
1954		Metallic whiskers CERN (Geneva) established Molecular beams Maser	Polio vaccine	Nuclear powered submarine Polypropylene discovered Owen Falls, Uganda; 204,800 million cubic metres man-made lake
1955		Anti-proton discovered Nuclear rotation	Structure of insulin	Oral contraceptive
1956		Theory of super-conductivity JINR (Dubna) set up Parity non-conservation Neutrino detected	Lysosomes discovered Forty-six chromosomes of man	Commercial nuclear reactor Solid-state radio receiver
1957	(International Geophysical Year) Theory of origin of elements	Theory of superconductivity	Interferon Oral polio vaccine	Suborbital space missions ICBM Jodrell Bank radiotelescope Hybrid sorghum ·
1958	Van Allen belts discovered	Mossbauer effect		

Date	Universe	Matter	Life	Technology
1959	Far side of Moon photographed		Zinjanthropus boisel Abnormal number of chromosomes: medical significance Three-dimensional structure of myoglobin	Nuclear-powered merchant ship Polypropylene Integrated circuits
1960	Magnetic anomalies found on ocean floor Exploration at depth of 11,700 metres of Mariana Trench	Large number of new particles	Homo habilis	Laser First weather satellite and Comsat First patent on proteins from petroleum
1961	Theory of ocean floor spreading Manned balloon experiments at 34,600 metres	Two neutrino hypothesis confirmed Eightfold way for particle physics	700 enzymes known Theory of gene regulation	Manned space flight Miniaturized radar sources Dwarf wheat
1962	X-ray sources discovered Mars fly-by Mariner 2 Venus probe	Helium group gas compounds Josephson's effect	Genetic code deciphered	First geodetic satellite Grande Dixence (Switzerland), 284 metres high
1963	Quasars 1,000-foot Arecibo radiotelescope	Solvated electron identified		Commercial colour television Scanning electron microscope
1964	(International Quiet Sun Year) Mariner 4 Mars probe	Omega minus discovered Confirming eightfold way Quark hypothesis Gunn effect Violation of CP conservation		Carbon fibre First hologram
1965	Microwave background radiation discovered Venera II and Venera III planetary probes Ocean floor spreading		Cyclic-AMP Somatic hybrid animal cell 3-D structure of lysozyme Four-carbon photosynthesis discovered Discovery of fossils more than 3 billion years old Homo erectus palaeohungaricus	Biodegradable detergents Dwarf rice Opaque-2 maize (high lysine)
1966			Insulin synthesized Repressor proteins isolated	Large-scale integration
1967	Pulsars		3-D structure of ribonuclease	Wankel engine in production
1968	Deep-sea drilling	Granular charge structure within proton	Synthesis of ribonuclease molecule	Audiovisual systems (EVR) introduced Hovercraft in service Krasnoyarsk (U.S.S.R.), 5,080 MW hydroelectric plant

23

Date	Universe	Matter	Life	Technology
1969	Moon rocks analysed First report of gravitational waves *Glomar-Challenger* expedition *Mariner 6* and *Mariner 7* photograph craters of Mars		RNA synthesized 3-D structure of insulin Structure of gamma globulin Australopithecus bosei	Man on the Moon Hybrid barley
1970	Solar neutrino flux measured		Chemical synthesis of a gene Reverse transcriptase Growth hormone synthesized Endocrine function of hypothalamus discovered Structure established of transfer RNA	Hybrid cotton
1971	Uhuru satellite detects many X-ray sources Moon exploration (Apollo, Lunakhod) and Mars survey (Mariner, Mars) continued	Wider energy range available for proton-proton collision at CERN's ISR	Experimental creation of a new species of *Drosophila* Triple hybrids Foetal diagnosis of chromosomic diseases General model for chromosomes of superior organisms proposed	Assuan dam inaugurated First polymer with nitrogen-nitrogen bond Hormonal insecticides A big plant in Scotland produces proteins from petroleum
1972	*Pioneer 10* flies to Jupiter Deuterium first observed in a sky object Radio source Cyg X-3 bursts Dating of most ancient rocks on Earth: 3.7–3.75 billion years	Superfluidity of ^{3}He below 1 mK 104, 105 elements discovered 'Unusual' chemical structures: synthesis of prismane	Organic matter found in rocks older than 2.6 million years Angyogenetic tumoral factor discovered Insulina-receptor insulated	NAL inaugurated at Batavia (400 GeV) Power laser Fourth-generation computers Virtual memories in commercial computers Nuclear energy powered pacemaker First breeder reactor at Shevchenko (U.R.S.S.)
1973	*Pioneer 10* sends photos of Jupiter *Pioneer 11* flies to Jupiter and Saturn *Skylab* and three manned vehicles in orbit: Sun and Kohoutek comet observed Two quasars observed whose red-shift is more than 3	Neutral current events New list of recommended atomic weights Big European bubble chamber of CERN	Discovery of a bacterium which could live in the (presumed) environment of Jupiter Tridimensional structure of a t-RNA Bidirectional growth of the chromosome of *Escherichia coli* Atomic resolution of double-helix structure Second artificial gene synthesized	Superconductive Nb_3Ge alloy at 22.4 K New titanium alloys Prototype of an electric generator based on superconductivity

Date	Universe	Matter	Life	Technology
1974	Helios flies to Sun Venus and Mercury photographed by *Mariner 10* *Mars 5* on Mars Jupiter's 13th satellite discovered Radio message sent from Arecibo observatory to M31 Methylammine discovered in interstellar space	106, 107 elements discovered LAMPF (meson factory) in function New heavy neutral particles discovered: signs of the existence of fourth (charmed) quark?	End of the third phase of International Biological Programme Conference on world population at Bucarest α-thalassemy due to genic deletion	First superplastic alloy discovered Silicon strips for solar cells Raw materials (manganese) from ore deposits in ocean floors
1975	Jupiter photo- graphed by *Pioneer 11* Apollo-Soyuz mission in orbit	Neutrons obtained by means of laser implosion		Big Tokamak designed for nuclear fusion

Science and technology

The terms 'science' and 'technology' refer to such a complex set of interwoven human activities that, as one would expect, they are used in a rather loose way in the writings of specialists, in official reports, as well as in the everyday language of the layman. Not only are these terms far from being univocal: they often are used singularly to indicate the whole set of activities, products, institutions and specialized communities, as we shall discuss in greater detail in Part III.

As a result of the rapid growth of science, pure and applied, in this century, there occurred a proliferation of terms to indicate various aspects of research related to the field. Of course, it would be impossible to review the immense literature on the subject, and we will limit our discussion to some recent comments and classifications.

Unesco's definitions of science and technology, the products of research and development (R & D), which we shall adopt are the following:[1]

1. Science may be defined as mankind's organized attempt, through the objective study of empirical phenomena, to discover how things work as causal systems. By means of systematic thought, expressed essentially in the symbols of mathematics, it brings together the resultant bodies of knowledge in an effort to reconstruct the world *a posteriori* by the process of conceptualization. Its purpose is not to invent but to comprehend. The sciences thus constitute an interlocking complex of attested fact and speculative theory, with the essential proviso that theories must be capable of being tested experimentally.

One might add that there is in fact a modern school of thought which tends to attribute to the concept of science an even wider definition, seeing it as embracing the whole range of intellectual and imaginative effort whose goal is to establish a consensus of rational opinion over the widest possible field.

2. Technology denotes the whole—or an organic part—of knowledge about: (a) scientific principles or discoveries; (b) industrial processes; (c) material and energy resources;

1. *A Selection of Working Definitions used in Science Policy*, Paris, Unesco, 1970 (Unesco doc. NS/ROU/207 prov.).

(d) methods of transport and communication, so far as it relates directly to the production or improvement of goods and services. Engineers, whose task it is to apply technology to development, are thus dealing with the conception, design and application of new forms of equipment, machines or installations, and with ensuring the most efficient and economic means of achieving defined objectives by such means.

Table 2 summarizes current definitions of various types of scientific research.

In a recent publication of the Academies of Science of the U.S.S.R. and of Czechoslovakia[1] the following definition is used:

Any definition is conventional and relative but a brief definition of science could still be attempted as a system of developing objective knowledge which is gained by the appropriate methods of cognition and is expressed in conceptions verified in practice: at the same time science is a sphere of the activity of people who are engaged in the production of new knowledge. Science incorporates simultaneously two opposing aspects (or two essences): the spiritual, exhibiting itself in the fact that science is a specific form or knowledge (cognition), and the material, exhibiting itself in the fact that science is a direct productive force.

And the following further comments are worth quoting:

While in the past revolutions in natural science and technology sometimes did and sometimes did not coincide in time, now they become a single united process of the scientific-technological revolution. . . . The scientific and technological revolution has led to a new relationship between science and technology. . . . Science becomes a direct part of the production process so that the process itself changes.

On the relationships between fundamental and applied research a report of the Organization for Economic Co-operation and Development (OECD) states:[2]

Within the spectrum of scientific activity it is becoming increasingly difficult to define precisely the zone of fundamental or basic research. Simply stated, fundamental research consists of investigations aimed at extension of scientific knowledge, of increasing our understanding of nature; it is therefore exploration of the unknown or of the insufficiently known. Applied research, on the

other hand, attempts to put to use the findings of basic research or even to discover new knowledge which might have immediate practical application. A few decades ago the difference between these two categories of investigation was clear and their methods distinct—the approaches of the 'pure scientist' and of the 'engineer'. The distinction is becoming less and less sharp and indeed the strength of contemporary science depends very greatly on intellectual co-operation between pure and applied scientists, while in fact a single individual can be at times 'engineer' and at times 'pure scientist' as he works on the same problem using different approaches or tackling different elements, or else facing different problems. Fundamental and applied science are perhaps most easily distinguished today in terms of motivation rather than differences of nature or technique—the aim of fundamental research is extension of knowledge for its own sake, that of applied research the utilisation of existing knowledge.

An original approach to the definition of science, technology and their relationship is the one of Derek J. de Solla Price.[3] Analysing the structures of publication in science and technology, the author develops a model, according to which

Science must consist of the scientific papers that are being cited, counted, and otherwise manipulated in such studies. I therefore propose, as a formal definition, to take as science that which is published in scientific papers.

And as for the difference with respect to technology:

. . . in this new sense is pointed to most clearly by the sociology of publication. The scientist, it was reiterated, is heavily motivated to publish—this is the key to all the inner springs of his drive to do science. In technology it is otherwise; the tradition, crudely speaking, is to conceal in order to have a new product or process before others. . . . One may put the whole thing in an aphorism I have used before; the scientist wants to write but not to read, the technologist wants to read but not write.

1. 'Man—Science—Technology'. A Marxist Analysis of the Scientific-technological Revolution, Prague, Academia, 1973.
2. Fundamental Research and the Policies of Governments, Paris, OECD, 1966.
3. D. J. de Solla Price, 'The Structure of Publications in Science and Technology', in: Factors in the Transfer of Technology, 1969, p. 91–104.

TABLE 2. Definition and description of the various types of scientific research

Types of research		Motive of the investigator	Degree of freedom of the Director of Research	Individual or team research	Method of financing	Prospective applicability of results	Scientific significance of results
		From the standpoint of the investigator				From the standpoint of the results obtained or expected	
Fundamental research	Pure research (free fundamental research)	Research directed toward fuller understanding of nature and discovery of new fields of investigation, with no practical purpose in mind	Choice of field, programme and method of work	Generally individual research	Funds allocated to the individual	Delay of practical application unpredictable	Results affect a broad area of science and often have a penetrating and far-reaching effect
	Oriented fundamental research	*Field centred research* (Exploiting new fields of investigation) Fundamental research focused on a given theme, generally connected with a natural phenomenon of broad scope, and often directed towards a well-defined objective	Choice of programme and method of work	Generally team research	Funds allocated to an institution or to a laboratory	Delay of practical application generally long·	Results affect a well-determined field of science and have a general character
		Background research Research directed towards increased accuracy of scientific knowledge in a particular field by gathering essential data, observations and measurements	Choice of method (and sometimes programme) of work	Generally team research	Funds generally allocated to an institution or to a laboratory, and often related to a research programme	Delay of practical application depends essentially upon field of research	Results are of empirical character and provide the necessary basic facts for the advancement of pure and applied sciences
Applied research	Agricultural research	Research directed primarily toward understanding and improving agricultural productivity (including animal husbandry, forestry and fisheries)	Choice of method (and exceptionally programme) of work	Generally team research	Funds allocated to an institution or to a laboratory, and related to a research programme	Delay of practical application generally short	Results generally affect a limited area and have a specialized character
	Medical research	Research aimed at understanding human diseases, maintaining and improving human health					
	Industrial research	Research aimed at increasing scientific knowledge in a specific field of man's industrial activity					
Experimental development		Systematic use of the results of applied research and of empirical knowledge directed toward the production and use of new materials, devices, systems and methods for agriculture, medicine, industry, etc., including the development of prototypes and pilot plants	Field and programme of work laid down by sponsor (sometimes also experimental design of research)	Generally team research	Funds generally related to specific development programme	Practical application generally immediate	Results affect a very limited area and have a narrowly specialized character

Source: A Selection of Working Definitions used in Science Policy, Paris, Unesco, 1970 (Unesco doc. DNS/ROU/207 prov.).

Whatever approach or definition is accepted, it must be conceded that in modern societies science and technology represent a continuum. No borderline between the two can any longer be drawn and, therefore, science is technology (see Part IV, Chapter 32). Only the institutional setting in which work is carried out still permits the use of these vanishing distinctions.

Scientific medicine

The impact of new fundamental knowledge about living systems has been immediate and far-reaching on the progress of medicine.

'In considerable measure, the history of biology is the history of attempts to cope with disease'—as stated in a recent report of the National Academy of Sciences of the United States.[1] Many disorders have fruitfully been viewed as 'nature's experiments' and, as such, have proved to be cardinal clues in elucidation of major fundamental phenomena. Thus, vitamin-deficiency diseases—for example pellagra, beriberi, sprue and scurvy—were the clues to the very existence of vitamins and, hence, to the coenzymes of metabolism; investigations of diabetes and glycogen-storage diseases revealed the hormonal control of carbohydrate metabolism and, indeed, the pathways of the metabolism; the prevalence of pernicious anaemia revealed the existence of vitamin B_{12} and of the unique biochemical reactions it makes possible; the requirement for agents to manage infectious diseases stimulated the discovery of antibiotics, and these, in turn, proved to be powerful tools in the elucidation of the mechanism of operation of the genetic apparatus and synthesis of bacterial cell walls; the dramatic changes in the volume, pH and salt concentrations of blood plasma in such disorders as infantile diarrhoea, pernicious vomiting, diabetic coma, and Addison's disease have been both the primary stimuli and the major 'experiments' in revealing the complex homeostatic mechanisms that control the volume, acidity and electrolyte composition of the body fluids of both the intracellular and extracellular compartments; the variety of cardiac disorders has revealed the fine mechanisms and neural control of the cardiovascular system; and the existence of sickle cell anaemia and other instances of altered haemoglobin structure were the first demonstration that a 'point' mutation results in a specific amino acid replacement in a protein, as well as the demonstration that the genetic code in man must be identical with that in the bacterial species in which it was first determined. In each instance, the knowledge so gained, abetted by the insights from other areas of biology, has resulted in expansion and improvement of the therapeutic armamentarium to the great benefit of those afflicted with the very disorders that served as clues.

The great advances of biology in this century, reflected with brief delay in medical practice, have been primarily the results of three analytical procedures of great finesse: biochemical research, which led to the identification of the function and structure of molecules essential for living systems; immune reactions, which through the study of how the body responds to the presence of foreign organisms or molecules led to the discovery of antibodies; genetic experiments, which, associated with cytological observations, made possible the discovery of the gene. Biochemistry, with its vast arsenal of sophisticated procedures for the separation and identification of the structure of biological molecules, lent to clinical research essential tools for diagnosis and treatment of disease. Immunology, with the discovery of the chemical nature of mechanisms for resisting infection and of the system for antibody formation, offered to medical practice a wide array of tools for preventing the insurgence of disease and for discovering the uniqueness of the individual. The development of the gene theory of inheritance provided a unifying conceptual frame for ordering the most diverse biological phenomena, which until that time appeared to be totally unrelated, and for offering a re-interpretation of the process of evolution of organisms. The momentous merging of genetics with biochemistry gave birth to the new discipline of molecular biology, which made possible the de-

1. *The Life Sciences*, Washington, D.C., National Academy of Sciences, 1970.

28

ciphering of the genetic code and the solution of the enigma of how proteins are constructed by the cell.

Reading through the column 'Life' of the preceding subsection one may have an idea of the pace of discovery in fundamental biology, which steadily increased during the last two decades. Many of those discoveries have already been translated into clinical advances, in the relief from pain, in the almost complete eradication of certain diseases, in the decline of infant mortality throughout the world, in the prolongation of life expectancies. Obviously, it would be impossible to encompass in a few pages the dramatic improvements in human health due to the parallel progress of biological and medical research; but a few selected examples may offer a measure of that success.

MALARIA

Malaria eradication has been one of the most successful global public health programmes ever undertaken. Of the 1,800 million persons who lived in malarious areas before this programme was started, 1,000 million have now been freed from the threat of the disease. The infection has been almost eliminated in most of the temperate areas, and national malaria programmes are in operation in almost all Asian and Latin American countries. In Africa, where special administrative and biological problems exist, a more gradual approach is being made, the present goal being malaria control rather than eradication.

According to a report of the World Health Organization:[1]

Danger of epidemics

The threat of malaria epidemics exists wherever antimalaria activities have succeeded in interrupting or drastically reducing transmission over several years and where the cessation of attack measures has permitted the re-establishment of a high potential of transmission. The maintenance of important foci of infection in the neighbourhood of such areas (for example, urban foci in Asia), the reduction or breakdown of surveillance, and delay or failure in applying effective remedial measures (as has occurred, for example, in Asian countries as a result of war and the lack or disruption of administrative support) may lead to rapidly spreading epidemics. The danger of epidemics can be contained only by a continuous and consistent effort, demanding competent and adequate services with administrative flexibility.

Eradication and control programmes

Malaria used to be endemic in 145 (69.4 per cent) of the 209 countries or territories for which information is available. As of 31 December 1973, the disease had been eradicated from 36 entire countries or territories containing 10.3 per cent of the total population of the originally malarious areas of the world.

During the year, eradication programmes continued in 47 countries that still have malarious areas, i.e., in 32.4 per cent of those originally affected. Malaria control activities were carried out in 45 countries (32 per cent), but only in 8 did they protect more than 50 per cent of the population, while in 23 they covered less than 10 per cent. No specific antimalaria measures were applied in 17 countries or territories (11.7 per cent).

By 31 December 1973, the population of the originally malarious areas of the world was 1,900.15 million, of which 786.77 million (41.4 per cent) were living in maintenance phase areas and 596.17 million (31.4 per cent) in areas in other phases of the eradication programme. Of these, 285.80 million were residing in consolidation phase areas, 306.41 million in attack phase areas, and 3.96 million in preparatory phase areas.

Among the 517.21 million people not yet protected by malaria eradication operations, 148.14 million (28.6 per cent) were benefiting from antimosquito measures or regular drug prophylaxis; 93.27 million (18.0 per cent) were not yet systematically protected but resided in areas where antimalarial treatment was available; while 275.80 million (53.3 per cent) were still completely unprotected.

In the year ending 31 December 1973 the maintenance phase areas increased by 2.2 per cent; at the same time, there was a decrease in areas in the other phases of malaria eradication, as no new programmes were started. A slight increase was observed in areas under intensive mosquito control or regular drug prophylaxis, while those not yet protected by any measures remained virtually unchanged.

1. 'Re-examination of the Global Strategy of Malaria Eradication', *Official Records of WHO*, No. 176, December 1969, Geneva.

TABLE 3. Malaria eradication and other antimalaria activities in the WHO regions, 31 December 1973

	Population (thousands)						
	African Region	Region of the Americas	South-East Asia Region	European Region	Eastern Mediter- ranean Region	Western Pacific Region	Total
Areas where malaria was never indigenous or disappeared without specific antimalarial measures	26,405	342,615	35,761	437,799	42,277	182,891	1,067,748
Originally malarious areas	234,478	198,259	834,551	357,866	205,376	69,624	1,900,154
Areas where malaria eradication is claimed (maintenance phase)	5,251	89,030	370,640	297,475	5,836	18,537	786,769
Areas where eradication programmes are in progress:							
consolidation phase	2,233	46,165	164,471	40,971	28,427	3,533	285,800
attack phase	775	46,755	165,721	5,696	77,048	10,416	306,411
preparatory phase	51	—	—	2,954	—	957	3,962
	3,059	92,920	330,192	49,621	105,475	14,906	596,173
Areas protected by exten- sive mosquito control measures	10,001	—	78,459	2,059	44,355	5,655	140,529
Areas where prophylactic drug administration is extensively used as a control measure	2,307	—	—	4,382	922	—	7,611
Areas where there is an organized effort to make antimalarial drugs easily available	5,038	16,199	11,923	136	34,790	25,184	93,270
Areas with no specific antimalarial measures	208,822	110	43,337	4,193	13,998	5,342	275,802
Total population	260,883	540,874	870,312	795,665	247,653	252,515	2,967,902

Source: 'The Smallpox Situation', WHO Chronicle, Vol. 26, 1972, p. 393–400.

Tables 3 and 4 show the malaria eradication situation at five-yearly intervals since 1958. It is evident that the programme as a whole expanded between 1958 and 1968, while there was some regression between 1968 and 1973 as a result of the strategy review and a change of medium-term objectives in some areas. The tables also show that in general the eradication programmes have made continuous progress as far as phase distri- bution is concerned.

In 1973 WHO assisted antimalaria activities in 70 countries, 38 of them having malaria eradi- cation programmes and the remainder control programmes. In four more countries, the Organiza- tion helped with the planning of malaria control programmes expected to become operational in 1974.

In the *African Region* the epidemiological situ- ation has remained largely unchanged, being characterized by a high level of malaria endemicity and intense transmission in most of the Region. Two islands have reached the maintenance phase: Mauritius, which has been entered in the register of areas from which malaria has been eradicated, and Réunion, where the relevant assessment was carried out in 1973.

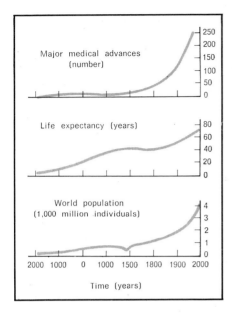

FIG 2. The similarity between the curves for world population growth, life expectancy and number of major medical advances, especially for the past few centuries, is certainly not a coincidence. It is calculated that, since 1900, the life expectancy in the materially most advanced countries has increased by twenty years (on average). (From J. Mehale, *World Facts and Trends*, New York, 1972.)

TABLE 4. Progress in malaria eradication and other anti-malaria activities, 1958–73

	Population (millions)			
	1958	1963	1968	1973
Areas with eradication programmes:				
maintenance phase	283	343	651	787
consolidation phase	64	354	346	286
attack phase	516	359	356	307
preparatory phase	—	49	13	4
TOTAL	863	1,105	1,366	1,384
Areas without eradication programmes but with limited control measures	396	397	367	148
Areas with no specific antimalaria measures				368
TOTAL	396	397	367	516
GRAND TOTAL	1,259	1,502	1,733	1,900

SMALLPOX

'From the earliest chronicles of ancient civilization, smallpox has written across the pages of recorded time an unparalleled history of death, blindness and disfigurement throughout the world.' So wrote L. Thapalyal in the magazine of the World Health Organization, *World Health*, October 1972.

After the disease had almost completely disappeared in the industrialized countries through compulsory mass vaccination, the World Health Organization launched a global strategy for eradicating smallpox from Asia, Africa and Latin America. A dramatic reduction has occurred throughout the world. The estimated annual incidence of the disease has declined from about 2.5 million cases in 1967 to less than 200,000 in 1972; the number of countries in which smallpox is endemic has decreased from 91 in 1945 to 7 in 1972. By 1974 smallpox has been eradicated from the Americas and from Indonesia; WHO experts foresee that the disease will disappear from the last foci in Bangladesh, Ethiopia, India and Pakistan by the end of 1975.

TUBERCULOSIS

With the advent of highly effective, relatively non-toxic and inexpensive anti-tuberculosis drugs it has been possible to develop treatment regimens, the efficiency of which is nearing the

TABLE 5. Tuberculosis mortality rates (per 100,000 population)

Country	1952	1957	Total number of deaths
Finland	57.7	38.0	402 (in 1968)
France	43.8	27.0	153 (in 1964)
United Kingdom	28.4	12.3	
Italy	27.7	20.6	
Switzerland	25.4	17.6	
Denmark	24.9	10.9	
Sweden	17.3	8.8	
Canada	17.1	7.1	25 (in 1969)
United States	15.8	7.5	27 (in 1969)
Netherlands	12.3	4.7	19 (in 1964)

Source: WHO.

100 per cent level. Thanks to the advent in recent years of a newer and better version of antituberculosis vaccine (BCG) and especially of the chemical compound Isoniazid, there has been a dramatic fall in mortality figures which, for example, during the ten-year period from 1954 to 1963, decreased in the United States by 85 per cent, in France by 67 per cent, and in Romania by 80.6 per cent. Table 5 gives similar statistics for the first five years after introduction of the drug.

Over a period of about one century, for example, mortality due to tuberculosis has fallen in Italy from over 500 to less than 20 per 100,000 inhabitants per year.

ENDEMIC TREPONEMATOSES

Yaws is a contact disease among children, characterized by crops of highly infectious and relapsing skin lesions in the first five to six years of the natural course of the infection. In adolescent and adult life, outbreaks of incapacitating hyperkeratosis occur on the palms and soles, and destructive mutilating lesions of subcutaneous tissues and bones develop in a large proportion of those infected. By contrast, endemic syphilis involves also mucous membranes, while pinta involves mostly the integument alone.[1]

On the basis of pilot studies of yaws in Haiti, endemic childhood syphilis in Yugoslavia and pinta in Mexico, mass penicillin campaigns

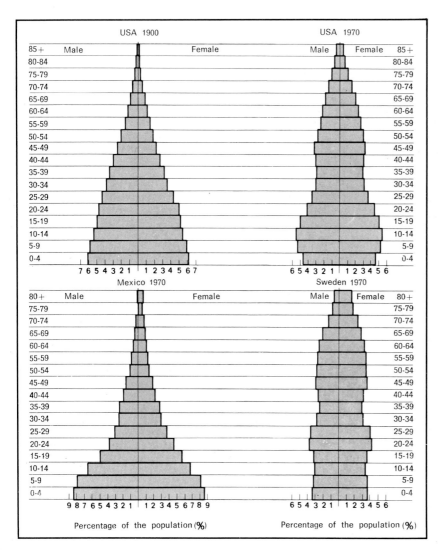

FIG. 3. With the progress of medicine (compare the United States in 1900 and 1970) and the spread of medical care (compare Mexico and Sweden, a developing and a developed country), life expectancy increases, broadening the top of the population's age pyramid.

TABLE 6. WHO treponematoses programme: prevalence and reduction of infectious endemic treponematoses in mass penicillin campaigns, 1954–65, in areas where sero-epidemiological studies were subsequently undertaken. All these programmes concern yaws except in Bosnia (Yugoslavia), where the campaign was directed against endemic syphilis

| Country or area | Period | Rural population involved (millions) | Initial treatment survey (ITS) | | No. of re-surveys | Last re-survey level of infectious yaws (%) |
			Population coverage at ITS (%)	Infectious yaws (%)		
Northern Nigeria	1954–65	2.65	83	4.2	1–7	0.02
Togo	1956–65	1.50	40	4.1	2–4	0.45
Midwestern Nigeria	1955–64	1.49	77	3.2	2–5	0.18
Western Samoa	1955–61	0.10	96	3.0	5–7[1]	0.005
Eastern Nigeria	1954–63	6.80	54	1.9	1–5	0.09
Western Nigeria	1956–63	1.90	59	1.8	2–5	0.21[2]
North-eastern Thailand	1952–60	8.40	50	0.7	2–5	0.09
Southern Thailand	1952–60	3.00	70	0.13	2–5	0.01
Philippines	1952–60	2.40	33	0.1	1–4	0.01
Yugoslavia	1948–54	0.83	80	0.4	1–8	0.00

1. Includes also a child survey and a sampling survey.
2. Includes also non-infectious cases.

Source: R. Guther, J. Ridet, F. Vorst, J. D'Costa and B. Grab, 'Methods for the Surveillance of Endemic Trepanomatoses and Seroimmunological Investigations of "Disappearing" Disease', WHO Bulletin, No. 46, 1972, p. 1–14.

were undertaken by health administrations in forty-six countries in the context of the WHO treponematoses programme. Table 6 shows the geographical distribution and extent of endemic treponematoses of childhood twenty years ago. Up to 1970, some 160 million people had been examined and some 50 million clinical cases, latent cases and contacts had been treated in these campaigns. In the first decade, attention was focused on programme application and on the control of disease.

Table 6 also summarizes some of the results obtained by the WHO treponematoses programme.

CHOLERA

The incidence of cholera in the 1960s as reported to WHO is shown in Table 7; it should be noted that these figures are not absolutely reliable.

As may be seen from the table, in 1961 the overall case fatality in the world was about 50 per cent (in earlier years it was even higher) while in 1970 it was only 17 per cent, i.e. one

in six. This change during ten years is mainly due to the use of modern rehydration therapy, and means that some 115,000 lives were saved. In well-organized hospitals or treatment centres, where facilities for care are available, the case fatality rate is below 1 per cent.

In the same period the fatality in cerebrospinal meningitis has fallen, due to the use of sulphonamides, from about 40 to 17 per cent in the 'meningitis belt' of Africa.

Strangely enough, it appears that no attempt has been made to estimate the number of lives that have been saved as a result of the introduction of new treatments for selected diseases. Tables 8 and 9, however, offer some interesting indication of the overall effect of improved sanitation, medical care, economic growth, and social improvements.

1. R. Guther, J. Ridet, F. Vorst, J. D'Costa and B. Grab, 'Methods for the Surveillance of Endemic Trepanomatoses and Seroimmunological Investigations of "Disappearing" Disease', WHO Bulletin, No. 46, 1972, p. 1–14.

33

TABLE 7. Incidence of cholera

Year	Cases	Deaths	Year	Cases	Deaths
1961	62,509	31,215	1966	33,779	4,886
1962	45,751	15,777	1967	24,167	3,399
1963	67,363	23,380	1968	28,463	4,324
1964	79,001	21,406	1969	33,041	5,833
1965	55,297	14,442	1970	45,011	7,882

TABLE 8. Changes in infant mortality rates in selected countries

Country	Infant mortality rate[1]	
	1950	1968–69
Poland	108	34.3
Yugoslavia	118.6	58.6
France	51	19.6
Switzerland	31.2	15.4

1. Deaths under 1 year of age per 1,000 live births.

TABLE 9. Changes in expectation of life at birth in selected areas

Areas	Expectation of live at birth (years)	
	1950	1970
Developed countries	64.6	70.4
Developing countries	41.7	49.6
West Africa	32.5	39.2
Japan	61.9	70.9
U.S.S.R.	61.7	70.3
United States/Canada	68.7	70.5
Europe	65.4	70.9

People living in the developing countries have seen the traditional pattern of disease and misery change within their own lifetime and life expectancy rise by almost twenty years (see Fig. 3). With the advent of preventive medicine, sanitary engineering and the use of 'miracle drugs' death rates dropped almost suddenly (by 50 per cent in one decade in certain cases) and population started to grow at unprecedented rates (in some cases over 3.5 per cent per year as compared to 1 per cent per year before the introduction of modern medicine and technologies, see Chapter 5).

Scientific agriculture

For most ... the outstanding technological achievement of this generation was the landing on the moon ... but for one billion Asians for whom rice is the staple food, the development of IR-8 and its dissemination throughout Asia is a more meaningful achievement. It is literally helping to fill hundreds of thousands of rice bowls once only half full. For those for whom hunger is a way of life, technology can offer no greater reward.

In these words Hasan Ozbekhan expressed the impact of a product of scientific plant breeding which brought about the 'green revolution'. And on this subject it is worthwhile to quote the words of Norman Borlaug, who was awarded the Nobel Prize for Peace for his contribution to the development of the new varieties of rice and other crops:[1]

There are no miracles in agricultural production. Nor is there such a thing as miracle variety of wheat, rice or maize which can serve as an elixir to cure all ills of a stagnant traditional agriculture. Nevertheless, it is the Mexican dwarf wheat varieties, and their more recent Indian and Pakistani derivatives, that have been the principal catalyst in triggering off the 'green revolution'. It is the unusual breadth of adaptation, combined with high genetic yield potential, short straw, a strong responsiveness and high efficiency in the use of heavy doses of fertilizers, and a broad spectrum of disease resistance that had made the Mexican dwarf varieties the powerful catalyst that they have become in launching the 'green revolution'. They have caught the farmers' fancy, and during the 1969–70 crop season 55 per cent of the 6 million hectares sown to wheat in Pakistan and 35 per cent of the 14 million hectares in India were sown to Mexican varieties or their derivatives.

This rapid increase in wheat production was not based solely on the use of Mexican dwarf varieties; it involved the transfer from Mexico to Pakistan and India of a whole new production technology that enables these varieties to attain their high yield potential. Perhaps 75 per cent of the results of research done in Mexico in developing the package of recommended cultural practices, including fertilizer recommendations,

1. N. E. Borlaug, *The Green Revolution: For Bread and Peace.* Reprinted by permission of Science and Public Affairs from *Bulletin of the Atomic Scientists,* June 1971, p. 6–9, 42–9. Copyright by the Educational Foundation for Nuclear Science.

were directly applicable in Pakistan and India. As concerns the remaining 25 per cent, the excellent adaptive research done in India and Pakistan by Indian and Pakistani scientists, while the imported seed was being multiplied, provided the necessary information for modifying the Mexican procedures to suit Pakistani and Indian conditions more precisely.

Just as in the case of the numerous victories obtained by medical practice based on scientific knowledge, the latter has provided the basis for increases in agricultural production and fish catch around the world. The combined efforts of geneticists, physiologists, entomologists, biochemists—and of many other specialists concentrating their efforts on cultivated plants, domesticated animals, timber and fish—laid the basis for a technical revolution in agriculture which is unparalleled in history. While important contributions have come from many different sources, such as universities, foundations and experimental stations, the Food and Agriculture Organization of the United Nations (FAO) has played a very significant role in co-ordinating the efforts to feed the growing population of the world and more particularly to spread modern agricultural techniques and practices in the developing countries.

Among the major technical innovations made possible by research undertaken during and immediately after the Second World War were the development of 'selective' herbicides and insecticides, the vast range of chlorinated hydrocarbon- and phosphorus-based pesticides; the application of ionizing radiations and chemical compounds to induce mutations in plants and in pests; the use of sonic methods in water exploration and in oceanic fishing; other great advances in land and water resource survey methods resulting from a combination of improved photographic equipment in aerospace technology, culminating in electronic remote sensing; improvement in farm machinery and methods of traction in seeding, fertilizer placement, cultivation, crop protection (including aerial spraying) and harvesting, all resulting in increased yields and much higher output per worker; 'factory farming' of livestock based on better disease control, automation and computerization.

Changing patterns of consumer demand—as is pointed out in a recent FAO report[1]—both in food products (freezing, freeze-drying, prepacking) and in relation to industrial or export crops—have generated research and changes in technology not merely in end uses but right back to the breeding of special varieties, suited to particular consumer requirements or manufacturing specifications. Canadian work on breeding oil rapeseed varieties with different fatty acid compositions, to meet varied and specific end-uses, is an outstanding recent example. This, as with certain other significant advances in plant improvement, has been made feasible not only by improvements in breeding techniques, but by teamwork between breeders, chemists and physicists using novel equipment and screening methods such as chromatography and automatic protein analysers.

The most spectacular achievement of scientific agriculture in recent years has been the 'green revolution' already mentioned. Over the period 1968–70 wheat production rose spectacularly in India and Pakistan. Using as a base the pre-'green revolution' crop year 1964–65, which produced an all-time record harvest in both countries, the production in Pakistan increased from the 1965 base figure of 4.6 million tons to 6.7, 7.2 and 8.4 million tons, respectively, in 1968, 1969 and 1970. West Pakistan became self-sufficient in wheat production for the first time in the 1968 harvest season, two years ahead of predictions. Indian wheat production has risen from the 1964–65 pre-'green revolution' record crop of 12.3 million tons to 16.5, 18.7 and 20.0 million tons during 1968, 1969 and 1970 harvests, respectively. India is approaching self-sufficiency and probably would have attained it by now if rice production had risen more rapidly, because, with a continuing shortage of rice, considerable wheat is being substituted for it.

If the introduction of new varieties of rice and wheat into the agricultural practices of Asian peoples achieved such spectacular results, one should not forget that these achievements were preceded by decades of scientific

1. *The State of Food and Agriculture, 1972*, Rome, FAO, 1972.

research on plant genetics and by a vast array of important improvements in the production of crops relevant to the agricultures of countries with temperate climates.

Skimming through the columns 'Life' and 'Technology' of the earlier subsection of this chapter, one sees that the original scientific discovery of hybrid vigour in maize was made in 1908 and that it took twenty-five years before this product reached the stage of commercial production of hybrid seed for improving the yields of the 'corn belt' of the North American continent. In that year early in the century, two American geneticists, E. M. East and G. H. Shull, discovered that if one crosses two inbred lines of corn one can

obtain hybrids having a much greater yield than either of the original strains, and remarkable uniformity. By choosing for hybrid production appropriately inbred lines carrying desirable traits, such as disease resistance or peculiar characteristics of the kernel, one could obtain high yielding corn especially suited to a variety of environmental conditions. Later D. F. Jones discovered the efficiency of four-way crossing, in which four different inbred lines are crossed two at a time and then the two resulting hybrids crossed again, to produce in the second generation commercial seed for corn production. As a result of this procedure corn yield per cultivated surface area increased by a factor of two and even three throughout the corn belt of the United States and later in other countries.

In the meantime the knowledge of hereditary mechanisms in plants and animals was progressing by leaps and bounds, so that a vast variety of genetic techniques could be exploited for improving the production of agricultural products. These included artificial induction of mutation, crossings between cultivated plants and their wild relatives, induced doubling of chromosome numbers (polyploidy), introduction of self-sterility genes, and so forth.

The 'green revolution' had an immediate impact of incalculable importance, not merely in raising production and productivity in the critical food-deficit areas of Asia (as the FAO report makes clear), but in disposing of two myths. The first of these was that research was either slow-yielding or low-yielding, or both; the second, that even if research produced potentially successful results, the farmers growing food crops were too subsistence-minded or ignorant to utilize them. Calculations by Griliches showed a rate of return of no less than 700 per cent on United States hybrid maize research as of 1955, and of 360 per cent on hybrid sorghum as of 1967, while Barletta had calculated a return of 750 per cent on wheat research in Mexico between 1943 and 1963 (see Table 10).

Although it is difficult to calculate a rate of return on the global effects of the spread of Mexican wheats or on the Asian rice improvement programme, because of the dual involve-

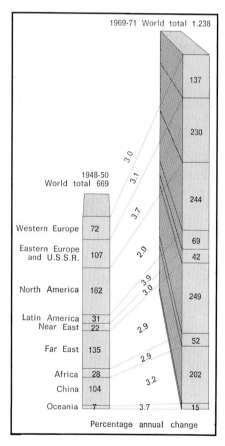

Fig. 4. Growth of world cereal production (in millions of metric tons) between 1948–50 and 1969–71. (From FAO, *The State of Food and Agriculture, 1972*, Rome, 1972.)

TABLE 10. Estimates of the social rates of return to investment in agricultural research

Type of study	Social rate of return	
	Returns at end of year above a 10 per cent discount rate[1]	Returns distributed internally[2]
	%	%
Particular United States farm products		
Hybrid maize research, public and private, as of 1955[3] and internalized over 1910–55	700	35–40
Hybrid sorghum research, public and private, as of 1967[3]	360	
Poultry research, public, 1960 and internalized over 1915–60[4]		
Feed efficiency	178	25
Total productivity	137	21
United States agriculture, 1949, 1954 and 1959		
Public and private agricultural research and extension adjusted for excess capacity[5]	300	
United States agriculture, 1938–63		
Public agricultural research and extension[6]		54–57
Adjusted for private research[6]		46–48
Agricultural research in Mexico		
Wheat research, 1943–63[7]	750	
Maize research, 1943–63[7]	300	
Total agricultural research in Mexico, 1943–63[7]	290	
Japanese agriculture, 1880–1938		
Predominantly investment in education, for example, in 1880, education 23.6 million yen, and agricultural research and extension 0.3 million yen; and in 1938, 185 and 21.5 million yen, respectively[8]		(35)

1. These returns are obtained by applying a 10 per cent discount rate to the flow of cost incurred over time accumulated and, also, to the flow of benefits obtained over time accumulated. The 10 per cent discount rate is assumed to be a reasonable proxy for the rate of return on alternative social and private investment. The use of estimate B, the internal rate of return, may attribute an inordinately high value to a dollar spent in the more distant past. For example, in the case of hybrid maize, the internal rate of

ment of international and national research and plant breeding programmes in each case, the incremental returns have to some extent been assessed. The cumulative value of the addition to wheat production in India between 1966 and 1969 is estimated to be U.S.$850 million (mainly attributed to high-yielding varieties and related inputs; the rice production increase for Asia (excluding China) over the same period is believed to be around $1,500 million. In contrast, the capital and recurrent expenditures necessary for the International Rice Research Institute (IRRI) in the Philippines, from its establishment in 1962 until 1970, were of the order of $20 million.

Too few people realize the immense progress achieved over the last twenty years in producing the historic 3 per cent annual growth rate in world agriculture. Had the world's population not rapidly increased, there would be no hunger in the world. Figure 4 shows that world grain production almost doubled between 1948–50 and 1969–71, from 669 million to 1,238 million tons, at the growth rate of 3 per cent per year already cited.

The applications of discoveries made in other fields of science have led similarly to major increase in the marine food catch (see Fig. 5). The possibility of identifying the presence of fish schools and of estimating their

return attributes a value of $2,300 to a dollar spent in 1910 in developing hybrid maize.
2. This is that rate of return which equates the flow of costs and flow of returns over time: it thus distributes the net benefits equally over the entire period measured in terms of the internal rate of return. Estimates A and B are different ways of interpreting the same set of cost and benefit facts.
3. Zvi Griliches, Research Costs and Social Returns: Hybrid Corn and Related Innovations, *Journal of Political Economy*, Vol. 66, 1958, p. 419–31.
4. Willis Peterson, *Returns to Poultry Research in the United States*, Ph.D. dissertation, University of Chicago, 1966.
5. Zvi Griliches, Research Expenditures, Education and the Aggregate Agricultural Production Function, *American Economic Review*, Vol. 54, 1964, p. 967–8.
6. Robert E. Evenson, *The Contribution of Agricultural Research and Extension to Agricultural Production*, Ph.D. dissertation, University of Chicago, 1968.
7. Nicolas Ardito-Barletta, *Costs and Social Returns of Agricultural Research in Mexico*, Ph.D. dissertation, University of Chicago, 1967.
8. Anthony M. Tang, Research and Education in Japanese Agricultural Development, 1880–1938. *Economic Studies Quarterly*, Vol. 13, 1963, p. 27–42, 91–100.
Source: The State of Food and Agriculture, 1972, Rome, FAO, 1972.

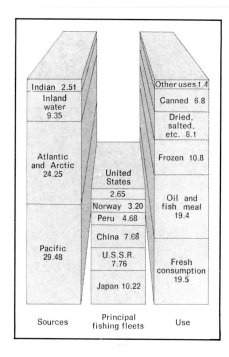

FIG. 5. World marine food catch (in million metric tons) showing major sources and use in 1973 (From J. McHale, *World Facts and Trends*, New York, 1972, updated).

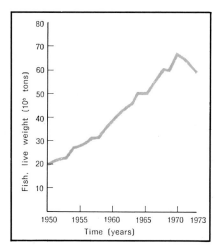

FIG. 6. Since about 1970, the figures for the world marine food catch (hitherto on the increase) began to drop disturbingly; this seems to be attributable not so much to actual 'ceilings' as to indiscriminate fishing. The situation requires planned cultivation of marine resources.

dimensions and movements improved steadily the efficiency of oceanic fishing to the point of having reached perhaps the maximum compatible with the reproduction and survival of fish populations. Today, refined physical methods permit a kind of scientific husbandry of fish populations.

Scientific engineering

The attempt to summarize in a few pages the developments of scientific engineering in the past few decades would be doomed to an even greater failure than in the case of medicine and agriculture. Our daily life is conditioned by innumerable technological products, each of which would call for comment: air transport, plastics, stainless steel, television, radar, copying machines, automobiles, nuclear energy, computers, detergents, transistors, communication satellites, radio-active tracer elements, synthetic fibers, high-strength materials, etc.; the list could be extended at will. Figure 7 shows an example of the immense impact of science-based technologies on the world in which we live today.

As John McHale puts it, in *World Facts and Trends*:[1]

In dealing with physical technology, we need to note the difference between preindustrial tool forms and specifically industrial tool forms. The former tended to single inventions locally operable by individuals or small groups and maintainable within the knowledge, limited energy, and material resource capacities of local regions, e.g., the sled, canoe, hammer, or plow. Industrial tools may be more specifically characterized as those which represent the coupling together of numbers of single tool forms into complex and large-scale systems which may not be wholly operated by individuals or small groups nor maintained by local resources alone, e.g., an airline, a telephone system, or a whole industry. They require the co-operative organizational energies of myriad individuals and access to the widest range of accumulated human knowledge and to the variety

1. J. McHale, *World Facts and Trends*, New York, Collier Books, 1972. Reprinted with permission of Macmillan Publishing Co. Inc. Copyright 1972 by John McHale.

of energies and materials obtainable only at the global scale.

Man has always assumed that an 'evolving' technology would be of the mythological robotic variety, formed in his own image. It is rather more difficult to observe the evolution of the airplane from single person/single engine with multiple-wing surfaces, to multi-engine/single wing, to propellerless jets of enormous size, speed, and 400 passenger carrying capacity—almost in one human generation.

It is as difficult to equate the evolution of the family of 'extended eyes'—from bulky, tripod, wetplate still cameras to microminiaturized tele-vision scanners spinning around the globe outside of the earth's atmosphere. We may better assess and control the further development of our tech-nologies by applying to them some bio-evolution-ary approach that we use in studying other life forms in the ecosystem.

As powered, renewed, and ultimately directed by human life, technology is as organic as a snail shell, a spiderweb, the carapace of a turtle, or an airborne dandelion seed. In many respects, it is now more ubiquitous as a functional component of the ecosystem than any other organic form except man himself.

The amounts of energy converted by machines; the materials extracted from the earth, processed, recombined, and redistributed in the technological metabolism; and the gross effects of such increased industrial metabolic rates on the ecosystem are now greater than the effects of many global popu-lations of other organic species.

The industrial revolution, i.e., the large-scale application of inanimate machine energies to productive use, marks the point where humanity was potentially freed from the constraints of agriculturally based, marginal, survival societies to forms based on possible machine-produced abundance.

It is the point also, we may note, when there occurred the specific and critical interdependency of the various world regions—through improved communications and transportation, and the need for globally available materials and markets. This is a major characteristic of the world industrial process which tends to be ignored, i.e. 'the peculiar interrelatedness that makes it impossible for cer-tain countries to talk in terms of national econ-omic growth'.

We will limit our survey to only a few indi-cators of major technological change: energy, materials, speed, information processing and large systems.

ENERGY

The world's energy production, as measured in tons of coal equivalent, has increased 3.5 times during the 1930–70 period, a figure which corresponds on average to about 3.2 per cent per annum. The average growth rate of total energy production in the United States was 2 per cent, only marginally more than the average population growth rate over the same period.

NEW MATERIALS

Industrial synthesis of carbon compounds at first used coal-tar as the major source of new products, such as dyes; but when oil, bitumi-nous rocks and other fossil resources became available in large quantities, hydrocarbons replaced coal-tar as the raw material for most synthetic materials (such as plastics). There is now practically no limit to the carbon com-pounds which can be synthesized. As an indi-cation of the economic importance of plastics, Table 11 shows the development of the manu-facture of plastics in the Federal Republic of Germany.

At the level of inorganic chemistry, how-ever, major technical and industrial advances have occurred, primarily because a vast series of new analytical tools became available. Elec-tron microscopes, X-ray diffraction analysis, microprobe analysers and many more new instruments or methods made possible a better

TABLE 11. Development of plastic manufacturing in the Federal Republic of Germany, 1958–69

	1958	1969	1969 as % of 1958
Percentage of total production of manufacturing industries	0.60	1.86	310
Value added in million D.M. at prices of 1962	7.51	47.92	635
Persons employed (in 1,000s)	64	153	238
Value added per person in 1,000 D.M. at prices of 1962	12	31	267
Capital input per person employed at 1,000 D.M. in prices of 1962	11	25	236

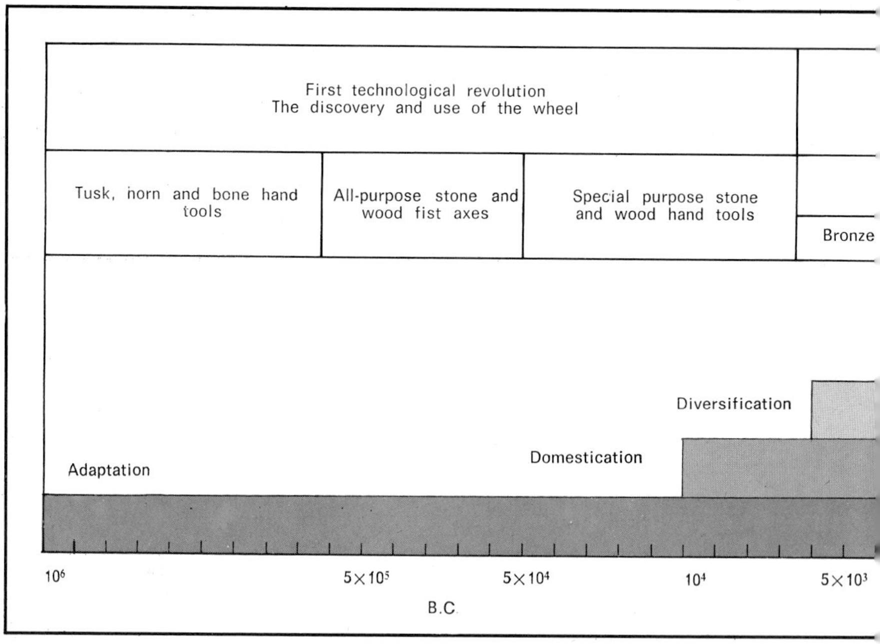

First technological revolution
The discovery and use of the wheel

Tusk, horn and bone hand tools | All-purpose stone and wood fist axes | Special purpose stone and wood hand tools | Bronze

Diversification

Domestication

Adaptation

10^6 5×10^5 5×10^4 10^4 5×10^3

B.C.

FIG. 7. The diagram illustrates the evolution of *Homo faber* and the increasing speed of his progress in regard to the subjugation and control of nature, territorial expansion, change of forms and methods of interaction with his environment. Such are the stages, following one another in ever faster succession, of the transformation of biosphere into noosphere. (From J. McHale, *World Facts and Trends*, New York, 1972.)

understanding of the relationships between the structure and properties of metal alloys: we now have many new alloys which have been 'designed' to meet specific industrial requirements. Similarly, the acquisition of detailed knowledge about the molecular structures of a vast series of chemical compounds affecting industrial development has played a significant role in many technological developments influencing our everyday life. To mention only one field, solid state physics made the advent of the transistor possible and this in turn catalysed the immense growth of electronic industry.

SPEED

The mobility of man, speed in transportation and the ensuing ability to explore the universe have increased at unprecedented rate particularly during the last decades. Correspondingly, our planet has shrunken as a result of man's increased speeds of travel and communication.

INFORMATION PROCESSING

Development in data processing as a result of the Second World War resulted in a new

awareness of the enormous potential of electronic computers, leading to research and development in this sector on a vast scale in subsequent years. With the introduction of the first commercial computing machines in 1953, the information processing industry began to grow rapidly.

LARGE SYSTEMS
AND THE INTERDISCIPLINARY
APPROACH

The problem of optimizing the efficiency of large organizations and systems has become one of almost universal concern. In order to control any system effectively, one must have a model of the system, which might be optimized with respect to suitable performance criteria. Model building is an essential feature of both 'operational research' and 'system engineering': both attempt to design or to improve upon complex systems involving men, machines, materials and money. Whereas operational research developed during the Second World War to help the decision-making process at the strategic and tactical levels and later evolved into a quantitative procedure for helping management in public and private agencies, the interest of control and system

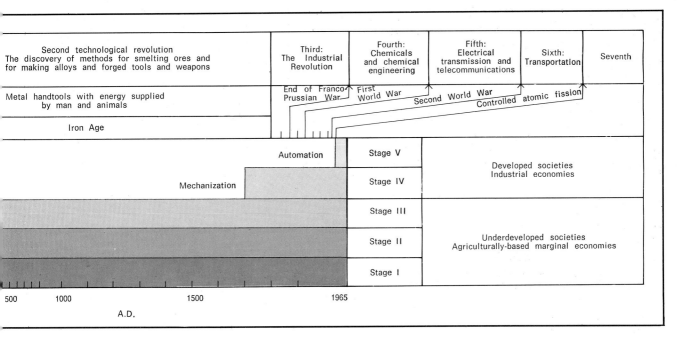

Second technological revolution The discovery of methods for smelting ores and for making alloys and forged tools and weapons	Third: The Industrial Revolution	Fourth: Chemicals and chemical engineering	Fifth: Electrical transmission and telecommunications	Sixth: Transportation	Seventh

Metal handtools with energy supplied by man and animals

End of Franco-Prussian War · First World War · Second World War · Controlled atomic fission

Iron Age

Automation	Stage V	Developed societies Industrial economies
Mechanization	Stage IV	
	Stage III	Underdeveloped societies Agriculturally-based marginal economies
	Stage II	
	Stage I	

500   1000   1500   1965

A.D.

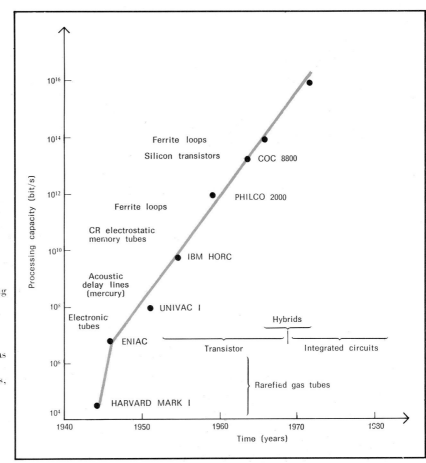

FIG. 8. The processing capacity of a computer, measured here by the operating speed of the machine (i.e. the data storage capacity—information units or bit—and the processing speed) is constantly increasing. This has made it possible for growingly complex systems to be run by computers: machines can handle an ever-larger number of elements, analyse possible combinations, and identify links between seemingly unconnected variables. (From United Nations, *The Application of Computer Technology to Development*, New York, 1971.)

Processing capacity (bit/s)

10^{16}

10^{14} Ferrite loops
          Silicon transistors   COC 8800

10^{12} Ferrite loops   PHILCO 2000

CR electrostatic memory tubes

10^{10} IBM HORC

Acoustic delay lines (mercury)

10^8 Electronic tubes   UNIVAC I   Hybrids

          ENIAC   Transistor   Integrated circuits

10^6 Rarefied gas tubes

10^4 HARVARD MARK I

1940   1950   1960   1970   1980

Time (years)

engineers in management systems has arisen as an extension of their experience in modelling and controlling the processes of dynamic engineering.

Problems arising in this zone of the management of large systems involving men, machines, materials and money call for an interdisciplinary approach since single specialists

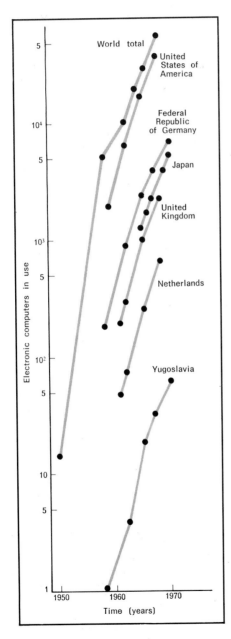

FIG. 9. The number of electronic computers (indicated on a logarithmic scale in the diagram) increases with the growth of the number of uses—scientific, technical and organizational.

TABLE 12. Computer development

1950	First generation
	First computers (United Kingdom and United States)
	Assembly language
1953	First commercial delivery
	Introduction of magnetic tapes
	Development of magnetic core memories
	Computers oriented towards data processing
	High speed alphanumeric (alphabetic and numeric) output available
1958	Second generation
	Introduction of operating systems
	First user language (FORTRAN)
	Multiplicity of problem-oriented languages (ALGOL, COBOL, etc.)
	Process control (machine tools, petrol cracking, etc.)
	Transistors (increased reliability, less air-conditioning)
	Hardware reliability high: ceases to be measured
	Large disc storage
1965	Third generation
	Computer systems (hardware-operating system)
	Integrated circuits (more reliability, improved performances/cost ratio)
	Multiprogramming systems
	Remote batch-job entry
	Time-sharing
	Interactive graphics
	Monolithic circuits
	Optical character reading
1974	Fourth generation
	Distributed systems of general purpose and decentralized satellite computers interacting through a data communication network
	Data base systems, integration of files, protocols for computer intercommunication
	Functional modules for data processing built as single electronic components (LSI— Large-scale integrated units; COC— Computer on a chip, Microprocessor)
	Conversational languages, data manipulators, simulators
	Software engineering

The first computers, designed for scientific computation, were built in university or research environments. With the first commercial deliveries, major application shifted to data processing.

will not have, in general, sufficient experience or wide enough knowledge to deal with the multiplicity of issues requiring resolution. At present, only preliminary steps have been taken towards the understanding and control of complex systems; models of national economies or of large industrial organizations that are of significance in managerial decision-making are only now emerging. The availability of large digital computers makes it possible to take into account, in a short time, immense masses of data, but the design of successful models is severely hampered by the inadequacy of techniques available for the analysis and synthesis of large time-varying systems and more particularly by the paucity of adequate data in this field.

In recent years major changes have occurred in the kinds of technology used, but the quantitative aspects of technological change in modern societies are even more impressive. Increases of several orders of magnitude have occurred in the size of productive units, in the dimensions of economies, and in the quantity and quality of performance. At the same time, the increase in mobility and in ease and speed of communication have allowed the rapid transfer of ideas from one country to others, the acceptance of new ways of life, and the multiplication of client demands. The profound transformations of modern societies resulting from the advance of science and technology have undoubtedly brought many benefits to mankind. This advance, however, is not an unqualified blessing. Unfortunately, technological progress has been marred in many cases by untoward effects: these aspects will be discussed in Chapter 5 and, in more general terms, in Part IV of this volume.

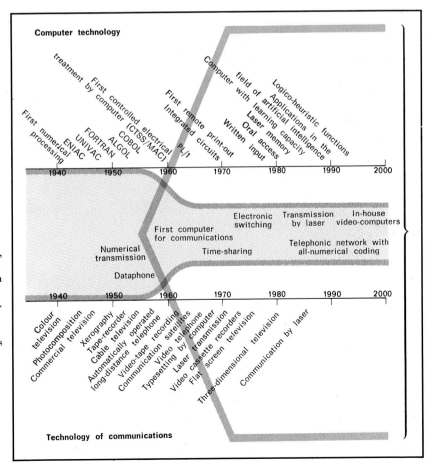

FIG. 10. In the second half of the century, the volume information flow has increased exponentially. Communication networks now cover the entire globe, unhampered by either time or distance. The importance of these services is growing, and the interval between the advent of social and technological changes and their repercussions has been shortened. This increases the interdependence between systems which were formerly autonomous, producing a marked change in the perception of the social and physical environment. (From J. McHale, *The Changing Information Environment: A Selective Topography*, Binghampton, 1971.)

4 Dangerous science

... the nuclear armories which are in being already contain large megaton weapons, every one of which has a destructive power greater than that of all the conventional explosives that had ever been used in warfare since the day gunpowder was discovered. Were such weapons ever to be used in numbers, hundreds of millions of people might be killed, and civilization as we know it, as well as organized life, would inevitably come to an end in the countries involved in the conflict.   *From a special committee report to the Secretary-General of the United Nations, 1967.*

Unless the strategic arms race between the superpowers ends, or even slows down, during the 1970s, the stockpile of nuclear warheads will rapidly grow, nuclear warhead design will continue, and new generations of increasingly sophisticated delivery systems will be developed and deployed. *Frank Barnaby, Director of SIPRI, 1972.*

The largest share of the total expenditures of the governments of the world for sustaining the scientific and technological effort goes to the production of weapons. In the period 1961–70 (inclusive), $200 billion were spent in military research; in the year 1970 alone, $25 billion were spent for this purpose. More than one-quarter of all the scientists and engineers of the world, including those working in industrial laboratories, devote their energies to military developments. This type of research, specifically directed at the destruction

of man and his buildings, crops, industries and environment, qualifies as dangerous science. The problem of war and military destruction remains the number one problem of the world, far exceeding in importance any of the other problems with which the United Nations and the peoples of the world are concerned (Table 13). While the military effort throughout history has been closely connected with technological advance, it is only with the Second World War that scientists, theoretical and experimental, have become personally involved in the development and use of weapons. The ingenuity of the best minds of the world has produced a staggering and frightful array of warfare devices, potentially capable of destroying the human species, and many others with it.

The rise of explosive power

The first large-scale mission-oriented research effort was the developing of the atom bomb. Thousands of scientists, engineers and technicians were mobilized over a period of four years in the 'Manhattan Engineering District Project'; $2,000 million were spent, and three atom bombs were produced, each of which had an explosive power 10,000 times greater than the most efficient chemical explosive then

TABLE 13. Gross national product (GNP), military expenditure and public expenditure on education and health, 1968 (absolute amounts in U.S. dollars and as percentage of GNP)

Major region	GNP		Military expenditure		Public expenditure on education		Public expenditure on health	
	Million U.S.$	Per-centage	Million U.S.$	Per-centage	Million U.S.$	Per-centage	Million U.S.$	Per-centage
World[1]	2,601,400	100	182,054	7.0	131,640	5.1	65,265	2.5
Africa	57,700	100	1,769	3.1	2,370	4.1	799	1.4
Northern America	931,800	100	82,379	8.8	56,510	6.1	23,788	2.6
Latin America	118,900	100	2,235	1.9	4,430	3.7	1,370	1.2
Asia[1]	286,700	100	8,302	2.9	10,660	3.7	1,460	0.5
Europe and U.S.S.R.	1,173,600	100	86,315	7.4	56,220	4.8	36,924	3.1
Oceania	32,700	100	1,054	3.2	1,450	4.4	924	2.8
(Arab States)	(28,800)	(100)	(2,000)	(6.9)	(1,340)	(4.7)	(400)	(1.4)
Developed countries	2,298,120	100	171,922	7.5	120,820	5.3	62,195	2.7
Developing countries	303,280	100	10,132	3.3	10,820	3.6	3,070	1.0

1. Not including People's Republic of China, Democratic People's Republic of Korea and Democratic Republic of Viet-Nam.

existing (1945). As shown in Figure 12, a second quantum jump in explosive power and maximum killing area was obtained with the hydrogen bomb: the increase was 1,000 times as great. The knowledge derived from what was at one time considered the most abstruse and useless of all the sciences, theoretical physics, brought about the most terrifying destructive device. The immense research and development effort carried out in the United States in time of war was repeated subsequently by the United Kingdom, the U.S.S.R., France and China.

According to a report published in 1967 by the United Nations,[1] there were, at that time, only seven countries that could afford, scientifically, technologically and financially, to join the club of the nuclear powers: Canada, Federal Republic of Germany, Japan, Italy, India, Poland and Sweden. The overall cost was estimated at $5,600 million spread over a ten-year period. With such an effort, each of these countries would acquire within five years ten to fifteen bombers as carriers and fifteen to twenty nuclear weapons. During the second five years the force would include 20 to 30 thermonuclear weapons, 100 intermediate range missiles and 2 missile launching submarines. The annual cost of maintaining bomb production would be about $5 million a bomb.

When estimating the costs of nuclear weapons, one should not forget indirect costs. Countries which decide to enter such a costly course are penalized further by the diversion of a specialized labour force away from peaceful and socially productive activities. Thus, the United Nations report states that to build the nuclear warheads would require 1,300 engineers and 500 scientists. If 50 missiles are added,

a peak labour force directly applied of 19,000 men would be needed, over 5,000 of them scientists and engineers with access to high speed electronic computers. Skilled personnel would include physicists, aerodynamic, mechanical and other engineers and a large number of production workers, including machine operators and welders. The suggested fleet of 50 bombers would require a minimum of from 1 to 2 million man-hours of skilled and unskilled labour for just the assemblage. The design and development stage would absorb an additional 2 million or more engineering man-hours, which would involve highly skilled efforts in aerodynamics, stress analysis, design work and flight testing.

1. *Effects of the Possible Use of Nuclear Weapons and the Security and Economic Implications for States of the Acquisition and Further Development of these Weapons*, New York, United Nations, 1967.

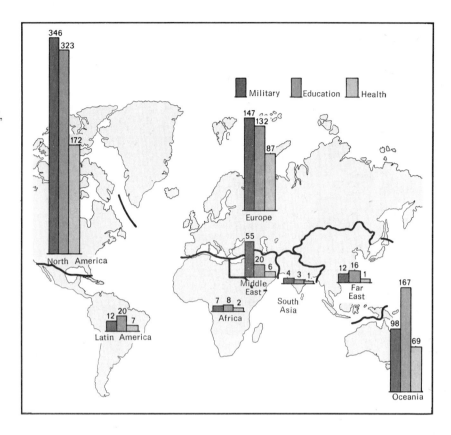

FIG. 11. Main public expenditures *per capita*, 1972 (expressed in U.S.$). Throughout the world, military expenditures greatly exceed those on health; in many parts, military expenses also exceed those for education. (From R. L. Sivard, *World Military and Social Expenditures, 1974*, Leesburg, Va., WMSA Publications, 1974.)

The temptation to become a nuclear power is unfortunately encouraged by the fact that more and more nuclear power plants are being built around the globe, not only among the advanced countries but also in the less developed ones. In most cases, it appears that nuclear power reactors will provide the cheapest, when not the only, means of producing electricity in the future. And since plutonium, an essential component of some nuclear bombs, is a by-product of the operation of a nuclear power plant, it will become available wherever electricity is produced by nuclear power plants. According to the 1972 yearbook of the Stockholm International Peace Research Institute (SIPRI):[1]

In 1971, 16 countries had 128 nuclear power reactors in operation with a total capacity of 35,000 MWe (1 MWe is equal to 1 million watts of electricity); in 1977, 32 countries will have 325 nuclear power reactors with a total capacity of 174,000 MWe. Plans put forward indicate that by 1980 installed nuclear power will exceed 350,000 MWe. As a by-product of this nuclear power production, considerable quantities of plutonium will be produced each year: about 13 tons in 1972, 65 tons in 1977, 130 tons in 1980. By 1980 about one-third of this plutonium will be produced in the present non-nuclear-weapon countries. This would, in theory, correspond to the production of some 100 nuclear weapons of nominal size per week in the non-nuclear-weapon countries. However, since the plutonium produced in nuclear power reactors cannot be used directly for nuclear weapons' manufacture without substantial investment in plants for chemical separation of plutonium-239 from plutonium-240, this may be somewhat misleading. Chemical separation plants exist in the five-nuclear-weapon states, but also in Argentina, Belgium, India, Italy, Spain and West Germany; Japan is constructing one.

To make the picture even gloomier, we must remember that availability of plutonium is not

1. 'World Armaments and Disarmaments', *SIPRI Yearbook 1973*, Stockholm, Almquist & Wiskell, 1973.

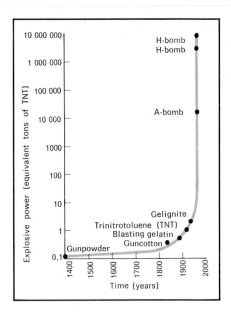

Fig. 12. Curve of the exponential growth of the power of explosive weapons, from the black powder used as explosive propellant from the fourteenth century onwards to the thermonuclear bomb. (From *SIPRI Yearbook*.)

a must for the production of nuclear weapons. Uranium enrichment, a process whereby the relative percentage of uranium-235 is increased over that of uranium-238 (the latter being the isotope prevailing in natural uranium ore), is the original procedure used for the production of the first atom bombs and is still used by the nuclear powers. Plants for the production of enriched uranium have not been proliferating primarily because of the immense cost and technological difficulties of building enrichment plants, which separate the two isotopes. But in recent times, in Europe, a new separation technique has been found, based on the use of batteries of gas centrifuges: it is said to be five times cheaper in running costs than the one based on gas diffusion. This may lead to a marked increase in the number of countries capable of producing enriched uranium that could be used in weapon production.[1]

The largest H-bomb ever exploded had a power equal to that of 57 million tons of the conventional TNT explosive; this is equivalent to a train loaded with TNT long enough to stretch all the way round the globe. As Robin Clarke recently pointed out:[2]

. . . the American scientist, Dr. Frank L. Klingberg, was asked during World War II to find some means of predicting how heavy Japanese losses would have to be before the Japanese surrendered. The question became academic when the atom bomb was dropped, but in the course of his work, Dr. Klingberg discovered how quickly lethality was escalating:

War	Date	Average deaths per day
Napoleonic	1790–1815	233
Crimean	1854–56	1,075
Balkan	1912–13	1,941
World War I	1914–18	5,449
World War II	1939–45	7,738
(Hiroshima)	6 August 1945	80,000

What level of killing and misery will be reached next time?

Chemical and biological weapons

Primitive types of chemical and biological warfare are to be found in the records of ancient wars, but it is only recently that the great progress of bacteriology and virology has increased the death potential of toxic substances and infectious micro-organisms.[3]

The number of substances that have been examined as candidate chemical warfare agents runs into hundreds of thousands—as stated in a recent report of the World Health Organization (WHO).[4] During the First World War, virtually every known chemical was screened, and much of the work done then was repeated during the Second World War; in addition, a high proportion of the new compounds that had been synthesized or isolated from natural materials during the inter-

1. *Economic and Social Consequences of the Armaments Race and its Extremely Harmful Effects on World Peace and Security*, New York, United Nations, 1971.
2. R. Clarke, *The Science of War and Peace*, London, Jonathan Cape, 1971.
3. *Chemical and Bacteriological (Biological) Weapons and the Effects of their Possible Use*, New York, United Nations, 1969.
4. *Health Aspects of Chemical and Biological Weapons. Report of a Group of Consultants*, Geneva, WHO, 1970.

war years were examined. Since the Second World War, the major laboratories engaged in such work have probably been carrying out a systematic check on all the compounds whose properties suggest utility in chemical warfare, however remote. The requirements that must be satisfied by candidate chemical warfare agents in respect of production cost, and physical, chemical and toxicological properties are severe, and the number of chemicals that have actually been used in chemical warfare or stockpiled in quantity for such use is small, probably not more than sixty.

From a military point of view, chemical warfare agents have been developed with three quite different tactical functions in mind: (a) 'lethal agents', used either to kill an enemy or to injure him so severely as to necessitate his evacuation and medical treatment; (b) 'incapacitating agents', used to put an enemy completely out of action for several hours or days, but with a disablement from which recovery is possible without medical aid; and (c) 'harassing agents', used to disable an enemy for as long as he remains exposed.

The properties of some of the most important agents are summarized in Table 14.

In the case of the development of biological warfare agents, the military research establishment aims at identifying and isolating micro-organisms, mostly viruses and bacteria, capable of infecting and producing diseases in the target—human, animal or vegetable. The same procedures that are used in the pharmacological and vaccine industries for producing strains of bacteria and fungi with high yields of antibiotic substances, or strains of viruses with low pathogenicity, can be applied for the production of micro-organisms characterized by enhanced virulence, increased resistance to environmental conditions and to antibiotic treatment, etc. In the laboratory, through isolation and production of high quality strains, crossing of different strains, and screening for the desired characteristics, the microbiologist has today in his hands an exceedingly powerful tool to 'improve' on existing diseases.

Table 15 contains a list of weapons that are considered most likely to be used because of their microbiological and epidemiological characteristics. While large-scale laboratory production of all the agents is feasible in most advanced microbiological laboratories, the drying, processing and delivery of most of them as military weapons require sophisticated technology. Some of them, however, such as smallpox virus, can be produced and used as weapons with relatively simple techniques.

The panel of experts advising the World Health Organization (WHO), has attempted quantitative estimates of the effects of possible small-scale airborne attacks on cities of 0.5–5 million population in industrially developed and developing countries. The postulated mode of attack consisted of one or a few bombers dispersing specific chemical or biological agents along a 2-km line, perpendicular to the direction of the wind.

On the basis of the particular assumptions employed, the following conclusions have been reached:

1. (a) Of the known chemical warfare agents, only the nerve gases, and possibly botulinal toxin, have a casualty-producing potential comparable to that of biological agents.

   (b) Under atmospheric conditions favourable to the attacker, an efficiently executed attack on a city with 4 tons of sarin (requiring some 15–20 tons of weapons) could cause tens of thousands of deaths in an area of about 2 km². Even in unfavourable conditions there could be thousands of deaths. If 4 tons of VX were used in such an attack, the casualties would not be appreciably greater in unfavourable meteorological conditions, but in favourable conditions this small attack would affect an area of about 6 km² and could cause anywhere between 50,000 and 180,000 deaths.

   (c) If a suitably stabilized botulinal toxin or a fine aerosol of VX (particles of 5 μ diameter) were developed and 4 tons were employed, several hundreds of thousands of deaths could result because of the greater coverage possible with such agents—12 km² for botulinal toxin and 40 km² for monodispersed VX aerosol. A larger total weight of weapons, perhaps two to three times

49

that needed for the agents in (b) above, would have to be used to deliver these forms of botulinal toxin and VX.

(d) If a biological agent such as anthrax were used, an attack on a city by even a single bomber disseminating 50 kg of the dried agent in a suitable aerosol form would affect an area far in excess of 20 km², with tens to hundreds of thousands of deaths. A similar attack with any one, of a number of other more labile biological agents could affect from 1 km² to more than 20 km², depending upon agent used, with tens to hundreds of thousands of casualties and many thousands of deaths.

2. Limited sabotage of a communal water supply with the typhoid fever bacillus, LSD, or a stable botulinal toxin, could cause considerable disruption and deaths in a large city, affecting tens of thousands of people.

3. Sabotage-induced or open attacks, causing the secondary spread of epidemics of yellow fever, pneumonic plague, smallpox or influenza, might under certain conditions ultimately result in many millions of illnesses and deaths.

TABLE 14. Some properties of selected chemical warfare agents

1	Sarin	VX	Hydrogen cyanide	Cyanogen chloride	Phosgene
2	Lethal agent (nerve gas)	Lethal agent (nerve gas)	Lethal agent (blood gas)	Lethal agent (blood gas)	Lethal agent (lung irritant)
3	Vapour, aerosol or spray	Aerosol or spray	Vapour	Vapour	Vapour
4	All types of chemical weapon		Large bombs	Large bombs	Mortars, large bombs
5	1,000 kg	1,000 kg	1,000 kg	1,000 kg	1,500 kg
6	100%	1–5%	100%	6–7%	Hydrolyzed
7	12,100 mg/m³	3–18 mg/m³	873,000 mg/m³	3,300,000 mg/m³	6,370,000 mg/m³
8 (a)	Liquid	Liquid	Liquid	Solid	Liquid
(b)	Liquid	Liquid	Liquid	Vapour	Vapour
9 (a)	1/4–1 hour	1–12 hours	Few minutes	Few minutes	Few minutes
(b)	1/4–4 hours	3–21 days	Few minutes	Few minutes	Few minutes
(c)	1–2 days	1–16 weeks	1–4 hours	1/4–4 hours	1/4–1 hour
10	> 5 mg-min/m³	> 0.5 mg-min/m³	> 2,000 mg-min/m³	> 7,000 mg-min/m³	> 1,600 mg-min/m³
11	100 mg-min/m³	10 mg-min/m³	5,000 mg-min/m³	11,000 mg-min/m³	3,200 mg-min/m³
12	1,500 mg/man	6 mg/man	—	—	—

Key: 1. Common name.
2. Military classification.
3. Form in which the agent is most likely to be disseminated.
4. Types of weapon suitable for disseminating the agent.
5. Approximate maximum weight of agent that can be delivered effectively by a single light bomber (4–ton bomb load).
6. Approximate solubility in water at 20° C.
7. Volatility at 20° C.
8. Physical state:
(a) at —10° C;
(b) at 20° C.

4. The numbers of potential casualties and deaths recorded in this report represent the possibilities arising out of a very small and limited attack already well within the capabilities of a number of nations, with the possibility that an ever-increasing number of countries will acquire similar capabilities. With technologically advanced weapons and a larger scale of attack, achievable without too much difficulty by militarily advanced powers, the magnitude of destructiveness attendant upon the use of chemical and biological weapons would be considerably increased.

Tables 16, 17 and 18 summarize the quantitative estimates of the WHO panel on the possible primary effects of limited biological warfare attack on unprotected civilian population groups.

Environmental warfare

During the Second World War we witnessed the horror of the destruction of Dresden, Hamburg and Tokyo through the use of incendiary bombs: judgements expressed later on such events considered them to be among the cruelest of all times. But the sophistication of

Mustard gas	Botulinal toxin A	BZ	CN	CS	DM
Lethal and incapacitating agent (vesicant)	Lethal agent	Incapacitating agent (psycho-chemical)	Harassing agent	Harassing agent	Harassing agent
Spray	Aerosol or dust	Aerosol or dust	Aerosol or dust	Aerosol or dust	Aerosol or dust
All types of chemical weapon	Bomblets, spray-tank	Bomblets, spray-tank	All types of chemical weapon		
1,500 kg	400 kg	500 kg	750 kg	750 kg	750 kg
0.05%	Soluble	?	Slightly soluble	Insoluble	Insoluble
630 mg/m^3	Negligible	Negligible	105 mg/m^3	Negligible	0.02 mg/m^3
Solid	Solid	Solid	Solid	Solid	Solid
Liquid	Solid	Solid	Solid	Solid	Solid
12–48 hours				—	
2–7 days	—	—	—	2 weeks for CS1: longer for CS2	—
2–8 weeks					
> 100 mg-min/m^3	0.001 mg (oral)	100 mg-min/m^3	5–15 mg/m^3 concentration	1–5 mg/m^3 concentration	2–5 mg/m^3 concentration
1,500 mg-min/m^3	0.02 mg-min/m^3	?	10,000 mg-min/m^3	25,000–150,000 mg-min/m^3	15,000 mg-min/m^3
4,500 mg/man	—	—	—	—	—

9. Approximate duration of hazard (contact, or airborne following evaporation) to be expected from ground contamination:
   (a) 10° C, rainy, moderate wind;
   (b) 15° C, sunny, light breeze;
   (c) —10° C, sunny, no wind, settled snow.
10. Casualty-producing dosages (for militarily significant injuries or incapacitation).
11. Estimated human respiratory LC_{50} (for mild activity: breathing rate approx. 15 litres/min).
12. Estimated human lethal percutaneous dosages.

Source: Health Aspects of Chemical and Biological Weapons. Report of a Group of Consultants, Geneva, WHO, 1970.

TABLE 15. Some potential biological weapons against man[1]

	Protection by vaccination against currently known strains	Effectiveness of serotherapy or chemo-therapy
A. *Viruses*	±	—
Adenoviruses (some strains)		
Arthropod-borne viruses (eastern, western and Venezuelan equine encephalitis; Japanese B; Russian spring-summer group; dengue; yellow fever; etc.)	+ to —	—
B virus and related herpesviruses	—	—
Enteroviruses (some members)	—	—
Influenza	±	—
Marburg virus (vervet monkey disease)	—	—
Sandfly (phlebotomus) fever	—	—
Smallpox	+	±
B. *Rickettsiae and bedsoniae*		
Psittacosis (ornithosis)	—	+
Q fever	?	+
Rickettsiaipox	—	+
Rocky Mountain spotted fever	±	+
Scrub typhus (tsutsugamushi fever)	—	+
Typhus (epidemic)	+	+
C. *Bacteria*		
Anthrax	±	±
Brucellosis	±	±
Cholera	±	+
Glanders	—	?
Melloidosis	—	?
Plague	±	+
Shigellosis	?	+
Tularaemia	±	+
Typhoid fever	±	+
D. *Fungl*		
Coccidioidomycosis	—	?
E. *Protozoa*	—	
Schistosomiasis (bilharziasis)		±
Toxoplasmosis	—	±

1. + = suitable or good
    ± = moderately effective
    — = unsuitable or poor
    ? = marginal or of questionable value, either because of the nature of the agent or because of insufficient knowledge.

Source: Health Aspects of Chemical and Biological Weapons. Report of a Group of Consultants, Geneva, WHO, 1970.

current dangerous science has enlarged and 'improved' the arsenal of weapons which supposedly are used for the destruction of objects but produce, at the same time, untold human suffering. There are four main groups of devices aimed at the destruction or modification of the environment in which men live, work or fight: incendiary weapons, herbicides and defoliants, special bombs and weather modification.

To quote a recent report prepared by a group of consultant experts, appointed by the Secretary-General of the United Nations:[1]

Incendiary weapons may be defined as weapons which depend for their effects on the action of incendiary agents. These in turn may be defined as substances which affect their targets primarily through the action of flame and/or heat derived from self-supporting and/or self-propagating exothermic chemical reactions; these reactions, for all practical purposes, are combustion reactions. Production of poisonous substances and certain other side-effects may also bring significant harm to the target.

The massive spread of fire is largely indiscriminate in its effects. When there is a difference between the susceptibility to fire of military and civilian targets, it is commonly to the detriment of the latter. The same applies to certain tactical applications of incendiaries, for the ability of these weapons to strike over an appreciable area, and the often close proximity of military and civilian targets, may also have consequences that are essentially indiscriminate.

Burn injuries, whether sustained directly from the action of incendiaries or as a result of fires initiated by them, are intensely painful and, compared with the injuries caused by most other categories of weapon, require exceptional resources for their medical treatment. Under war conditions only a few of the people exposed to more extensive napalm burns survive to the period of real convalescence, which is long and difficult. Permanent loss of function, disfigurements and severe scarring are frequent. Disabilities, impairment

1. *General and Complete Disarmament—Napalm and Other Incendiary Weapons and All Aspects of their Possible Use*, New York, United Nations, 1972.

TABLE 16. Numbers of persons at risk from an attack with chemical or biological weapons on the most densely populated areas of typical urban targets

| Population of town | Industrially developed country | | | Industrially developing country | | |
| | Type of area[1] | | | Type of area[1] | | |
	I	II	III	I	II	III
500,000	50,000	75,000	120,000	180,000	200,000	250,000
1,000,000	60,000	100,000	180,000	100,000	160,000	250,000
5,000,000	150,000	300,000	500,000	60,000	150,000	250,000

1. The types of area selected are the most densely populated areas of the sizes specified below in the different kinds and sizes of town:
Area I extends from an attack line 1 km upwind of the town centre to 2 km downwind of the town centre, and is 2 km wide.
Area II extends from 1 km upwind of the town centre to 9 km downwind of the town centre, and is 2 km wide.
Area III extends from an attack line 10 km upwind of the town centre to 10 km downwind of the town centre and is 2 km wide.

Source: Health Aspects of Chemical and Biological Weapons. Report of a Group of Consultants, Geneva, WHO, 1970.

of sight or hearing, and grave emotional disorders are often the consequences of this. Plastic repair and reconstruction of the damage is very difficult and painful, and may have only limited effect. When judged against what is required to put a soldier out of military action, much of the injury caused by incendiary weapons is therefore likely to be superfluous. In terms of damage to the civilian population, incendiaries are particularly cruel in their effects.

Napalm and other incendiary weapons owe their effect not only to heat and flame, but also to the toxic effects of carbon monoxide and other combustion products. This is true particularly for people sheltering in confined spaces, but may also be significant when they are in the open. Asphyxiating effects may also be important, particularly in connexion with the massive use of incendiaries. Here it is the action of dense smoke and the depletion, through fire, of atmospheric oxygen that are primarily responsible. The toxicity of white phosphorus may also be an important consideration for the injuries sustained from weapons utilizing this material. Napalm weapons often contain white phosphorus, and this may further aggravate the burn injuries they cause.

Attempts have been made to use incendiaries to damage crops, forests and other features of the rural environment. Although there is a lack of knowledge of the effects of widespread fire in these circumstances, such attempts may lead to irreversible ecological changes having grave long-term consequences out of all proportion to the effects originally sought. This menace, though largely unpredictable in its gravity, is reason for expressing alarm concerning the massive employment of incendiaries against the rural environment.

The rapid increase in the military use of incendiary weapons, especially napalm, during

TABLE 17. Effects of an attack along a 2 km line using up to 4 tons of chemical agent requiring 15–20 tons of weapons

Chemical agent	Area affected[1]
Sarin[2]	2 km² in area I
VX	Area I (=6 km²)
Botulinal toxin (if adequately stabilized)[3]	Intermediate between areas I and II (approx. 12 km²)
VX, dispersed as a monodisperse 5 μ diameter aerosol (if technically feasible)[3]	Area III (approx. 40 km²)

1. See Table 16 for definitions of areas I, II and III.
2. Agents less toxic than sarin would produce effects only in the immediate vicinity of the explosion of a weapon.
3. A larger weight of weapons, perhaps 2–3 times as great, might be required to deliver the botulinal toxin or to produce the fine VX aerosol because of the greater amount of hardware required for such purposes.

Source: Health Aspects of Chemical and Biological Weapons. Report of a Group of Consultants, Geneva, WHO, 1970.

the past thirty years is but one aspect of the more general phenomenon of the increasing mobilization of science and technology for war purposes. New weapons of increased destructiveness are emerging from the research and development programmes at an increasing rate, alongside which the long upheld principle of the immunity of the non-combatant appears to be receding from the military consciousness.

The destruction of vegetation through the

TABLE 18. Estimated possible primary effects of limited (single bomber) biological warfare attack on unprotected civilian population groups[1]

Disease caused by agent	Pattern of attack (type of areas)[2]	Downwind carriage of lethal or incapacitating concentrations		1		2	
		Approximate extent	Approximate time	D	I	D	I
Venezuelan equine encephalitis (similar effects can be expected from Rift Valley fever, chikungunya and O'nyong-nyong)	I	1 km	5–7 min	100	10,000	100	12,000
Tick-borne (far Eastern) encephalitis (similar effects can be expected from Japanese encephalitis and yellow fever)	I	1 km	5–7 min	2,500	10,000	3,000	12,000
Influenza, antigenically modified, non-virulent strain	I	1 km	5–7 min	50	10,000	50	12,000
Epidemic typhus	II	5 km	30 min	2,500	30,000	3,000	37,000
Rocky Mountain spotted fever	II	5 km	30 min	2,000	30,000	2,500	37,000
Brucellosis	II	10 km	1 h	150	27,000	200	50,000
Plague	II	10 km	1 h	6,500	27,000	12,000	50,000
Q fever	III	> 20 km	> 2 h	50	60,000	100	90,000
Tularaemia	III	> 20 km	> 2 h	4,500	60,000	7,000	90,000
Anthrax	III	≫ 20 km	≫ 2 h	24,000	60,000	35,000	90,000

1. Approximately 50 kg of dried powder containing 6×10^{15} organisms are assumed to have been aerosolized to form a band 2 km long at right angles to the wind direction under type F meteorological conditions.
2. See Table 16 for definitions of attack patterns (areas) I to III.
3. The figures are rounded estimates, based on the number of individuals at risk and taking into account such factors as an assumed attack rate of 50 per cent, the case fatality rates, and for some of the agents the use of antibiotics as described under 'Remarks'.
   D = deaths; I = incapacitated, including deaths.

use of chemical compounds that interfere with normal physiological processes of plants is a new type of environmental warfare which has been experimented and developed over a period of about ten years. No assessment by experts officially appointed by the United Nations (such as those previously quoted in the case of nuclear and incendiary weapons) has been made yet of the effects of herbicides and defoliants. We can rely, however, on the

Number of casualties[3]								Remarks
Population group[4]								
3		4		5		6		
D	I	D	I	D	I	D	I	
300	30,000	400	35,000	200	20,000	100	10,000	
7,800	30,000	2,500	35,000	6,000	20,000	3,000	10,000	
50	30,000	100	35,000	100	20,000	50	10,000	
3,000	125,000	19,000	85,000	15,000	65,000	14,000	60,000	Use of antibiotic therapy considered to cause reduction of mortality by 70 per cent and 10 per cent for economically developed and developing communities, respectively.
7,500	125,000	11,500	85,000	9,000	65,000	7,500	60,000	Same as above for epidemic typhus.
600	150,000	500	100,000	400	80,000	350	75,000	
38,000	150,000	55,000	100,000	44,000	30,000	41,000	75,000	Same as above for epidemic typhus.
250	250,000	150	125,000	150	125,000	150	125,000	No calculations made for after two hours: part of infective cloud would persist downwind.
19,000	250,000	30,000	125,000	30,000	125,000	30,000	125,000	Same as above for epidemic typhus. No calculations made for after two hours; part of infective cloud would persist downwind.
100,000	250,000	95,000	125,000	95,000	125,000	95,000	125,000	No calculations made for after two hours: part of infective cloud would persist far downwind. Use of antibiotic therapy considered to cause reduction of mortality by 50 per cent and 5 per cent for economically developed and developing countries, respectively.

4. The population groups are defined as follows:
    1. Urban population of approximately 500,000 in economically developed country.
    2. Urban population of approximately 1,000,000 in economically developed country.
    3. Urban population of approximately 5,000,000 in economically developed country.
    4. Urban population of approximately 500,000 in developing country.
    5. Urban population of approximately 1,000,000 in developing country.
    6. Urban population of approximately 5,000,000 in developing country.

Source: Health Aspects of Chemical and Biological Weapons. Report of a Group of Consultants, Geneva, WHO, 1970.

TABLE 19. Application of herbicides in the Viet-Nam war by year (millions of gallons)

	1962-July 1965	August-December 1965	1966	1967	1968	1969	1970	1971	Total
Orange	NA[1]	.37	1.64	3.17	2.22	3.25	.57	.00	11.22
White	NA[1]	0	.53	1.33	2.13	1.02	.22	.01	5.24
Blue	NA[1]	0	.02	.38	.28	.26	.18	.00	1.12
TOTAL	1.27	.37	2.19	4.88	4.63	4.53	.97	.01	18.85

1. NA = Not available.

TABLE 20. Estimated acreage sprayed one or more times, 1965–71[1]

Vegetation type[2]	Total in SVN in 1953		Number of times sprayed August 1965 to March 1971				Total sprayed one or more times	
	Millions of acres	Percentage	Millions of acres				Millions of acres	Percentage
			1	2	3	4+		
Inland forest	25.91	62.4	1.72	0.62	0.22	0.11	2.67	10.3
Cultivated land	7.80	18.8	0.20	0.04	0.01	0.00	0.26	3.2
Mangrove forest	0.72	1.7	0.14	0.07	0.03	0.02	0.26	36.1
Other	7.07	17.1	0.31	0.07	0.02	0.00	0.39	5.5
TOTAL	41.50	100.0	2.37	0.80	0.28	0.13	3.58	8.6

1. Does not include coverage of missions before August 1965 (1.27 million gallons) and missions after that date for which location information is incomplete (1.1 million gallons), representing about 12.5 per cent of the total gallonage accounted for.
2. Inland forests include those areas classed as dense forest, secondary forest, swidden zones, bamboo forests, open dipterocarp, *Lagerstroemia* and Leguminosae forests. 'Other' includes pine forests, savannah and degraded forests, grasslands and steppes in higher elevations, dunes and brushland, grass and sedge swamps and areas of no vegetation (urban areas, roads, water courses, etc.). Classification and area figures follow Bernard Rollet (1962).

report prepared for the National Academy of Science of the United States by an international group of specialists.[1]

The herbicide programme in Viet-Nam began in an experimental way in 1961, became operational in 1962, and then grew rapidly to peak in 1967, before declining somewhat in 1968 and 1969. Over the period, about one-seventh of the land area of South Viet-Nam has been treated with herbicides, most of them sprayed from low-flying C-123 cargo aircraft that made more than 19,000 individual spray flights between 1962 and 1969. About 90 per cent of the herbicide was dropped on forest land and about 10 per cent on crop land.

The committee of the National Academy of Science conducted as thorough as possible an inventory of the herbicide operations in Viet-Nam, as a basis for assessing the effects of these operations on vegetation, soils and people. The number of gallons sprayed on the area is shown in Table 19, and the areas sprayed once, twice and more times appear in Table 20. About 88 per cent of the herbicide missions were designated for defoliation, about 9 per cent for crop destruction and the remaining 3 per cent were directed at base perimeters, enemy cache sites, waterways, and

1. *The Effects of Herbicides in South Vietnam.* Part A: *Summary and conclusions*, Washington, D.C., National Academy of Science, 1974.

lines of communication. According to the report:

Death and damage to vegetation caused by herbicides can have many different consequences: loss of potential production at a stage before the growth becomes economically valuable; loss of commercial products such as timber, grain and fruit; lack of young plants and of seed necessary to maintain the 'system', the latter type of effect being particularly important in native vegetation. The committee studied herbicide damage to three major vegetation types of South Vietnam: the inland forest, the mangrove forest, and (permanently) cultivated land. . . . With the exception of extensively sprayed mangrove forests, aerial photographs showed that vegetation cover of some type returns to most areas six months to a year after they had been sprayed. . . . The fact that vegetation of some type generally returned promptly suggests, however, that there was no permanent inhibition of plant growth because of adverse conditions in the soil.

A large proportion of the mangrove forests was sprayed with herbicides and was more heavily affected by the spraying than any other vegetation type in South Vietnam. Of the approximately 720,000 acres of South Vietnam that were covered by mangrove (representing about 1.7 per cent of the total area of the nation), about 260,000 acres, or 36 per cent, were sprayed. One spray usually killed all mangrove trees; large contiguous areas were devastated, and there has been little or no recolonization of mangrove trees in extensively sprayed areas, except along the margins of some of the canals that drain these swamps. . . . An estimate based on a model suggests that, under present conditions of use and natural regrowth, it may take well over 100 years for the mangrove area to be reforested. With a massive reforestation program, the forest could probably be restored in approximately 20 years if sufficient money and seed resources were available.

No conclusive evidence was found of association between exposure to herbicides and birth defects in humans; but the material studied was not adequate for definite conclusions.

The bomb is code-named BLU-82/B and bomber crews have nicknamed it either 'daisy-cutter' or 'cheese-burger'. The bomb explodes just before touching the ground. 'Normally the resulting radial explosion does not dig a crater', Westling noted, 'but uproots and blows away the trees and other obstacles in the very heart of a very dense jungle, creating a perfect clearing of the size of a football pitch.' The resulting landing space can be used immediately by helicopters. In June 1970, no less than 160 of these bombs were dropped in Viet-Nam; after that they continued to be dropped at the rate of several a week. The explosion of the bomb is so powerful that all animal, plant and human life within a radius of not quite 1,000 metres is annihilated by the shock wave. The corresponding lethal area is thus about 300 hectares.

Finally there is the possibility of using weather control for military aims. For many years meteorologists have been studying how to influence weather in order to bring under control its vagaries. If we could make rain at will or stop it when it pours in excessive amounts, if the course of storms and hurricanes could be deflected over uninhabited areas, or if we were able to disperse fog, many inconveniences would disappear and major economic benefits would ensue. Obviously, the same scientific and technological advances could be used in waging war and indeed, in many countries, the military are in control of meteorological research, weather forecasting and experiments in rain-making. The hearings at the United States Senate showed that a classified rain-making programme was conducted in South-East Asia from 1967 to 1972, which employed air-dropped silver and lead iodide seeding units to increase normal monsoon rain. Purpose of the programme was that of: softening road surfaces; causing landslides along roadways; washing out river crossings; maintaining saturated soil conditions beyond the normal time span.

The results of the programme cannot be expressed in precise quantitative terms. Using empirical and theoretical techniques based on units expended and the physical properties of the air mass seeded, however, it has been possible to estimate that the rainfall was increased in limited areas up to 30 per cent above that predicted for the existing conditions. 'Subjectively', the report concludes, 'it is believed that this rainfall was heavier than that which would have fallen normally and that it did contribute to slowing the flow of supplies into South Vietnam along the Ho Chi Minh trail.'

Antipersonnel
and computerized warfare

At the very beginning of this subsection, it must be stated that while we could rely on official documents of the United Nations or on analyses made by scholarly sources such as the Stockholm International Peace Research Institute when we discussed nuclear, chemical and biological, or environmental warfare, we shall have to refer only to reports of research workers or of specialized journalists in commenting on the scientific and technological developments discussed hereunder. Indeed, the following quotations could be criticized for their subjectivity; they should be discussed in this context, however, for the recent developments referred to represent a further escalation in the exploitation of scientific advance for evil purposes. It should be stated further that the information on new weapon systems comes almost entirely from American sources, this being due to the liberal attitude prevailing in that country about information in general and even about news concerning military developments. It is certain that the direct involvement of scientists and engineers in the military enterprise occurs in other countries as well, and particularly among the great powers. The use of the euphemism 'defensive measures' does not detract much from the responsibility of those who partake in such technical innovations.

The arsenal of deadly devices resulting from science-based military research recently has undergone further expansion. While nuclear weapons and chemical and biological warfare may find their use in all-out war, the risk of which appears currently to be dwindling, the immense potential of recent advances in electronics, solid state physics, coherent optics and automation is being exploited for a vast variety of new military technological developments.[1]

High-speed computers, advanced electronic devices and the laser have already brought about major changes in non-nuclear weapons in recent years. Milton Leitenberg, an expert in the study of war and peace formerly attached to SIPRI, has given the following brief survey of some of them:[2]

1. Large computer-controlled air defence networks—for example, the United States semi-automatic ground environment, and NATO's air defence ground equipment—with large early-warning, over-the-horizon radars for ballistic missile warning and forward emplaced radar networks for anti-aircraft defence. Radar has been credited with maintaining a rate of product improvement doubling its capabilities every second year since the end of the Second World War.
2. Smaller field, mobile, radar-controlled anti-aircraft guns, often mounted on tracked tank-like vehicles.
3. Surface-to-air anti-aircraft missiles, with ranges as high as 70 miles, and analogous shipborne weapons. (Some are nuclear armed.)
4. Self-guided target-seeking air-to-surface missiles; some 'home' on opposition radar, others have active search guidance.
5. Avionics: electronics and airborne computers play a near complete role in advanced combat aircraft—navigation, reconnaissance, bad-weather operations, engaging opposing aircraft, fire control, weapon guidance. It should be said that such devices become increasingly expensive, more difficult to repair and suffer from longer and longer fractions of their lifetime in a malfunctioning state due to their complexity.
6. Airborne antisubmarine warfare has undergone an enormous development, with long-range, long-duration patrols, expendable sonobuoy systems, other buoy telemetry, airborne dipped sonars, infra-red and magnetic anomaly surveillance.
7. Development of new materials and advances in the aeronautics industry in production methodology for aircraft and spacecraft.
8. Advanced weapon guidance, using lasers for targeting of many kinds of ordnance in field weapons and ground-support aircraft.
9. Increased deployment capabilities: first airlift, now rapid deployable air bases.
10. Greatly increased use of the helicopter.
11. Night-time target acquisition and fire-control devices. These can also penetrate forest cover.
12. Radars for artillery and mortar location.
13. Increased use of napalm.

1. M. Leitenberg, 'The Dynamics of Military Technology Today', *International Social Science Journal*, Vol. XXV, No. 3, 1973, p. 336–57.
2. M. Leitenberg, 'The Present State of the World's Arms Race', *Bulletin of Atomic Scientists*, January 1972, p. 15–21.

14. Computer-aided logistics management.
15. Increased exercise of command and control via various electronic instrumentation.
16. Nuclear power for naval vessel propulsion.
17. The military use of space is itself an entire group of activities. Satellites are used for military communications, for photographic reconnaissance in visible, infra-red and ultra-violet spectra, for electronic intelligence and for navigation and guidance. There are classified meteorological satellite programmes and geodetic survey satellites, which supply information for ICBM guidance. There are also large satellite-tracking radar and computer networks and operational missile-borne satellite destruction systems maintained in readiness both by the United States and the Soviet Union. Much manned space effort is directly oriented towards strategic military uses. The new United States space shuttle is the direct descendant of the Dyna-Soar orbital space bomber programme cancelled in 1963.
18. The oceans have also seen and are yet to see further invasion by entirely new military support systems. Bottom-mounted acoustic surveillance systems may soon be mounted on sea-mounts or table-guyots in midocean. Similar systems have been deployed since the mid-1950s off the east coast of the United States. The United States NR-1 and the Dolphin are precursors of the next generation of United States ballistic missile and 'hunter-killer' nuclear submarines, and have depth capabilities of 6,000 to 10,000 feet. The United States DSSP (Deep Submergence System Project) and DSRV (Deep Submergence Rescue Vehicles) programmes are essentially for the development of vehicles and equipment to inspect, install, repair or serve bottom-mounted surveillance of weapon systems. ASWEPS (Anti-Submarine Warfare Environmental Prediction System) has peppered the ocean surface and depths with various sensor and buoy systems. Some are bottom-mounted, located as deep as 20,000 feet and powered by nuclear SNAP (System for Nuclear Auxiliary Power) packages. The United States plans one atmosphere-manned bottom station in the coming years at depths of 1,000 feet and more. Project Rocksite hopes to adapt techniques for dry tunneling under the sea to great depths from under the continental land mass itself.

Finally, the remotely piloted vehicle (RPV) is the next 'miracle killer' for the immediate future, as reported by Robert Barkan, a former electronic engineer now with the Pacific Studies Centre in East Palo Alto, California:[1]

A fancy model aeroplane flown by a pilot on the ground outmanoeuvred and 'shot down' the best U.S. fighter plane. But this aircraft was no toy; rather it is the most spectacular success in a deadly earnest project to reduce the cost—in money and men—of modern air warfare.

Anti-aircraft defences are getting tougher, the costs of modern warplanes are rising rapidly, and the American public is becoming increasingly reluctant to risk pilots' lives. The answer to these problems is the RPV—the robot aircraft flown by pilots on the ground who watch the action on a television set. Advances in electro-optical sensors, data transmission links, microelectronics, computers, and display systems have prompted the United States Air Force to spend millions of dollars to develop RPVs for bombing, air-to-air combat, reconnaissance, and radar jamming. RPVs are the 'hottest idea in the Defense Department today', according to the authoritative *Armed Forces Journal.*

The scientist and the military establishment

The museum of horrors we have briefly visited in this chapter is undoubtedly the product of a large fraction of the scientific community. In the course of the last twenty-five years defence and science have fed one another. The armaments race, the purpose of which is political, has turned into a scientific and technological race. Of all governmental departments or ministries the most 'scientific' is the one responsible for military developments. Furthermore, military escalation is largely determined by an internal, self-sustaining acceleration prompted by science. Military competition, in fact, is a peculiar one: the development of a new weapon system calls for the most advanced research, pure and applied; during the time necessary for getting the system ready for use, the frontier of knowledge advances, and the military of one side will wonder

1. R. Barkan, 'The robot airforce is about to take off', *New Scientist*, 10 August 1972.

whether his competitor (the potential enemy) might not have already made use of the most recent breakthroughs for his own new military development; to be safe, newly produced scientific knowledge and technological know-how will be used promptly for more sophisticated weapons. Thus the spiral of the armaments race finds at its base a scientific and technological catalyst.

The scientific component of military developments was assessed by economists at the Stockholm International Peace Research Institute in an attempt to measure the rate of military product improvement:[1]

As there was no obvious measure—such as destructiveness—that could be used, they looked at the real cost over the years of a number of weapons: four specific types of aircraft, the attack submarine, the aircraft carrier and the destroyer. Having allowed for world price increases, they found that the cost of each of these weapons increased over two decades by between 6.6 per cent per year for the aircraft carrier to 18 per cent per year for the air force fighter plane. These increases give some indication of the product's increased performance. In comparable prices, for instance, the air force fighter cost $110,000 in 1945 but $6.8 million in 1968. From this the SIPRI team were able to make a startling conclusion:

The figures for the seven weapons . . . suggest an average increase in performance of something over 10 per cent a year. This implies a doubling every seven years, and a twentyfold increase over thirty years. Civil goods do not increase in performance or capability in this way. The performance or capability of a present-day car is not twice that of a 1962 model, or twenty times that of a 1939 model. If calculations were made on the same basis . . . for a typical collection of consumer goods, they would show very little rise at all.

The SIPRI study went on to single out the major cause of this vast rate of weapons improvement. SIPRI showed that the research input in the military field was at least twelve times greater than in the civilian field. For instance, in the United Kingdom, which in 1964–65 operated the most research-intensive military industry in the world, $62.2 were spent for every $100 worth of military equipment bought. The comparable figure for civilian industry was $4.9 for every $100 of manufacturing output. Similar figures were found for the United States and for France: $54 compared to $7.5 for the United States and $51 compared to $1.9 for France. This work also shows the research cost of being a nuclear power. The next highest country in the league was non-nuclear Canada; she spent 'only' $20.4 on military research and development for every $100 of military procurement, compared with $1.3 for manufacturing industry as a whole.

Faced by such an appalling situation, many members of the international scientific community have expressed their concern not only for the dangers to the world as a whole, deriving from military developments of scientific advances, but also for the role of the scientist who no longer can claim his neutrality. In Parts III and IV we will have the opportunity to discuss the ethical implications of scientific endeavour and the problem of the divided loyalties of the scientist. But at this point it seems desirable to record the stand taken by some scientists, as an indication of the increasing awareness of the need for action on their part.

An American chemist, Franklin A. Long,[2] who has been concerned for many years about the interactions between science, technology and society, wrote:

What specifically can scientists do? First they must develop a greater sense of social responsibility for all aspects of the application of science to technology. This means that individual scientists must analyse the implications of their own work. It implies that there must be educational programmes to give scientists and engineers a firmer understanding of the potential impact of science and technology, as well as explicit practice in participating in the kinds of interdisciplinary technology assessments which are increasingly needed.

Scientists also have a major role to play in fostering international discussions of these grave problems. Given the international character

1. 'World Armaments and Disarmaments', *SIPRI Yearbook 1972*, Stockholm, Almquist & Wiksell, 1972.
2. F. A. Long, *Science and the Military. Civilization and Science, in Conflict or Collaboration?*, Amsterdam, Elsevier-Excerpta Medica-North Holland, 1972, p. 115–30.

of science it is not surprising that the international Pugwash movement was started by scientists, brought together from many nations by the dreadful prospects of nuclear war.

Above all, for all of these efforts, we need more scientific activists, more people sufficiently persuaded that these problems are so important that they personally must give much of their time to analysing them.

A Dutch industrial scientist, H. B. G. Casimir,[1] in his address to the second General Conference of the European Physical Society (of which he was president), stated in October 1972:

As I said, I am unable to clearly define my position. Still less can I present a solution. I should like, however, to formulate a few simple and straightforward recommendations.

1. A physicist should realize that being a physicist does not put him in a position beyond all responsibility, even if he is dealing with abstract academic subjects. . . .
2. Since we are unwilling and probably unable to stop the development of science and technology and since we are quite obviously unable to design a comprehensive masterplan for science and society, our best chances for gaining some control over the ominous spiral lie in a plurality of controls, in an independence of opinion of the several participating groups, and in openness.
3. In particular, universities should maintain their independence versus industry and industry should respect this independence.
4. Relations between the military and the universities—if they exist at all—should at the very least follow similar rules as those between industry and universities.
5. It would be desirable to have also a more clear cut separation between industry and the military. . . .
6. The time lag between fundamental science and technological applications makes it difficult to assess possible consequences and to define responsibilities. If academic scientists working on basic problems would refuse to collaborate with industry or with the military from today onwards this will hardly influence technical development for quite some time to come.

Yet I want to make one more recommendation. It is my personal opinion that, at the present time and in view of the alarming uses of science-based technology in warfare, no scientist in an academic position should of his own free will be active in or advise on military technology. As I said, this is a personal opinion but I know it is shared by many of my colleagues and an impressive number of students. . . .

And a Belgian scientist and international civil servant, Yvan de Hemptinne, made the following proposal in the course of the 1972 Pugwash Conference:[2]

When more than a quarter of the world's research scientists and technologists are working towards the perfection of destruction techniques aimed at killing people and wiping out mankind from the surface of our planet it is time to control the mechanism which mobilises such effort and to put on the brakes.

How could we bring the aforementioned mechanism under democratic control?

First of all, by bringing to light the way in which it works—one would do better to say the way in which it is going out of control—at the moment. . . .

As a second step, it would then be necessary to subordinate the choice of priorities and the allocation of resources flowing into military research to the veto of the highest science and technology policy-making organ of the various national governments. . . .

The third step would be for each government to decide to reduce progressively and proportionally its military research expenditure and the number of research workers employed in this field. . . .

Thus could the pace of the arms race gradually be slowed down, and, in the long run, be brought to a halt. To reverse the order of attack on the problem, as has been done since the arms race began, is in fact trying to put the cart before the horse. This has led to nothing in the past. It will lead to nothing in the future.

First of all we must master and then break the motor of the arms race. And then the arms can be left to rust away.

Statements like these are clear signs that the mood of active scientists is changing, that they have become aware of the great responsibilities falling upon them, and that they begin to realize that the survival of the scientific enterprise itself is at stake. Science and all its

1. H. B. G. Casimir, 'Physics and Society', *Kosmos*, Uppsala, Fysiska Institutionene, 1972, p. 105–11.
2. Y. de Hemptinne, 'Military Research and its Impact on World Peace', *Proceedings of the 22nd Pugwash Conference on Science and World Affairs*, Oxford, 1972, p. 390–7.

national and international institutions are part of the world socio-political system; to remedy the negative impact of science—it is maintained today in many circles—socio-political systems need to be reformed. This might be true, but it must be conceded that over the last thirty years or so, in the technologically advanced countries at least, the average scientist too often has accepted financial support for his research without due consideration of what such acceptance might mean. Whether support was coming from governmental sources or from industry, the average scientist was actively searching for this aid as long as his research interests could benefit therefrom. Besides the academic scientist, working on basic research problems, there have been hundreds of thousands of scientists and engineers who accepted engagement in governmental or industrial enterprises clearly devoted to the production of increasingly sophisticated tools of destruction. As implied in the statements reported, the scientist can no longer contend that his activity is one of the noblest of man (in that it is tantamount to the pursuit of truth) and at the same time share the primary responsibility for the production of destructive weapons, an effort incompatible with human dignity and decency.

This is a major dilemma, whether the scientist's primary loyalty should be to the human race or to his country. But the dilemma goes beyond the responsibility of the scientist, in that what is at stake are the acceptable limits of human action. If, in the course of the Nuerenberg trials and on other similar occasions, it has been recognized that citizens should refuse to obey when orders received are contrary to human conscience, there should be equally good reasons for the refusal of the scientist to co-operate with governmental authorities or industry when the aims of his work are clearly against the interests of mankind. In the case of the scientist, conscientious objection to military research ought to be expected even more promptly than from the ordinary soldier ordered to perform inhuman actions, since one would think that the research worker has received a deeper and more thorough education than the average military man.

One may object that, in reducing the involvement of the scientist in the scientific enterprise, the problem of the employment of thousands of scientists and engineers would arise. Indeed this is a problem that cannot be solved in a short time; but if the scientific communities of each country were to take an open stand on the issues here discussed, a more genuine internationalization of scientific effort would result (see Parts III and IV), and practical solutions to alternative employment for scientists refusing to co-operate in military research and development could be found.

The proposed course of action is admittedly difficult, painful and long. Scientific communities would experience advances and setbacks, as in any other major programme for social action, for many people would be ready to accuse the responsible scientist of working against the interests of his country. But it is precisely on this point that the validity of the stand of the scientist will be measured, for his refusal to co-operate in the military enterprise is tantamount to exerting powerful pressure against the maintenance of national sovereignties and in favour of international government. One important step in such direction would be the lifting of secrecy from all scientific activities.

Secrecy

The corruptive effects of secrecy on science were commented upon with acuity in 1938 by J. D. Bernal. In recent times, surprisingly enough, the subject has found only occasional attention, while in reality it would probably call for extensive study. It is worthwhile to quote Bernal's words on the subject:[1]

The growth of modern science coincided with a definite rejection of the idea of secrecy. It is nowhere expressed more clearly than by Réaumur in his book, L'Art de convertir le fer forgé en acier. Here he published openly the principles of steelmaking which he had discovered experimentally, though it had been a secret in the trade for two or three thousand years. For this he justifies himself

1. J. D. Bernal, *The Social Function of Science*, London, Routledge, 1939.

in the following terms, which are worth quoting at length:

'Reproaches [from those] completely opposed, & to which I find myself flattered to reply, are those made subsequent to the Assemblies of the Academy that I have just cited; there are persons who found it strange that I had published some secrets not to be revealed; others would have wished that these had been entrusted to Companies that would have made use of them, & in working for their own account would thus have worked for the general good of the Kingdom. The emotions giving rise to the first way of thinking are not sufficiently noble that one could even glorify oneself by feeling the directly opposite [sentiments]; are they not even against natural equity? and it is quite sure that our discoveries are so much our own that the Public have no right to them, that [the Public] have not a right to them in some measure? we should all, it is our first duty to, support the well-being of Society; that there lacks, when it is possible to contribute something, that there lacks, when the only cost would be to speak out—there is lacking a fundamental duty, and in the most odious of circumstances?

'It is true that there are complaints, forever long, of the little reaction from the Public, from whom comes so little praise but of which it is quite capable; a secret, guarded as it may be, is deemed miraculous—once divulged, it is considered to be nothing more than something already known before. The slightest pre-knowledge, the least significant of traces, the slightest analogies are served up as proof. It is this which has provided the pretext to various Scholars to withhold their knowledge; & others to wrap [the knowledge] they seemed to be communicating in such a way as to make the buyer of the pleasure pay dearly. Once these complaints justified, and the injustice of the Public established as certain and generalized as certain Authors contend, would one be permitted to withhold the part [of the information] that could be useful to him? Would the Physician be in the right by refusing succor to patients in urgent cases if there would be no recognition forthcoming, & who even might suffer ingratitude thereby? Are the advantages of the intellect less significant than those of the body? Is not knowledge, justly appreciated, the most real of goods? I shall say even more, that not to publish one's research as clearly as one can—revealing but a portion of it, and wishing that the rest be guessed—this is, in my view, assuming the responsibility for the time wasted by one's Readers. I wish that Mankind would admire not at all those who seem to have sought more to have themselves esteemed than to be useful. . . .

'But to respond to the second objection to which I have alluded, there are some people who have not approved the public revelation of the discoveries exposed in these Transactions; they would have willed these reserved for the Kingdom; that we would have emulated the examples of mystery, hardly praiseworthy in my view, set for us by a few of our Neighbours. We owe ourselves first to our Country, but we owe ourselves also to the rest of the world; those working to perfect the Sciences and Arts should even consider themselves to be citizens of the world entire. After all, if the research in these Transactions meets with the success that moved me to undertake it, there is not a Country capable of drawing as much benefit from it as the Kingdom; it could in the future pour its own fine steel, currently procured in foreign Countries. Even here, I would suppose that we would not fail, as we are too often wont, to take advantage of what we have at home; and I suppose that we would not abandon some establishments as lightly as they have been organized.'

In this Réaumur shows himself at the same time a true scientist and a true patriot. The two principles which he enunciates—that the work of the discoverer belongs to society and that those who work for science and art are citizens of the world—have been the guiding principles of the relation of science and society from then on and are only now once again in danger.

The idea of National Science is, of course, as old as modern science itself; the Royal Society, the Académie des Sciences, the Prussian and Russian Academies were all founded for the purpose of fostering national talent in science and also quite explicitly for the improvement of national trade and manufactures. But in early days these dangers were obviated by the much greater freedom with which scientists lived and worked in countries other than their own and by the prestige value which patronage of science conferred on the rulers of the states, a prestige dependent on open publication. The danger now is that in Governments obsessed by an autarchic economy and preparations for totalitarian war the value of science should appear in its narrowest economic form. Research into making substitutes both for industrial materials and foodstuffs derived from abroad has already, and not only in Germany, assumed immense importance. As this would be unnecessary in a rational world economy, it represents an unfortunate diversion of scientific ingenuity. Now that the control of scientific research, as that of the Government itself, is so much in the hands of large monopoly firms, the pressure to divert it to these ends may be too much for the free and open tradition of science to resist.

The situation today is much worse than the one that Bernal foresaw thirty-nine years ago Secrecy has permeated practically all levels of scientific research, even the most abstruse, and, what is worse, few scientists seem to realize how much the integrity of the scientific establishment has deteriorated during the last decades, and therefore very few fight against the dangerous trend. Not only do we find in most countries the institutionalization of security measures in many laboratories (e.g. need of special clearances for entering a scientific establishment, the distinction between different grades of 'classified research'), but even at the level of so-called 'pure' and academic science, the research worker no longer keeps a position of complete openness towards his colleagues, as in the past, but often hides his techniques, his results and even his thoughts because he fears that other colleagues may announce a discovery, even a minor one, before he does. Such debasing conduct goes under the heading of the so-called 'protection of early stages of basic research'. In fact a recent report of the Committee on Science for the Promotion of Human Welfare of the American Association for the Advancement of Science[1] contained the following statement:

Privacy is also necessary for the protection of the early stages of pure research in situations in which the highly competitive nature of modern science makes the research worker wary of premature disclosure. The demand for priority publication often results in the publication of uneven and incomplete work. If the scientific community would demand that publication be withheld until the work was more complete and elegant, this would encourage and protect the right to secrecy, that is, to privacy during the early developments of a research lead. It is also possible that secrecy which is connected with competition among research groups—whether they are working within nations divided by hostilities, in parallel industries, or in the research institutes of different universities—may lead to productive excursions into research which would not have been undertaken had the work of rival institutions been known. Phantom rivalry, that is, the unverifiable attribution of some research direction to a competitor whose actual work is protected by the walls of secrecy, may advance rather than impede discovery.

The fact that such an authoritative scientific society as the American Association for the Advancement of Science would endorse such a statement proves without possible doubt that the level of integrity of today's scientists has reached a very low level. The story of one of the major discoveries of this century, the identification of the chemical structure of the hereditary material as recounted by one of its major contributors, James Watson, in *The Double Helix* (see Part I, Chapter 2), is a poignant testimony that today even the most 'pure' research activity is not substantially different from that of any other less respected activities where secrecy, competition, rivalry, spying, conceit and social recognition prevail.

The fact registered by R. K. Merton and R. Lewis[2] that also great scientists in the past, such as Newton, were concerned with establishing the priority of their discoveries does not make the mores of today's competition any more respectable. The fact that there are many sinners does not justify my committing sin. Furthermore, the relatively recent phenomenon of the institutionalization of the scientific enterprise render rivalry and competition in scientific research not much different from those occurring in business.

What matters—it seems to me—is not the rate of advancement of this or that scientific discipline but rather that such advancement occur with full respect of intellectual honesty, i.e. with no barriers whatsoever to the communication of scientific data among research workers. If we accept the validity of competitive methods based on secrecy, as industry does, we can just as well forget about the proclaimed ideals of science: instead, let us all become businessmen—we would make more money!

1. 'Secrecy and Dissemination in Science and Technology—A Report of the Committee on Science in the Promotion of Human Welfare', *Science*, No. 163, 21 February 1969, p. 787–9.
2. R. K. Merton and R. Lewis, 'The Competitive Pressures (I): The Race for Priority', *Impact of Science on Society*, Vol. XXI, No. 2, 1971, p. 151–61.

We must recognize the fact that, particularly in the course of the last few decades, secrecy has undermined the respectability of scientific enterprise. It is high time that we re-examine our goals and our methods if we wish to reassert the human values of science. We should study carefully the deleterious effects of secrecy, demonstrate that the value of secrecy for government and industry is to a large extent a myth, and outline a programme for the progressive elimination of secretive practices in scientific research. Such proposals may sound Utopian but, as we shall examine in Parts III and IV, internationalizing the scientific effort—which would have the necessary corollary of the abolition of secrecy—could well be the only course of action compatible with the survival of our species.

5 The limits of usefulness

Science, as we have seen in the last two chapters, has made innumerable contributions to the production of technological innovations that have drastically altered our way of living. But has such change been for the better or worse? With some qualifications, the living conditions of the average man in industrially developed countries are certainly better than they were two centuries ago. It appears to some that now the 'law of diminishing returns' has set in, if we consider the whole gamut of complex interactions that those innovations have brought to bear on modern man and his communities. Each new product or technique was developed with a view to improve health, reduce human toil, offer a more varied life, or raise the economic level of individuals and countries. As viewed through the circumstances in which each innovation was produced, they were undoubtedly useful. When considered as one component of a complex of conditions which render life agreeable, innovations can show negative aspects, especially discernible in high level technological societies.

Even a cursory glance at the right-hand column of the list of scientific and technical landmarks (Chapter 3) brings out the validity of the previous statement. Just a few items from the list will bring to mind positive and negative sides: automobile (speed and freedom in transportation; road accidents, air pollution); aeroplane (shortening of distances; effective bombing, air pollution); radio, radar, computers, transistors (fast communications, safe travel; limitations to privacy, computerized warfare); pesticides and fertilizers (increased agricultural production; pollution, destruction of wild life); and so on. Two examples deserve special attention, however, because of the magnitude of the problems they pose to the living generations of man; the population explosion and careless technology.

Population explosion

Which action is more admirable and noble than saving the life of our fellow human beings? Which profession is better respected than that of the physician, whose mission it is to return to healthy life those who are ill? Which scientific discoveries are more generally appreciated than those that make possible technological developments that limit human sufferings? In the course of the last three decades we have witnessed how the introduction of effective control of infectious diseases in tropical countries has brought about a dramatic decrease in death rates, but this at the same time has sparked the major crisis of our planet: the explosion of the population of our species.

TABLE 21. Estimates of world population and rates of increase, 1750–1960

Year	Population (millions)	Average annual percentage increase from preceding date	Approximate number of years required for population to double at computed rate of increase
1750	791		
1800	978	0.4	175
1850	1,262	0.5	140
1900	1,650	0.5	140
1950	2,486	0.8	85
1960	2,982	1.8	39

Source: Human Fertility and National Development. A Challenge to Science and Technology, New York, United Nations, 1971.

TABLE 22. Estimates of population in more developed and less developed regions, 1750–2000

Year	Population (millions)			Percentage of world population	
	World total	More developed regions	Less developed regions	More developed regions	Less developed regions
1750	791	201	590	25.7	74.3
1800	978	248	730	25.6	74.4
1850	1,262	347	915	27.7	72.3
1900	1,650	573	1,077	34.7	65.3
1950	2,486	858	1,628	34.5	65.5
1970	3,632	1,090	2,542	30.0	70.0
2000	6,494	1,454	5,040	22.4	77.6

Source: Human Fertility and National Development. A Challenge to Science and Technology, New York, United Nations, 1971.

The story can be told briefly, using as the main reference a report prepared for the United Nations Advisory Committee on the Applications of Science and Technology to Development.

Estimates of world population prior to the beginning of the nineteenth century, though subject to considerable margin of error, reveal that over past centuries and millenniums the tempo of population growth has been very slow. By the middle of the eighteenth century a change in trend began—a steady acceleration in population growth—which has continued to this day and which now constitutes a major world problem. Table 21 illustrates the nature of this change.

Estimates of past and current population in the developed and developing regions of the world, along with estimates of the future population up to the year 2000 based upon the United Nations most recent population projections are shown in Table 22.

That populations grew slowly is established by the fact that it took at least a million years for human numbers to reach the 1,000 million mark, about the beginning of the nineteenth century. Prior to that time, that the population grew sporadically, with periods of gain followed by eras of actual decline, is established by recorded history. At least a quarter—perhaps a third—of the population of Europe died within hardly more than a year when the Black Death struck from the east in 1348.

Since 1800, an entirely new pattern of growth has emerged. The second 1,000 million arrived on the scene not in a million years but in hardly more than 130 years. A third 1,000 million was added in only thirty years and the fourth increment materialized in 1975, or just fifteen years after the attainment of the third 1,000 million.

FIG. 13. The growth of world population, according to United Nations demographic projections, is illustrated here by a goblet-shaped diagram whose width represents the total population of the world for the corresponding year indicated on the goblet's stem. This shows very clearly the rapid population expansion in the period 1960–2075. At around 2075, the figures begin to be stabilized (in Europe, for instance, 698 million inhabitants around 2075; in Africa, 2,338 million inhabitants in about 2120). It should be pointed out, however, that these projections do not indicate how (through what changes or upheavals—social, economic, political) this stabilization will occur. (Diagram by F. Guillot, modified.)

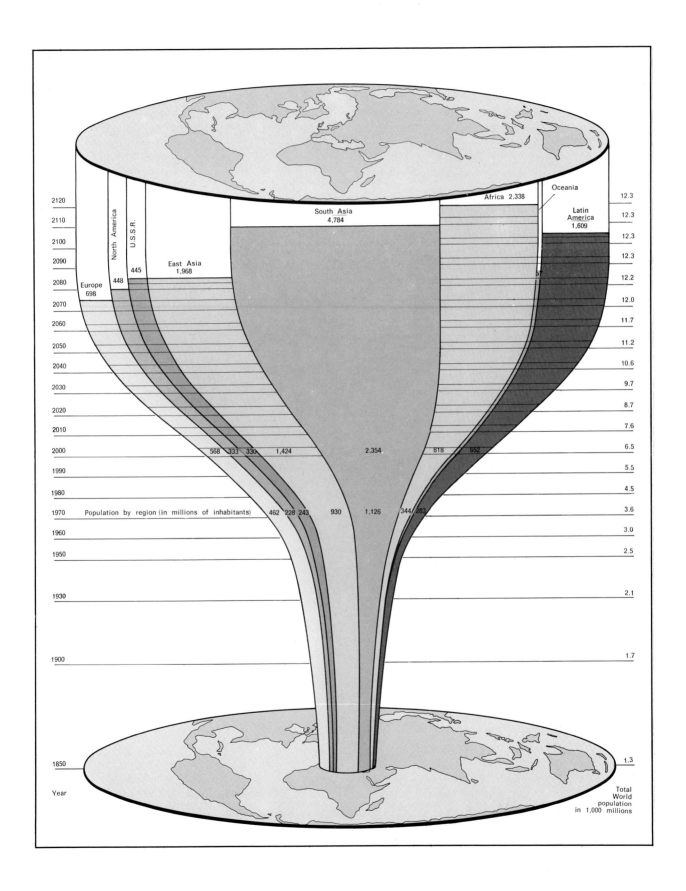

Europe
698

North America

U.S.S.R.

448

445

East Asia
1,968

South Asia
4,784

Africa 2,338

Oceania

Latin
America
1,609

2120   12.3
2110   12.3
2100   12.3
2090   12.3
2080   12.2
2070   12.0
2060   11.7
2050   11.2
2040   10.6
2030   9.7
2020   8.7
2010   7.6
2000   6.5

568 333 330   1,424   2,354   818 652

1990   5.5
1980   4.5
1970   3.6

Population by region (in millions of inhabitants)   462 228 243   930   1,126   344 283

1960   3.0
1950   2.5
1930   2.1
1900   1.7
1850   1.3

Year

Total
World
population
in 1,000 millions

At the beginning of this century, world population totalled a little over 1,500 million. In the ensuing seventy years, the world's population increase has amounted to 2,000 million or twice the total attained by 1800 since the beginning of mankind. If the present trend continues for thirty years more, world population will be well past the 6,000 million mark by the end of the century.

Of the more than 3,500 million inhabitants of the world in 1970, 2,500 million live in the world's developing regions, and about 1,000 million in the world's more developed regions. About three-quarters of the increase since 1900 has occurred in the developing regions. At present, the more developed regions—comprising less than a third of the world population—are contributing only a fifth of the total population increase.

If the assumption of the United Nations' projections are borne out, the population of the developing regions can be expected nearly to double by the end of the century, while that of the more developed regions will increase by only one-third. The projected population figures shown for the year 2000 in Table 22 are based on the United Nations medium variant which assumes that current rates of fertility will decline quite considerably in most of the world's developing regions. If this assumption is not borne out, and if fertility were to remain at present levels, the world's population would reach about 8,000 million, or more than double, by the end of the century. In that

event, the population of the developing regions alone would total over 6,000 million by the year 2000.

The annual population growth for the two regions since 1750 is shown in Table 23. This table presents in very condensed form the genesis of the current world population crisis: (a) the accelerating rate of the world population growth; (b) the shifting pattern of population increase vis-à-vis the more developed and the less developed regions.

Up to the first half of the nineteenth century, annual increments in the more developed regions amounted to no more than 1 or 2 million inhabitants. The rate of population growth was no more than one-half of 1 per cent. In the second half of the nineteenth century, the more developed countries entered into a new demographic phase. Initially, mortality began to decline at a slowly accelerating rate, this being the result of the industrial revolution, of advances in medicine and hygiene, and of better nourishment. Indeed, living conditions in mines and factories during that time were so bad as to arouse the indignation of some writers and politicians. But, on the whole, the average conditions of the industrialized regions of Western Europe substantially improved after approximately 1850, when compared to what had happened earlier in rat-infested cities and countryside slums. From then until the end of the century, annual population increments more than doubled, the rate of increase rising to about 1 per cent. With

TABLE 23. Estimated average annual population growth in more developed and less developed regions, 1750–2000

Period	Absolute annual increase (millions)			Relative increase (per thousand)		
	World total	More developed regions	Less developed regions	World total	More developed regions	Less developed regions
1750–1800	4	1	3	4	4	4
1800–50	6	2	4	5	7	5
1850–1900	8	5	3	5	10	3
1900–50	17	6	11	8	8	8
1950–2000	80	12	68	19	11	23
1960–70	65	11	54	20	11	24

Source: Human Fertility and National Development. A Challenge to Science and Technology, New York, United Nations, 1971.

the larger population base, this rate of growth produced at the turn of the century an annual population increment of about 11 to 12 million people.

The understanding of the nature of infectious diseases and a remarkable improvement in sanitation were responsible, early in this century, for lowering further the death rates. European death rates, which fluctuated around 22–24 per thousand in 1850, decreased to around 18–20 per thousand and went as low as about 16 per thousand in some countries. In Scandinavia, the drop was from 20 per thousand in 1850 to 16 in 1900. Approximately one generation after the beginning of the new phase, in the industrialized countries, one other trend appeared clear: birth rates began to decrease. This was the beginning of the so-called 'demographic transition' which has followed industrialization from the beginning of 1900 in Western Europe and in North America more recently.

For the less developed regions, average annual additions to the population are estimated to have numbered only 3 to 4 million up to the beginning of the present century, with growth rates generally of less than one-half of 1 per cent per annum. In the first half of the twentieth century, an accelerating trend in population growth in the less developed regions emerged. The average annual increment rose to 11 million, and the rate of increase approximately trebled: from a third of 1 per cent to nearly 1 per cent. But after the Second World War, a major demographic trend began in these regions; the death rate decreased from an estimated average of 28 in the 1940s to 18 per thousand per year in the 1960s. This decline was due primarily to the rapid diffusion of modern drugs, particularly antibiotics, and of advanced public health measures, exported from Europe and North America to other continents.

Victory over historical scourges such as malaria, yellow fever, cholera, smallpox and other infectious diseases, as we reviewed in Chapter 3, have saved the lives of hundreds of millions of children over the last twenty-five years, thus making it possible for them to have a much extended life expectancy. This trend has continued, and for the decade 1960–70 an annual increment of more than 50 million population is estimated in these areas, with the rate of growth reaching 2.4 per cent annually. With the numbers of the base population steadily increasing, about 70 million persons will be added annually to the populations of these regions.

Estimates of birth rates and death rates in the past are even more uncertain than those relating to population numbers. Based on an analysis of the sketchy information available, the United Nations has estimated models of the vital statistics of the last two centuries in the more developed and the less developed world regions that are consistent with what is conjectured about population sizes and growth rates.

These postulate that in the present developed regions of the world death rates above thirty per 1,000 population were common before the middle of the nineteenth century. In the second half of the nineteenth century and the beginning of the twentieth century, mortality steadily declined and at an accelerating rate. By the 1960s, an average death rate of about nine had been attained in these regions as a whole. The birth rate began to decline in the second half of the nineteenth century. In a little more than a century, it fell on the average to about one-half its previous level—from about thirty-nine around 1850 to nineteen or twenty at the present time. Little further change in the average level of fertility is expected by the end of the century for the developed regions as a whole. Though further improvement in health conditions is anticipated, the death rate in these regions is expected to remain rather stable until the end of the century, because of the increasing proportion of older persons in the population.

The trends of vital rates for the world's developing regions have been markedly different from those of the more developed countries. Average death rates well above thirty per thousand are believed to have prevailed in these regions, even into the twentieth century. There is evidence of large fluctuations resulting, *inter alia*, from wars, famine and epidemics. It was not until after the Second World War that a truly remarkable decline in mortality took place in these regions; the death rate decreased from an estimated average of twenty-eight in the 1940s to eighteen in the 1960s. Perhaps because of the extreme rapidity with which the death rate declined around the middle of this century, the second phase of the demographic transition that occurred spontaneously in the now developed regions during the nineteenth century has not taken place. The average birth rate is still estimated around forty per thousand, having shown no percep-

tible downward trend from its earliest estimated levels. The gap between birth and death rates has thus dramatically widened, bringing the rate of population growth in these regions to an all-time high (see Table 24).

The reduction of mortality is almost certain to persist, provided that present methods of improving health and prolonging life continue to be applied effectively. The continuing gap between the birth rate and the death rate has

FIG. 14. World hunger map. (From *Time*, map by P. Puglisi, archives of Mondadori.)

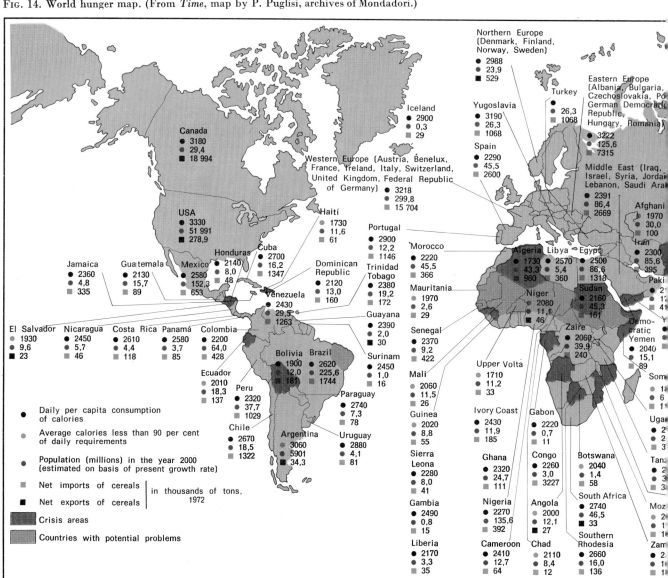

not only accelerated population growth, it has also profoundly affected the age-distribution of the population. Over 40 per cent of the population in the developing regions are children under 15 years of age. Because of the extreme youthfulness of these populations, crude death rates may possibly fall to a level as low as or even lower than prevailing in the developed regions long before health and survival conditions become comparable with those in

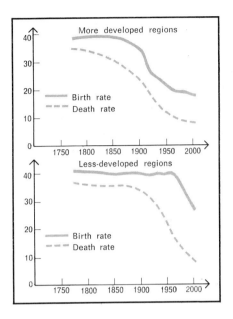

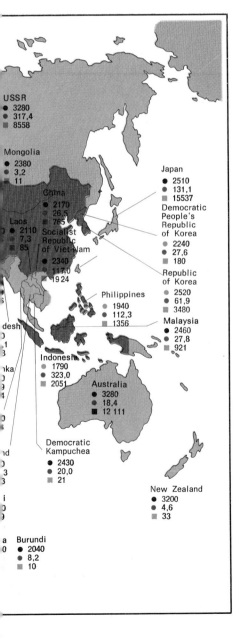

Fig. 15. Birth and death rates (per thousand) in the economically advanced and developing countries respectively (United Nations data.)

the developed regions. The United Nations' projections foresee the death rate falling from fifteen per 1,000 population at present to eight towards the end of the century.

Another consequence of this youthful age-distribution is that the fertility potential of the developing regions is very high. In less than two decades hence, the number of women aged 20–29 in these areas will have approximately doubled. Currently, the growth rate of the developing regions—more than 2 per cent per annum—doubles their population in about thirty years. This rate of growth produces an eightfold increase in a century. Such a rate cannot continue for long. Curtailment of population growth is inevitable in the long run, either as the result of disasters or of decreases in the birth rate commensurate with those that have already occurred and are expected to occur in the death rate. Current estimates indicate how great is the 'transition' yet to be made in the developing regions. The major uncertainty pertains to the speed at which fertility may decline during the decades that lie immediately ahead.

73

The control of death, thanks to the use of antibiotics, vaccines, DDT and the entire panoply of modern public health measures, brought suddenly to bear on the populations of the developing world has caused an unprecedented growth of their populations. This appears to be the major factor preventing their economic and social development. This is the cause of untold hunger, poverty and despair for hundreds of millions of human beings. Is it worthwhile to save the life of a new-born child through the use of the products of scientific advance if the child is inevitably condemned to malnutrition, no access to effective schools, unemployment in adulthood, and perhaps the victim of rioting or war?

Careless technologies and environmental problems

There hardly is a need today to stress the ecological problems brought about by industrial developments, particularly after the Conference on the Human Environment convened by the United Nations in Stockholm in June 1972. But it is worthwhile to point out that such problems are today of primary importance not only for the industrialized countries but for the less developed countries as well. In both cases problems have arisen because the benefits which were expected to accrue from a certain technological development were not weighed against its negative effects.

One-dimensional approaches were followed whereas a multidimensional assessment of the consequences should have been made before, and not after, the relevant decisions were taken.

A volume was published in 1972 bearing the title *The Careless Technology—Ecology and International Development*, edited by M. Taghi Farvar and John P. Milton. The volume epitomizes the problems under discussion. We shall use some excerpts from the book to underline the fact that the usefulness of the applications of science is not unlimited: the needs of men and societies are complex and require multifaceted evaluation before taking action. In the book, forty-seven case histories of technological developments are presented and discussed.

In each of them the selected health or nutritional measures (nine examples), irrigation and water development (eleven), procedures for intensification of plant productivity (fourteen) and animal productivity (seven), or advanced technologies (three), produced the benefits expected but also unexpected negative effects.

In his summary of the conference of which the volume presents the proceedings, Barry Commoner writes:[1]

We in the 'advanced' nations are proud of our science and technology. One of the striking features of modern life is a deep and widespread faith in the efficacy of science and the usefulness of technological progress. The United States is widely envied as an advanced nation not merely for its magnificent material wealth, but because it possesses the marvellous instruments—science and technology—which are the continuing source of more wealth. It is these wealth-producing instruments which less developed nations are eager to acquire.

There is now at least one good reason to question this faith—environmental pollution. It is my contention that environmental pollution reflects the failure of modern science to achieve an adequate understanding of the natural world, which is, after all, the arena in which every technological event takes place.

The roster of the recent technological mistakes in the environment which have been perpetrated by the most scientifically advanced society in the history of man—the United States of 1968—is appalling:

We used to be told that radiation from the fallout produced in nuclear tests was harmless. Only now, long after the damage has been done, do we know differently. The bombs were exploded long before we had even a partial scientific understanding that they could increase the incidence of harmful mutations, thyroid cancer, leukemia and congenital birth defects.

We built the maze of highways that strangles almost every large city and filled them with hordes of automobiles and trucks long before it was learned—from analysis of the chemistry of the air over Los Angeles—that sunlight induces a complex chain of chemical events in the vehicles' exhaust fumes, leading eventually to the noxious accumulation of smog.

1. Barry Commoner, in: *The Careless Technology. Ecology and International Development*, edited by M. T. Farvar and J. P. Milton.

74

TABLE 24. Population, rate of increase, birth and death rates, area and density for the world, major areas and regions: selected years[1]

Major areas and regions	Estimates of mid-year population (millions)							Annual rate of population increase (%)		Birth rate (°/oo)	Death rate (°/oo)	Area (km²) (000's)	Density[2]
	1950	1955	1960	1963	1965	1970	1972	1963–72	1965–72	1965–72	1965–72	1972	1972
World total	2,486	2,713	2,982	3,162	3,289	3,632[3]	3,782[3]	2.0	2.0	34	14	135,906	28
Africa	217	241	270	289	303	344	364	2.6	2.6	47	21	30,320	12
Western Africa	64	71	80	85	90	101	107	2.5	2.5	49	24	6,142	17
Eastern Africa	62	69	77	82	86	98	103	2.5	2.6	47	22	6,338	16
Northern Africa	51	58	65	71	75	87	92	3.0	3.1	47	17	8,525	11
Middle Africa	25	27	29	31	32	36	38	2.1	2.2	45	24	6,613	6
Southern Africa	14	16	18	19	20	23	24	2.4	2.4	41	17	2,701	9
America[4]	328	368	412	441	460	511	533	2.1[5]	2.1[5]	29	10	42,083	13
Northern America[4]	166	182	199	208	214	228	233	1.3[5]	1.2[5]	18	9	21,515	11
Latin America	162	186	213	232	246	283	300	2.9	2.9	38	10	20,567	15
Tropical South America	84	96	112	122	130	151	160	3.0	3.0	40	10	13,700	12
Middle America (mainland)	35	41	48	53	57	67	72	3.4	3.4	44	10	2,496	29
Temperate South America	27	30	33	35	36	39	41	1.8	1.8	26	9	4,134	10
Caribbean	17	18	21	22	23	26	27	2.3	2.2	35	11	238	113
Asia[6,7]	1,355	1,487	1,645	1,754	1,833	2,056	2,154	2.3	2.3	38	15	27,655	78
East Asia[6]	657	715	780	822	852	930	962	1.8	1.8	32	14	11,757	82
Mainland region	536	586	640	675	700	765	792	1.8	1.8	33	15	11,129	71
Japan	83	89	93	96	98	103	106	1.1	1.1	18	7	370	287
Other East Asia	38	40	47	51	54	61	64	2.6	2.5	35	10	258	248
South Asia[7]	698	772	865	931	981	1,126	1,191	2.8	2.8	44	17	15,898	75
Middle South Asia	481	528	588	632	665	762	806	2.7	2.8	44	17	6,771	119
South-East Asia	173	193	219	236	249	287	304	2.8	2.9	45	16	4,621	66
South-West Asia	44	51	58	63	67	77	82	2.9	2.9	44	16	4,506	18
Europe[6,7]	392	408	425	437	445	462	469	0.8[5]	0.8[5]	17	10	4,936	95
Western Europe	122	128	135	140	143	149	151	0.8[5]	0.7[5]	16	11	995	151
Southern Europe	109	113	118	120	123	128	131	0.9[5]	0.9[5]	19	9	1,315	99
Eastern Europe	89	93	97	99	100	104	106	0.8[5]	0.8[5]	17	10	990	107
Northern Europe	72	74	76	78	79	81	82	0.6[5]	0.6[5]	17	11	1,636	50
Oceania[4]	12.6	14.1	15.8	16.8	17.5	19.4	20.2	2.1[5]	2.1[5]	25	10	8,510	2
Australia and New Zealand	10.1	11.4	12.7	13.5	14.0	15.4	16.0	1.9[5]	1.9[5]	21	9	7,955	2
Melanesia	1.8	1.9	2.2	2.3	2.5	2.8	2.9	2.5	2.5	42	18	524	6
Polynesia and Micronesia	0.7	0.8	0.9	1.0	1.1	1.2	1.3	3.2	3.1	40	9	30	44
U.S.S.R.	180	196	214	225	231	243	248	1.1	1.0	18	8	22,402	11

1. Unless otherwise specified all figures except in 'Area' are estimates of the order of magnitude and are subject to a substantial margin of error.
2. Population per square kilometre of area. Figures are merely the quotients of population divided by area and are not to be considered as either reflecting density in the urban sense or as indicating the supporting power of a territory's land and resources.
3. Adjusted for discrepancies in national data on international immigration and emigration; unadjusted totals are 3,635 million for 1970 and 3,788 million for 1972.
4. Hawaii, a state of the United States of America, is included in Northern America rather than Oceania.
5. Rate reflects combined effect of natural increase and migration.
6. Excluding the U.S.S.R., shown separately.
7. The European portion of Turkey is included with South Asia rather than Europe.
Source: United Nations Demographic Yearbook 1972, New York, United Nations, 1973.

For more than forty years massive amounts of lead have been disseminated into the environment from automotive fuel additives; only now has concern developed about the resultant accumulation of lead in human beings at levels that may be approaching toxicity.

Insecticides were synthesized and massively disseminated before it was learned that they kill not only insects, but birds and fish as well, and accumulate in the human body; and that they also stimulate the development of insecticide-resistant pests and kill off natural predators and parasites.

Billions of pounds of synthetic detergents were annually drained into U.S. surface waters before it was learned—more than ten years too late—that such detergents are not degraded by bacterial action and therefore accumulate in water supplies. Nor were we aware until a few years ago that the phosphates added to improve the cleansing properties of synthetic detergents would cause overgrowths of algae, which on their death pollute surface waters.

In the last twenty-five years the amount of inorganic nitrogen fertilizer used on United States farms annually has increased about twelvefold. Only in the last few years has it become apparent that this vast elevation in the natural levels of soil nutrients has so stressed the biology of the soil that harmful amounts of nitrate have been introduced into surface waters.

The rapid combustion of fossil fuels for power, and more recently, the invasion of the stratosphere by aircraft, are rapidly changing the earth's heat balance in still poorly understood ways. The effects of these processes may drastically influence the climate and the level of the sea.

And for the future, if we make the catastrophic blunder, the major military powers are prepared to conduct large-scale nuclear warfare, even though no one knows how our societies could survive.

Each of these problems is a technological mistake, in which an unforeseen consequence has seriously marred the value of the undertaking.

For the case of the developing countries, Barry Commoner comments:

In my view, the ecological mistakes that have been reported in this conference reflect a grave and systematic fault in the overall approach which has thus far guided most international development programmes. Although each new technology that is introduced into an underdeveloped country impinges on a complex natural system, we have generally failed to take into account the effects of this technological intrusion on the properties—indeed, the very stability—of the system as a whole. Given the complex feedback networks in even the simplest ecosystem, these intrusions have nearly always led to unforeseen effects, often sufficiently deleterious to counterbalance significantly the good derived from the intended programme. In some cases, such ecological backlash has destroyed the effectiveness of the intended programme itself. There are many examples of this general failure:

1. Nearly every irrigation project reported to the conference has been followed by outbreaks, some of them disastrous, of waterborne diseases of human (in particular schistosomiasis) or of animals (such as tsetse-borne trypanosomiasis of cattle). Some irrigation projects, such as the pervasive system of dams on the Nile, have induced large-scale geophysical changes which have, in turn, reduced the agricultural potential of the region. Small-scale irrigation projects, as in Israel, have generated new insect pest problems.

2. Nearly every reported instance involving the introduction of chemical control of agricultural pests in newly developed agricultural areas has been characterized by serious ecological hazards. With awesome regularity, major outbreaks of insect pests have been induced by the use of the modern contact-killing insecticides—by stimulating the development of resistant strains and destroying the natural predators which ordinarily regulate the densities of pest populations.

3. Case histories of technological improvements in animal husbandry and fisheries, while less numerous, yield the same picture: unexpected hazards resulting, for example, from ecological interactions in the food chain (as in Peruvian fisheries) or in host-parasite relationships (as in the several tsetse fly problems in Africa).

4. The introduction of modern health programmes has in some cases involved specific intrusions on human physiology with untoward effects on aspects of nutrition and health.

Thus, while nearly all of the projects described in the conference were conceived as specific technological advances—the construction of a hydroelectric plant, the development of an irrigation system, enhancement of crop yields by chemical control of insect pests—they were in operational fact powerful intrusions on large-scale geophysical and ecological systems. Most of the difficulties which have been recounted result from the failure to recognize this basic fact.

Such failures are exemplified by many of the specific case histories which we have heard. Lake

Kariba is the largest man-made body of water in the world; surely its creation must be regarded as a huge and intricate ecological operation. Yet Scudder reports that '. . . no ecological surveys of the lake basis or the relocation areas were initiated, let alone completed, prior to the decision to proceed with the construction of the dam in 1955'. Van der Schalie reports that he had great difficulty in obtaining support for a study of the ecological consequences of the Aswan High Dam designed '. . . to obtain the data necessary to show whether the prediction that the new dam might prove to be a liability rather than an asset' was correct. Perhaps the most authoritative evidence is that provided by Riney's account of FAO experience:

Much of the assistance which has been given to developing countries has used criteria found within the teacup perimeter of various single disciplines. Within this limited horizon decisions have often been well meaning and seemingly logical, but catastrophic in their ultimate effect on the environment. The danger comes from not realizing that the virtuous activity of giving outside advice often triggers the most profound effects on the environment extending far outside the scope of the single discipline responsible for the original advice. The decisions in fact quite often guarantee failure rather than success.

I believe that we must conclude that the ecological failures recounted at this conference are not the random accidents of progress. They are rather, evidence that (a) regardless of its conceived purpose, the introduction of new technology into developing countries is always an ecological operation which must be expected to affect the complex network of physical and biological processes that characterize natural systems; and that (b) with rare exceptions, development programmes were planned, put into operation and sometimes closed down in failure, before their ecological consequences were appreciated.

I chose to quote the comments of Barry Commoner because, during the last decade, he has been one of the most systematic and outspoken critics of present-day scientific enterprise. He is not alone. Views similar to those of Commoner are shared by an increasing number of scientists, young and old. Today, it is much easier to find in the world literature books or articles criticizing science and its products than writings on the benefits human beings continue to enjoy thanks to advances in medicine, agriculture and engineering.

Are such criticisms justified? What alternative courses of action have been proposed? While a more thorough discussion of such issues will be presented in Part IV, at this stage we wish to limit our analysis to a more specific question, directly related, however, to those problems.

Must science have a stop?

The mechanism of research is clear. The march of science proceeds onwards, and the boundaries of knowledge are expanding. The prime mover of the scientist is curiosity, as we have seen. He is interested in little else than explaining the universe, matter, and life. As we shall examine in Part II, the pace of discovery in these three principal domains is fast, and has been accelerating during the last few decades. If the advancement of knowledge were not accompanied by applied developments, there would be (theoretically) no problem. Yet throughout the short, but increasingly intense, history of science, Bacon's proposal, 'the enlarging of the bounds of Human Empire, to the effecting of all things possible', has guided man's action.

The search for wealth, prestige or the fear that the 'enemy' might become more powerful in military terms than one's own country, all things possible have been and are being affected—often with total disregard for man's needs.

A growing number of pensive men began wondering whether the phenomenon makes sense, whether technological developments are not the inevitable consequence of scientific discovery, whether the survival of the species is not at stake because of runaway technology, whether life in overcrowded conditions is human at all. And since each step, in the succession science→technology→social disruption→human alienation, seems to be governed by an inexorable relationship of cause and effect with the preceding step, the question has been raised whether at least a moratorium, and possibly a global freezing, of scientific research should not be enforced. It has even been suggested that Nobel Prizes for scientific achievements should no longer be awarded inasmuch as they represent an incentive for ac-

77

tivities which may turn out to be dangerous for man.

Commenting on the misdeeds of careless technology, Barry Commoner writes:[1]

Each stems directly from misconceptions which are engendered by a specific fault in our system of science and in our understanding of the natural world. This fault is reductionism, the view that effective understanding of a real, complex system can be achieved by investigating the properties of its isolated parts. The reductionist methodology, which is so characteristic of much of modern biological research is not an effective means of analysing the vast natural systems which are threatened by modern technology. Water pollutants stress the total ecological web which ties together the numerous organisms that inhabit rivers; their effects on the whole natural system are not adequately described by laboratory studies of pure cultures of separate organisms. Smog attacks the self-protective mechanism of the human lung; its noxious effects on man are not accountable by the influence on a single enzyme or even single tissue. If, for the sake of analytic detail, molecular constituents are isolated from the smashed remains of a cell or single organisms are separated from their natural neighbours, what is lost is the network interrelationships which crucially determine the properties of the natural whole.

Indeed, I must record my own very strong conviction that the rescue rope offered to developing nations by modern science and technology is intrinsically unsound. I believe that science generally, and biology in particular, have been dominated by an intensely reductionist approach; scientific analyses engendered by this reductionist approach are a poor guide to the understanding of those realms of nature which are stressed by modern technology. This fault is the cause, not only of the ecological failures in international development, but also of the grave and unwitting deterioration of the environment in advanced nations, for example, the United States. These are strong claims, but the evidence is, I believe, of sufficient force to support them.

That modern science is a dangerously faulty foundation for technological interventions into nature becomes evident if we apply the so-called 'engineering test'—that is, how well does it work in practice? Science represents our understanding of the natural world in which man must live. Man, like any other living thing, can survive only in a given set of environmental conditions. At the same time, like all living things, man influences the very environment on which he depends, and

human survival hangs on the maintenance, at a suitable ecological balance, of this reciprocal interdependence. Since man consciously acts on the environment through technology, the compatibility of such action with human survival will, in turn, depend on the degree to which our technological practices accurately reflect the nature of the environment. We may ask, then: Is the understanding of nature which science gives us an effective guide to technological action in the natural world?

It is my feeling that such pessimistic views of science are basically wrong, for scientists like Barry Commoner obviously seem to be unaware of the fact that what is being described as reductionism is precisely the powerful analytical procedure through which science has achieved its successes. Moreover, the detractors of reductionism have no alternative methodology to suggest, one that would permit further advancement of knowledge more compatible with the needs of men and society. The fact that in order to obtain novel insights into physical or biological phenomena one has to follow the reductionist approach does not mean that, at the time when the scientist attempts to express a synthetic overview of complex systems, he denies and rejects the validity of the very method which made possible his breaking through previously unexplored frontiers that lead to new conceptions. The vast natural systems are complex indeed, but the history of science shows clearly that an understanding of such complexity is reached only through the analysis of isolated components and not by having recourse to semimythical entities as entelechy or *élan vital*.

We must, in other words, be modest and realize that even the most outstanding and elegant breakthroughs of science are partial and temporary. This is why we complain today of the misdeeds of careless technology. It is careless precisely because inadequate care was taken in evaluating the pros and cons of decisions that were reached hastily, long before all the necessary data were at hand. Too often the scientist has offered to the powers of the world the results of his labour as if they were absolute truths. Too often the scientist has been bashful and has not asked to participate

1. Commoner, *op. cit.*

with full responsibility, on equal terms with the politician, in major decisions.

Science cannot be stopped, since the curiosity and creativity of men knows no laws or imposed rules. But the manner in which the products of science are going to be used can, indeed must, be governed by less haphazard procedures than the ones prevailing until now. Serious attempts should be made to foresee untoward side-effects of technology and ethical principles and political decisions should, when necessary, limit and even forestall clearly dangerous developments. Technological innovations can no longer be accepted blindly but need to be evaluated in the broad social interest. As we shall discuss in Parts III and IV the humane management of technology is a very complicated process. Numerous attempts are under way in this direction and have been recently reviewed and discussed in the volume *Society and the Assessment of Technology*.[1]

Perhaps we, the people of the latter part of the twentieth century, have been at fault in not giving adequate consideration to the warning that Francis Bacon was giving long ago:

It were good therefore that men in their innovations would follow the example of time itself; which indeed innovateth greatly, but quietly, and by degrees scarce to be perceived. For otherwise, whatsoever is new is unlooked for; and ever it mends some, and pairs other; and he that is holpen takes it for a fortune, and thanks the time; and he that is hurt, for a wrong, and imputeth it to the author. It is good also not to try experiments in states, except the necessity be urgent, or the utility evident; and well to beware that it be the reformation that draweth on the change, and not the desire of change, that pretendeth the reformation. And lastly, that the novelty, though it be not rejected, yet be held for a suspect; and, as the Scripture saith, 'that we make a stand upon the ancient way, and then look about us, and discover what is the straight and right way, and so to walk in it'.

1. *Society and the Assessment of Technology*, Paris, OECD, 1973.

Part II

*Current trends
in scientific
research*

The universe

6   The solid Earth

Origin of the Earth

Any discussion of the origin of the Earth in effect poses four main problems: how the Earth originated, when it happened, when the Earth assumed its present-day characteristics, and what happened in that interval of time.

Our present knowledge of the origin of the universe and of the solar system is discussed in Chapter 12. The sequence of events leading to the formation of the Earth, according to the generally accepted model, runs as follows. Around the Sun a large and cold cloud of ionized gas (plasma) and dust began to condense to form 'grains', the remains of which may be the chondrules found today in many meteorites; the aggregation of these grains produced embryonic planets, and further aggregation or collision gave rise to protoplanets of which present day asteroids may be examples; further impacts and aggregation, together with gravitational collapse and heating phenomena, finally produced the planets and their satellites.

Radiometric dating of meteorites and of some lunar samples both date the end of these processes at 4.6 billion[1] years ago. This date is commonly accepted as the date of the origin of the Earth and solar system. Moreover, the presence in meteorites of ^{129}Xe derived from the decay of ^{129}I (half life, 16 million years) and

of $^{131\text{-}136}$Xe derived from the decay of ^{244}Pu (half life, 76 million years), together with fission tracks from ^{244}Pu, is interpreted as evidence that the interval of time between the formation of these elements by nuclear synthesis and the formation of solid bodies able to preserve their fission products cannot exceed 100 million years. The presence of small quantities of ^{244}Pu in terrestrial rock confirms that nuclear synthesis was still occurring less than 5 billion years ago. Had it been earlier, this isotope would have decayed in such a way as to be no longer detectable.

The oldest terrestrial rock (recently found in Greenland) is 3.9 billion years old. Nothing is known of the first six hundred million years of the Earth's history. In fact, because all the oldest minerals on Earth seem to be about 3.6–3.9 billion years old, it seems that at that time the Earth may have been involved in a thermal event which reset all radiometric clocks and obliterated all traces of former events. Whether this 'thermal event' was some kind of heating phenomenon that heated a formerly colder Earth or simply the end of a longer heating period begun with the heat released during accretion and infall of celestial bodies captured by accreting Earth, is still a matter of debate. Searching for a cause of the

---

1. 1 billion$=10^9$ or 1,000 million.

thermal event outside the Earth itself, it has become important, therefore, to ascertain whether Earth has always had the Moon as its companion or if the latter entered terrestrial orbit later.

There are three main hypotheses as to how the Moon originated: (a) the Moon was once part of the Earth; (b) the Moon was once an independent body which later entered terrestrial orbit; and (c) the Moon was an independent body which originated at the same time as the Earth and has ever since been its satellite.

The *Apollo* samples of the Moon's surface show that the abundance and distribution of elements are very different from those of the Earth. Lunar dust has an age of 4.5–4.6 billion years. These facts exclude hypothesis (a), which was also rejected for mechanical reasons by H. Jeffreys in the early 1930s. It should be added that the supporters of this idea originally thought that the Pacific Ocean was the scar caused by the separation of the Moon from the Earth, but marine geology has now explained the origin of this oceanic basin in terms of movements of the Earth's crust.

Hypothesis (b) could explain why all the oldest minerals on Earth have nearly the same age. The thermal event which happened at that time might have been caused by the Moon entering terrestrial orbit. The Moon would have been much nearer the Earth than it is now, causing very high tidal energy that would have melted the surface of both the Earth and the Moon itself. The hypothesis is supported by astronomical observations which show that there has been a progressive increase in the average distance between the Moon and the Earth, and by palaeontological observations (on Precambrian fossils living in the intertidal zone) which suggest that the height of tides has diminished over time. But radiometric dating of rocks in the lunar maria makes it difficult to explain lunar magnetism as the result of a unique event such as a close approach of the Moon to the Earth.

Hypothesis (c) now seems, therefore, the least open to criticism although the chemical differences between the Earth and the Moon—two bodies which originated in the same region of the solar nebula—have yet to be explained. Deeper knowledge of the general characteristics of the Moon is still required.

As has been mentioned, the first event of which there is evidence on the Earth is the thermal event which occurred about 3.9 billion years ago. The oldest sedimentary rock is 3.2 billion years old and, during the 1960s, the remains of living organisms were found embedded in it. We do not know yet whether there was a primordial atmosphere and hydrosphere before the thermal event (see Chapter 8). The presence of sedimentary rock means that 3.2 billion years ago there was a hydrosphere in which deposition occurred and probably an atmosphere which gave rise to weathering and erosion, but neither the 'air' nor the 'sea' were necessarily the same as today. Detrital particles of minerals, which in the present atmosphere would be quickly oxidized, are known in alluvial deposits to be some 2 billion years old; primary ferric deposits are known in marine sediments of the same period. There is no evidence of free oxygen in the atmosphere till 1.8 billion years ago, when 'red beds' of nonmarine origin appeared.

There are two models of the composition of the primary atmosphere and hydrosphere. According to the first, the atmosphere formerly consisted mainly of hydrogen, methane and ammonia; according to the second, it consisted of those gases that now flow out of volcanoes, such as CO_2, nitrogen and water vapour. It seems clear, however, that the primary atmosphere and hydrosphere, from which are derived today's atmosphere and hydrosphere, were produced by degassing and dewatering of the Earth itself. A better knowledge of the thermal history of the Earth seems to be required to understand the details of how the atmosphere and the hydrosphere evolved. If the Earth and its protoatmosphere were once very hot, condensation of the hydrosphere could account for the great quantity of CO_2 found in limestone. Condensation would have produced a hydrosphere extremely rich in CO_2, and limestone would have been the most abundant early sedimentary rock. The deposit of carbonate, however, seems to have occurred at a nearly constant rate during geological time, the oldest sedimentary rock not being limestone.

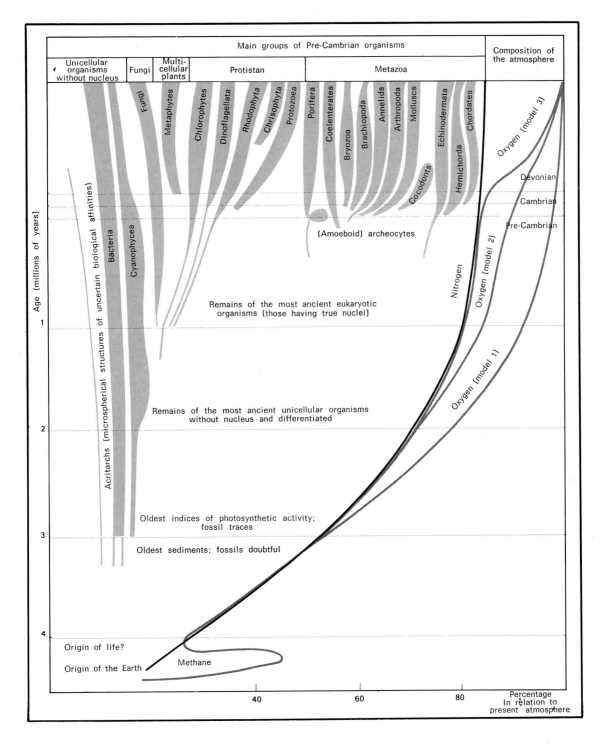

FIG. 16. Evolution of the atmosphere and the first
organisms. Of the three possible curves for the oxygen
present in the atmosphere, the most probable,
according to the data at present available, is number 2.
(From A. G. Fischer.)

In short, by the middle 1970s the problem of how the Earth originated was still one of the great unsolved questions in the earth sciences. Nevertheless, the problem—which once belonged to the realm of theoretical speculation—is now being attacked in a quantitative way and from several points of view. Although a coherent and conclusive model has not yet been accepted, there are several promising research trends in this field. Space exploration may provide information on the chemistry and mineralogy of planets and satellites, where evidence of early conditions has not been obliterated by later geological phenomena as happened on Earth, on the behaviour of cosmic plasma outside and inside the magnetosphere, and on the nature of micrometeorites which may be fossil remains of the material from which the Earth was formed. An especially interesting goal of space research might also be the exploration of the asteroids since these may represent an intermediate stage in the formation of the planets.

Studies of plasma (see Chapter 18) will be particularly important in understanding the behaviour of matter in the solar nebula from which planets and satellites originated, as will studies on collisions at high speed between solid bodies. A better understanding of how early forms of aggregate matter rotated round a central body in the solar nebula may be provided by studies of Keplerian motion in a viscous medium. The application of computers to classical celestial mechanics and to the study of resonance phenomena, which seem to have played a decisive role at the time of the formation of the solar system, will also be very valuable. Research on the magnetization of meteorites to ascertain the presence and intensity of a magnetic field at the time of the origin of the solar system seems to be promising, too. Finally, all studies directed at finding superheavy elements such as ^{244}Pu and their fission products will be extremely helpful in understanding how and when nucleosynthesis occurred.

Models of internal structure

The deepest holes in the Earth's crust are 8 km deep in the continental crust and 6 km deep in the oceanic crust (the latter being the depth reached in the Deep-sea Drilling Project in 1970). The rest of the Earth is inaccessible to direct observation and its properties have to be deduced indirectly from other measurements.

All models or theories of the Earth's interior constitution have to be consistent with at least three known parameters: its mass M, its volume V (calculated two centuries ago) and the coefficient y of the moment of its intertia. The recalculation of the latter in 1963, with the help of data from the orbits of artificial satellites, modified the previous value of 0.334 to 0.331, and thus led to the modification of all former models of the Earth's interior.

The simplest of these models, for example the Legendre-Laplace and the Roche density models, considered the distribution of only one property as a function of the Earth's radius. Now five parameters have to be considered: density ρ, pressure p, gravity g, bulk module k, and rigidity μ. The last two parameters emerged together with seismology during this century, since the velocity of seismic waves in the Earth is a function of k, μ and ρ. Seismology made possible the calculation of k/ρ and μ/ρ. But the problem is how to calculate k, μ and ρ separately with independent methods. In the past ten years great progress has been possible through experiments on rocks and various materials at high pressures and temperatures, the further development of the finite strain theory showing a correlation between k and p, and the development of explosion seismology and seismological array stations. Furthermore, after theoretical calculations of the 'free oscillation' periods of the Earth in 1959 for various models of the Earth's interior, a great interest arose in recording such oscillations. In fact, free oscillations made the direct calculation of density possible. After the great 1960 Chilean earthquake and the 1964 Alaskan earthquake, more than 100 free oscillation periods were determined. Each one is a new condition that all models of the interior of the Earth have to satisfy.

At the beginning of the 1970s a model of the Earth was proposed which summarizes the data collected with the help of large aperture seismic observatories, and of large under-

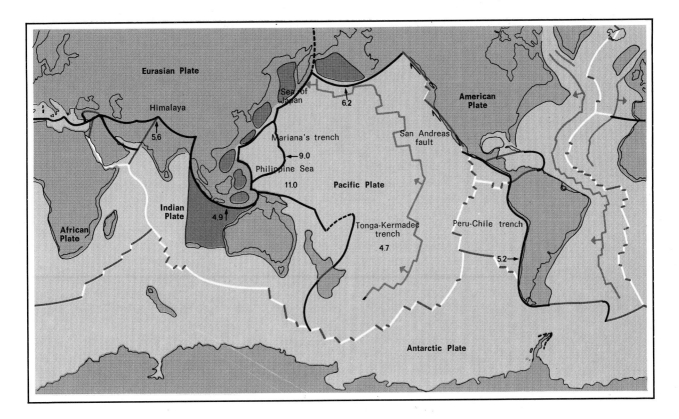

FIG. 17. Plate tectonics has developed into a highly scientific theory, with geometric and kinematic bases providing a convincing explanation of the nature of earthquakes, volcanoes and mountain chains. The basic hypothesis is that the various different plates (spherical caps of lithosphere fitting into one another to form a series of blocks covering the Earth) are in a state of constant movement (the figures in the diagram indicate the speed of movement in cm/yr; the lines represent the various types of edges between contiguous blocks), while the dimensions of the Earth remain approximately unchanged.

ground nuclear explosions. The model also took into account data from long-period surface waves and the free oscillation of the Earth. It seems to be consistent, also, with laboratory studies on ultrasonic measurements, and with the results of high pressure and shock wave experiments on a great variety of rocks and materials. According to this model the Earth's interior may be subdivided into:

1. The crust, 30–40 km thick under the continents and 10 km thick under the oceans, its lower boundary being marked by the Mohorovičić Discontinuity.

2. A high velocity layer, 0–50 km thick, which together with the crust constitutes the lithosphere and with the layers 3, 4, 5 and 6 listed below constitutes the mantle of the Earth.

3. A low velocity layer, partly molten, 100 km thick, called the astenosphere.

4. Another high velocity layer, relatively homogeneous, from 150 to 400 km deep.

5. The transition zone, from 400 to 800 km deep, in which several abrupt increases of seismic wave velocity occur.

6. The lower mantle (from 800 to 2,900 km deep) where density and velocity of seismic waves increase with depth, its lower boundary being the Wiechert-Gutemberg Discontinuity.

7. The outer core (from 2,900 to 5,150 km deep) which seems to be liquid and have the properties of molten iron with sulphur or silicon (motion in it may be responsible for the Earth's magnetic field).

8. The inner core, from 5,150 km deep to the centre which (according to indirect evidence) may be solid. Its density is 13 g/cm^3.

The core and mantle

The Earth's core is the region under the Wiechert-Gutemberg Discontinuity, which occurs at a depth of 2,890–2,900 km and marks a sharp increase of density. The outer part of the core is certainly liquid, since S seismic waves do not travel through it. At a depth of 5,150 km a sharp increase of P wave velocity marks the boundary of another region with higher density, the inner core.

Knowledge of the composition of the core is hindered by our poor understanding of the phenomena that led to its formation. In fact, even though there are several lines of evidence to follow up and several properties which can be measured or extrapolated, final conclusions are dependent on exactly which theory of how the Earth itself originated is used.

By analogy with iron meteorites and in accordance with the way in which the chemical elements are distributed on the Earth and other astronomical bodies, it used to be assumed that the core of the Earth was composed of an iron-nickel alloy. In the 1960s, shock wave experiments on iron-nickel alloys at typical pressures which would exist in the core confirmed that a pure iron-nickel core would be too dense. Later, the 'bulk' speed of sound waves in the core was estimated to lie between 4.88 and 5.18 km/s while the values for iron-nickel alloys in the same experimental conditions lay between 3.1 and 3.7. A pure iron-nickel core is therefore much too dense and has too small a bulk sound speed to explain geophysical observations.

Comparing the p, k and ρ variations (see page previous) with the results of experimental petrology, it is possible to deduce that the main core component is an iron-nickel melt, with lighter alloying elements included in the outer core. The latter could be silicon or sulphur: both would decrease the core density and increase the bulk sound speed. Determinations of the density of iron-silicon alloys in core conditions are consistent with a core containing 14–20 per cent of silicon; further, the bulk sound speed of such an alloy is consistent with the data. But a sulphur-iron core would have almost the same physical properties. More laboratory experiments on sulphur-iron samples seem to be needed before a choice can be made.

The liquid outer core is the region in which, following the Bullard-Elsasser dynamo theory, the Earth's magnetic field is produced: motion in the liquid metallic (i.e. electrically conductive) outer core causes it to act as a self-exciting dynamo and to produce the Earth's observed magnetic field. This theory is now quite generally accepted, even though there are several different theories about the forces that set the outer core in motion and even though the dynamo theory is difficult to work out in detail. Studies of the Sun, in which surface motions as well as magnetic fields are observable, would be very helpful.

The mantle of the Earth is the region between (a) the Mohorovičić Discontinuity (10–40 km beneath the Earth's surface) at the boundary between crust and mantle and (b) the Wiechert-Gutemberg Discontinuity (2,890–2,900 km beneath the Earth's surface) at the boundary with the core. The mantle itself can be divided into two main regions, upper and lower. The boundary between these is not sharp, and coincides with a transition region (between 400 and 800 km deep) in which several abrupt increases of seismic velocity occur. A convenient boundary could be 700 km, below which earthquakes do not occur.

The mantle contains nearly 68 per cent of the mass of the Earth and is also the source of energy from which the dynamic processes of the Earth's crust are driven. For this reason, a great effort was made in the 1960s to learn more about it and particularly about its upper part. Fundamental in the promotion and co-ordination of the effort has been the international research programme called the Upper Mantle Project, launched early in the 1960s, and out of which, early in the 1970s, was born the Geodynamics Project.

The upper mantle now seems to be composed of four main layers: a high velocity layer (0–50 km thick), a relatively homogenous layer (from 150–200 km to 400 km deep), and the transition zone from 400 to 800 km deep. The presence in the upper mantle of a low velocity zone makes the interpretation of seismic data in terms of mineralogy more complicated, but it is accepted that olivine, pyroxene and garnet are the most important minerals together with, in the upper part of the upper mantle, spinel and amphibole. The low velocity zone (called the astenosphere by geologists) is thought to be the result of partial melting or dehydration of the mineral content. The transition zone, in which several sharp increases of velocity and density occur, seems to be a region in which pressure induces a change of the structure of olivine into a spinel-type structure; this happens 400 km beneath the surface. Deeper down, the fourfold oxygen co-ordination of silicon is transformed into a denser sixfold form.

Data on the lower mantle are far less abundant than on the upper one. The lower mantle seems to be a relatively homogeneous zone with a density and velocity consistent with a denser ionic packing than MgO, $MgAl_2O_4$ or garnets. Shock wave data, seismic data, free oscillation data, finite strain theory and the seismic equation of state are all helping to provide more knowledge of the deeper mantle.

The crust

The crust is the thin outer shell of the Earth, extending from the surface down to a depth of 30–40 km beneath the continents or to no more than 10 km beneath the oceans. The lower boundary of the crust is marked by a region in which there is a sharp increase of seismic wave velocity called (as already mentioned) the Mohorovičić Discontinuity or, more briefly, Moho.

The crust is the most heterogeneous layer of the Earth, but it is possible to distinguish the upper sialic or granitic layer from the lower sialic or basaltic layer. The former is the crust generally observed on continents; the latter forms the ocean floor where the sialic layer is absent. The lower layer of the continental crust is known only from seismic data and not from direct observation. At a depth of 15–20 km, seismic velocities are somewhat higher than in the upper part, thus indicating higher density. This is consistent with both a chemical constitution similar to the oceanic crust and one similar to the continental one but with minerals formed at a higher pressure.

The crust has been the most studied portion of the Earth. This is particularly true of the continental crust and the orogenic or mountainous belts, the typical areas studied in classical geology. But solutions to the main problems of continental geology now come from the sea. The deeper knowledge of the oceanic crust which was acquired mainly in the 1960s gave rise to the theories of ocean floor spreading and plate tectonics.

Early in the 1970s, as more and more evidence accumulated in favour of the plate tectonics model, the mountain belts came to be regarded as former plate boundaries and the concept of geosynclines was carefully re-examined. The geosynclines, thick deposits of sediments from which (it used to be thought) folded mountain chains had been formed, seem now to be the deposits formed on the continental shelf and the continental rise. Seen in this light, the great subsidence which is characteristic of geosyncline deposits, and the subsequent folding and thrusting, seem to be the results of the interactions between two plates. The peculiar characteristics of insular arcs and of continental margins, also, with their typical seismic and volcanic activity, can now be explained in terms of the interaction of two neighbouring plates. As a consequence of the dynamics of the Earth's crust, the oceanic crust is being continually renewed along the mid-oceanic ridge and consumed in the trenches, while continental masses are slowly expanding by marginal addition of the geosyncline sediments mixed with fragments of oceanic crust involved in the process of mountain building.

Classical geology is now receiving increasing help from aerial and orbital photogeology, development of remote sensing techniques that will make orbital stations very important research vehicles, seismology, a more profound knowledge of the oceanic environment which makes the interpretation of sedimentary structures and the palaeontological content of sedimentary rocks less difficult, geochemistry, and laboratory experiments which allow matter to be studied—even when it cannot be directly observed from internal heat studies and from studies on palaeomagnetism.

Dynamics of the crust

The Earth's crust appears now to be more mobile than was once imagined, and the long struggle between followers of the fixed and the mobile crust hypotheses is gradually being settled. This dispute had separated Earth scientists into two camps: on the one hand, geologists who worked mainly within stable areas and therefore believed that the Earth's crust moved almost exclusively in the vertical direction; and those, on the other hand, who did research on recently corrugated zones (such as the Alps) and therefore believed in large-scale horizontal displacement. And continental drift, the theory proposed in 1912 by Wegener according to which all emerging lands were in the past united in a single continent (Pangea) and only subsequently reached their present positions, required such huge displacements that not even those who believed most ardently in the Earth's mobility could accept it without further evidence.

By the end of the 1950s, two trends appeared which were to modify this situation drastically. Research on palaeomagnetism which in 1956 led some scientists to postulate a certain degree of mobility in poles and continents was one of these. And oceanographic research which led others to propose the existence, in every ocean, of a very long and winding underwater ridge, already known in one ocean from the Challenger expedition as the mid-Atlantic Ridge.

Geologists then turned increasingly to the study of the ocean and more and more refined methods of research appeared, marking a shift in emphasis from the study of the single ocean ridge to that of the entire ocean so that international co-operation was increasingly needed. The existence of the ridge was at first acclaimed as a proof of the fixed crust hypothesis coupled with the idea that the Earth had expanded over geological time. Thus the continents would have separated without noticeable movements on the surface of the globe, and the mid-Atlantic Ridge would be the great scar along which the laceration developed. But this idea was immediately countered by another which postulated mobility, that of the spreading ocean floors, proposed in 1961–62. Accord-

ing to this theory, the mid-Atlantic Ridge would be the great scar along which the laceration developed. But this idea was immediately countered by another which postulated mobility, that of the spreading ocean floors, proposed in 1961–62. According to this theory, the mid-Atlantic Ridge is a rift system along which material from the Earth's mantle rises to become fresh crust; since new crust is continuously generated in the middle of the ocean, in an Earth of constant size the bordering continents must drift away.

Meanwhile, studies of palaeomagnetism continued and confirmed a theory first put forward in 1906 that in the past the Earth's magnetic field had undergone repeated inversions. New radiometric methods made it possible to construct first a relative and then an absolute chronological scale of these inversions. On this base a brilliant new proof of the spreading of the ocean floor was proposed in 1963 which took into account the discovery, made in 1960, of magnetic anomalies running in parallel bands along the northeastern Pacific shores: if new crust is continuously generated and if, meanwhile, the Earth's magnetic field changes its polarity, the crust should contain symmetrical bands of material of alternating magnetic polarity. An aerial magnetic survey performed in 1966 showed that magnetic anomalies of this kind existed on the Reykjanes ridge, south of Iceland. Radiometric dating of the ocean floor was done, and it was soon possible to verify that the oldest crust was in fact the farthest from the ridge, and to give an absolute age to such bands of normal and inverted magnetization.

The absolute magnetic stratigraphy thus achieved provided a direct means of dating oceanic crust after a survey of magnetic anomalies: in every ocean, patterns similar to those of the Reykjanes ridge have been found. The whole of the Earth's crust is by now divided into a number of plates capable of moving relative to each other, in order to accommodate the new crust generated along the ridges. If the Earth does not expand, and if new crust is generated somewhere, it must also be consumed elsewhere. Subsequent studies of deep earthquakes revealed, in the overlapping zone of two plates, a downward movement of the crust which would eventually rejoin the mantle.

The systematic drilling of the ocean's floors within the framework of the Deep-sea Drilling Project (started in 1969) provided a huge amount of material and, above all, confirmed that nowhere is the ocean crust older than 150 million years. This rules out the expanding Earth hypothesis, since such an expansion would have started much earlier and some fragment of the oldest crust would have been found by now. Spreading ocean floors and continental drift thus seem to be sufficiently proved, while the geodynamic model of plate tectonics is still under test.

Today research is aimed at learning what forces and mechanisms drive the plates and how traditionally accepted continental structures can be re-interpreted in the light of new discoveries. Mountain building, volcanism and earthquakes all seem to be effects of a single cause—the dynamics of the Earth's crust. Thus the fracture of the single continent named Pangea and the drift of its fragments to produce the face of today's world may be only an episode in the much longer history of the dynamics of the Earth's crust.

Further reading

AL'VEN, Ch. O Proischoždenii Solnecnoj sistemy. *Budušee Nauki.* Moskva, Znanie, 1971.

CAILLEUX, A. *L'anatomie de la terre.* Paris, Hachette, 1968.

CLAYTON, K. *Earth's crust.* London, Aldus Books, 1967.

EIBY, G. A. *Earthquakes.* London, Muller, 1967.

EICHER, D. L. *Geological time.* Prentice Hall, 1968.

GASS, I. G.; SMITH, P. J.; WILSON, R. C. L. (eds.). *Understanding the earth.* Artemis Press, 1971.

HAMILTON, E. I.; FARQUHAR, R. M. (eds.). *Radiometric dating for geologists.* London, Interscience, 1968.

KALESNIK, S. V. Obščie geograficeskie zakonomernosti Zemli. Moskva, Znanie, 1970.

KEEN, M. J. *An introduction to marine geology.* London, Pergamon Press, 1969.

LISICYN, A. P. Burenie Dna okeanov. *Nauka i Čelovečestvo.* Moskva, Znanie, 1970.

LUKASEV, K. I. Gorizonty geochimii. *Budušee Nauki.* Moskva, Znanie, 1968.

MAXWELL, A. E. (ed.). New concepts of sea floor evolution. *The sea,* vol. IV. New York, Wiley Interscience, 1971.

MESČERJAKOV, Ju. A. Dychanie Zemli. *Nauka i Čelovečestvo.* Moskva, Znanie, 1965.

NAGATA, T. Magnitnoe pole v proslom. *Nauka i Čelovečestvo.* Moskva, Znanie, 1965.

PHINNEY, R. A. (ed.). *The history of the earth's crust.* Princeton, Princeton University Press, 1968.

PIEL, G. (ed.). Gondwanaland revisited: new evidence for continental drift. *Proc. Amer. Phil. Soc.,* vol. 112, no. 5, 1968.

RAUZER-ČERNOUSOVA, D. M. Mikropaleontologija i istorija Zemli. *Nauka i Čelovečestvo.* Moskva, Znanie, 1968.

RITTMANN, A. *Volcanoes and their activity.* New York, Wiley Interscience, 1966.

SHEPARD, F. P. *The earth beneath the sea.* Baltimore, John Hopkins Press, 1967.

SIDORENKO, A. V. Geologija v 2000 gody. *Budušee Nauki.* Moskva, Znanie, 1971.

SMITH, B. L.; JOHNSON, H. (eds.). *The megatectonics of continents and oceans.* Rutgers, Rutgers University Press, 1970.

STRANGWAY, D. W. *History of the earth's magnetic field.* New York, McGraw-Hill, 1970.

TAKEUCHI, H.; UYEDA, S.; KANAMORI, H. *Debate about the earth.* San Francisco, Freeman & Cooper, 1970.

TAZIEFF, G. O vulkanologii. *Nauka i Čelovečestvo.* Moskva, Znanie, 1965.

TRESNIKOV, A. F. Arktika i ee izucenie. *Nauka i Čelovečestvo.* Moskva, Znanie, 1967.

TUCKER, R. H.; COOK, A. H.; IYER, M. H.; STACEY, F. D. *Global geophysics.* London, English University Press, 1970.

VARIOUS AUTHORS. Deep-seated foundations of geological phenomena. *Tectonophysics,* vol. 7, no. 5–6, 1969.

——. *The ocean.* A Scientific American Book. San Francisco, W. H. Freeman, 1970.

——. The world rift system. *Tectonophysics,* vol. 8, no. 4–6, 1969.

WHIPPLE, F. L. Earth, moon and planets. Oxford, Oxford University Press, 1968.

WOOD, J. A. *Meteorites and the origin of plants.* New York, McGraw-Hill, 1968.

7 The Earth's oceans

Origin of the oceans

The problem of the origin of the oceans is really two problems: that of the forming of oceanic basins and that of the origin of the oceans themselves. The first is intimately connected with the evolution of the Earth's crust (Chapter 6), the main problem being how and when the ocean's water was formed.

The presence in sedimentary rocks 600 million years old of fossil forms similar to organisms living today in the sea seems to indicate that the composition of sea water has not changed dramatically since then. But the age of the Earth is 4.6 billion years. What happened to the oceans in their first 4 billion years? Is the present ocean a remnant of a primordial ocean and has ocean water changed drastically, both in quantity and in quality, since its formation?

Recent research into the abundance in the universe of the noble gases has now thrown some light on these questions. Stars whose composition is thought to be representative of the early make-up of the universe at the time the Earth was formed are rich in noble gases. The Earth has become depleted of these chemicals, particularly the lighter ones. This is to be expected if a thermal event caused the Earth to heat up soon after its formation (see Chapter 6). At that time many of the gases in the original atmosphere would have been driven off into space, lighter gases escaping more easily than the heavier ones. It seems that most of the Earth's argon, with atomic weight of 40, has been lost in this way and the same must therefore be true of lighter gases such as water vapour, which has a molecular weight of only 18. In other words, almost none of the original atmosphere and oceans is now preserved on the Earth.

This consideration is now put in doubt by a number of scientists who, interpreting the results of lunar and meteorite studies, reach other conclusions. *Ad hoc* experiments developed on the lunar surface and studies on the composition of the outer skin of lunar samples show that noble gases' relative composition of the Sun (as shown by the solar wind's composition) is different from that of the Earth's atmosphere (the latter being depleted of lighter gases). At the same time, the analysis of noble gases included in crystals deep inside meteorites has shown a composition rigidly similar to that of the atmosphere of the Earth. Researchers have concluded that the difference between the solar and terrestrial noble gases' relative composition is not a consequence of some kind of event which happened on the earth some time after its formation but of the different positions of the Earth and Sun in the solar nebula.

The ultimate conclusion of this line of reasoning is that our present atmosphere is the end-product of the evolution of the first 'astronomical' atmosphere acquired by the Earth in the accretion period. This means also that the thermal history of early Earth would be quite different from that formerly supposed. After a maximum reached at the time of the formation of the Earth's outer core (the sole shell of the Earth now liquid, see Chapter 6), the temperature would have remained at a level sufficiently low to maintain the original abundance of volatile compounds. These would be preserved first in a dense atmosphere and later in the earliest kind of ocean.

As previously indicated, many scientists think that the heating up of the accreting Earth became so intense that the lighter gases escaped into space. Thus they believe that, as they do now, the light gases would have added sufficient new material to fill the oceans many times over.

This finding suggests that the volume of the oceans has increased substantially. Indeed, if volcanic activity has remained constant over time, the volume of the oceans must have grown at a steady rate. The distribution of marine rocks in the past seems to indicate, however, that the marine environment has not undergone dramatic changes; this depends, rather, on how the average ocean depth—the mean difference in height between the continents and the oceans—has changed over time. It now seems that volcanic activity accompanied the differentiation of the Earth's surface into continents and oceans, and led to simultaneous increases in·the volume of the oceans.

The ocean waters must have originated from the Earth's mantle. That this is possible has been proved by studies of stony meteorites, bodies which can reasonably be assumed to have a similar chemical composition to the mantle. Such meteorites contain an average water content of some 0.5 per cent. The oceans, however, have a total mass of about 0.035 per cent of the Earth's mantle and hence the water content of the mantle could have accounted for more than ten times the volume of water which is found in the oceans today.

Composition of the oceans

In one sense, the oceans are remarkably homogenous. The composition of sea water with respect to its eleven major constituents, which make up 99.9 per cent of the dissolved salts, is very nearly uniform. Yet some of the dissolved gases and particulate matter, as well as the other minor constituents such as certain nutrient salts and organic components, may be very variable both in time and space, and vertically as well as horizontally.

Oceanographic expeditions during the past three decades have greatly extended our knowledge of these variations. International and regional co-operation has been a feature of oceanographic research for many years but, with the International Geophysical Year (IGY) and the birth of the International Oceanographic Commission (IOC) late in the 1950s and early in the 1960s, an intensive international effort has marked the past decade. For example, much of our information concerning the Indian Ocean came from the International Indian Ocean Expedition of 1960–65, and recent research efforts have greatly added to our knowledge of the polar regions (see Figure 18). As a result, much more data are now available on changes in the ocean's composition at all depths and during different seasons. Although some preliminary conclusions can be drawn from this work, much more observation remains to be done.

Sea water consists of 96.5 per cent water and 3.5 per cent salts, organic compounds, particulate matter and dissolved gases. In the past salinity or chlorinity measurements were made chemically. These techniques have now been largely replaced by methods which involve the measurement of a physical parameter such as conductivity or refractive index. The relationships between these measurements and salinity or chlorinity as determined by the older methods has been precisely established, so that both old and new data are comparable. The new techniques are quicker, more precise and often automatic.

The salinity of surface water in the world's seas and oceans varies from the 5 $^o/_{oo}$ (five parts per thousand) found in the Gulf of Bothnia to more than 40 $^o/_{oo}$ in the Red Sea. In the open ocean, however, the variation usually falls between 33 and 37 $^o/_{oo}$, with an average of 35 $^o/_{oo}$. The average is somewhat higher in the northern hemisphere than in the southern, and markedly higher in the Atlantic than in the Pacific; this is due to many different causes, such as the flow of very salty water from the Mediterranean into the Atlantic and the transfer of water through evaporation into the Pacific from the Atlantic. Further, salinity reaches a maximum at subtropical latitudes, at 30° N. in the northern hemisphere and between 20 and 30° S. in the southern hemisphere. Salinity reaches its lowest values in the polar regions, where it also undergoes large seasonal variations.

The variation with depth is no less complex. First, there is a maximum of salinity in all three world oceans at about 100 m. Second, there is an 'intermediate depth salinity minimum' between 500 and 1,000 m down, attributable mainly to the spreading of 'Antarctic Intermediate' water. This spreads to different latitudes in the three oceans—about 30°–40° N. in the Atlantic, no further north than 150° N. in the Pacific and not north of 10° S. in the Indian Ocean. Below the 1,000 m line, salinity increases all the way to the bottom in the Pacific proper (north of 55° S.). In the Atlantic another salinity maximum is formed between 1,000 and 2,000 m down, and a similar maximum is formed in the Indian Ocean up to about 15° S. The most important cause of these maxima is thought to be the sinking of saline water in high latitudes during the winter. In the Atlantic there is then a further decrease in salinity toward the bottom, chiefly because of the northward spreading of the 'Antarctic Bottom' Water. Overall, the way salinity changes with time depends largely on the evaporation-precipitation relationship, variations in the pattern of ocean currents and, in the polar areas, the melting of ice in the summer and the freezing of water in the winter.

Attention is also still focused on the minor constituents of sea water, a field which has been aided by the development of autoanalysers capable of making continuous measurements of the concentration of one or several of these constituents. Strontium, for example, has assumed greater importance in the past decade because of its use as a nuclear pollution index. Research on oxygen and carbon dioxide, important in biological studies, continues and new interest has been shown in some of the largely unknown trace elements.

Ocean structure and circulation

Observations of the physical, chemical and biological characteristics of the world's oceans have led to the conclusion that an ocean is not a homogeneous body of water but consists of a number of more or less distinct water masses, each originating from a specific area. More observations are required, and attention is now being diverted to both vertical and horizontal mixing at the boundaries of water masses with different physical and chemical properties.

Ocean currents have been studied for a long time. They have assumed, however, greater practical importance recently, especially after the dramatic wreckage of large oil tankers, as there is now a clear need to forecast the distribution of sea pollution caused by the traffic of ever larger ships crossing the ocean at faster and faster speeds and carrying greater quantities of oil. Several experiments with tracer materials have been carried out to learn more about how a floating body moves and is influenced by ocean currents and wind. The results of these and many other experiments show that meteorological phenomena have a considerable effect on the ocean's surface movements and that their pattern is much more complex than that revealed by average figures. Consequently, daily observations and forecasts of the actual patterns of ocean currents, such as those available for weather forecasting, seem highly desirable.

During the 1960s much was learned about deep-water movements which, together with superficial currents, make up the ocean's general circulation, but the real speeds and rates of transport in deep water are still poorly quantified. At the beginning of the 1960s a series of observations showed that deep-water movements were significantly stronger and more variable than expected. Oceanographic buoys fixed to the ocean floor and supporting (at various depths) a series of current meters showed, together with observations of the movements of objects floating at predetermined depths, the presence of medium scale eddies. This was an expected effect of the rotating Earth on a large water mass. Several large-scale studies of the dynamics of such features, which seem to be key structures in the general ocean circulation, are planned for the future.

The study of individual currents, such as the Gulf Stream, the Somali current and the equatorial undercurrent, has been very useful in understanding more about the fluid dynamics in the ocean. Fruitful studies on the vertical mixing of surface waters cooled in winter have also been carried out in the Gulf of Lions (Bay of Biscay).

Laboratory experiments and computer analysis of the observational data coming in at increasingly rapid rate are expected soon to explain a number of characteristic phenomena, such as internal waves, and to provide a global picture of the ocean's circulation. In this task, however, more observational data are needed.

Atmosphere/ocean relationships

The action of the wind causes waves and the wind-driven oceanic circulation. The transfer of latent heat of evaporation provides energy for weather systems and for global wind circulation. The water evaporated from the ocean and precipitated on land is the main source of the world's fresh water. And salt particles injected into the atmosphere from sea water act as condensation nuclei and seem to be quantitatively sufficient to account for all the sodium chloride found in river water.

These large-scale phenomena are ultimately due to microprocesses such as the transfer of heat and momentum from ocean to air and vice versa, the evaporation of water from the sea's surface, the way droplets of sea water are whipped up by the wind into the atmosphere, and the mixing of atmospheric gases (such as oxygen and carbon dioxide) in sea water. In the past decade, understanding all the phenomena that act at the interface between the ocean and the atmosphere has become one of the most important goals of both oceanogra-

phy and meteorology. Ultimately, a complete understanding of either will lean heavily on the ocean/atmosphere interaction.

This interaction can be studied from a climatological point of view on the basis of average data. Oceanographic expeditions have provided much data in the last fifty years about the average behaviour of the ocean and the atmosphere and their relationships. The more observations that are made, the better the description of phenomena such as the wind-driven oceanic circulation or the mean oceanic heat balance becomes, and variations in time or space can be more accurately plotted. A number of national and international efforts are now being made to obtain more data, mainly in the lesser known regions such as the oceans of the southern hemisphere.

The study of momentum transfer is very important, mainly from an oceanographic point of view, because it explains both wave motion and wind-driven circulation. During the past decade classical methods of the study of momentum transfer, based on indirect measurements, were critically reviewed. It has been shown that direct measurements are needed; new equipment for this purpose is now becoming available.

Heat transfer and the related water vapour flux from sea to air are very important for both meteorology and oceanography. The latent heat of water vapour evaporated from the sea surface is the main source of energy of motion in the lower levels of the atmosphere. Conversely, as evaporation takes place, sea water cools and its salinity is increased; the increase in surface density which results provides the driving mechanism for the deep ocean circulation. Here again classical measurements no longer seem to be very helpful. New techniques are now available which provide a direct estimate of the flux of water vapour and of the actual heat transfer. They have been tested on land and are now being applied at sea.

As far as the transfer of gases and salt nuclei is concerned, intensive studies are being developed. The exchange of gases across the sea-air boundary and their diffusion in the upper layers of the ocean is inadequately known. The exact mechanism driving salt nuclei, which later plays an important role in water vapour condensation over the sea, from the ocean to the atmosphere, is not well understood although its importance is obviously very great.

Thus in order to understand the exact mechanisms of sea-air interaction, direct, synoptic and continuous measurements must be made. This kind of research needs either a sustained effort by one or two countries. Examples are the joint United States, Canada and Barbados project BOMEX (Barbados Oceanographic and Meteorological Experiment, see page 104); an international effort such as the 1968–69 Joint North Sea Wave Project (JONSWAP, scheduled again for this decade) in which a great number of wave spectra were obtained to improve classical wave theory; or the Cooperative Investigation of the Caribbean and Adjacent Regions (CICAR) in which, under the co-ordination of the Intergovernmental Oceanographic Commission, fifteen countries are involved.

Continental water

Inland seas, salt marshes, rivers and lakes —the continental waters—comprise about 2.8 per cent of all the world's water. Most of this (about 2.2 per cent) is frozen in glaciers.

Thus not more than 0.6 per cent of the world's water is available as continental fresh water. The chemical and physical properties of this water, its occurrence and its circulation are the domain of hydrology, a science which embraces all aspects of water from the moment of precipitation from clouds over land to its flow into the sea.

Continental water, vitally important for domestic, agricultural and industrial uses, also occasionally causes immense physical destruction. Its presence is of crucial concern to any human settlement and its study, though mainly restricted to hydrology, also includes aspects of meteorology, climatology, hydrogeology, applied geology, soil conservation and regional planning.

Physiologically man needs a minimum of about one litre of fresh water per day, but the average consumption is some hundreds of litres per day. Much of the fresh water used by man is used for irrigation, and the demand for this is now increasing fast to keep pace with the exponential growth rate of population. Recently, this has provoked renewed interest in the problems of fresh-water management, conservation and production. For this reason, the major problems faced by hydrology in recent years have been the survey of the real amount of fresh water available, research on new sources of fresh water, and the management of the quantity available particularly in arid lands. The conservation of water quality must now be added as a fundamental problem, mainly in the developed countries.

The problem of continental water is therefore a typically interdisciplinary one. Furthermore, because of its size and the political implications connected with the multinational distribution of the global water supply, the problem requires extensive international cooperation. The major effort in this direction has been the International Hydrological Decade (IHD), an international programme organized by Unesco with the scientific advice of the ICSU Committee on Water Research (COWAR), launched in 1965. The programme ended in 1974.

The tasks of IHD are to plan and co-ordinate scientific research on the various aspects of the fresh-water problem, to establish the global water balance, to prepare various kinds of hydrological maps for many countries; to study the behaviour of groundwater, the influence of man on the hydrological cycle and the forecasting and prevention of floods.

The basic objects of research in this field are to (a) measure global and local amounts of precipitation, evaporation, evapotranspiration and infiltration in soil; (b) measure the volume of fresh water stored in glaciers and ice caps; (c) measure the volume of water discharged by rivers into the sea, and their solid load; (d) evaluate the volume of water stored in underground reservoirs and their variations in connexion with regional needs; (e) study movements of water in underground reservoirs; (f) evaluate the superficial consequences in terms of subsidence of overexploitation of an underground reservoir; and (g) develop engineering hydrology to improve the use of water in irrigation and in flood control.

Marine and continental ice

More than 2.1 per cent of the world's water is frozen in the form of glaciers and the total volume of the world's ice is estimated to be at least 24 million km³. Many glaciers are present on high mountains at all latitudes but 96 per cent of the world's ice occurs in two great ice caps, the ice sheets of Antarctica and Greenland; together they have an area of nearly 10 per cent of the world's dry surface area. The antarctic ice sheet extends to 12 million km² in area and in some places is more than 4 km thick. The Greenland ice sheet has an area of 1.8 million km² and a thickness of more than 3 km. This means that if the characteristics of the greatest glaciated area are to be accurately known, a continent larger than Europe or Australia must be explored under very arduous conditions. Such studies involve many disciplines: glaciology, climatology, meteorology, the physics of ice, thermodynamics, geophysics, geomorphology and geology.

The greatest boost to the study of glaciers came in the 1960s from the study of the largest ice mass in the world, the antarctic ice sheet, and from the study of the polar areas in general. Antarctic research is a typically international field; international co-operation, stimulated by the 1959 Antarctic Treaty, is now organized by the Scientific Committee on Antarctic Research (SCAR) of the International Council of Scientific Unions (ICSU).

The inland region of Antarctica was virtually *terra incognita* until the 1950s. Research work began early in the 1950s with the Norwegian-British-Swedish expedition in Queen Maud Land and the French expedition in Adélie Land. As in other fields of the earth sciences, much research was carried out during the International Geophysical Year (1957–58). Only in the past decade, however, has the scientific effort assumed the proportions needed to explore a continent. Together with the build-up of permanent settlements of small communities of scientists, new research tools capable of measuring the nature, behav-

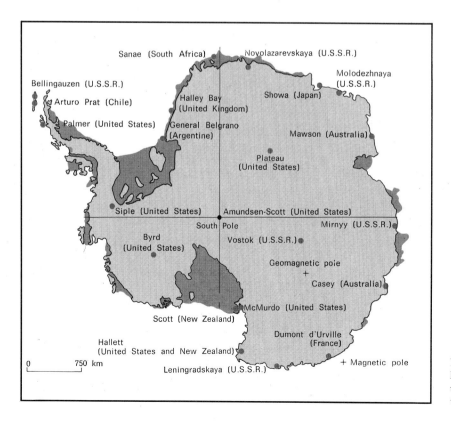

FIG. 18. Map of the Antarctic, with location of the main scientific bases set up by the countries which are parties to the Antarctic Treaty.

iour and age of the ice sheet are now at work in Antarctica.

The first task was to measure the thickness of the ice cover, its variations and consequently the morphology of its interface with the land surface. By the early 1970s, a profile nearly 50,000 km long had been taken using seismic and gravimetric methods. Gravimetric prospecting, begun late in the 1960s, proved to be a great improvement over previous, seismic profiling techniques because of its greater speed. A new development in this field is airborne radar profiling. The radar operates on a frequency band which can penetrate the ice cover and is reflected not only by the rocky bottom but also by some of the reflecting horizons in the ice mass itself. The greatest advantage of the new method is its speed. This makes it possible, for example, to repeat measurements at the same place at various times to study the movements of an ice mass. Improvements in the method should make it possible soon to define topographic details of the bottom of the ice sheet, its thickness and internal structure.

The second task of antarctic research has been to investigate the internal structure and movements, over time, of the ice mass. Here ice coring is very helpful. Several holes have been drilled completely through the ice, not only in Antarctica but also in Greenland. The cores obtained allowed temperature, density and other physical properties to be measured, the structure of the ice at various depths (and therefore at various pressures) to be determined, and the age of the ice to be estimated. Particularly interesting, from both a chronological and palaeoclimatological point of view, are measurements of the $^{18}O/^{16}O$ ratio in those ice samples. The isotopic composition of precipitation depends, in fact, on air temperature at condensation time. Summer precipitations are richer in ^{16}O than winter precipitations, and so are precipitations during warmer years. Thus the $^{18}O/^{16}O$ ratio can be used to pin-point the relative times of seasonal variations and to identify a particularly cold or warm period. Since the ice cores reach down to ice at least 100,000 years old, the oxygen isotope ratio can be used as a palaeoclimatological indicator and thus contributes to our knowledge of how the Earth's climate evolved in the recent past.

As far as polar studies are concerned, two methods of observation other than inland settlements have been established: satellites and ice islands. Meteorological satellites can give a synoptic view of the meteorological situation and its evolution. At the same time, photography by satellite provides a continuous record of the ice cover and of the position of icebergs. Ice islands are thick bodies of floating ice, main source of which seems to be the Ward Hunt Ice Shelf near Ellesmere Island in the Arctic Ocean, which drifts westward around the north polar area. Although known for a long time, the first icefloe was not occupied, and then only temporarily, until the 1950s; this was the Fletcher Ice Island, or T–3. Later, several other ice islands were occupied by both United States and Soviet scientists who used them as meteorological and oceanographical research stations.

Of the ice floating in the sea, at least 99 per cent is sea ice or ice formed by the freezing of sea water. Its salinity depends mainly on the rate of freezing of sea water, and therefore on the temperature. During the 1960s much research was carried out on the physical properties of sea ice, and the mechanisms of its formation.

Further reading

AMERICAN GEOPHYSICAL UNION. *Antarctic snow and ice studies*. Washington, 1972.

An oceanic quest: the international decade of ocean exploration. National Academy of Sciences, 1969 (Pub. 1709).

AYALA-CASTANARES, A. Lagunas costeras: un symposio. Mexico, Libreria Universitaria, 1970.

BRANCAZIO, P. I.; CAMERON, A. G. W. (eds.). *The origin and evolution of atmosphere and oceans*. New York, Wiley, 1964.

DAVIS, R. A. *Principles of oceanography*. Reading, Mass., Addison-Wesley, 1972.

GROSS, M. G. *Oceanography: a view of the Earth*. Englewood Cliffs, Prentice-Hall, 1972.

KAMENKOVIČ, V. M. *Osnovy dinamiki okeana*. Leningrad, Gidrometeoizdat, 1972.

KUSTO, Z. M. Proščanie s poverchnost' ju. *Nauka i Čelovečestvo*. Moskva, Znanie, 1967.

KVJATKOVSKIJ, I. A. *Okean i korabl'*. Leningrad, Gidrometeoizdat, 1972.

LEGRAND, L.; POIRIER, G. *Théorie des eaux naturelles*. Paris, Eyrolles, 1971.

MERO, J. L. *The mineral resources of the sea*. Elsevier, 1965.

MUGA, B. J.; WILSON, J. F. *Dynamic analysis of ocean structures*. New York, 1971.

QUAM, L. O. (ed.). *Research in the Antarctic*. American Association for Advancement of Science, 1971.

REID, J. L. (ed.). *Antarctic oceanology*. Washington, American Geophysical Union, 1971.

REMSON, I.; HORNBERGER, G. M.; MOLZ, F. J. *Numerical methods in surface hydrology*. New York, Wiley-Interscience, 1971.

RILEY, J. P.; SKIRROW, G. *Chemical oceanography*. New York, Academic Press, 1965.

SHEN, H. W. (ed.). *River mechanics*. Fort Collins, Colo., Water Resources Pub., 1971.

STASKEVIČ, A. P.; TJURIN, A. M.; TARANOV, E. S. *Gidroakusticeskie izemerenija v okeanologii*. Leningrad, Gidrometeoizdat, 1972.

STRAHLER, A. N. *Introduction to physical geography*. Chichester, Wiley, 1970.

STRINGER, E. T. *Foundations of climatology; an introduction to physical dynamic, synoptic and geographical climatology*. San Francisco, Freeman, 1972.

——. *Techniques of climatology*. San Francisco, Freeman, 1972.

TREŠNIKOV, A. F.; BARANOV, G. I. *Struktura cirkuljacii vod Arkticeskogo bassejna*. Gidrometeoizdat, 1972.

VARIOUS AUTHORS. *The ocean*. San Francisco, Freeman & Co., 1969.

VILELA, C. R. *Hidrogeologia*. Tucuman, Universidad National de Tucuman, 1970.

WEYL, K. *Oceanography, an introduction to marine environment*. New York, Wiley, 1970.

YOSHIDA, K. (ed.). *Studies on oceanography*. University of Washington Press, 1965.

8

The Earth's atmosphere and beyond

Meteorology and the lower atmosphere

The Earth's climate depends ultimately on our planet's relationship with the Sun but it varies over the Earth's surface in a way which depends both on a region's latitude and on the presence of large topographical features such as oceans and mountain ranges. The study of the Earth's weather concerns the short-term, local variations that are introduced into this picture by, for example, the presence of air masses of different origin, cloud cover, and smaller, local topographical features. But although weather is a more local phenomenon, its genesis may occur over several weeks or months, and over distances of several thousand kilometres: variations in the weather may be introduced over a whole continent by developments which occur over an adjacent ocean, a neighbouring continent or even the polar regions in each hemisphere. For this reason weather forecasting demands a detailed knowledge of both local conditions and of a general weather picture covering a complete continent or even the whole hemisphere.

Research in this field has been marked by two significant events over the past decade: the development of new observing systems, particularly weather satellites which can now be equipped with a whole range of observing equipment which existed only in theory ten years ago; and the use of the computer, both for analysing meteorological data, and for preparing and testing models of the Earth's atmospheric circulation which until recently could be tested only by crude physical methods such as the study of circulation patterns in slowly rotating liquids. Together, these advances have led to a much better understanding of the weather and promise an improved ability to forecast weather conditions, particularly over medium- and long-term periods.

After the 1958 *Vanguard 2* and the 1959 *Explorer 7*, the first successful operational meteorological satellite was *TIROS 1*, launched in 1960; it carried into space television cameras to photograph the cloud cover of the Earth; the following eight *TIROS*, in addition to television cameras, carried infra-red sensors with the aim of measuring the temperatures radiated by the Earth's surface. *TIROS*' view of the Earth, however, was confined to the sunlit portion of the planet—it was not able to photograph the night-time cloud cover. The first satellite able to do this was *Nimbus* launched in 1964. *Nimbus* carried a television camera and an infra-red scanning radiometer to provide day and night coverage. The success of *Nimbus* was a big advance in meteorological satellites and in weather-observing systems. Next came the

101

TOS programme in which two satellites were simultaneously used to provide a global daily coverage from a high altitude; after their launch in 1966, the TOS satellites were renamed ESSA and nine further satellites were launched. Finally, ·early in the 1970s the United States ITOS satellites and the U.S.S.R. METEOR satellites were launched; ITOS means Improved Tiros Operational Satellite

and is an enlargement of the TOS programme. The first such satellite was launched in 1970 and carried television cameras and infra-red radiometers for day and night coverage. Television and infra-red data are directly transmitted to users all around the world. At the same time the *Nimbus* programme is continuing: *Nimbus 4*, launched in 1970, is able to provide a vertical profile of the atmosphere between the satellite and the Earth's surface; it also gives a quantitative distribution of ozone and water vapour; because of its polar orbit *Nimbus 4* scans the whole Earth twice a day.

Weather satellites have given meteorologists a more synoptic view of weather phenomena and their evolution; they have also provided an enormous mass of data in those regions of the Earth in which it would be very difficult

FIG. 19. By juxtaposing photos taken by the satellite *ESSA 9*, these pictures of the Northern and Southern Hemispheres were obtained, giving us an overall view of the way the earth's atmospheric systems are distributed. (*Source*: United States Department of Commerce.)

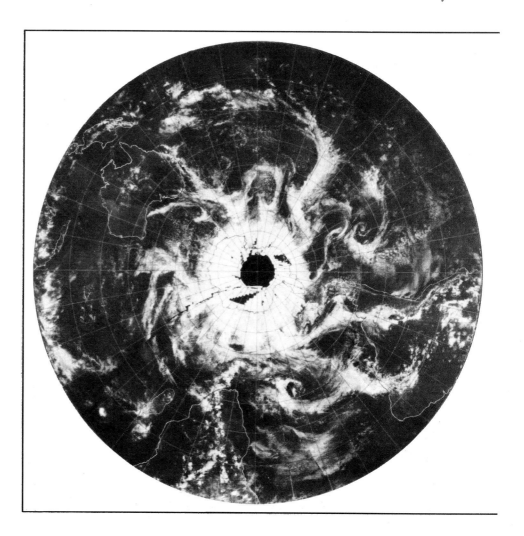

to build ground stations (very high mountains, deserts, oceans and so on) but which are often critical for weather behaviour.

Ground stations, however, are no less important than the satellites. The satellite network that will be operational in the 1970s is only one part of the necessary global weather observing system, although it is of fundamental importance.

A global weather observing and forecasting system is now under development—the World Weather Watch (WWW). Sponsored by the World Meteorological Organization (WMO), WWW began in 1968 and faced serious organizational and technical problems: how to provide a sufficient number of surface and upper air stations; how to assemble the observations fast enough to allow the elaboration of data;

and how to process such an enormous amount of data? To answer these three questions, the Global Observing System (GOS), the Global Telecommunication System (GTS) and the Global Data Processing System (GDPS) were planned. The GOS is largely based on existing surface and upper-air stations, but in order to satisfy the aims of the programme the existing network must be greatly improved to provide more upper-air observations; a major problem is that of obtaining sufficient information about weather behaviour on the oceans, particularly the southern ones for which a certain number of additional weather ships were planned. All the data have to be communicated, through regional and national meteorological centres, to the three world meteorological centres at Melbourne, Moscow and

Washington. The total traffic scheduled is enormous: 700,000 groups of coded data and 2,500 charts of processed data per day. World meteorological centres are designed to provide global analyses and forecasts and then to communicate them to regional and national meteorological centres.

Early in the 1970s WWW was not fully operational but it seemed likely to improve drastically our knowledge of the dynamics of the atmosphere and of the weather.

Parallel to WWW was another growing global programme, the Global Atmospheric Research Programme (GARP), which aims to improve knowledge of the general circulation of the atmosphere and the processes involved in it. GARP is sponsored by the International Council of Scientific Union (ICSU) and by WMO. Because many atmospheric phenomena originate from tropical conditions, which are not well understood, the first GARP experiment took place in the tropics; named GARP Atlantica Tropical Experiment (GATE), it began in 1974 and will benefit from the experience of the United States and Canadian experiment, begun in 1969, called the Barbados Oceanographic and Meteorologic Experiment (BOMEX) which was the major meteorological experiment of the 1960s: twelve oceanographic ships including the Floating Laboratory Instrument Platform (FLIP), an oceanographic buoy, twenty-eight aircraft, satellites and 1,500 people were involved. Many of the results of BOMEX are now under evaluation and it is quite clear that they will provide a basis for further studies of tropical atmospheric phenomena.

The ever-growing mass of observational data requires a parallel development of processing systems, particularly in the use of larger and faster computers. Numerical models of the atmosphere are the end-point of observational data: a mathematical model simulates the behaviour of the atmosphere while the observational data provide the input data and the output is a weather forecast. Numerical modelling of the atmosphere promises much but in the middle of the 1970s a wider knowledge of the behaviour of the atmosphere to improve the models, and still faster computers to process the data, were needed.

Climatology

A climate is a statistical average of the weather in a certain area over a given period of time. The area observed may be all the Earth or its largest geographical units (macroclimate) or smaller units such as a lake and the land around it or a town and its neighbourhood (mesoclimate); when the area is very small, such as the upper boundary of trees in a forest, or the first three metres above the ground of an airport, the microclimate is concerned.

The macroclimate is largely conditioned by the radiation balance of the planet and its atmosphere: how much solar energy is incoming, how much is outgoing, how the energy is transformed, and how it is distributed. Here the position of the Earth in its orbit around the Sun, the orientation of its axis of rotation with respect to the Sun and the distribution of land, water and of reflective surface (such as snow) are very important. All these determine point to point differences in the atmosphere's temperature and consequently its pressure; such a situation gives rise to the general circulation of the atmosphere, which is then modified not only by the rotation of the Earth but also by the irregular distribution of land, water and features such as mountains.

An understanding of all the factors which affect the climate may lead to improved knowledge of climates in the past. Further, several independent lines of research are now producing new data on past climates—the isotopic analysis of deep cores of arctic and antarctic ice (see page 98) and of the composition of fossil shells, the micropalaeontological and sedimentological analysis of marine sediments, and the mathematical analysis of astronomical cycles, are all interesting research areas which may produce a clearer picture of the evolution of past climates on the Earth and at the same time may help us to understand today's climate better.

Microclimatology is also important from a practical point of view: a knowledge of the microclimate (which may differ greatly from the meso and macroclimate in the same area) is obviously very relevant to agriculture and urban or regional planning. Microclimatologists are now developing new instrumentation to measure different properties of the atmosphere at different levels and models which make it possible to predict what effect new structures such as buildings are likely to have on the microclimate.

Aeronomy and the upper atmosphere

The object of aeronomy is to study the upper atmosphere, a region which extends from nearly 30 km above the surface of the Earth to the outer limit of the atmosphere. The weather observed on the surface of the Earth develops well below this limit; the top of cumulonimbus, the vertical storm clouds, very rarely exceeds 12 km in height. The upper atmosphere is in some ways comparable to the interior of the Earth; both are very near to the environment in which man lives but have very different properties and are too far away to be reached directly. However, in the past ten years the upper atmosphere has been reached by rockets and satellites and much more data is now available.

A relatively complex temperature pattern divides the atmosphere into several main regions in which temperature either decreases or increases with height: above the troposphere and the stratosphere, which compose the lower atmosphere and in which temperature respectively decreases and remains nearly constant, there are the mesosphere (in which temperature decreases) and the thermosphere in which temperature increases to a maximum value at 300 km. Chemically, the upper atmosphere is divided into the homosphere (up to 85 km), in which the mean composition is nearly independent of altitude, and the heterosphere, in which the composition changes with height.

The properties of the homosphere have been measured many times with rockets and artificial satellites and latitudinal and seasonal changes are well established. The variations in density and temperature in the region between 30 and 90 km high are respectively about 20 per cent and 10 per cent of the mean value at that altitude; the seasonal change consists of an increase of temperature in winter. The region above 90 km but below 200 km is not so well known as the homosphere but

more data are available on the region above 200 km; they have been provided by the study of atmospheric drag on artificial satellites: in fact the period of revolution of a satellite decreases proportionally to the drag force exerted by the atmosphere which is, in turn, proportional to the air density. In the past fifteen years measurements of atmospheric drag have provided a powerful method for studying the upper atmosphere. For example, it has been shown that its properties are highly dependent on solar activity; when the latter is intense the density deviates by a factor of 100 from the mean value and the temperature changes by hundreds of degrees. These changes result from changes in the absorption of ultra-violet radiation and by the energy that the solar wind, a stream of particles coming from the Sun, carries to the Earth.

An international study of the properties of the upper atmosphere was included in the International Year of the Quiet Sun (IQSY) and a working group of the Committee for Space Research (COSPAR) co-ordinates research in this area.

The magnetosphere

The study of the Earth's magnetic field and its interactions with the solar wind (the stream of particles flowing from the Sun, see below) was greatly stimulated by developments in space in the 1950s and 1960s. Both artificial satellites and spacecraft bound for other celestial bodies have helped to map the Earth's field, which has turned out to be very different in shape from the simple dipolar field (similar to that of a straight bar magnetic) which had been suggested on the basis of terrestrial measurements and studies. In fact, the lines of force of the Earth's field are bent round it by the impact of the solar wind, creating a cavity (called the magnetosphere). In this respect the Earth behaves very much like a bullet travelling through the air, with a shock front ahead of it and a long and turbulent tail behind it. The magnetosphere extends to about 10 Earth radii on the day side and to about 30 radii on the night side.

One of its main features, the existence of 'belts' of trapped particles, mainly electrons,

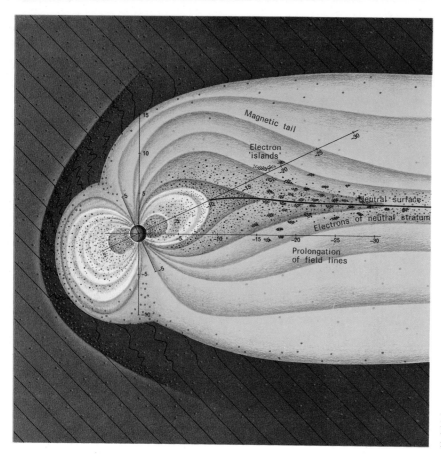

FIG. 20. Diagram of interaction between the solar wind and the Earth's magnetic field.

was discovered by some of the earliest artificial satellites (*Explorer 1* and *Explorer 2*, *Sputnik 3*). They are sometimes called the Van Allen belts, from the name of one of their discoverers, and until recently they were thought to be two, one inside the other, but further measurements have demonstrated that the electrons form a single zone. The density of the trapped particles is not constant but depends, among other things, on the local time on the Earth below; there is also a sharp fall in the concentration of electrons in equatorial latitudes (the so-called 'knee effect').

Both the magnetosphere and the radiation belt are very strongly influenced by solar activity. Measurements made by artificial satellites during magnetic storms have shown that the beginning of a storm is accompanied by a sharp rise in the magnetic field and by a contraction of the magnetosphere, which comes closer to the Earth. After the storm's

main phase the magnetosphere expands back past its previous boundary, then oscillates back and forth. In the storm's main phase the magnetic field decreases; the reason for this is the penetration in the magnetosphere of high energy protons (about 100 keV) coming from the Sun.

The study of this interaction between the solar wind and the magnetosphere has important implications for life on the Earth: the magnetosphere normally acts as a shield against charged radiation coming from space. To further research on the magnetosphere satellites especially designed for its study are being actively developed. Among them are the so-called highly eccentric orbit satellites, whose elongated elliptical orbits are able to cross the magnetosphere's boundaries again and again, during different phases of the Sun's activity. Once more, as during the International Geophysical Year of 1957–58, which saw the launching of the first artificial satellites, a large international effort devoted to the study of the magnetosphere, the International Magnetospheric Study scheduled for 1976–77, will see the launching of more sophisticated space vehicles, among them a pair of satellites, the so-called 'mother-daughter' pair, specially designed for the mapping of the spatio-temporal structure of the magnetosphere. In the same year other satellites will be placed in solar orbit to study the solar wind and its connexions with the Sun's surface activities. Correlations between the measurements of the solar probes and the magnetospheric probes will be of the greatest scientific value.

The Earth-Sun relationship

All the energy available on the Earth comes from the Sun (except for nuclear energy and that of the tides). Man has always been conscious of the vast amount of radiation coming from the Sun as light: it was not until recently, however, that it was realized that the Sun also emits other forms of radiation, both electromagnetic and corpuscular, which are potentially dangerous to all forms of life. The corpuscolar radiations are, at least partially, shielded by the magnetosphere; where, in the polar regions, the Earth's magnetic field lines reach down to the Earth, the flux of high energy electrons causes the striking phenomenon of aurorae. The electromagnetic radiations are partly shielded by the atmosphere, except for those with a narrow range of wavelength which are able to pass through the 'windows' in the atmosphere. Since the solar radiations reach the Earth as a whole and influence a wide range of phenomena spanning the whole globe (for instance, radio and television broadcasts), it was natural that the study of the Earth–Sun relationship became an important part of man's first great international scientific programme, the International Geophysical Year.

In 1964–65, however, another programme was launched that was devoted solely to the study of solar-terrestrial interactions—the International Year of the Quiet Sun (IQSY). It was timed to include the solar activity minimum in the Sun's eleven-year cycle. Scientists from seventy countries co-operated in this research effort to understand the effects of the Sun's activities on terrestrial phenomena.

This second programme made full use of the available space technologies, including solar orbiting satellites. Among the programmes deriving from IQSY is the Proton Flare Project, which keeps a constant watch on the Sun to signal, and even predict, events which could endanger astronauts travelling through space, unshielded from radiation.

The study of the Sun has taken full advantage of orbiting space platforms such as *Skylab*, launched in 1973, in which three consecutive crews of scientists-astronauts worked in co-operation with astronomers on the Earth to keep watch on the Sun's activities. The next minimum point in the Sun's cycle, in 1976–77, will see another international effort devoted to the study of the Earth–Sun relationships, namely the International Magnetospheric Study (see above). Meanwhile studies are being carried on all over the globe, and especially in Antarctica which, being close to the south pole, provides a valuable scientific vantage point. A whole chain of scientific stations exists there, manned by scientists from many countries, which work in co-operation with scientific stations elsewhere. A concept which has become very important is that of studying phenomena simultaneously at two different locations on the Earth which are linked by a line of force of the Earth's magnetic field (conjugate points). Thus measurements are made and data collected in a network which spans the whole Earth and knows no borders.

Further reading

BUGAEV, V. A. Tri sistemy meteorologiceskich sputnikev. *Budušee Nauki*. Moskva, Znanie, 1971.
——. Sputniki i Sluzba pogody. *Nauka i Čelovečestve*. Moskva, Znanie, 1972.
CURTILLOT, V.; LE MOUËL, J.-L. Le champ magnétique de la terre. *La recherche*, vol. VI, no. 59, 1975, p. 720.

DZERDZEEVSKII, B. L.; POGOSYAN, K. P. *General circulation of the atmosphere*. London, Keter, 1971.

HAMMOND, A. L. Global meteorology (I): experiments in the tropics. *Science*, vol. 174, no. 4006, 1971, p. 278.

——. Global meteorology (II): numerical models of the atmosphere. *Science*, vol. 174, no. 4007, 1971, p. 393.

KOUTCHMY, S. Le vent solaire. *Atomes*, vol. 24, no. 270, 1969, p. 678.

MUSTEL', E. R. Solnecnaja aktivnost'i niznie sloi zemnoj atmosfery. In *Nauka i Čelovečestve*, Moskva, Znanie, 1968.

RATCLIFFE, J. A. *Introduction to the ionosphere and magnetosphere*. Cambridge, Cambridge University Press, 1972.

ROTHMULLER, I. J.; BEER, T. Conjugate point phenomena. *Science Progress*, vol. 60, no. 238, 1972, p. 205.

SEVERNYJ, A. B. Vspyski na solnece i magnitnye polia. *Nauka i Čelevečestve*. Moskva, Znanie, 1970.

SKURIDIN, G. Space physics: the new outlook. *Soviet science review*, vol. 3 III, no. 2, 1972, p. 107.

——. *Space physics and space apparatus*. Moscow, Znanie, 1970.

9

The planetary
system

The Sun

The Sun, a typical star of average mass and luminosity, is the source of much information on the properties of stars in general. But, in one respect, it is unique: it is the star around which revolves a very complex planetary system of nine planets and thirty-two satellites, plus an almost countless number of smaller bodies, the asteroids, between the orbits of Mars and Jupiter. Is a planetary system a common feature of stars, or is it exceptional? This is a very important question since it is only on planets of a given kind that life can flourish in the universe. But how did the solar system form, and under what conditions? Once this question is answered, it may be possible to estimate the probabilities that other stars, too, have solar systems of their own.

The source of all the energy falling on the Earth, the Sun, is powered by thermonuclear fusion reactions which until recently were thought to be completely understood. But scientists who since 1968 have been trying to detect the flux of neutrinos escaping from the Sun's core have now uncovered some surprising information: the flux is lower, by a factor of ten, than that expected on the basis of the thermonuclear reactions thought to occur in the Sun. Either the properties of the neutrino are not completely understood, and it decays

in the eight minutes it takes to make the journey from the Sun to the Earth and thus rarely reaches the detectors set up in mines and caves to record it without interference from cosmic radiation, or the Sun's internal temperature is much higher than we thought. As in many fields, this problem illustrates the very important relationship which exists between elementary particle physics and astrophysics. The problem will require more research in both fields before a solution is found.

Besides neutrinos, the Sun emits a vast range of radiation, from radio waves to X-rays, which can be detected with the appropriate instruments, as well as charged particles which form the cosmic rays of solar origin. All these radiations reach the Earth in a very short time, and play an important role in the high atmosphere, influencing all kinds of phenomena, from aurora to magnetic storms (see above, Chapter 8). For these and other reasons, such as the danger of high-energy charged particles to space travellers and possibly passengers on high-flying aircraft, a close watch is maintained on the Sun by Earth-based observatories, X-ray telescopes mounted on satellites, and orbiting solar observatories which are able to monitor continuously the visible activities on the surface of the Sun (the photosphere) and sound the alarm should a high intensity solar flare develop. As is well known, the Sun goes through a cycle of activity which lasts roughly eleven years from maximum to maximum; as already mentioned, during the last period of minimum activity a large international research effort was launched, the International Year of the Quiet Sun (IQSY), in 1964. The next large international effort will be the International Magnetospheric Study, scheduled for the year 1976–78 (Chapter 8). Very important solar researches were performed by the three *Skylab* missions (1973–74), during which solar phenomena were monitored closely from space in an extended wavelengths range. Tens of thousands of solar pictures were brought to Earth and their study will keep solar scientists busy for years.

Current objectives of solar research include the study of magnetic phenomena on the Sun's surface, and their relationships with other phenomena such as flares, protuberances and sunspots. How the lines of force of the Sun's magnetic field extend into space and shape the form of the corona, a region of ionized atoms (plasma) which is visible only during solar eclipses, has become another important research area. The ionization of the atoms in the corona is so high, as proved by the wavelength of the light that it emits, that its temperature must be one to several million degrees. This poses a very interesting problem, since the temperature of the photosphere is 'only' 4,000–6,000 degrees: how can a body at a lower temperature heat up another to a higher temperature? An answer is sought in terms of acoustic waves originating in the convective region of the Sun, below the photosphere; as the density of the corona becomes lower and lower, the outflow of acoustic energy agitates the atoms in the plasma more and more rapidly since they collide less often, thus giving birth to the higher temperatures needed to dissipate energy at the same rate. The corona has no perceptible limit, and the plasma moves outwards from the Sun at great speed; it reaches the Earth with a speed of 300 km/s and flows past its magnetic field, producing a shock wave which closely resembles that of a supersonic flow around an object in a high velocity medium. The Earth is luckily shielded from this by its magnetic field, but planets without a magnetic field are not. The solar wind has been measured by space probes in interplanetary space and by astronauts on the Moon. Thus the Sun extends its influence well beyond its visible limits, and is responsible for many complex phenomena in the space between the planets and on the planets themselves: a typical example are the radiation belts (or Van Allen belts) of trapped particles, discovered by one of the very first artificial satellites (*Explorer I*), which are known to exist so far only around the Earth and Jupiter: these latter have been explored by the two *Pioneer* spacecraft (*Pioneer 10* and *Pioneer 11*) during their fly-bys of Jupiter, respectively in December 1973 and December 1974. They resulted to be at least 10,000 times more intense than the Earth's, thus representing a real danger not only to live astronauts, who would be certainly killed by the dose of radiation absorbed passing through them, but also to the delicate electronic circuits of the space probes. Those, however, survived and are now bound towards the outer planets, still transmitting data.

The outer planets

The outer planets are Jupiter, Saturn, Uranus, Neptune and Pluto. Of the latter, very little is known. The four planets Jupiter to Neptune account for about 99 per cent of the mass of the planetary system, and Jupiter in particular, the largest planet, carries most of the angular momentum of the planets, besides having a very rapid rotation about its own axis, producing a 10-hour 'day'. Its composition may be very close to that of the material from which the whole system was formed. Other interesting features of Jupiter are its strong radio emissions in the decimetric and decametric band (the latter being apparently modulated by its second satellite, Io), its Van Allen belts and its rapidly rotating surface, on which there apparently floats a permanent mark, the Great Red Spot. For these reasons Jupiter is a primary target for exploration among the outer planets. The first spacecraft ever aimed at Jupiter, *Pioneer 10*, left the Earth in March 1972, and flew past Jupiter on 3 December 1973 grazing it by a mere 136,000 kms, after having crossed unscathed the dangerous asteroid belt. It was followed by a sister spacecraft, *Pioneer 11*. They carry thirteen instruments, designed to gather information about Jupiter, its magnetic field, temperature, trapped radiation belts and atmosphere, and the nature of the interplanetary space they fly through. A number of colour pictures of Jupiter and its satellites have been taken and transmitted to Earth, a huge feat of electronic engineering applied to space astronomy. The second half of the 1970s, because of a favourable alignment of the outer planets which will not repeat itself for another 180 years, offers the chance of visiting two or more of the outer planets with a single spacecraft, taking advantage of the gravitational pull of a planet to accelerate the spacecraft on to the next one: this is the basic concept of the so-called 'Grand Tour' missions, which have however been cut back for financial reasons. A possible Grand Tour mission might have visited Jupiter, Saturn and Pluto in 1976–77; another one Jupiter, Uranus and Neptune in 1979. However, the only mission of this kind now planned is a fly-by of Saturn by *Pioneer 11* in 1979. The concept of gravitationally assisted navigation requires, among other things, as exact a knowledge as possible of the masses of the target planets. This is by no means an easy task, especially in the case of planets such as Pluto, whose mass is known only with great uncertainty through long celestial mechanics calculations based on the perturbations provoked in the orbits of nearby planets. The radar observation of the satellites of Jupiter and Saturn, employing more powerful radars than those in use today, might help in determining more accurately their masses; the best way to do this is, however, by careful monitoring of radio-tracking data from deep space probes such as *Pioneer 10* and *Pioneer 11*.

Of the other outer planets, Saturn probably has the same chemical composition as Jupiter, which may be somewhat similar to that of the Sun; Uranus and Neptune may be deficient both in hydrogen and helium. The hydrogen-helium ratio in Jupiter may be very important as a test of cosmological theories. Too little is known of Pluto to say anything about its composition.

An interesting fact: the *Pioneer 10* spacecraft, which is bound outside the solar system and towards the Taurus constellation (it should reach it in 8 million years), carries a plaque designed to give information about the Earth and its dwellers to any hypothetical living beings it should meet in its course.

The asteroids

The asteroid belt, between the orbits of Mars and Jupiter, which may include as many as 50,000 celestial bodies with diameters ranging from 1 to 800 km and countless others with diameters less than 1 km, has always intrigued astronomers because, according to Bode's law, a planet should be found in its place. Are the asteroids then the remnants of a planet which broke up for unknown reasons, or are they pieces of a planet in the course of formation from interstellar dust by the process of accretion? The problem is a very important one, since the asteroids, which divide the inner planets from the outer ones, could represent a stage in the formation of a planet from a disc of dust; this is the general mechanism by which all planets (and most of their satellites) are believed to have been formed. The disc is unstable and breaks up in a series of separate, smaller clouds, all rotating in the same plane around a central mass (the Sun).

For these reasons a mission to the asteroids has been proposed as a future space programme; because of their low mass, the landing of a space vehicle on an asteroid such as Ceres (the biggest asteroid) should be very easy, and the departure from it would not require the burning of much fuel. On a more practical plane, unmanned space vehicles which cross the asteroid belt, such as *Pioneer 10* and *Pioneer 11*, have registered very few impacts with microasteroids and have not even sighted a big one. They should not represent, therefore, a big danger for space travellers.

Not all asteroids, however, have circular or near circular orbits: some have eccentric orbits which periodically bring them nearer to the Earth. These orbits are studied both from photographs which go back to the last century (the photographic camera is the best way to discover asteroids, since they leave a track on the background of fixed stars) and by modern radar-tracking methods; one of the aims of these studies is to discover relativistic effects in the asteroids' orbits (precession of perihelion) as a check of general relativity against competing theories. Icarus and Eros were studied in this way during their recent close approaches to Earth.

Asteroids are believed to be the source of meteorites falling on the Earth. They thus provide scientists with the opportunity of studying extraterrestrial material and analysing it for chemical composition, abundance of elements, isotopic ratios and radio-activity induced by the cosmic ray flux during its trip through space. Meteorites can be considered as a very cheap form of space probe which provides much useful information. The search for them makes use of radar or photographic smthods, using automatic cameras at fixed destances to form a network of observing tiations. From the pictures taken by all the cameras which register the fireball of an incoming meteorite, its path of descent can be determined, and the meteorites can be searched for in a small area near its predicted point of impact. The Lost City meteorite of 1970 was discovered in this way by the Smithsonian Observatory Prairie Network a few days after its fall. It is very important to recover a meteorite as soon as possible after its fall in order to record the radio-activity from short-lived isotopes contained in it. Research on meteorites has also revealed fission tracks from nuclei of superheavy elements within them (see pages 135 ff.) and traces of organic material of extraterrestrial origin, mainly amino acids. These traces have recently been found in the Murchison, Murray and Orgueil meteorites and, together with the discovery of interstellar organic molecules, have led to many speculations on possible forms of biological evolution in space and in the primaeval atmospheres of the planets.

The inner planets

Dramatic advances were made, during the 1960s, in our knowledge of the inner planets thanks to the use of space probes, especially in the case of Mars and Venus, while Mercury was left largely alone: the first mission aimed at Mercury, a joint Venus-Mercury fly-by, left the Earth only in November 1973 and reached Mercury and its environment in March 1974, after having photographed Venus en route. Its solar orbit carries it near Mercury every six months, so a second fly-by occurred in September 1974 and a third one in March 1975. Mercury was shown to be a moonlike planet, very much cratered, but with features typical of its own, such as long 'scarps'.

An unmanned approach to a planet usually involves two main tasks: first a fly-by mission of reconnaissance, after which the space probe becomes an artificial planet of the Sun, then a deep entry mission into the atmosphere of the planet, with a possible landing package of instruments, or an orbiter mission with a possible landing package. Venus has been the target of several missions, fly-by (*Mariner 2, 5* and *10*), deep entry (*Venera 4, 5, 6, 7* and *8*) orbiter and landing (*Ven 9* and *10*); Mars has been the target of fly-by missions (*Mariner 4, 6* and *7*), orbiter missions (*Mariner 9*; *Mars 4* and *5*), orbiter and landing (*Mars 2, 3, 6* and *7*). The next step in the exploration of Mars will be the orbiter-landing mission *Viking 1* and *2*, with the landing section capable of analysing the Martian soil and transmitting back its findings, with perhaps an emphasis on the search for possible biological material. The requirements for a landing craft on the surface of Venus are much more severe, since it has to withstand a pressure of 90 atmospheres and a temperature of 475° C, according to the *Venera* missions' findings. The conditions on the surface of Mars are likely to be much less extreme and easier to withstand, at least as far as pressure and temperature are concerned, but the 1971 *Mariner* and *Mars* missions have encountered very strong winds, with speeds up to 270 km/h, which raise vast dust storms and may be a danger for the stability of landing craft especially if provided with large arrays of solar cells.

The study of a planet by space probes, often supported by optical and radar observations, has a twofold purpose: the study of its atmosphere and of its surface in the hope of understanding its evolution in the framework of the evolution of the planetary system. Atmos-

FIG. 21. The structure of the atmosphere of Venus, detected from the Earth by radar and studied by the instruments on board the various space probes which have approached the planet, presents many unresolved problems. It is thought, for instance, that the prevailing carbon dioxide has accumulated as a result of the 'hot house' effect; water may have been lost in the upper troposphere through reaction with minor constituents. The diagram shows atmospheric characteristics of the lower strata; variations in temperature of the exosphere, between minimum and maximum, are indicated.

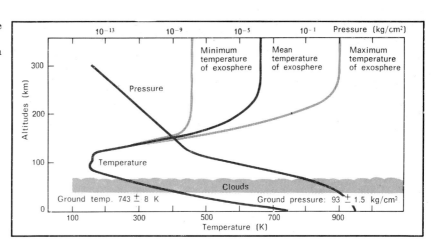

pheric studies are also very interesting because they are closely related to the study of the Earth's atmosphere and its motions, while surface studies of planets are similarly related to the geological evolution of the Earth.

In the case of Venus, which has an opaque and dense atmosphere composed mainly of CO_2, atmospheric studies have to rely on *in situ* measurements made by space probes, while studies of surface features once limited to low resolution surveys made by radar were initiated in October 1975 by *Venera 9* and *10*, which transmitted the first pictures of Venus' surface.

In the case of Mars, much can be learnt about its atmosphere by the study of clouds and their motions viewed through telescopes, such as those of the Planetary Patrol network especially dedicated to the study of planets, while its very thin and transparent atmosphere allows high resolution photographic mapping of the surface, which has produced hitherto unimagined detail, such as deep irregular valleys, big volcanic craters, great smooth plains and other surface features very different from both those of the Earth and the Moon. The polar caps of Mars, until recently considered to be frozen CO_2, are very likely to be of water ice or of both water ice and CO_2 (which is the main constituent of the Martian atmosphere); and the presence of water means that some form of life is possible. While the surface study of Mars is still in its infancy and may have to wait for samples of Martian soil brought back

either by manned missions or automatic space probes, space flight has not yet found any sign of a magnetic field. This means that Mars has no melted liquid outer core like the Earth's or Jupiter's to act as a dynamo and that therefore the solar wind (see above) flows almost unimpeded around Mars. Neither does Venus have a magnetic field, but it has a highly conducting ionosphere, which shields it: there is a bow shock at about 1.3 Venus radii. Venus may have a liquid core, but it rotates too slowly (in 245 days) to create an appreciable field. Mercury also rotates too slowly, and is also probably solid throughout, so it should not have any appreciable magnetic field; the small field measured by *Mariner 10* is, very likely, a fossil field. On the planetary scale, the Earth and Mars are likely to be sister planets.

The analysis of pictures taken by *Mariner 10* in February 1973 shows that the atmosphere of Venus rotates in four earth days only, corresponding to the optical observations, that is sixty times the rate of rotation determined by radar in 1965 for the bulk of the planet.

The Earth-Moon system

The first manned landing on the Moon in July 1969 represented the culmination of a decade of human effort in space, which had begun with Yuri Gagarin's first flight in 1961. But, as mission followed mission, the human and technological aspects of space flight were

113

FIG. 22. The exploration of the Moon has, since the 1960s, been the main objective of both American and Soviet space programmes. In the complete illustration is the Moon photographed in July 1971 by astronauts of the *Apollo 15* mission; the South pole of the Moon is at the top, the North pole at the bottom. The dark circular patch in the centre is the Mare Crisium, which can also be seen in the composite version, made from views taken in November 1973 from *Mariner 10*.

superseded by the scientific motivations which led men to explore another celestial body in the hope of understanding its evolution and its relationships to its companion body, the Earth. So what had begun as a thrilling adventure became more and more a routine mission whose purpose was that of placing scientific instruments at selected places on the Moon, and of carrying back samples of lunar soil to be studied in the Earth's laboratories.

In this process much was learnt about both the Moon and the Earth, but many new questions were raised, and it will take many years to solve all the problems posed by the new data. New missions to the Moon, beyond those already planned, will certainly become necessary and the establishment of scientific bases on the lunar surface is no longer a science fiction dream. New generations of launchers and spacecraft may become necessary for this purpose, and it is to be hoped that, considering the great financial effort involved, an international space programme can be planned in order to avoid duplication and waste, but also as a mean of promoting co-operation between the nations of the Earth. A first step in this direction resulted in the rendezvous of American and Soviet spacecraft in Earth orbit, the *Apollo-Soyuz* test project carried out in July 1975.

Since the lunar exploration programme is still continuing (after American-manned missions of the *Apollo* series ended in December 1973) with Soviet automatic probes of the *Luna* series, its scientific results are far from complete and the study of the gathered materials is continuing in many laboratories throughout the world (it is in this aspect of space flight that international co-operation has often worked best).

The preliminary results are as follows. The big, dark lunar maria are made up of basalt rock, similar to terrestrial lava but completely dry and without free oxygen. This means that the Moon had a melted surface in its early days (from 4.5 to 3.3 billion years ago). On the melted material a crust developed, which has been extensively broken up by the impact of heavy meteorites, with great ejections of material. It is still not clear whether the Moon melted completely down to the core. Some evidence for this can be gathered from the magnetic field 'frozen' into the lunar material which was measured by the *Apollo* instruments; it should have been there for 3 to 4 billion years.

A molten outer core of ferrous material, of course, is needed to explain a magnetic field through a dynamo effect; on the Moon, however, there is very little iron. Some measurements on the heat flow at the surface are surprisingly high, so high that the Moon ought to be completely molten to account for them. But they may be due to local concentrations of radioactive materials, just as some higher magnetic measurements are perhaps due to concentrations of magnetic materials. A very puzzling discovery was that of mascons, gravitational anomalies indicating the presence of extra mass (mascon stands for mass concentrations). They

are located in mare regions of circular shape, like the Mare Imbrium, and not within mare of irregular shape like the Mare Tranquillitatis. There is no generally accepted theory for mascons but they may be related to the fall of very large meteorites.

Other puzzling problems are represented by 'red spots', parts of the Moon which do not cool as rapidly as the rest when the Sun no longer heats them. There is evidence of seismic activity, especially when the Moon is nearer to the Earth; when hit by a meteorite or excited by the impact of a space craft aimed at it on purpose the Moon 'trembles' much longer than the Earth, as a rigid body would do; there is evidence that most of the seismic energy comes from a single zone, about 800 km below the surface. So the Moon may not be completely dead, geologically speaking, after all.

Still unsolved is the problem of the Moon's origin. It was probably already orbiting the Earth 4 billion years ago, but nothing is known about the earlier period. On the other hand, tidal friction is such that the Moon is now receding from the Earth so rapidly that it may have been close to the Earth's surface at the time of its formation. This is as yet an unsolved problem. One hypothesis is that there were several moons around the Earth at one time, together with much debris, and that one moon, the biggest, captured the others. There is, in fact, evidence of low velocity (2–4 km/s) impacts on the Moon, as would be expected in this case. There are countless other problems, such as the origin of craters and rilles, which will be clarified only after many more missions to the Moon, in areas so far neglected such as the far side of the Moon.

Further reading

ASHBROOK, J. Findings from Mercury's transit. *Sky and telescope*, vol. XXXIX, no. 6, 1970, p. 20.

BAHCALL, J. N. Neutrino from the Sun. *Scientific American*, vol. CCXXI, no. 1, 1969, p. 28.

CAMERON, A. G. W. Le système solaire. *Atomes*, no. 270, 1969, p. 643.

DOBROVOLSKY, O. Comets as cosmic measurement devices. *Soviet science review*, vol. III, no. 1, 1972, p. 49.

Dossin, F. Spectra of comet Tago-Sato-Osaka. *Sky and telescope*, vol. XXXIX, no. 3, 1970, p. 148.

Eshleman, V. R. The atmospheres of Mars and Venus. *Scientific American*, vol. CCXX, no. 3, 1969, p. 78.

Goldreich, P. Tides and the Earth-Moon system. *Scientific American*, vol. CCXXVI, no. 4, 1972, p. 42.

Hartmann, W. K.; Yale, F. G. Mare Orientale and its intriguing basin. *Sky and telescope*, vol. XXXVII, no. 1, 1969, p. 4.

Jérome, D. Y.; Lancelot, J.; Perrier, G. Le bilan des expéditions lunaires. *La recherche*, vol. II, no. 10, 1971, p. 203.

Kopal, Z. *The solar system*. Oxford, Oxford University Press, 1973.

Koutchmy, S. Le vent solaire. *Atomes*, no. 270, 1969, p. 648.

Kuiper, G. P. Lunar and planetary laboratory studies of Jupiter, I and II. *Sky and telescope*, vol. XXXXIII, no. 1, 1972, p. 4, no. 2, 1972, p. 75.

Kuzmin, A. D.; Marov, M. Ya. L'atmosphère de Venus. *Sciences*, no. 76, 1972, p. 55.

Leighton, R. B. The surface of Mars. *Scientific American*, vol. CCXXII, no. 5, 1970, p. 26.

Marov, M. Ya. Vénus. *La recherche*, vol. V, no. 50, 1974, p. 927.

Pneuman, G. W. The chromosphere-corona transition region. *Sky and telescope*, vol. XXXIX, no. 3, 1970, p. 148.

Rasool, S. I.; De Bergh, C. La planète Mars. *La recherche*, vol. I, no. 1, 1970, p. 25.

Rasool, S. I.; Encrenaz, Th. Les planètes géantes. *La recherche*, vol. II, no. 1, 1971, p. 317.

Rawlins, D. The mysterious case of the planet Pluto. *Sky and telescope*, vol. XXXIX, no. 3, 1970, p. 144.

Reeves, H. L'origine du système solaire. *La recherche*, vol. VI, no. 60, 1975, p. 808.

Sagan, C. The solar system. *Scientific American*, vol. CCXXXIII, no. 3, 1975, p. 22.

Skuridin, G. Space physics: the new outlook. *Soviet science review*, vol. III, no. 2, 1972, p. 107.

Watts, R. N. Findings of a sample of lunar material. *Sky and telescope*, vol. XXXXII, no. 6, 1971, p. 346.

——. *New science in the solar system*. London, New Science Publications, 1975.

——. Three spacecrafts study the red planet. *Sky and telescope*, vol. XXXXIII, no. 1, 1972, p. 14.

——. Soviet exploration of Mars. *Sky and telescope*, vol. XXXXIII, no. 2, 1972, p. 91.

10

Stars

Life and death of a star

A star is essentially a huge nuclear fusion reactor, of the kind man is trying with much effort to duplicate on Earth on a much smaller scale, so far without success. Its fuel is in most cases hydrogen, which 'burns' in a series of nuclear reactions transforming it into helium and heavier elements. The helium nucleus (also known as an alpha particle) is particularly stable, and therefore the reaction is exothermic, giving out energy as electromagnetic radiation (photons) and neutrinos.

Stars are believed to be born from vast clouds of hydrogen present in space, which contract under the gravitational attraction of the atoms forming the cloud. In the process the cloud heats up, much as does the gas in a piston engine during compression, and the temperature eventually rises high enough to start thermonuclear reactions, which transform matter into energy (radiation). The outward pressure of radiation balances the inward pull of gravitation and, given a reasonable mass for the star, a steady state results: the star then slowly burns its fuel for billions of years. When the hydrogen is nearly spent, the star contracts and starts falling towards its centre, its temperature increasing still further; this may start other thermonuclear reactions which are strongly temperature-dependent,

but when those have exhausted their fuel, the process of implosion starts again.

What happens after this contraction has started depends mainly upon the mass of the star: if it is less than 1.4 solar masses (the so-called Chandrasekhar limit), the contraction is halted by the rise of internal pressure when a central density of 10^6 g/cm^3 is reached. The star then becomes a cold 'white dwarf' with a typical radius of 10,000 km; such stars are barely visible even with the more powerful telescopes. If the star is more massive than this, the stellar core is squeezed to such high densities that violent nuclear processes occur, causing a huge explosion which sends a shock wave outwards at speeds comparable with that of light: an explosion of this kind is called a supernova explosion, and is visible in the sky as an extremely brilliant star, which burns itself out in a very short time on the stellar scale. Three supernovae explosions were observed in historically recent times, one in 1034 by Chinese astronomers, one in 1572 and one in 1604. Supernovae explosions can often be seen in extragalactic nebulae through optical telescopes.

Neutron stars, pulsars and black holes

On a summer night in 1967 a young graduate

student at the Cambridge University Mullard Radio-astronomy Laboratory was scanning the sky with a new radiotelescope designed to measure rapid variations in celestial radio-sources when she noticed a series of strong, regular bursts of radiation coming from one position in the sky. The radiotelescope she was using was of the fixed kind, which depends upon the Earth's rotation to sweep the 'beam' through the sky. She had therefore to wait until the next night to obtain a confirmation of her findings: the source was there again. All activity at the laboratory was then concentrated upon this object, which was nicknamed, only half jokingly, LGM (from the initials of 'little green men') because such a regularly pulsating signal could well have been the product of an extraterrestrial civilization signalling its existence to mankind (such signals had been looked for in vain before). However, the explanation of these objects—now called pulsars—turned out to be less mysterious and to fall entirely within the framework of known physical laws as they apply to stellar evolution.

What is left when a supernova explosion has died out is a strange and supermassive object, with a radius of about 10 km and a density of about 10^{14} g/cm³: it is a neutron star, since the pressure has forced the electrons 'inside' the protons, thus producing neutrons. The existence of such weird objects, invisible even with the most powerful telescopes, was postulated on theoretical grounds by nuclear physicists about forty years ago. But when the existence of pulsating radio stars (pulsars) were announced the neutron star was re-examined to see if it could explain the peculiarities of the pulsar's emission. It was found that a rotating neutron star emitting a beam of radio waves matches the pulsar's characteristics almost perfectly. It is now generally believed that pulsars are neutron stars, relics of supernova explosions. One pulsar has been found in the middle of the Crab Nebula where Chinese astronomers had witnessed a 'guest star' in A.D. 1054. This particular pulsar has a period of 33 ms, the shortest known for pulsars, and has been found to be pulsating, with the same period, even in the optical, radio and X-ray regions of the spectrum.

The period of pulsars is not strictly constant, but has been found to increase slowly and steadily; in terms of the rotating neutron star model, this means that the neutron star is rotating increasingly slowly, probably because of some kind of drag in its atmosphere. Neutron stars are very complicated objects, and provide a challenge to many different kinds of physicists because they are an example of matter under very extreme conditions. According to one model, a neutron star has a central core composed of elementary particles, an inner shell made out of neutrons and protons and an outer shell made up of heavy nuclei which act as a crust; sudden variations in the pulsar's periods are interpreted as 'starquakes' in this crust. The internal temperature of a neutron star is about 10^{10} K but its density is so high that the neutrons form a superfluid similar to liquid helium near absolute zero. Another important feature of a neutron star is its strong external magnetic field, which may be as high as 10^{12} gauss and which presumably rotates with the star, emitting magnetic dipole radiation; this means there must be a plasma-filled region, a magnetosphere, near the star. Suggestions have been made that the strong magnetic field acts like a gigantic particle accelerator and is responsible for accelerating the elementary particles which bombard the Earth from space and are known as cosmic rays.

The existence of neutron stars exactly as predicted has reinforced speculation on the existence of a third class of dead star, apart from the white dwarfs and the neutron stars: the 'black holes'. General relativity predicts, in fact, that if a star is so massive that the gravitational pressure completely overcomes the nuclear, very short-range repulsive forces, the collapse of the star will continue until it is reduced to an object of zero volume and infinite density, which is called a singularity. Such a weird object would be completely invisible because it absorbs even the radiation emitted by itself (except for gravitational waves which are very difficult to detect: see Chapter 12). The exact limit of mass above which a star cannot settle into a neutron star but is forced to become a black hole is not known exactly; it is speculated that the critical factor may be not the mass but the angular momentum of the star, and that stars which rotate more slowly are more likely to end up as black holes. Black holes are being looked for particularly in binary systems of stars in which one of the stars is invisible. The search is very difficult since a black hole can be discovered only by indirect means, for instance by the radiation emitted by particles falling into the 'hole'. Such radiation is likely to be in the X-ray band, so that among the first candidates for the role of black holes were the X-ray sources discovered in 1962 by the use of rockets and balloons carrying X-ray telescopes. Since then X-ray detectors have been placed aboard astronomical satellites but a 'black hole' had still to be unambiguously identified. The most likely candidate is the X-ray source Cyg X-1, in the constellation of Cygnus. This is considered to be a binary star where a highly collapsed object is sucking matter from the companion star, a red giant. Where the matter falls on the collapsed object, it heats up by compression and emits X-rays. Other similar sources are known in the sky. It must be said, however, that not all scientists agree that a black hole is needed to explain celestial X-ray sources, and that more conventional explanations are also given.

The interstellar medium

Until 1963 only three molecules were known to exist in interstellar space, namely CH, CN (cyanogen) and CH^+; atomic hydrogen was thought to be its major constituent, together with large fluxes of ultra-violet radiation and cosmic rays, which rendered impossible the formation of larger molecules. A low density of dust grains, 10^{12} smaller than that of atomic hydrogen, was also thought to be present and to have the effect of reddening the light passing through space. But the use of new radio-telescopes, capable of gathering and focusing radiation in the centimetric and millimetric bands, which correspond to the rotational and vibrational modes of many molecules, soon gave unexpected results.

In 1963 the hydroxyl radical (OH) was found by a research group at Massachusetts Institute of Technology, in 1968 ammonia (NH_3) and water molecules were found at Berkeley, in 1969 formaldehyde (H_2CO) at the National Radio-astronomy Observatory

in Virginia. Then the discoveries became more frequent; up to the end of 1971 various groups working in the United States and in Australia had found carbon monoxide (CO), molecular hydrogen (H_2), hydrogen cyanide (HCN), cyanoacetylene (HC_3N), methyl alcohol (CH_3OH), formic acid (CHOOH), formamide (NH_2CHO), silicon oxide (SiO), carbonyl sulphide (OCS), acetonitrile (CH_3CN), isocyanic acid (HNCO), methyl acetylene (CH_3C_2H), hydrogen isocyanide (HNC), acetaldehyde (CH_3CHO) and thioformaldehyde (H_2CS). Subsequent discoveries included hydrogen sulphide (H_2S), sulphur monoxide (SO), methylammine (CH_3NH_2) and other compounds, but the list is certainly destined to grow as more and more interstellar molecules are discovered.

Meanwhile the OH radical has been detected in external galaxies, too. This has been possible because the intensity of the OH emission is particularly strong, suggesting that a kind of interstellar maser process is responsible for the emission. The emission seems to be associated with infra-red sources of small angular diameter which may be protostars, that is stars in the process of formation from a hydrogen cloud. Dust grains in space are thought to play a major role in the synthesis of interstellar molecules in the extreme conditions of the space environment. In this way, a first link is established between interstellar molecules and the evolution of stars.

A second, even more interesting, link can be established between interstellar molecules and the evolution of life: a number of the organic molecules found in space are among the 'building blocks' from which amino acids have been synthesized in the laboratory. In fact, these amino acids were also found in the Orgueil, Murray and Murchison meteorites, suggesting that interstellar space provides a unique environment, with its very low temperature and its abundant ultra-violet radiation, for the synthesis of prebiotic compounds. It has been pointed out, however, that the densities of the organic molecules in space are so much lower than those used in the laboratory that no conclusion can be correctly drawn from experiments performed under such different conditions. There might be, however, a direct

119

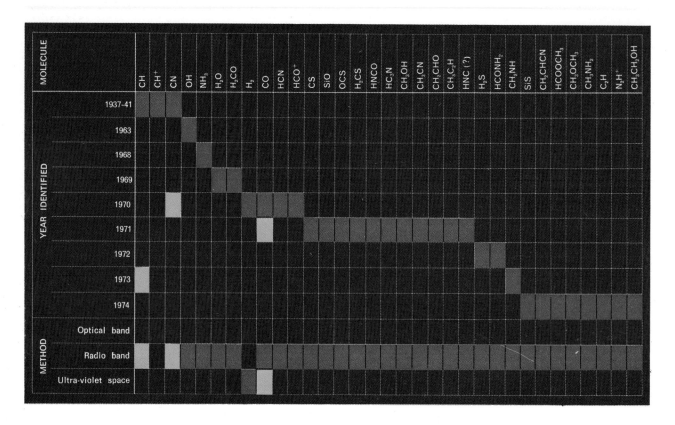

Fig. 23. Summary table of interstellar molecules identified by the end of 1974, principally by means of radiotelescopic observation. Those identified with certainty are shown in dark, those with probability in light shades. (From *La Recherche*, March 1975.)

The Galaxy

All objects visible to the naked eye, except for one in the northern hemisphere (the Andromeda galaxy) and two in the southern hemisphere (the Magellanic Clouds), belong to the Galaxy, a huge stellar disc with a diameter of 100,000 light years and a thickness of 1,700 light years, except for a bulb in its centre which is 16,000 light years thick. It is a spiral galaxy, as was discovered by radio-astronomers listening to the 21.11 cm wavelength emission of neutral hydrogen contained in the spiral arms; neutral hydrogen comprises to about 2 per cent of the whole galactic mass. While the spiral arms are flying outward, hydrogen clouds have been discovered infalling on the galactic nucleus.

Radio-astronomy and other non-optical techniques such as infra-red astronomy are much used in the study of the Galaxy because

connexion between the evolution of stars and biological evolution: in the first phase of a star's formation an atmosphere may develop which includes organic molecules. This atmosphere may be retained fully or partially by the developing planets, while it disappears in the star as soon as thermonuclear reactions start. The process of biological evolution may then start in the planetary atmospheres, or in the oceans deriving from them. This is why a direct investigation of the atmospheres of Mars and Venus by space probes could give some understanding of both stellar and biological evolution, as these planets may have retained their primaeval atmospheres, unlike the Earth's atmosphere, which has been modified by biological processes such as photosynthesis.

its centre is practically obscured by interstellar dust. In 1968 an infra-red source was found in the constellation Sagittarius, in the direction of the centre, near the position of a radio-source (Sag A) discovered in 1951; the infra-red emission is extremely powerful. In 1971 the small X-ray astronomy satellite *Uhuru* found an X-ray emission coming from the galactic centre; in 1975 another satellite, *Ariel 5*, confirmed this finding. The source is variable and flares up, at irregular intervals, becoming very bright.

What kind of process is involved and what is the cause of the strong infra-red and X-ray emissions which seem to accompany it? Speculations include gravitational collapse, by which up to 45 per cent of the total mass can be converted into energy. This hypothesis is often advanced to explain the violent phenomena and the very intense energy emissions apparently occurring in the nuclei of Seyfert galaxies and in quasars. Another possibility is that of matter-antimatter annihilation, which allows 100 per cent of the matter to be transformed into energy, provided there is sufficient antimatter available; there is no sign, however, of the existence of so much antimatter in the Galaxy, certainly not in the cosmic radiation which bombards the Earth with an apparently isotropic flux and which is therefore believed to be of galactic origin. Be that as it may, it is evident that even our Galaxy, the Milky Way, so familiar to anyone who has looked at the night sky, is not such an ordered and peaceful system as was once believed.

Further reading

AMBARCUMJAN, V. A. O jadrach galaktik. *Nauka i Čelovečestvo*. Moskva, Znanie, 1969.

BARRETT A. H. Radio signals from hydroxyl radicals. *Scientific American*, vol. CCXIX, no. 2, 1968, p. 50.

BERGMANN, P. G. *The riddle of gravitation*. New York, Scribner's, 1969.

BOK, B. J. The spiral structure of the Galaxy, I and II. *Sky and telescope*, vol. XXXVIII, no. 6, p. 392, vol. XXXIX, no. 1, p. 21.

CHARADZE, E. K. Ob'ject izucenija—nasa Galaktika. *Nauka i Čelovečestvo*. Moskva, Znanie, 1970.

CHIU, M. Y. Pulsars, radiophares de l'espace. *La recherche*, vol. II, no. 8, 1971, p. 13.

COWSIK, R.; PRICE, P. A. The origin of cosmic rays. *Physics today*, vol. XXIV, no. 10, 1971, p. 30.

GIACCONI, R. X-ray stars. *Scientific American*, vol. CCXVII, no. 6, 1967, p. 36.

GINZBURG, J. L. Cosmic rays astrophysics. *Scientific American*, vol. CCXX, no. 2, 1969, p. 50.

GORENSTEIN, P.; TUCKER, W. Supernova remnants. *Scientific American*, vol. CCXXV, no. 1, 1971, p. 74.

GREEN, L. C. Ordinary stars, white dwarfs, neutron stars. *Sky and telescope*, vol. XLI, no. 1, 1971, p. 18.

. Pulsars today, I and II. *Sky and telescope*, vol. XL, no. 5, 1970, p. 260, no. 6, 1970, p. 357.

. Starquakes: have they been observed? *Sky and telescope*, vol. XLI, no. 2, 1971, p. 76.

GREENBERG, J. M. Interstellar grains. *Scientific American*, vol. CCXVII, no. 4, 1967, p. 107.

GURSKY, H.; VAN DEN HEUVEL, E. P. J. X-ray emitting double stars. *Scientific American*, vol. CCXXXII, no. 2, 1975, p. 24.

HERBIG, G. H. The youngest stars. *Scientific American*, vol. CCXVII, no. 2, 1967, p. 30.

HEWISH, A. Pulsars. *Scientific American*, vol. CCXIX, no. 4, 1968, p. 25.

KRAFT, R. P. Exploding stars. *Scientific American*, vol. CCVI, no. 4, 1962, p. 54.

KUCHOWICZ, Br. The fifth state of matter: superdense matter and neutron stars. *Science progress*, vol. LVI, no. 224, 1968, p. 531.

LITVAK, M. M.; ZUCKERMANN, B. M. Interstellar masers. *Sky and telescope*, vol. XL, no. 5, 1970, p. 267, no. 6, 1970, p. 345.

MIRZOJAN, L. V. Problema dozvezdnoj materii. *Budušee Nauki*. Moskva, Znanie, 1971.

NEUGEBAUER, G.; LEIGHTON, R. B. The infrared sky. *Scientific American*, vol. CCXIX, no. 2, 1968, p. 50.

NOVIKOV, N. D. O konečnoj sud'be zvezd. *Budušee Nauki*. Moskva, Znanie, 1968.

PECKER, J. C. *La nouvelle astronomie*. Paris, Hachette, 1971.

. *Les observatoires spatiaux*. Paris, Presses Universitaires de France, 1969.

PIKEL'NER, S. B. Spiral'nye vetvi galatkik i ich magnitnoe pole. *Budušee Nauki*. Moskva, Znanie, 1968.

RUFFINI, R.; WHEELER, J. A. Introducing the black hole. *Physics today*, vol. XXIV, no. 10, 1971, p. 30.

SHKLOVSKIJ, I. S. Puls'ary. *Budušee Nauki*. Moskva, Znanie, 1971.

SNYDER, L. E.; BUHL, D. Molecules in the interstellar medium, I and II. *Sky and telescope*, vol. XL, no. 5, 1970, p. 267, no. 6, 1970, p. 345.

THORNE, K. S. Gravitational collapse. *Scientific American*, vol. CCXVII, no. 5, 1967, p. 88.

    The search for black holes. *Scientific American*, vol. CCXXXI, no. 6, 1974, p. 32.

WEBER, J. The detection of gravitational waves. *Scientific American*, vol. CCXXV, no. 1, 1971, p. 74.

WEEKES, T. C. *High energy astrophysics*. London, Chapman & Hall, 1969.

11   Galaxies

Optical galaxies

Until recently the only known objects outside the Milky Way, the optical galaxies are still the major source of information to the astronomer. For this reason new and bigger optical telescopes are being built in a number of countries. The biggest so far is the 6 m telescope at Zelenchukskaya (U.S.S.R.), while others were built in Chile and Australia, to view the southern sky which contains many fascinating celestial objects. Many of these projects are so costly that they are being funded by international co-operation (such as the 4 m telescope at Ceno Tololo, Chile, the 3.9 m Angloaustralian telescope in Australia and the 3.6 m European telescope at La Silla, Chile).

The main problem in extragalactic astronomy, aside from the cosmological problem of the red shift of light coming from distant galaxies (see below), is that of understanding the evolution of galaxies and of correlating this evolution with the morphological classification scheme made by Hubble (the so-called Hubble diapason). Roughly speaking, the Hubble sequence runs from irregular galaxies to spiral galaxies and elliptical galaxies. The evolutionary sequence presumably follows the same pattern; according to a

theory advanced in 1964 the proportion of uncondensed gas decreases as the galaxy evolves, and so a younger galaxy should contain more uncondensed gas. A model according to which a galaxy is formed from a large, spherical and slowly rotating mass of gas was also put forward in 1964, and it accounts for at least some of the observed facts.

Another problem consists in estimating the total mass of the galaxies by counting the number of visible galaxies and multiplying by their estimated mass; this count has yielded a matter density that is considerably lower than that calculated from cosmological models (see Chapter 12). In clusters of galaxies the total mass of the cluster can be estimated from dynamics considerations and is always higher than the sum of the individual masses of the galaxies apparently composing the cluster, as estimated from the sum of the masses of stars composing puzzle. Naturally, this last estimate is strongly dependent on the masses of the stars composing the galaxies, which in turn can be estimated from the mass-to-light ratio of the stars (since the light coming from a given star is its only directly measurable physical parameter). Stars of very large mass and low luminosity (neutron stars or even 'collapsed objects' such as black holes) could make up for this 'missing mass'.

Finally, there is the problem of the hierarchical structure of the universe; can the clusters of galaxies be regrouped in superclusters, or are they uniformly distributed in all directions? This important problem has a direct bearing on estimating the mass of galaxies, since the latter is done by statistical methods, that is by counting the number of galaxies visible in a sample section of the sky and by extrapolating this count to the whole sky. This method is admissible only if the distribution of galaxies is uniform, and not if there is superclustering.

Radiogalaxies

The main theoretical problem posed by radiogalaxies has been how to explain their enormous energy emission which is known as synchrotron radiation because it is similar to the radiation produced in synchrotrons (particle accelerators) by charged particles accelerated in magnetic fields. The strongest radiosources among the galaxies are the elliptical giant ones, or those consisting of a very bright nucleus and a wisp projecting from it (an outstanding example is M 82, where M stands for Messier catalogue). Another strong radiosource is M 87, where the radio emission comes from two jets of material shooting out at great speed in opposite directions from the nucleus of this exploding galaxy. It has been suggested that violent events are taking place in the nuclei of radio-emitting galaxies, and this suggestion was corroborated by observation in 1963.

The development of radio-astronomy, like that of optical astronomy, has been greatly dependent on the construction of bigger instruments such as the Jodrell Bank 250 ft (76 m) radiotelescope, which was overtaken in size in 1971 by the 100 m radiotelescope at Effelsberg (Federal Republic of Germany). At the same time, an intelligent use of smaller instruments, often combined together as radio-interferometers, has provided instruments of smaller gathering power but often with better resolution and hence able to 'see' radiosources in greater detail.

The distance of a radiosource cannot be measured unless it can be identified with an optical object, since the synchrotron emission of radiogalaxies contains no spectral lines which might reveal a red shift and hence, through Hubble's law, lead to a distance determination. The first radio red shift, relative to the 21 cm line emitted by hydrogen, was measured only in 1973 for the quasar 3C 286 and found to be 0.69. Radiotelescopes have a much greater diameter than optical telescopes and are thus able to gather more radiation and 'see' farther in the Universe than optical telescopes. It is therefore not always possible to associate an optical object with a radiosource, particularly because radiotelescopes use longer wavelengths than optical instruments and hence are not able to pin-point radiosources with great accuracy. To improve the resolving power of radiotelescopes, a sophisticated method has recently been developed, called very long base interferometry (VLBI), which makes use of simultaneous observations by instruments in places as far apart as the U.S.S.R., the United States and Australia. If the information from each instrument is synchronized by atomic clocks it can later be combined to give an 'image' of the radiosources which is much more detailed than would be that provided by a single telescope. In this way, the identification of radiosources with optical counterparts has been made easier.

Research of great potential cosmological importance has been done by counting the density of radiogalaxies at different distances from the Earth, yielding the so-called log S-log N relation, which is a crucial test of cosmological theories. This count has been made many times by various scientists, using ever-improving techniques. The results seem to favour the evolutionary models of the Universe as opposed to the static ones, but have not yet provided a clear-cut test between the two main evolutionary models, the so-called big bang theory and the steady state theory, although they tend to support the former.

Seyfert and compact galaxies

Among the galaxies with a very bright nucleus are the Seyfert galaxies, called after the astronomer C. K. Seyfert who was the first to describe them. Their nuclei are very small, and their spectra have very broad emission lines, which correspond to a spread in velocity of the emitting region equal to about 3,000 km/s. Other emission lines show that matter within the Seyfert galaxies exists in a high state of ionization, as was shown for NGC 4151 in 1968 (NGC stands for new general catalogue). In this case too there may be ionized gas spurting from the nucleus at high speed, and in fact some Seyfert galaxies are also radiosources, and their nuclei may also emit strongly in the infra-red. A typical feature of Seyfert galaxies is their variability on a time scale of years, or even months, as demonstrated in 1967.

Another kind of peculiar galaxy to have been catalogued independently are the compact galaxies. They occur in small groups, and some fall in the class of Seyfert galaxies, as in the case of VV 144 (VV stands for Vorontsov-Velyaminov catalogue). Among the more interesting cases, VV 172 is a chain of galaxies first studied in 1963. In this chain of galaxies the second member of the quintet has a much higher red shift than the other members, that is, it should be much farther away if one applies Hubble's law. But the five galaxies seem to form a physically bound system, and to assume that one member of the quintet is a galaxy in the background which by chance falls into the line of sight of another group of galaxies seems more and more difficult to believe as more examples of these strange chains are found. Another example is the so-called Stephan's quintet. These, and other examples in which quasars are involved, suggest that Hubble's law may not be applicable to all celestial objects, especially those with compact nuclei and high brightness and in which violent events seem to occur resulting in the emission of many different kinds of radiation.

Quasars

The discovery of the quasi-stellar objects (or quasars) ranks among the greatest scientific discoveries of the 1960s. It was made by several astronomers who were trying to pin-point the positions of the radiosources contained in the third Cambridge catalogue, using a radio-interferometer at the California Institute of Technology. They sought to associate the radiosources with visible objects contained in the photographs taken with the Schmidt camera of Mount Palomar. It was thus discovered that many of the radiosources did not correspond to galaxies, but instead to starlike objects of bluish light, which were termed 'radio stars' in 1960, when they were still considered as stars; their optical spectra did not correspond, however, to those of any known star, and the mystery persisted.

In 1964 it was shown that their spectra could be easily understood if one postulated for them a red shift similar, but usually greater, than that of the more distant galaxies; the first objects identified as quasars, 3C 273 and 3C 48, had red shifts respectively of 0.158 and 0.367 but by 1965 the spectrum of 3C 9 had been interpreted on the basis of a 2.012 red shift. In 1970, in the radiosources of the fourth Cambridge catalogue, a quasar, 4C 0534, was found with the huge red shift of 2.877; in 1973 two quasars were found with red shifts greater than 3: OH 471, with a red shift of 3.4 and OQ 172, with a red shift of 3.53. It was realized in 1965 that some blue stellar objects, although not sources of radio waves, had many of the characteristics of quasars, that is blue light, small angular diameters and a huge red shift; for this reason they were dubbed 'interlopers', but is was soon evident that those radio-quiet objects (quasi-stellar objects) were much more common than the quasi-stellar radiosources. The name quasars was officially accepted in 1971 as covering both kinds of object.

The main problem posed by these strange objects is how to interpret their red shifts: if it is cosmological in nature, and is caused by their outwards movement as part of the expansion of the Universe as postulated by Hubble's law, they are extremely distant objects and about as bright as 10^{13} suns! The source of this huge amount of energy is unlikely to be the thermonuclear processes which power the Sun and the other stars; one must resort to the gravitational collapse of enormous masses, matter-antimatter annihilation or the creation of matter and energy in 'white holes', as postulated by some later versions of the steady state theory. Quasars thus pose enormous problems in astrophysics.

If these difficulties are to be avoided, the cosmological interpretation of the red shift must be abandoned and another explanation found. Tentative suggestions include the effect of a very strong gravitational field which reddens the light (by the so-called Einstein effect), the ejection of quasars from nearer objects (galaxies) with a high velocity which would produce the red shift by the Doppler effect, and yet others such as the action of a 'gravitational lens' which would focus the light coming from other galaxies. But none of these ideas has proven to be entirely satisfactory. Evidence against the cosmological interpretation of the red shift for quasar 3C 279 was adduced at the Massachusetts Institute of Technology (MIT) in 1971. It was shown that the two jets of gas spurting from it in opposite directions would have to travel at ten times the speed of light if the red shift were of cosmological origin. Since the absolute velocity of light is postulated by the special theory of relativity, and is generally believed correct, it is unlikely that any massive object anywhere in the Universe travels faster than the speed of light unless the physical laws valid on Earth have to be modified to account for phenomena in distant parts of the Universe. This position has been taken by two scientists at Cambridge University, who devised in 1971 a theory in which the mass of a particle depends on its location in the Universe in order to account for the red shift phenomenon (see above). The strongest argument against the cosmological interpretation of the quasar's red shift has been a picture in which a quasar (Markarian 205) is seen to be linked to a galaxy by a bridge of hydrogen gas: the red shifts of the two objects differ by a factor of ten. For these reasons, it seems likely that the unresolved paradox of the quasars will continue to puzzle astrophysists for many years to come. A first clue towards the solution of the puzzle consists perhaps in the fact that brilliant quasars were found to be entoured by diffuse objects resembling elliptical galaxies; the hypothesis was made that quasars are nothing but brilliant nuclei of very distant galaxies, which are too faint to be seen except for the most powerful telescopes employing the more refined electronic image intensifiers. Many astronomers are now working on this clue.

Further reading

AMBARCUMJAN, V. A. Ob evoljucii galatktiki. *Nauka i Čelovečestvo*. Moskva, Znanie, 1965.

BURBIDGE, E. M.; BURBIDGE, G. Peculiar galaxies. *Scientific American*, vol. CCIV, no. 2, 1961, p. 50.

BURBIDGE, E. M.; LYNDS, C. R. The absorption lines of quasi-stellar objects. *Scientific American*, vol. CCXXIII, no. 6, 1971, p. 22.

GREEN, L. C. Quasars six years later. *Sky and telescope*, vol. XXXVII, no. 5, 1969, p. 290.

GREENSTEIN, J. L. Quasi-stellar radiosources. *Scientific American*, vol. CCIX, no. 6, 1963, p. 54.

HARP, H. C. The evolution of galaxies. *Scientific American*, vol. CCVIII, no. 1, 1963, p. 70.

HEESCHEN, D. S. Radiogalaxies. *Scientific American*, vol. CCVI, no. 3, 1962, p. 41.

HODGE, P. W. Dwarf galaxies. *Scientific American*, vol. CCX, no. 5, 1964, p. 78.

HOYLE, F. *Galaxies, nuclei and quasars*. London, Heinemann, 1965.

KELLERMAN, K. I. Intercontinental radioastronomy. *Scientific American*, vol. CCXXVI, no. 2, 1972, p. 72.

LEQUEUX, J. *Physique et évolution des galaxies*. Paris, Dunod, 1968.

McNALLY, D. Quasi-stellar sources. *Science progress*, vol. LII, no. 207, 1964, p. 426.

PAGE, J. *Étoiles et galaxies*. Paris, Marabout Université, 1966.

PAGE, T. L. Galaxies and quasars in Prague, I and II. *Sky and telescope*, vol. XXXIV, no. 6, 1967, p. 372; vol. XXXV, no. 1, 1968, p. 16.

PECKER, J. *La nouvelle astronomie*. Paris, Hachette, 1971.

REES, M. J.; SILK, J. The origin of galaxies. *Scientific American*, vol. CCXXII, no. 6, 1970, p. 26.

SANDAGE, A. R. Exploding galaxies. *Scientific American*, vol. CCXI, no. 5, 1964, p. 38.

SCHATZMANN, E. *Structure de l'univers*. Paris, Hachette, 1968.

SCHMIDT, M.; BELLO, F. The evolution of quasars. *Scientific American*, vol. CCXXIV, no. 5, 1971, p. 54.

SHAKESHAFT, J. R. Radiogalaxies and quasars. *Science progress*, vol. LIII, no. 212, 1965, p. 512.

SHAPLEY, H. *Galaxies*. Cambridge, Harvard University Press, 1961.

STROM, R. G.; MILEY, G. K.; OORT, J. Giant radio galaxies. *Scientific American*, vol. CCXXXIII, no. 2, 1975, p. 26.

TOOMRE, A. and J. Violent tides between galaxies. *Scientific American*, vol. CCXXIX, no. 6, 1973, p. 38.

WELIACHENS, L. La dynamique des galaxies spiroles. *La recherche*, vol. VI, no. 55, 1975, p. 318.

WEYMANN, R. J. Seyfert galaxies. *Scientific American*, vol. CCXX, no. 1, 1969, p. 28.

12

The origin
of the Universe

Observational evidence

Like a policeman trying to solve an important crime, the scientist interested in the origin of the Universe must base his deductions on what scanty evidence he has gathered and cannot commit the crime again to prove his theory. Thus, in cosmology, the very term 'observational evidence' has a different meaning from that used in experimental science, where theories can be tested by laboratory experiments. The birth of the Universe is not an experiment reproducible in the laboratory, and is therefore outside the scope of accepted physical laws which postulate, among other things, the conservation of mass-energy, electric charge, baryon number and lepton number; all these laws would be violated in the creation of a Universe *ex nihilo*, either in the form of a single act of creation or of the continuous creation of matter. The simultaneous creation of equal amounts of matter and antimatter, postulated by some scientists, could maintain some conservation laws, but the conservation of mass-energy would be violated in any case. For this reason, some scientists prefer a model in which the Universe has neither an origin nor an end, but is always there, not static but pulsating like a gigantic engine between a phase of expansion and a phase of contraction. We are now living in a phase of expansion, which has been going on for about 10,000 million years (this is often referred to as 'the age of the Universe').

There are three main proofs of the expansion of the Universe. The first is related to the simple fact that the sky is black at night, the so-called Olber's paradox: simple mathematical reasoning shows that, in a static and infinite Universe, the light coming from an infinite number of stars should be infinitely bright. There are two ways out of this impasse: either the Universe is finite or it is expanding. The second proof is related to the red shift of galaxies, through Hubble's law which states that the velocity of a celestial object (as determined by its red shift) is proportional to its distance. This is precisely what should happen in an expanding Universe, which is often pictured as the surface of an inflatable balloon, where the distance between fixed points on the surface increases (if measured on the surface), as the balloon is inflated, according to the same type of law.

The third proof of expansion in the Universe came almost by chance in 1965. Two radio engineers had been charged by Bell Telephone Company to inquire about the causes of a background noise in an antenna built for communication with artificial satellites at Andover, Maine. When the noise turned out to be of celestial origin, they contacted astrophysicists at nearby Princeton University, who had in fact been looking in vain for exactly the same kind of 'noise', but without success: this had been foreseen around 1948 as the cool remnant (about three degrees above absolute zero: 3 K) of the 'primaeval fireball' from which the Universe originated, according to the 'hot big-bang' theory of the origin of the universe. It is called 'fossil radiation' or, more usually, 'microwave background' radiation, because it was discovered in the microwave part of the radio spectrum. The predictions of the big bang theory are that the background radiation should have the typical bell-shaped form of thermal emission from a 'black body', if the intensity of the radiation is plotted against its wavelength, corresponding to a 3 K temperature, and should be perfectly isotropic. The first prediction is well fulfilled in the microwave part of the spectrum but at smaller wavelengths (in the infra-red spectrum) it is difficult to make accurate measurements from the Earth because of atmospheric absorption. Thus detectors must be carried above the atmosphere by rockets or balloons.

In 1968 various research groups made balloon and rocket flights to measure the background radiation in the infra-red part of the spectrum, and a relatively high intensity was measured, corresponding to an 8 K black body curve, higher than the predicted 3 K. These results still await confirmation; other, more recent, aircraft and balloon flights have recorded different temperatures, and consequently it has been suggested that this radiation originates from 'local' infra-red sources, or even from crepuscular light in the high atmosphere. It is likely that astronomical satellites equipped with infra-red detectors can clarify the point. The isotropy of the background radiation was checked in 1967 and later, and the condition is fulfilled to a given degree of accuracy.

Relativity and gravitation

Gravitation is the physical force or interaction which dominates at very large distances in the Universe, and is therefore most important in cosmology. The Einstein equations of general relativity are so far the best theory of gravitation available, and most cosmological models are based on their solutions. General relativity, however, is by itself much more difficult to check experimentally than special relativity, since it involves great distances and effects which are generally very small and difficult to measure accurately. In recent years, however, the progress of space research and of electronic instrumentation devised in order to communicate with space vehicles has offered an opportunity of testing the predictions of general relativity in actual experiments, performed over astronomical distances.

In 1968 the large (120 ft diameter) antenna of the Haystack Radar at the MIT Lincoln Laboratory was employed for a 'fourth test' of general relativity (the three 'classical' ones are: advancement of Mercury's perihelion; deflection on light in the gravitational field of the Sun; and the gravitational red shift of photons). A radar beam was directed at Mercury and Venus when those planets were in superior conjunction with the Sun, thus making it graze the Sun's edge, and its echo was timed very accurately, in order to measure the delay in the round trip of the radar waves against that predicted by the theory of general relativity; the agreement was checked, within a 10 per cent limit of accuracy. This was a marked improvement over the roughly 20 per cent accuracy of the classical tests, but still insufficient to discriminate between Einstein's classical relativity predictions and those of the so-called scalar-tensor theory, a major variant of general relativity; scientists at Princeton University have been, since 1966, developing a kind of solar telescope designed to detect any alteration in the shape of the Sun which would explain, at least partially, the results of the 'classical tests' in terms of their modified theory.

A further check was therefore carried out by scientists at Pasadena's Jet Propulsion Laboratory in 1970, using the spacecrafts *Mariner 6* and *Mariner 7* in solar orbit, and later with *Mariner 9* in orbit around Mars. They were ordered by the scientists, through the 250 ft Goldstone antenna of the Deep Space Network, to emit a radio signal, and the delay in the round trip of the radio waves was again measured, this time to a 2 per cent accuracy. Other experiments with celestial radiosources,

FIG. 24. According to the theory of the expanding universe, it began swelling from the primordial fireball, which represents zero years in time (the 'Big Bang'), about 10,000 million years ago. The galaxies we can see are found in the zone limited by the horizon, and their light has reached us along a curved path. (Copyright: Field Enterprise Educational Corporation.)

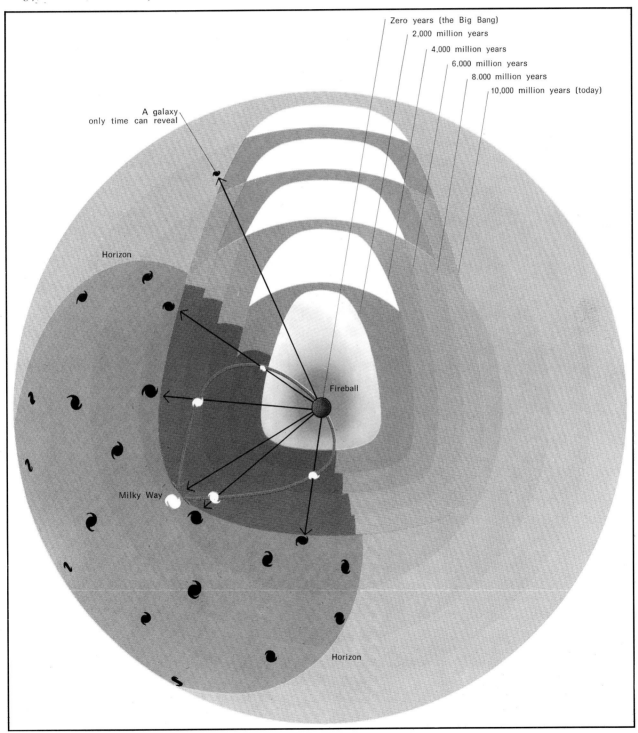

occulted at times by the Sun, reached an accuracy of 1 per cent. The results were more favourable to Einstein's theory than to the scalar-tensor theory.

Gravitational wave detectors, first developed at the University of Maryland, where positive results were claimed for in 1969 are also being developed at Moscow University and at other European and American universities which make use of cryogenic equipment in order to reduce thermal noise, and should be very sensitive. To date, however, nobody has been able to duplicate the original Maryland results, and they are much doubted. Other instrumentation is being developed to be put aboard artificial satellites, in order to measure the curvature of space predicted by general relativity. Space stations, of the *Saljut* and *Skylab* kind, are almost ideal laboratories in which to make tests of general relativity in a gravity-free environment, and will probably be employed to this end, together with specialized satellites, such as the Relativity Satellite being prepared at Stanford University.

Laser light is often used in this research because it permits distance measurements to be made with very high accuracy, through the use of interferometric methods. It is to be expected that the laser retroreflectors placed on the Moon by the American astronauts and the Soviet automatic probes will offer a chance of measuring the Earth-Moon distance with such a high degree of accuracy that the whole Earth-Moon system can be used as a huge gravitational wave detector.

On the theoretical plane, mention must be made of a theory which has been developed since 1964 by two scientists at Cambridge University's Institute of Theoretical Astronomy; its main departure from general relativity lies in the fact that it suggests that not only does mass vary with velocity, as in general relativity, but also with the space-time location of the body. In this theory anomalous red shifts are explained as a consequence of the mass variations; but since anomalous red shifts can perhaps be explained by other means (a photon-photon interaction, according to a suggestion formulated in 1972 by French and Australian scientists), the value of this extension of general relativity lies more in

the fact that is appears to relate microscopic to macroscopic phenomena, by means of symmetry laws (conformal invariance) which appear to be valid both in the microscopic and the macroscopic world, thus following the trend first set by Einstein in his attempts to create a unified field theory which could embrace all natural phenomena.

Cosmological models

Once one accepts the fact that the Universe is expanding as a whole, three main models can be used to explain this fact: the steady state model of Bondi-Gold-Hoyle, in which new matter is continuously created to make up for the loss of density due to the expansion; the big bang model of Alpher-Bethe-Gamow, in which the universe is expanding from a 'primaeval fireball' of very high temperature, getting cooler as it expands; and the cycloidal model, in which an expansion is followed by a contraction which brings the Universe through a state of very high density, and then by another expansion. All three models give rise to predictions which are very difficult to check against the observed facts, because of the large margin of error in the observations. These are largely of a statistical nature, that is they consist in counting, by means of telescopes and radiotelescopes, the number of galaxies and radiogalaxies in a given zone of the sky and then extrapolating this result to obtain an average density of the universe, or its density at a given distance.

Such observations have been made carefully over and over again in the past ten years by astronomers and radio-astronomers of many countries, and the results tend to favour the big bang theory over the steady state theory, which postulates a uniform universe not only in space (the 'cosmological principle') but also in time (the 'perfect cosmological principle'). A parameter which could help to make a choice between the three models is the so-called 'deceleration parameter', or the slowing down of the rate of expansion; its observed value is again dependent on statistical observations, the results are not sufficiently accurate to be able to discriminate between the three models.

An argument often used by opponents of the steady state theory is that it violates the conservation laws which underlie all known physical laws; to this the followers of that theory reply that the big bang theory merely removes this violation to an instant in the past (the origin of the Universe). Since, however, the initial state of the universe should be a 'singularity', that is a state of infinite density, it is far from clear whether one can speak of a violation of physical laws, which would break down in any case in a singularity. It has been recently shown, using the theory of general relativity, that matter of very high density collapses unavoidably in a state of singularity; this has been used as an argument against the cycloidal model, since a contraction of the universe would necessarily bring it in a state of very high density, and then into a singularity, with the result that physical laws would be completely unable to predict its state after the singularity (or before the singularity, if one tries to extrapolate back in time to the situation existing before it).

The argument has recently been advanced that, since general relativity is as yet a classical and not a quantum theory, these singularities could be eliminated if one took into account the quantum fluctuations of the gravitational field. This is now a very active research field. Today, microphysics and macrophysics intermesh very closely in cosmology especially since the discovery of very high density objects such as neutron stars (pulsars) and very high energy sources (compact galaxies and quasars), in which the effects of gravitation and of nuclear forces compete with each other. It has been shown by British scientists that, even under broad assumptions, any model of the Universe is bound to collapse to a singularity, even if quantum conditions are taken into account. Collapse at the end seems thus inescapable.

Origin of the elements

One of the basic facts to be explained by any cosmological model is the relative abundance of the chemical elements in the Universe. Evidence on this point comes from the following sources: the Earth's crust, meteorites, cosmic rays, the Sun and other stars (through absorption spectroscopy), gaseous nebulae and the interstellar medium (by spectroscopy in the radio band). In recent years samples of the Moon's crust have been brought back to Earth by astronauts and automatic vehicles, and automatic sampling of the crust of other planets, especially Mars, is likely in the future.

The evidence gathered so far can be summarized as follows: in the Galaxy, about 91 per cent of the atoms are hydrogen atoms, 8 per cent are helium atoms and less than 1 per cent heavier atoms. The 1 : 11 ratio between helium and hydrogen is thus a crucial test of any cosmological model, which must be able to account for it. The first calculations made for the big bang model gave a higher percentage of helium than the correct one, and the model had to be refined in order to arrive at the right ratio. According to this model the Universe started from a fireball of very high temperature which cooled by expanding very rapidly; during the cooling the elements were created, first from elementary particles and then through nuclear fusion reactions. The initial stage of the big bang can be understood only in terms of particle physics and nuclear physics, and thus this model is strongly dependent on the knowledge attained in these fields. For instance, a recent suggestion made by Soviet astrophysicists that neutrinos might be very important in the early stages of the Universe's evolution, playing a role in converting neutrons into protons, might lower the helium abundance expected from the big bang theory.

As for the heavier elements, it has been shown that they cannot have been synthesized in the big bang, and it is now generally accepted that they are produced in the interior of stars by nuclear processes, during the star's evolution. Since there are so many stars, including the Sun, which can be studied at different stages in their evolution, the processes of chemical element synthesis in the stars are much better known than those hypothesized in cosmological models of the early stages of the universe, and observed abundances are well explained. But even in this field unexpected things can happen, as when the progress of neutrino detection instrumentation revealed in 1970 that the Sun emits far fewer neutrinos than was expected from theory (Chapter 9). Since the Sun is obviously the easiest star to investigate, the new observations are very puzzling. One way out of this dilemma is to assume that the Sun is twice as old as previously thought, and thus has a much higher core temperature, yielding fewer neutrinos. However, the Sun is thought to be so well understood that other astrophysicists prefer to modify the laws of elementary particle physics by assigning a small, but finite, mass to the neutrino, which might then decay into other particles during the Sun-Earth trip and thus not register on the neutrino detectors which are put underground in order to shield them from massive particles. Any progress in the physics of the neutrino might thus become very important for astrophysics, and especially for the study of stellar evolution; but it might also have important implications for the cosmological models, since neutrinos are thought to play an important role in the early stage of the Universe's evolution.

Isotopic abundances in lunar and meteoritic samples have also been used to determine their ages, through the knowledge of the half lives of the different isotopes. This method has been extremely useful in the dating of the lunar samples brought back by different Soviet and American missions.

Further reading

ALFVÉN, H. Antimatter and cosmology. *Scientific American*, vol. CCXVII, no. 4, 1967, p. 106.

——. Kosmologhja i fizika. *Nauka i Čelovečestvo*. Moskva, Znanie, 1972.

——. Plasma physics applied to cosmology. *Physics today*, no. 2, 1971, p. 28.

——. Worlds-antiworlds: antimatter in cosmology. San Francisco, Freeman, 1966.

BERGMANN, P. G. The riddle of gravitation. New York, Scribner's, 1968.

BONNOR, W. The mystery of the expanding universe. London, Macmillan, 1964.

MERLEAU-PONTY, J. *Cosmologies du XXe siècle*. Paris, Gallimard, 1965.

——. Les bases de la cosmologie moderne. *La recherche*, no. 2, 1970, p. 143.

MERLEAU-PONTY, J.; MORANDO, B. *Les trois étapes de la cosmologie*. Paris, Laffont, 1971.

PEEBLES, P. J. E.; WILKINSON, D. T. The primeval fireball. *Scientific American*, vol. CCXVII, no. 6, 1967, p. 28.

SACHAROV, A. D. Simmetrija vselennoj. *Budušee nauki*. Moskva, Znanie, 1968.

SCIAMA, D. La renaissance de la cosmologie. L'observation. *La recherche*, no. 2, 1970, p. 149.

SCIAMA, D. W. *Modern cosmology*. Cambridge University Press, 1971.

TREDER, H. J. Gravitazija. *Nauka i Čelovečestvo*. Moskva, Znanie, 1970.

ZEL'DOVICH, Ya. B. 'Gorjacaja' model vselennoj i teorija Fridmana. *Nauka i Čelovečestvo*. Moskva, Znanie, 1967.

Matter

13

The elementary particles

The experimental background

Most experiments performed in elementary particle physics consist in accelerating particles in a particle accelerator and making them hit a target. The particles are either simply scattered by the target (elastic collisions) or new particles are created in the process (inelastic collisions); the latter are by far the more important since they often provide information about the most basic processes in the natural world, although the former have a significant, if non-limited, interest.

The creation of new particles of mass m requires the availability of the equivalent energy, according to Einstein's formula $E=mc^2$, where E is the energy and c the speed of light. Thus the search for new particles of larger mass requires the construction of more powerful particle accelerators, many of which were built in the 1950s and 1960s. The most important are the 28 GeV (1 GeV$=10^9$ electron volts) machine of CERN (Geneva, 1959), the 33 GeV machine of Brookhaven (United States, 1960), the 76 GeV machine of Serpukhov (U.S.S.R., 1967), the 400 GeV machine of Batavia (United States, 1972) and the 300 GeV machine in construction at the CERN laboratories in Geneva. The trend is evident but high energy accelerators are very costly and can be afforded only by highly industrialized nations. This fact has led, on one hand, to the search for less costly and more effective methods of accelerating particles, such as the collective effect accelerators and electron ring accelerators, which are now being actively developed, and on the other hand to a fostering of international co-operation, both within international scientific organizations such as CERN at Geneva and JINR at Dubna (U.S.S.R.), and among scientists of different countries.

One method which has been developed to make full use of existing particle accelerators is that of causing beams of particles to collide with each other head on, instead of colliding with a fixed target. This can be done in the storage rings', long tubes of almost circular shape in which a high vacuum is maintained and in which beams of particles are conserved for hours circulating freely until they collide head-on; the particles travel clockwise and counter-clockwise in the same tube in the case of particles and antiparticles (proton and antiprotons, or electrons and antielectrons) while in the case of identical particles (mainly protons and pustons) they travel in tubes intersecting at fixed points ('interesting storage rings').

Particle accelerators are built for fundamental research but they are now finding increasing numbers of applications not only in particle, nuclear and atomic physics, but also in chemistry, engineering and medicine: the use of pion, neutron and heavy meson beams has recently been proposed in the radiation therapy of cancer, instead of the X-rays, gamma rays and beta rays so far used, over which they may have distinct advantages. New accelerators are being built with an emphasis on beam intensity instead of beam energy, such as the so-called meson factories, capable of producing intense pion and ion beams which are very suitable for radiotherapy. Examples are the LAMPF machine at Los Alamos (United States), the SIN machine at Villigen (Switzerland) and the TRIUMPF machine at Vancouver (Canada).

Large though the energies achieved in accelerators may be, nature provides even more energetic particles in cosmic rays from space. New particles have been found which are the products of collisions between cosmic rays and the nuclei of atoms in the upper atmosphere; the search continues but the new trend is to install particle detectors in mines and tunnels, in order to study rare neutrino events without the noisy background of primary cosmic radiation (neutrinos will penetrate the Earth's crust over great distances while other radiations are quickly absorbed at the surface of the Earth). Solar neutrinos are being looked for in a gold mine in South Dakota, while experiments in the Mont Blanc tunnel and a Utah salt mine recently gave a first clue of the existence of a particle which had been postulated on theoretical grounds, the W boson, a possible quantum of the weak interaction, one of nature's four basic forces (see below). Such experiments require great skill and electronic devices of high complexity capable of selecting out the required particle from a background noise of other radiations: very few 'events' (particle reactions) are counted each day. The situation is in marked contrast with that existing in particle accelerators, where the particles produced are so numerous that computers are installed 'on line' to keep count of them, and to select the desired 'events'. Computers are also used to control the accelerators themselves, which are becoming more and more automated, and to protect scientists using them from dangerous radiations produced in the machines. For this reason a super-powerful

accelerator of 1,000 GeV, planned by the U.S.S.R., will be completely automated and has been called the 'cybernetic accelerator'. Particle detectors have also been developed to cope with the accelerators' improvement: they include both liquid hydrogen and heavy liquid bubble chambers, scintillation counters, Cerenkov counters, spark chambers, streamer chambers, proportional chambers, multiwire chambers and others. Superconducting magnets have been developed in order to give higher magnetic fields at lesser costs, and computers are widely used in accelerators' centres both to control the accelerators and to process the huge amounts of data obtained by them.

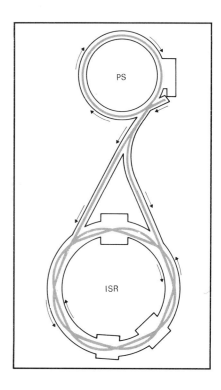

FIG. 25. The complex of accumulator rings with crossed beams (ISR) at CERN is composed, as shown in the diagram, of a ring 300 m in diameter containing two vacuum tubes intersecting one another at eight points. At these points, proton beams (moving through the two tubes in opposite directions) collide. (From CERN.)

Interactions

The four basic interactions (or forces) existing in nature are, in order of decreasing strength, the nuclear 'strong' force, the electromagnetic force, the nuclear 'weak' force and the gravitational force. By far the most important for elementary particle physics is the first one, since it is believed to be responsible for the very existence of elementary particles. Particles which participate in the strong interactions are called 'hadrons', as opposed to 'leptons', which do not. Hadrons can be divided into baryons and mesons, according to the value of their spin; baryons have spins values of $\frac{1}{2}$, $1\frac{1}{2}$, $2\frac{1}{2}$ etc., while mesons have integral spins (0, 1, 2, 3, etc.).

The basic problem of strong interaction physics is that of understanding the mass spectrum of hadrons, and this is still an unresolved problem. The electromagnetic interaction is well described by quantum electrodynamics, the most successful theory in particle physics, and has been confirmed by many experiments. The weak interactions produced a big surprise in 1956, when it was discovered that a fundamental physical parameter called parity (connected with left-right symmetry) was apparently not always conserved in weak interactions. A bigger suprise, however, was in store and in 1964 it was found that these interactions can also violate a yet wider symmetry, the CP symmetry, which is the product of C symmetry (consisting in the replacement of every particle by its antiparticle) and the P (parity) symmetry. Since the product of CP and yet another symmetry, the T symmetry (in which the direction of the flow of time is reversed) is held to be an exact or absolute symmetry by the so-called CPT theorem, the only way of maintaining CPT if CP is violated is that of postulating a violation of T in the opposite direction (others think instead that CT is violated, and not CP).

The violation of CP symmetry was discovered in the disintegration into two pions of the K^0_L meson, the 'long-lived' component of the K^0 meson, a particle which acts as if it were made up of two component particles, a short-lived one and a long-lived one. This decay is relatively rare, but a violation of T

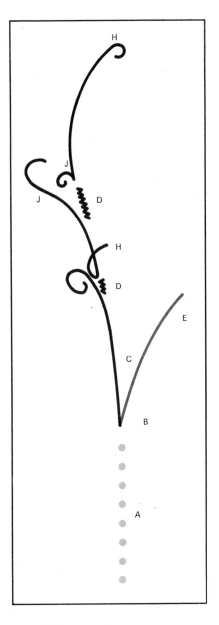

FIG. 26. Diagram of a 'neutral current' event obtained in the Gargamelle bubble chamber at CERN. An antineutrino (A) interacts (B) with a negative electron, producing another antineutrino (E), also another electron (C): an exceptional reaction indicative of a neutral current. The reaction also produces two high energy Υ-rays (photons indicated by D), which leave no photographic trace.

implies a violation of the time-honoured principle that basic physical laws (as opposed to the statistical laws of thermodynamics) are independent of the direction of time's flow. From these experiments, which have since been confirmed many times, it appeared that the laws underlying weak interactions are not time symmetric, at least not always. This would be such a profound modification in accepted physical thinking that the presence of a fifth 'superweak' interaction has been hypothesized in order to avoid a violation of T. Today, many scientists are looking for reactions in which time-reversal symmetry can be directly tested, without reference to the CPT theorem. All experiments performed so far, however, have confirmed T-symmetry. The problem is therefore still unsolved, and it is a very basic one. Are the laws of nature always time symmetric or not? Particle physics may eventually provide an answer.

The fourth basic interaction, gravitation, has so far been considered as unimportant in particle physics because it is so weak. Recently, however, there have been speculations that it may become important wherever matter has a very high density, for instance in the centre of neutron stars, where it can compete in strength with the nuclear or strong interaction. Moreover, attempts are being made, with encouraging first results, to relate it to the electromagnetic interaction, in order to eliminate some of the mathematical difficulties concerned with infinite quantities (divergences) in quantum electrodynamics. This research is related to the same problems which led Einstein to formulate his unified field theories, even if the mathematical framework is now different. Other attempts are being made to unify the electromagnetic and the weak interactions in an all-embracing scheme, with promising results; these efforts have been greatly encouraged by the discovery, in bubble chamber pictures, of reactions between particles of a kind specifically predicted by those theories, the so-called 'neutral current events', of which there was no evidence until the year 1973–74. Should it become possible to find some relationships between two or more of the basic interactions, it would become easier to answer the most fundamental ques-

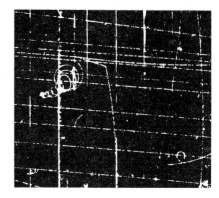

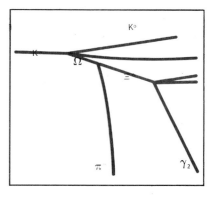

Fig. 27. An image of historical importance: the bubble chamber photograph (below, corresponding diagram) showing the first example of Ω^- particle to be definitely identified.

tion in particle physics: why are there four different interactions, and why are there so many particles?

Classifications and symmetries

The discovery of the Ω^- particle, seen as a track in a bubble chamber photograph taken in 1964 at the Brookhaven National Laboratory (United States), was one of the great scientific achievements of the 1960s. Since the particle's properties had been predicted in advance, this discovery is often compared to the discovery of a new chemical element on the basis of Mendeleyev's periodic table of elements; but what is the 'periodic table' of elementary particles?

At the 1959 International Conference on Elementary Particles held in Rochester (United

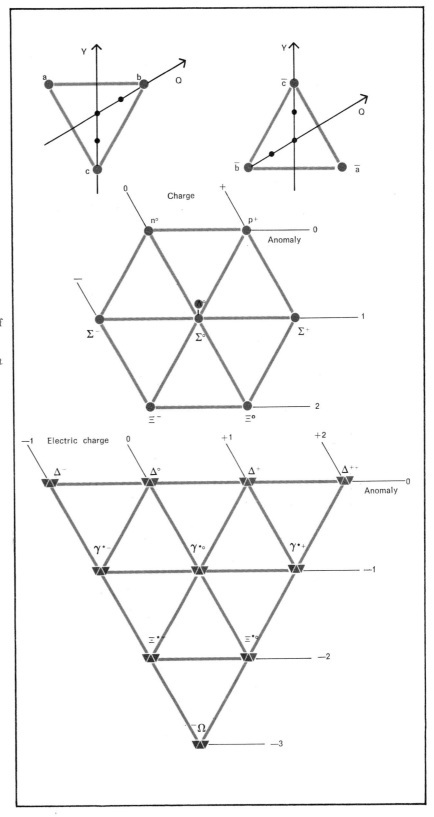

FIG. 28. The two triangles at the top of the diagram represent, in the plane Q, Y (electric charge, hypercharge) the triplet of the quark (left) and the triplet of the antiquark (right). Since the quarks have fractional charges it is possible, from these triangles, to construct the meson and baryon octets. The hexagon in the centre shows the eight baryons with 1/2-spin. Below, the baryon octet obtained, as we see, by combination in a variety of ways.

States), Japanese scientists from the University of Nagoya suggested that an empirical classification of particles on a two-dimensional plot, whose main axes were two 'internal quantum numbers', isotopic spin and hypercharge, had a deep theoretical reason behind it: the (approximate) invariance of strong nuclear interactions with respect to a new symmetry, called SU(3) symmetry. This suggestion was taken up and developed by other scientists, one of whom gave a form of this new symmetry the picturesque name of the 'eightfold way', referring to an aphorism of Buddha because the particles appear in the plots in patterns of eight called 'octets'. Other patterns include ten particles, and are called 'decuplets': it is at the bottom of a tenfold pattern that the Ω^- particle finds its place. It was the only unknown member of the pattern, and its properties were predicted on the basis of the properties of the other nine members of the group. The discovery of the Ω^- particle, exactly as predicted, caused great excitement because this suggested that all the properties of elementary particles might be explained on a purely theoretical basis: SU(3) symmetry is nothing but an extension of the SU(2) symmetry which expresses the fact that strong nuclear forces in the nucleus are the same for a neutron and for a proton.

Extensions of SU(3) were then investigated in the hope of including all elementary particles in an all-embracing scheme but this hope was frustrated, both because SU(3) symmetry is only approximately true and because higher and related symmetries are violated even more violently. But a lesson was learned in this attempt to classify the elementary particles: it is the weaker interaction which violates the symmetries of the stronger one, and since particles such as hadrons participate in all the four basic interactions, it is no wonder that SU(3) symmetry, which is relevant only to strong nuclear interactions, is violated. It is a basic law of nature that every symmetry gives rise to a conserved quantity, which is called 'charge': thus one speaks of electric charge, baryonic charge, leptonic charge and so on; if the particle is in motion one has a 'current', the most familiar of which is the electric current. In the course of research on SU(3)

symmetry, it was found that currents also fit into an SU(3) scheme and that, interestingly, electromagnetic and weak currents of hadrons can be described by simple formulae which include only terms related to strong interactions. This is called the algebra of currents, and is another hint that the strong, weak and electromagnetic interactions may be related to each other in a wider symmetry scheme which is still undiscovered.

An alternative approach to the classification of elementary particles is one which considers them as composites of each other, without any real 'elementary' particle. This is known as the 'bootstrap' theory, from the phrase 'lifting oneself up by one's bootstraps'. This approach derives from a mathematical treatment of scattering experiments, in which particles are made to collide with a target, called S-matrix theory, and also makes use of Regge theory. In the latter particles are represented as points on a line in a graph whose axes are mass squared and angular momentum; each line (known as a Regge trajectory) represents a family of particles, only some of which are known by experiments, while the search for others is going on in the accelerator laboratories. Particles of higher energy are known as 'resonances'; they should not, however, be considered as 'particles' but rather as excited states of the particle on the trajectory which has the lower mass, similar to the excited energy levels of an atom. The dream of every elementary particle physicist is to find an equation which describes the excited levels of mass for elementary particles, similar, at least in principle, to the Schrödinger equation for atomic levels. But a new problem is already being debated: are 'elementary particles' really elementary, or are they composites of even more 'elementary' subparticles? And, if so, where, if at all, are the really indivisible components of the atom to be found?

Quarks and partons

SU(3) symmetry is most naturally explained, from a physical point of view, if one postulates the existence of just three fundamental subparticles, called quarks. They should have a half integer spin and, most strangely, a fractional electric charge: one-third or two-thirds of an electron's charge. One baryon, such as neutron or proton, should be made up from three quarks, one meson from a quark and an antiquark. Quarks would then be the real 'elementary particles', from which ordinary matter is made up; they are being intensively looked for among particles deriving from nuclear reactions caused by high energy beams in particle accelerators and high energy cosmic rays. The quarks' signature ought to be a long track (that of a massive particle) but a thin one (because of its small charge) in a bubble chamber; so far no quark has been detected with absolute certainty, although the hunt is still going on. Since existing particle accelerators can create particles up to a finite limit of mass, the fact that no quarks have been found so far suggests that they might be very heavy, perhaps more than five proton masses. Thus a proton (whose mass is slightly less than 1 GeV, in energy units) should be made up from three particles whose total mass would amount to 15 GeV or more. This is clearly a weird situation, but is analogous to the situation already known for nuclei, where the mass of the nucleus is slightly less than the sum of the masses of its constituent nucleons (protons and neutrons). The missing mass (called the mass defect) goes into the binding energy which glues the nucleons together. The mass defect of a nucleon is instead enormous (about fourteen times its free mass!) and the binding energy is correspondingly huge; this is why quarks are so difficult to find.

But while the search for free quarks has met with little or no success, experiments performed since 1968 at Stanford University with the help of a 3 km long, linear electron accelerator, have produced results which are best explained if one postulates the existence, in nucleons, of three or more scattering centres, nicknamed 'partons'. The experiments, strikingly similar in purpose to Rutherford's scattering experiments with alpha particles in 1911, included the scattering of 20 GeV electrons by hydrogen nuclei in a liquid hydrogen target. The electrons, being leptons, are not influenced by 'strong' nuclear forces, and are thus able to penetrate deeply into the nucleons. Were the Stanford physicists seeing quarks in a bound state? Are partons quarks? The identification is not straightforward, since a nucleon, in the quark model, can be made up not only from three quarks but from an arbitrary number of quark-antiquark couples, which makes the problem mathematically extremely complex. During 1974 and 1975 new particles were discovered which may consist of quarks endowed with an additional conserved property, or 'quantum number', nicknamed 'charm'. This may point out the existence, for elementary particles, of still a wider symmetry than SU(3), namely SU(4).

The problem of the structure of elementary particles, such as nucleons, is at the forefront of elementary particle research today, even if some hold that there are no real 'elementary particles' and that a different approach to the problem of the constitution of matter may be necessary.

Another concept which developed from the Stanford experiments is scale invariance of physical laws, that is invariance with respect to a dilation or a contraction in physical units. This is a wider symmetry than that postulated by relativity, and may also play an important part in cosmology. Research in this area is being actively pursued by scientists from many countries.

Further reading

BARAŠENKOV, V. S. Sečenja bzaimodejstvija elementarnyh častic. Moskva, Znanie, 1966.

BARAŠENKOV, V. S.; BLOHINCEV, D. I. Problemy struktury elementarnyh častic. Dialektika i sovremennoe estestvoznanie. Moskva, 1970.

BLOHINCEV, D. I.; BARAŠENKOV, V. S.; BARBAŠOV, B. M. Struktura nuklonov. Uspehi Fizičeskix Nauk, vol. LXVIII, no. 3, 1969, p. 417.

CHEW, G. F.; GELL-MANN, M.; ROSENFELD, A. H. Strongly interacting particles. Scientific American, vol. CCX, no. 2, 1964, p. 74.

CLINE, D. B.; MANN, A. K.; RUBLIA, C. The detection of neutral weak currents. Scientific American, vol. CCXXXI, no. 6, 1974, p. 108.

DEVONS, S. Elementary particles: simple or complex? Science progress, vol. LII, no. 208, 1964, p. 543.

FORD, K. W. *The world of elementary particles*. London, Blaisdell, 1963.

FORD, S. D. Electron-position annihilation and the new particles. *Scientific American*, vol. CCXXXII, no. 6, 1975, p. 50.

GOUIRAN, R. *Particules et accélérateurs*. Paris, Hachette, 1966.

GOURDIN, M. Les symétries des particules élémentaires. *La recherche*, vol. III, no. 21, 1972, p. 243.

KENDALL, W.; PANOFSKY, W. K. The structure of the proton and the neutron. *Scientific American*, vol. CCXXIV, no. 6, 1971, p. 60.

MATTHEWS, P. T. *The nuclear apple*, London, Chatto & Windus, 1971.

POLKINGHORNE, J. Relativistic quantum theory and the S-matrix. *Science progress*, vol. LIX, no. 236, 1971, p. 551.

RYAN, C. Weak interactions of elementary particles. *Science progress*, vol. LX, no. 238, 1972.

SCHWARZ, J. H. Dual resonance models of elementary particles. *Scientific American*, vol. CCXXXII, no. 2, 1975, p. 61.

VARIOUS AUTHORS. *Philosophical problems of elementary particle physics*. Moscow, Progress Publishers, 1968.

14   The nucleus

Nuclear models

Since nuclear physics began many models have been proposed to describe the complex structure of the atomic nucleus, since its complete description starting only from the basic interaction between two nucleons is a task of formidable mathematical complexity. These models fall into two main categories, the independent particle models and the collective models, according to whether one considers the motion of the nucleus as a whole. The two approaches are based on completely different assumptions, yet it is a remarkable fact that both can account for many properties of atomic nuclei.

Recent research has shown, however, that a considerable unification of these models can be achieved if one considers an atomic nucleus as made up from a 'core' of nuclear material around which rotate one or more 'valence nucleons'. This model resembles that of the atom itself, with the nuclear 'core' being analogous to a noble gas in atomic theory. The independent particle model is called the 'shell model' of the nucleus because nucleons are held to belong to definite energy 'shells', like the electrons which circle around a charged nucleus. The number of nucleons on a shell is limited by the Pauli exclusion principle, as in atoms; a nucleus whose shells are completely filled is particularly stable, and unlikely to decay. This simple picture is able to account for the very stable 'magic' nuclei (those with 2, 8, 20, 28, 50, 82 and 126 neutrons or protons) if one considers the so-called spin-orbit interaction, which gives rise to only small effects in atoms but ones which become very important in nuclei.

Recent refinements in the shell model involve the grouping of valence nucleons in pairs and quartets acting as single particles, which are very important in the study of the excited energy states of nuclei. Some nuclei, for instance ^{24}Mg, exhibit what has been called a molecular configuration, with two groups of 12 nucleons moving relative to each other like two atoms in a diatomic molecule; other states of the same nucleus are like triatomic or tetratomic molecules. The shell model has been shown to be able to account for the shape of atomic nuclei, with magic number nuclei (closed shells) corresponding to spherical shapes and nuclei with unfilled shells to other, more complicated, shapes. Scattering experiments performed with high energy electron and protons have shown that the shell structure concept is valid not only in the outer part of the nuclei (where nucleons are less tightly bound) but also in the inner part of the core, where nucleons are tightly bound by the combined effect of the nuclear force and

FIG. 29. Scale shows distribution of nuclear orbits, each orbit having its own specific energy. Some of the resulting orbits are grouped. The number of nucleons of the same type present in the nucleus, once a certain shell and those of lesser energy are 'filled up', is indicated for each shell. (From University of British Columbia.)

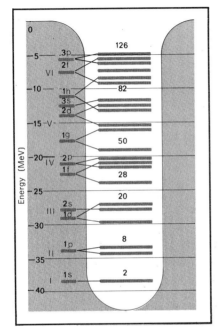

the Pauli principle (a nucleon cannot be excited to a nearby orbit if this is already occupied).

The shell model of the nucleus is generally accepted today, and is being continuously refined by scientists from many countries. Attempts have been made, however, to derive the properties of nuclei, such as the ^{186}Os, ^{188}Os, ^{190}Os and ^{192}Os isotopes, solely from the nature of the force between two nucleons as deduced from scattering experiments and from other reasonable assumptions. These calculations are very difficult because of the great number of particles involved, and require the use of great mathematical skill and high-speed computers. This line of approach to nuclear structure is very promising and represents a new level of sophistication in nuclear theory.

Nuclear spectroscopy and levels

The study of energy levels in nuclei involves the excitation by means of particles and ions accelerated by particle accelerators and the measurement of the energy of the radiation emitted during the return of the nuclei to their

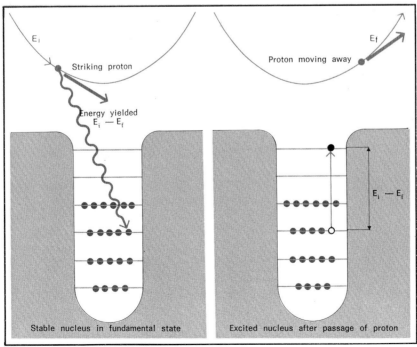

lower energy state. The study of the decay processes, which give rise to the emission of gamma radiation rather than optical, ultra-violet or X-radiation (as is the case for atomic energy levels) has been greatly facilitated in recent years by an improvement in gamma ray detectors of high sensitivity. They consist of crystals of pure germanium treated with lithium, and belong to the class of semiconductor detectors. The increase in accuracy of measurements performed with these new detectors with respect to those previously used (sodium iodide crystals coupled to a photomultiplier) varies from 10 to 100, thus opening up a whole new range of nuclear phenomena for study, and making it possible to study the fine structure of nuclear levels.

On the theoretical side, symmetry considerations based upon the use of group theoretical methods, similar in principle to those already used in atomic and particle physics, have come to play a very big part. One starts from the basic fact that the nuclear force between a neutron and a proton is the same if the neutron is replaced by a proton or the proton by a neutron; this fact is expressed by the isospin, or SU(2) symmetry (Chapter 13). Extensions of this symmetry are the SU(4) symmetry, postulated by E. Wigner in 1937, and the SU(3) symmetry of modern elementary particle physics. Both have been applied to the study of nuclear levels, even though they only apply approximately (for one thing, they neglect the Coulomb repulsion between protons). In 1972 experiments performed with heavy ion beams seemed to support a 'quartet' model of nuclei in which two protons and two neutrons act as a single valance nucleon: the quartet model could provide a physical basis for the SU(4) symmetry.

Another mechanism by which an excited nucleus can de-excite itself, rather than by the emission of gamma radiation, is by emission of a corpuscular radiation (alpha or beta particles). A new type of radio-activity, proton radio-activity, was discovered in 1971 for an excited state of ^{53}Co, which emits protons with a half life of 243 ms, which is very long by nuclear standards. The discovery was made at Oxford University and subsequently confirmed at the University of California; scientists at Moscow University then gave a theoretical explanation of this finding on the basis of present knowledge of the nucleus, demonstrating that this is the only isotope with a mass less than 80 for which such a phenomenon can occur. This is a proof both of the sophisticated level of our knowledge of nuclear matter and also of the degree of international co-operation now existing in this field.

Nuclear reactions

The study of nuclear reactions depends heavily on the use of particle accelerators specially designed for nuclear research, such as heavy ion linear accelerators, heavy ion cyclotrons and high flux particle accelerators ('meson factories'). High energy particle accelerators, however, have also been used in the search for supertransuranic nuclei such as element 112 and other, still heavier, nuclei. A tungsten target was bombarded with 24 GeV protons from a typical particle accelerator; after a cooling period of several months to reduce radio-activity, the sample was dissolved and analysed by chemical means. Since element 112, on the basis of the periodic table of elements, should be very similar to mercury in its chemical properties, the sample was analysed for the presence of mercury, and a spontaneous fission activity was observed in the mercury fraction. From this a half life of at least several weeks was deduced for element 112, but more experiments are still needed to confirm this result.

Reactions induced by heavy ions are very important both for the verification of nuclear models and the synthesis of superheavy elements, which then decay by the process of fission. An important discovery about fission was recently made by a Soviet group working at Dubna: the process is often a two-step one, because many heavy nuclei have two metastable configurations (one more stable than the other); these are called fission isomers, and the two configurations correspond to two different shapes of the nucleus. It is hoped that, through a better understanding of nuclear fission processes, it will be possible to increase the yield of neutrons in a nuclear chain reaction, with substantial increase in the power

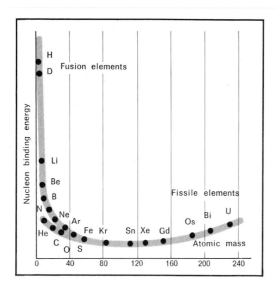

FIG. 30. Representation of the energy necessary for cohesion of the principal atomic nuclei: the higher the binding energy per nucleon, the better its quality as a nuclear fuel.

emitted in a nuclear fission reactor (especially in breeder reactors, where the flux of neutrons is needed to convert material in the blanket surrounding the core into fissile material).

Another type of research which makes use of particle and ion accelerators is the study of nuclei having an excess number of neutrons: with beams of high energy protons bombarding a suitable target the nuclei ^{8}He, ^{11}Li, ^{12}Be, ^{14}B, ^{17}C, ^{19}N and ^{21}O have been obtained; by bombarding a target of ^{232}Th with ions of ^{22}Ne accelerated to 7 MeV, 24 neutron-rich isotopes were synthesized and identified, from ^{24}Ne to ^{26}Ne, from ^{20}F to ^{25}F, ^{21}O to ^{24}O, from ^{17}N to ^{21}N and from ^{13}C to ^{18}C. The theoretical problem of the stability of nuclei with an excess of neutrons is important also for astrophysics, since it is relevant to the problem of nuclear matter made up solely from neutrons, as in neutron stars. One of the more sophisticated experiments in this field is being performed at CERN, in Geneva, by means of the ISOLDE apparatus (isotope separator on line) connected to the 600 MeV synchrocyclotron. The nucleus under study is that of mercury, and the techniques employed include 'optical pumping' (usually exploited in lasers),

and other very advanced methods. Isotopes under study vary from ^{205}Hg to ^{183}Hg; around ^{185}Hg the shape of the nucleus is seen to depart abruptly from its predicted form, and it probably changes from a spherical to a deformed shape. This is very unexpected, since Hg has an almost closed shell of protons.

Heavy ion accelerators are also used in the search for other possible supertransuranic elements, such as elements 126 and 164, predicted by the shell model of the nucleus to be stable against fission, but for which no known method of production exists in nature. This research is still continuing, but no definite results have been obtained as yet.

Applied nuclear physics

While the exploitation of the nuclear fission reaction is today an established branch of engineering, which aims to produce electric power in nuclear fission reactors, the feasibility of a nuclear fusion reactor for the purpose of energy production has yet to be demonstrated. The advantages of an eventual fusion reactor over its fission counterpart are so many, however, that research on it is being conducted throughout the world. The advantages are mainly that there is a much greater availability of nuclear fuel for a fusion reactor and that radio-active waste produced would be less dangerous. Furthermore, fusion reactors, unlike fission reactors, are not likely to produce materials that could be used in nuclear weapons.

There is no theoretical reason why controlled nuclear fusion (as opposed to uncontrolled fusion in H bombs) should not be attained; the major difficulty is that of containing a sufficiently hot plasma (see page 172) to trigger the D-D or D-T fusion reactions (D is ^{2}H, T is ^{3}H). Deuterium is contained in sea water in the proportion of 1 atom in 7,000, while tritium can be obtained from an initial D-D reaction. In order to contain the plasma, which is too hot to be contained by ordinary matter, magnetic confinement has been proposed and attempted in many different ways. However, the plasma must be dense enough, hot enough and contained for a given time in order to ignite a nuclear fusion reaction: these three conditions must be simultaneously fulfilled, and the quantitative requirements are accurately known (the so-called Lawson criterion states that the product of all three must reach a certain minimum). Although any two of these three conditions have been simultaneously fulfilled, there is as yet no nuclear fusion machine that can fulfil all three at the same time. The experimental apparatus which has come closest to the goal so far is the T-10 machine of the Soviet Tokamak type; other apparatus which has produced outstanding results are the 2XIIB mirror type (United States) and the Scylla IV of the theta pinch type (United States). Machines of the Tokamak type are actively being developed in the United States and Europe; another kind of machine, the Stellarator originally developed in the United States, is now being studied in the U.S.S.R. Research on nuclear fusion is thus being carried out in a truly international spirit which will help in overcoming the tremendous difficulty of the task.

Reaching the values required by Lawson's criterion would only be a first step on the road towards a practical fusion reactor. There are tremendous engineering problems, especially in heat transfer and in materials science, to be overcome. Another approach to plasma heating and confinement, which produced extremely good results in 1973, is that of laser heating. A beam of laser light of very high intensity is aimed at a target of frozen deuterated hydrogen. Its temperature rises so steeply that fusion reactions are triggered before the hydrogen plasma has time to escape. A more refined version of this experiment, developed independently by Soviet, American and French scientists, aims at the implosion of a spherical target by laser beams surrounding it from all sides and fired simultaneously at it:

the implosion combines with the heating to give very high density and temperature, as required by Lawson's criterion. The two approaches can be combined in such a way as to create a dense and hot plasma by laser heating, and then to contain it by magnetic fields, as in the machines already mentioned. This now looks like a very promising approach to the problem of controlled nuclear fusion.

Further reading

ARTSIMOVICH, L. The road to controlled nuclear fusion. *Nature*, vol. CCXXXIX, no. 5,366, 1972, p. 18.

BARANGER, M.; SORENSEN, R. A. The size and shape of atomic nuclei. *Scientific American*, vol. CCXXI, no. 2, 1969, p. 59.

BÈS, D. R.; SZYMANSKI, Z. The shape of the atomic nucleus. *Science progress*, vol. CCXVIII, no. 55, 1967, p. 187.

COHEN, B. *The heart of the atom.* New York, Doubleday, 1967.

DÉTRAZ, C. Les réactions nucléaires. *La recherche*, vol. VI, no. 56, 1975, p. 418.

EMMET, J. L.; MUCKOLLS, J.; WOOD, L. Fusion power by laser implosions. *Scientific American*, vol. CCXXX, no. 6, 1974.

GMITRO, M. Les modèles nucléaires. *Atomes*, vol. XXV, no. 272, 1970, p. 13.

IRVINE, J. M. Nuclear fision. *Science progress*, vol. LXII, no. 248, 1975, p. 501.

ISOBELLE, D.; RIPKA, G. La structure du noyau atomique. *La recherche*, vol. V, no. 46, 1974, p. 543.

KUZNETSOV, V. The synthesis and study of nuclei having an excess of neutrons. *Soviet science review*, vol. III, no. 4, 1972, p. 207.

LIDSKY, L. M. The quest for fusion power. *Technology review*, vol. LXXIV, no. 3, 1972, p. 10.

WEISSKOPF, V. *Physics in the twentieth century.* Cambridge, Mass., MIT Press, 1972.

15　Atoms and ions

Transuranic elements

The search for transuranic elements in recent years has followed two different directions: the search for elements with atomic numbers from 100 to 107, with lifetimes sharply decreasing with the increase of atomic number, and the search for 'supertransuranic elements', predicted by nuclear shell models, which lie on the so-called 'islands' or 'valleys' of stability, the first of which should be the element with atomic number 112. The first search has proved successful and, by using heavy ion accelerators, all the elements up to atomic number 107, with the exception of element 106, have been synthesized in the laboratory. However, there is still no unambiguous evidence of the supertransuranic elements, although they have been looked for in cosmic rays and in extraterrestrial material, such as meteorites and lunar samples, using refined means of chemical etching to show up the tracks left from their spontaneous fission.

Most work on the transuranic elements has been carried out at the Lawrence Radiation Laboratory in Berkeley, California, and at the Joint Institute for Nuclear Research, Dubna (U.S.S.R.). In 1964 and 1967 the Dubna group announced the synthesis of elements 104 and 105, produced artificially by heavy ion bombardment; the same elements were produced in 1969 and 1970 in Berkeley, and some controversy on the priority of the discoveries ensued, because the two groups used different methods to identify the elements, the Soviet group using spontaneous fission and the American one alpha radio-activity. In February 1971 the Dubna group, which has very advanced facilities, announced preliminary results (from ten to twelve events) which could indicate production of element 107. The controversy over the priority in the discovery resulted in the fact that there is no officially accepted name for elements after element 103 (lawrencium); for element 104 the names proposed are kurchatovium, by the Dubna group, and rutherfordium, by the Berkeley group; for element 105 the names proposed are respectively hahnium and bohrium. However, since the lifetimes of transuranic elements decrease so sharply with atomic number, they are becoming more and more difficult to detect and measure, so that priorities are very difficult to establish. This same controversy arose on the subsequent (1974) discovery, by the same two groups, of element 106. For this reason the naming of new elements will probably be abandoned by international authorities such as the International Union of Pure and Applied Chemistry (IUPAC) and the International Union of Pure and Applied Physics (IUPAP).

New elements will thus be identified solely by their atomic number.

Atomic spectroscopy

The study of the energy levels in atoms has been greatly helped in recent years by the use of new techniques borrowed from other fields of physics. Pure, energetic ion beams, originally developed for the study of nuclear structure, are used in the beam foil technique: excited atomic states are produced by an ion beam as it passes through a thin metal foil, and the excited atoms are made visible by the light they emit during their subsequent decay to the ground state. Knowing the speed of the beam, it is possible to measure the lifetimes of the atomic states, and hence the transition probabilities, simply by measuring the length of the path travelled by atoms, by time of flight techniques. Such methods have been recently used to measure the lifetimes of excited atomic states in neon, argon and iron; the last measurement has direct astrophysical implications, since the iron present in the Sun's photosphere is in an excited state, and the new measurements imply that it should be ten times more abundant than previously calculated, thus resolving a big discrepancy with the observed abundance. This is yet another example of the close relationship existing between atomic and nuclear physics on one side and astrophysics on the other.

The beam foil technique has also been used to test the applicability of atomic theories, both relativistic and non-relativistic, to different atoms. Hydrogen-like and helium-like heavy ions, having only one or two electrons left, were studied with the aim of testing the predictions of non-relativistic quantum mechanics: they turned out to be correct. Also tested in this case was the applicability of Dirac theory, which is a relativistic quantum theory, but not a quantized field theory like quantum electrodynamics (QED); this last, very refined theory was really only needed to explain one phenomenon of atomic physics, the Lamb hyperfine splitting in the levels of atomic hydrogen. But at the University of California in Berkeley another phenomenon has now been revealed by the ion beam foil technique which cannot be explained by Dirac theory but needs the full power of QED to be employed for its explanation: the decay of the 2^3S_I state of argon.

A second powerful method recently used to investigate the hyperfine structure of hydrogen involves laser light (see Chapter 19). The measurement of the exact frequency of a spectral line is vital in hyperfine spectroscopy; unfortunately, the motions of atoms in a gas give rise to a thermal broadening of the lines, since atoms travel in all directions and the light emitted by them is influenced by Doppler effects. A method called saturation spectroscopy has been devised which makes use of a tunable laser source to identify all atoms which travel in directions perpendicular to the laser beam, and hence do not suffer from a Doppler effect. If the frequency of the laser beam corresponds exactly to that of an atomic transition, light will be absorbed up to a point where a secondary, reflected beam of the same frequency will not be absorbed because all possible transitions have been saturated by the first. This is the signal that an exact, unbroadened, frequency has been achieved. By this method the Lamb shift has been measured in excited states of atomic hydrogen and in hydrogen-like ions, for instance in ionized oxygen. The advantage of using heavy ions lies in the fact that the magnitude of the Lamb shift depends strongly on the charge at the atomic nucleus, allowing easier measurements.

Exotic atoms

An important tool both in atomic and nuclear physics is the study of atoms in which an electron has been replaced by a heavier, negatively charged particle. This makes it possible to test existing atomic and nuclear theories, since an important parameter (mass) can be varied to test experimental results against those predicted by the theory. Such atoms are nicknamed 'exotic' atoms; the first one to be identified, in 1952, was the pionic atom, in which a negatively charged pion had replaced an electron. In 1953 muonic atoms were discovered at Columbia University, New York, and in 1967 kaonic atoms were discovered at the University of California, Berkeley; finally in 1970 sigmic and antiprotonic atoms were discovered at CERN in Geneva.

The study of such atoms requires the use of a powerful particle accelerator whose beam is made to collide with a target in order to produce a secondary beam, suitably filtered, of the wanted particles: pions, muons, kaons, sigmas, antiprotons, etc. Such secondary beams are then slowed down so that their particles are captured by atoms in a target, where they replace an electron; atoms then decay to a ground state as the exotic particle falls down the permitted energy levels, emitting X-rays (instead of the usual optical and ultra-violet radiation) because of the exotic particle's larger mass. Measurement of these X-rays by suitable detectors provides the test of theories of electric charge distribution in the nuclei (very important in nuclear physics) and of quantum electrodynamics at very small distances, since the exotic orbits are much closer to the nucleus than the electron orbits, again because of the exotic particle's bigger mass; in fact, they spend some of their time within the nucleus, unlike electrons, and some of them interact with it through the nuclear strong force (Chapter 14) as well as the electromagnetic force.

From the X-ray measurements X-ray spectra are obtained, similar in principle to normal atomic spectra, but usually complicated by the presence of the strong interaction. From these spectra information has been gathered on the mass of the antiproton (it was shown to be equal to that of the proton to within an error of 1 in 2,000), on the structure of the surface of the nucleus, and on the distribution of particles within the nucleus; these studies are continuing with the exotic atoms already mentioned, and there is hope that new kinds of exotic atoms will be obtained from new experimental facilities, such as the high flux accelerators, known as 'meson factories', which are being built in the United States, Europe and the U.S.S.R.

Free radicals

The study of chemical radicals, and in particular of free radicals, that is of neutral molecules with a broken covalent bond and an unpaired electron, was greatly accelerated in the 1960s by the use of magnetic electron resonance techniques (EMR). This consists in measuring the energy shift which occurs when the unpaired electron flips its spin from parallel to antiparallel with respect to an applied external field. When this happens different radicals produce different absorption lines and the study of the absorption peaks, and the 'hyperfine coupling constants' of the unpaired electron with nearby nuclei of nuclear spin other than zero, makes it possible to infer important facts about the structure and nature of the radicals. This technique has been used to solve important theoretical problems: the splitting of the absorption lines caused by a magnetic nucleus is related to the interaction between the wave function of the unpaired electron and the nucleus itself. Hence from the EMR spectrum one can deduce the distribution of the unpaired electron in a free radical, and this can be checked against the results calculated via the molecular orbital method (MO).

This technique of direct observation and identification of short-lived radicals is possible only if radicals are produced in relatively high concentration in the cavities of EMR apparatus (either through strong irradiation or by a rapid mixing of two or more solutions). Basic advances in the past decade were due to two methods which allow the identification of low concentrations of free radicals in reaction systems, namely spin trapping and chemically induced dynamic nuclear polarization (CIDNP). The spin trapping technique implies the trapping of a reactive free radical, which is very short-lived, through an addition reaction with compounds which produce a more stabler radical, detectable by EMR. Compounds used to this end are nitrous and nitroso compounds and nitrile oxides, which all produce relatively stable nitroxyradicals. Biological molecules have been investigated in this way; for instance, by attaching a 'spin label' to proteins, information can be deduced about their 'tertiary' structure.

The other technique, CIDNP, consists in the observation of an increased absorption or of an emission spectrum when the products of a very fast reaction, implying as intermediates some reactive radicals, are immediately observed by nuclear magnetic resonance (NMR) spectroscopy. This phenomenon was first observed in 1967, and has been recently theoretically explained as an effect of a strong polarization of some nuclear spins caused by the unpaired electron while the molecule exists as a free radical. On this basis reaction mechanisms (see Chapter 17) involving free radicals have been deduced for a great variety of reactions in which unpaired electrons have never been directly observed.

New techniques have thus provided a vast amount of new information in the chemistry of radicals, both in pure and applied chemistry and in other related fields such as biology. Future developments in this field are likely in biochemistry, where the intervention of free radicals produces both desirable and undesirable effects. Smog, to cite an important problem of modern cities, contains many free radicals, as NO, NO_2, ozone, etc.; penetrating radiations can also break chemical bonds and create free radicals; living in an atmosphere of high pressure oxygen, as may aquanauts and astronauts, can also give rise to free radicals in the process of respiration. Antioxidizers may perhaps be used to control these and other effects where free radicals are involved, perhaps even the combustion processes of petrol, where less noxious anti-knock additives could lead to cleaner exhaust fumes.

Further reading

HUGUES, V. W. Muonium atoms. *Scientific American*, vol. CCXIV, no. 4, 1966, p. 93.

JØRGENSEN, C. K. *Lanthanids and elements from thorium to 184.* New York, Academic Press, 1973.

METZ, W. D. Beam foil techniques: new approach to atomic lifetimes. *Science*, vol. CLXXVI, no. 4033, 1972, p. 394.

NIX, J. R. Predictions for superheavy nuclei. *Physics today*, vol. XXV, no. 4, 1972, p. 30.

ORGANESYAN, Y. The synthesis of element 105. *Soviet science review*, vol. II, no. 1, 1972, p. 23.

PRYOR, W. A. *Free radicals*, New York, McGraw-Hill, 1966.

——. Free radicals in biological systems. *Scientific American*, vol. CCXXIII, no. 2, 1970, p. 70.

SEABORG, G. T. *Man made transuranium elements.* Englewood Cliffs, Prentice-Hall, 1963.

WEISSKOPF, V. F. The three spectroscopies. *Scientific American*, vol. CCXVIII, no. 5, 1968, p. 15.

16

The structure of molecules

Theories of chemical structure

The most important problem in theoretical chemistry since 1928 has been to integrate Schrödinger's wave equation for a molecule or for more than one molecule. This amounts to the solution of the dynamics problem of how a number of bodies (charged nuclei and electrons) interact together; the exact solution is known only in the case of two bodies (the one nucleus and the one electron which make up the hydrogen atom). All molecular problems must therefore be solved by approximate methods which make use of simplifying assumptions. The first of these is the Born-Oppenheimer approximation, in which the nuclei are assumed to move much more slowly than the electrons, since they are so much more massive, and can therefore be treated separately: they are considered as fixed in the electron wave equation. The second step is to consider each electron in a molecule as independent of the others, and subject only to the fields of the nucleus and to an average field due to all other electrons. This second field can be calculated in a number of ways, the most common of which describes an electron molecular wave function as a linear combination, with appropriate coefficients, of electronic atomic wave functions and solutions of atomic Schrödinger equations (centred on one nucleus only). This method is known as molecular orbital-linear combination of atomic orbitals (MO-LCAO).

Once the average field is calculated, it can be used in the Schrödinger equation for one electron: the solution of this equation is then used again to calculate the average field in order to obtain a better approximation. The process is repeated until the wave functions reach a point where they cannot be bettered by further calculations; the field is then said to be self-consistent, and the method is known as the self-consistent field method (SCF), or Hartree-Fock method from the name of the physicists who devised it.

However, the mathematical complexity of the problem is so great that alternative methods, which make use of empirical parameters to be inserted in the calculations, are often used. One such approach, that of Heitler and London, which considers molecules as made up of atoms interacting through the coupling of pairs of electrons with opposite spins, is applicable only to simple molecules, and has now been more or less abandoned.

This situation has changed drastically in the past ten years through the use of high-speed computers in the numerical integration of the Hartree-Fock equations. The complexity of the problem rises sharply with the number of interacting nuclei and electrons: the num-

153

ber of separate equations to be integrated increases approximately with the fourth power of the number of interacting bodies. For instance, a calculation done in 1965 on a molecule of biological interest involved the computation and storage of the results of 2,400 million separate integrals: this could not have been done without the use of high-speed computers. At the end of the calculations, which may take several days, the computers display on a screen the shape of the molecular orbitals (or wave functions) squared, which gives the density of the electron cloud between the different nuclei of the molecule.

All the chemical properties of a molecule can, in principle, be calculated from the Schrödinger equation, through the Hartree-Fock method or its many variants, each labelled by a code name; even reaction rates can be calculated *a priori*, at least in some cases. Chemistry by computer is thus becoming a more and more widespread subject, and new compounds can even be discovered by the use of theoretical methods; their existence is then verified in the laboratory, under the conditions predicted by the theory.

Methods of determining molecular structure

The structure of a molecule can be determined at many different levels. Besides the empirical formula and the molecular weight, which are the basis for any further knowledge, the determination of the way atoms are bonded allows one to write down a planar idealized picture called the structural formula. To such a planar picture, one often needs to add the configuration of an atom or group of atoms, that is its position in relation to a given molecular plane. The actual shape of a molecule can be depicted only by a three-dimensional stereochemical formula, which shows the arrangement of the atoms in space and possibly the most stable among various possible arrangements. For asymmetric molecules, a further distinction can be made between optical isomers. One can then proceed even further, and try to describe the electronic structure of the molecule.

No single method can provide all this structural information, except in a few favourable cases of rather simple molecules. Over the past decade new chemical methods have been added to the classical ones. The advent of 'soft chemistry' methods has made it generally easy to obtain suitable derivatives of unknown compounds, either to obtain information on functional groups or to confirm a proposed structure. However, the real breakthrough in structural analysis has been the development of physical methods, a few of which have grown so much in importance that they are now almost indispensable tools for chemists.

The most effective method for the elucidation of the structure of molecules of crystalline compounds is X-ray diffraction. It is an absolute method, establishing the arrangement in space of all atoms, since X-rays are scattered by the electrons of an atom, so that the regions of high electron density can be associated with the positions of the atoms.

Recent improvements in the technique have greatly enhanced its effectiveness, allowing for the fast collection and processing of diffraction data. Difficulties in detecting diffractions from lighter atoms (such as hydrogen) can be overcome by neutron scattering techniques.

The next most important physical methods are the spectroscopic techniques based on selective absorption of electromagnetic radiation by molecular systems. The electronic spectra (ultra-violet and visible) and the vibrational spectra (infra-red and Raman) are still widely used; a recent improvement is the use of laser radiation to obtain Raman spectra.

The nuclear magnetic resonance (NMR) spectra of compounds containing 'active' atoms such as ^1H, ^{13}C, ^{19}F, and ^{31}P give detailed information on the number of such atoms as well as on their 'local structure'. NMR is an extremely powerful tool for structural analysis.

These techniques are now complemented by mass spectrometry, a non-optical method based on fragmentation of molecules by electron impact and analysis of the fragmentation pattern. Since mass spectrometry is a sensitive and fast technique it can be effectively coupled with an efficient and rapid separation method such as gas chromatography by which the mass spectra of all the components of a vapour can be recorded separately in less than one hour.

Numerous other physical properties can also be measured to study the special structural features of a molecule. Of these, optical rotation (especially in its more complete forms of optical rotatory dispersion and circular dichromism) discloses subtle stereochemical differences between optical isomers of asymmetric molecules. Electron spin resonance (ESR), a magnetic phenomenon like NMR, provides information on the local structure of paramagnetic molecules and free radicals.

Photoelectron spectroscopy (PES) gives direct information about the distribution of electronic densities and the energy of orbitals. In an indirect way this technique can also reveal the oxidation state of a metal in a compound; from this point of view it is similar to Mössbauer spectroscopy with which the nuclear resonance of various nuclei can be studied using radio-active atoms incorporated in a crystalline structure as γ-ray sources of very precise wavelength; the structure must be capable of uniformly absorbing the recoil energy of the γ-ray emission. The Mössbauer effect has made it possible to study the nature of chemical bonds, the electric charge distribution around the nuclei and the structure of the molecule.

There is little doubt that unknown chemical structures will be determined in the near future by routine procedures involving gathering of the largest possible amount of spectroscopic data and processing them by computer techniques.

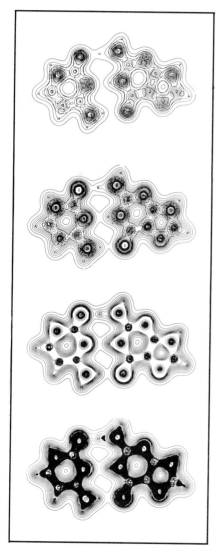

Fig. 31. Electronic density charts for two interacting molecules (guanine and cytosine). The schemata were obtained from parallel sections, then transferred vertically by 0.2 atomic units each (from top to bottom).

New chemical structures

Research over the past twenty years in the field of the organic chemistry of aliphatic, carbocyclic, aromatic and heterocyclic compounds has cleared up so many of the traditional problems of organic chemists that they have sought new perspectives and new interests. As a result a new subject of study, organic physical chemistry, has sprung up, while organic chemists interested in synthesis have devoted their activities, particularly recently, to the preparation both of natural molecules with extremely complicated molecular structures (such as the corrines, including vitamin B_{12} and the chlorophylls), and of geometrically unusual structures. Compounds belonging to this last class are those whose molecular structure probably represents the limit of the imagination of chemists to organize the tetrahedric valencies of the carbon atom into

155

situations that sometimes look more like a bet between the chemist and nature than a substantial experimental or theoretical contribution to chemical knowledge.

The synthesis of the cubane has now been successful while so far attempts to synthesize the tetrahedrane have failed; the prismane and 'Dewar's benzene' have also been synthesized. It has thus been shown that these structures are new and well defined chemical individuals which are isomers and not mesomers of Kekulé's benzene.

The concept of aromaticity, which was previously known only in the presence of $4n+2$ π-electrons in benzenoid systems, has now been extended to many non-benzenoid carbocyles, and several aromatic compounds called annulenes (with 10, 14 and 18 π-electrons) have been proposed.

A great deal of experimental evidence which shows that certain compounds containing $4n$ π-electrons are not aromatic has also been collected.

Several aromatic knot or ansa-systems (*ansa*, in Latin, means handle) called cyclophanes have been synthesized and their spectroscopic properties thoroughly studied; catenanes and rotaxanes are also examples of ansa-systems.

Chemistry of organometallic and co-ordination compounds

Rapid developments have been made in the past decade in the field of the co-ordination chemistry of metals in low oxidation states. This has been due mainly to the fact that new and sophisticated techniques (see page 155) for studying these compounds became much more widely available in research laboratories. For example, infra-red and Raman spectroscopy have played a decisive role in the development of the chemistry of the carbonylic compounds of transition metals. These techniques made it possible to draw rapid conclusions about molecular symmetry and the nature of the bond between metal and ligand. However, their importance is now decreasing, especially in examining complex molecules, because the time required for an X-ray diffraction study

is decreasing with the use of new diffractometers coupled with computers. Other techniques, such as NMR and mass spectrometry, which until a few years ago were peculiar to organic chemistry, are now widely employed in the study of organometallic and co-ordination compounds. Photoelectron and Mössbauer spectroscopy are also being increasingly used.

At present two research areas are likely to develop rapidly in the future, both theoretically and practically: the study of reactions in an homogeneous phase catalysed by complexes and the study of atmospheric nitrogen fixation by complexes.

Homogenous catalysis by means of transition metal complexes has produced new methods of organic syntheses and provided a comparative data base for studies on heterogenous catalysis. Many carbonylic complexes can be used as catalysts but cobalt carbonyl and its triphenylphosphinic derivatives are still the most commonly employed in industrial synthesis.

Ziegler-Natta catalysts, which permit polymerization reactions at low pressures, are very complicated. Recent studies indicate, however, that the active centre of polymerization is the transition metal (the Ti^{+3} ion in scheme 2), while the alkylating agents act essentially as co-catalysts.

The chemistry of transition metal complexes in which molecular nitrogen is the ligand has advanced rapidly, largely because these complexes represent a good model for studying biological nitrogen fixation as well as chemiadsorption and hydrogenation on the catalytic surfaces of several metals. The number of these complexes known now is very high, and the nitrogen can usually be easily replaced by hydrogen:

$$Ph_3Co(N_2)H \underset{H_2}{\overset{N_2}{\rightleftarrows}} Ph_3CoH_3.$$

In a few cases the nitrogen can be activated in this way, being then reduced to ammonia by hydrogen at low pressures.

Interest in research on biological processes has thus opened up a new area in which the reactivity of comparatively simple inorganic molecules—which have been synthesized and can be used to model biological systems—is studied.

Chemical structure and the behaviour of matter

The most recent studies of systems of associated atoms and statistical studies of the whole molecule have given rise to mathematical models which can predict the chemical, physical and physico-chemical behaviour of substances in solid, liquid and gaseous states, as well as the conditions under which transitions among the different phases occur. Often, such characteristics as internal energy, enthalpy, entrophy, free energy, specific heat, compressibility, viscosity, surface tension, thermal conductivity and the diffusion coefficient can be estimated in this way for both pure substances and mixtures.

Empirical correlations, which are really simplifications of theoretical models, have only a narrow range of applications and, generally, cannot be used for extrapolation. The theoretical correlations are more general, but the results are imprecise because of the approximations needed to simplify the calculations. Today, semi-empirical studies, in which experimental quantities are correlated with generally valid theoretical data, seem to be the most useful.

There are several calculation methods, most of them based on the application of the law of corresponding states. Of the latter, methods for calculating the physical properties and behaviour of substances on the basis of the atoms, the group of atoms and the different types of bonds present in a molecule have developed fast because they have important applications. A general survey of these methods has been prepared for gases and liquids and for crystals, liquids and glasses. The results obtained are valid enough for technological applications if the substances considered are not polar. For strongly polar substances, and for substances forming hydrogen bonds, the results are not always satisfactory. There are no valid general techniques of calculation, at present, for mixtures of polar substances.

Recently, the use of computers in theoretical chemistry (see Chapter 17) has greatly simplified techniques for calculating molecular structure. For simple molecules, with few atoms, very precise distances, angles and bond energies of the molecule can be obtained theoretically by a completely *a priori* calculation. Moreover, this can be done for all the different activation states of the molecule.

A branch of research which is likely to develop in the future is the possible means of progression from the properties of a molecule to those of an aggregate of molecules by means of *a priori* calculations. This type of calculus requires that the molecular interaction potentials can be expressed mathematically. Valid theoretical expressions for the attractive component of molecular interaction potentials have been known for some decades but for the repulsive term there are still only empirical expressions devoid of any theoretical content.

New experimental techniques, such as 'molecular beams', may soon provide a deeper knowledge of intermolecular forces. If molecular interaction energies could be calculated theoretically, we would then be able to predict the behaviour of matter at both the microphysical and the macrophysical level.

Structure of molecules of biological interest

Biological compounds are classified either according to chemical criteria (lipids, carbohydrates, proteins) or according to their biological function (vitamins, hormones, enzymes). The continuous improvement of methods for investigating the structure of organic compounds has now led to the elucidation of the structure of most small and medium-sized biological molecules. Present knowledge of these molecules often extends to their stereochemistry, that is to the arrangement of their atoms in space.

On the contrary, the detailed structure of large biological molecules (including most important biopolymers) is still largely unknown. The elucidation of the structure of a few important biopolymers has, however, been made possible by X-ray diffraction studies. The determination in 1953 of the double-helix

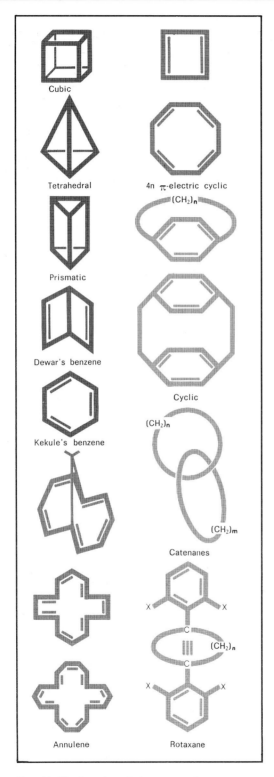

Cubic

Tetrahedral

4n π-electric cyclic

Prismatic

$(CH_2)_n$

Dewar's benzene

Kekule's benzene

Cyclic

$(CH_2)_n$

Catenanes

$(CH_2)_m$

Annulene

Rotaxane

$(CH_2)_n$

Fig. 32. Explanation of the names of many organic chemical compounds synthesized in the laboratory in recent years is made clear by examination of their molecular structures.

structure of deoxyribonucleic acid (DNA) is currently thought to mark the beginning of modern molecular biology. Since then, the detailed structures of other biopolymers have been outlined and refined by X-ray techniques (see Chapter 18).

Once a number of these structures was established, the way was open to a better understanding of the nature of the secondary forces which hold together atoms belonging to different units along the polymer chain or to units on adjacent chains. Among such forces, hydrogen bonds between electropositive hydrogen and electronegative atoms (O, N and S) play a major role in determining the shape of the polymer chain, by stabilizing it intramolecularly (as in the α-helix of proteins or in the V-helix of the carbohydrate amylose) or intermolecularly (as in the 'pleated sheet' structure of fibroin and in β-keratin). Helical chains can wind up further to form more stable double or triple helices, held together by covalent bonds (disulphide bridges), by hydrogen bonds or merely by van der Waal interactions between non-polar groups. Van der Waal interactions have been shown to play an important role in stabilizing the secondary and tertiary structure of biopolymers. In globular proteins, the non-polar groups are often situated on the internal surface of the macromolecule. In solution, such groups are held there by the so-called 'hydrophobic interactions', which are caused by the tendency of the water molecules to segregate from the aqueous environment any hydrophobic moieties of the polymer.

Since the first successful determination of secondary structures, it has become apparent that the shape of an isolated macromolecule can suggest its bulk physical properties: helical chains are used by nature like springs to build elastic structures; intermolecularly hydrogen-bonded elongated chains cannot be stretched and are used as structural supports: branched chains often form reserve materials whose moieties are readily accessible to enzyme attack.

The interest in the relationships between the shape and function of biological molecules has stimulated a number of theoretical approaches to predicting the secondary and also the tertiary structure of biopolymers. Calculations of the overall attractive and repulsive interaction between non-directly bonded atoms along the chain have produced computer-built models that often turn out quite similar to the structures experimentally determined by X-ray diffraction. These theoretical calculations have in turn proved useful in X-ray diffraction, since it is easier to compare experimental diffraction data with those calculated for reasonable models than it is to work out a completely unknown molecular structure. Several groups of X-ray crystallographers are at present working on the structure of biological macromolecules and the structure of a large number of biopolymers amenable to this approach—those that can be crystallized or have some degree of order in the solid or in the gel state—will probably be established within a few years.

Knowledge of the molecular structure of an ordered state does not usually provide information on the actual shape of the molecule in its natural aqueous environment. Although spectroscopic methods have thrown some light on special structural features (such as degree of 'exposure' of aromatic amino acids in proteins, the geometry of some important hydrogen bonds and the nature of specific interactions between enzymes and other molecules), present knowledge of the shape of biological macromolecules in solution is rather limited. NMR spectrometry is expected to provide the most reliable information on the secondary structure of biopolymers in solution, as well as information (at the atomic level) on physicochemical interactions betweens biopolymers and other molecules.

Chemical structure and pharmacological activity

Studies on the relationship between the structure and the activity of a drug are undertaken so that its original chemical structure may be changed either to improve its activity or to decrease its unwanted side-effects. This work therefore depends on some knowledge of the active chemical group involved, and sometimes of the configuration of the receptor on which the original drug acts and on which the modified one should act.

Since structural variations bring about new physical properties and alter the reactivity of a molecule, even an apparently minor modification in the chemical structure may result in major changes in its pharmacological properties and may uncover biological effects which were masked by unwanted effects in the original material. Moreover, as changes in the molecular configuration must not alter any of the actions of a drug, it is sometimes possible to develop a congener with more suitable properties than the original drug.

Good recent examples of correlation between chemical structure and biological activity are provided by antibiotics. A typical example is that of semisynthetic penicillins derived from a structure which is common to all penicillins, namely 6-aminopenicillonic acid, and which, unlike the mother product, can be taken orally and has a wider antibacterial spectrum. Rifamycins are also a relatively new antibiotic produced by the natural fermentation of *Streptomyces mediterranei*. The compound called Rifamycin B has proved to be the most active. It was found that the compound became more active in water solution and with ageing. The increased antibacterial activity arose in water solution when a glycolic acid was split off from the original compound by spontaneous hydrolysis (Rifamycin S). The new drug proved to be more active but also more toxic.

Hydrogen reduction of this compound produced Rifamycin SV, a widely employed antibiotic for treatment of gram-positive infections. Rifamycin SV is more active, less toxic, more stable in solution and more soluble in water than Rifamycin B.

Attempts were then made to increase oral absorption and effectiveness against gram-negative bacteria. More than 600 new compounds were synthesized: they included hydrazones of Rifamycin SV of which a piperazine derivative was found to have the desired therapeutic properties. These semisynthetic Rifamycins are now being investigated for their effectiveness in blocking 'reverse transcriptase', an enzyme specific of some tumour-inducing viruses (see Chapter 21).

Data on the relationship between structure and function has also been used to synthesize antimetabolites, some of the most widely used anticancer drugs. Antimetabolites are substances which differ only slightly from the fundamental molecules of cellular metabolism. Their chemical similarity to molecules occurring naturally in the cell means that they become incorporated into the structure of nucleic acids, altering their properties irreversibly. On the other hand, such antimetabolites can also bind themselves to essential enzymes, inactivating them and causing cell death. The search for anticancer drugs has also been focused, for example, on the structure of azoyperite (nitrogen mustard); variations on the original structure have produced a series of drugs with specific properties and a relatively low toxicity.

Such research has produced many hundreds of cytotoxic drugs. Unfortunately, it has not yet been possible to isolate either a molecule or a biochemical pathway which leads specifically to the formation of tumour cells. This is why tumour-specific drugs are not yet available. But research in this general area is likely to become more and more important in the future. Because small variations in structure can lead to wide variations in pharmacological activity, the possibilities seem almost unlimited. But what is needed is a better theoretical understanding of the problem so that specific drugs can be produced 'to order' and not by chance.

Further reading

ATHERTON, N. M. Electron spin resonance of organic free radicals in solution. *Science progress, Oxford*, no. 36, 1968, p. 179.

BARNER, R. *Organic chemistry of biological compounds*. Englewood Cliffs, N.J., Prentice-Hall, 1963.

BOVEY, F. E. Nuclear magnetic resonance. *Chemical and engineering news*, 1963, p. 30.

BREY, W. S. *Physical methods for determining molecular geometry*. New York, Reinhold, 1963.

VON BUNAU, G. Application of computers to M.C. theory and chemical kinetics. *Angew. Chem. Int. Ed.*, no. 11, 1972, p. 393.

COTTON, F. A. *Chemical applications of group theory*. New York, Interscience, 1971.

COTTON, F. A. A.; WILKINSON, G. *Advanced inorganic chemistry*. New York, Interscience, 1966.

CRIEGEE, R. Structure insolite in chimica organica. *Chimica e industria*, no. 53, 1971, p. 277.

DICKERSON, R. E.; GEISS, I. The structure and action of proteins. New York, Harper & Row, 1969.

DYER, J. H. *Applications of absorption spectroscopy of organic compounds*. Englewood Cliffs, N.J., Prentice-Hall, 1963.

GRAY, H. B.; HAIGHT, G. P. *Basic principles of chemistry*. New York, Benjamin, 1967.

HENRICI-OLIVÉ, G.; OLIVÉ, S. Non-enzymatic activation of molecular nitrogen. *Angew. Chem. Int. Ed.*, no. 8, 1969, p. 630.

HERBER, R. H. Mossbauer spectroscopy. *Scientific American*, vol. 223, no. 4, 1971, p. 86.

LAMBERT, R. J. The shape of organic molecules. *Scientific American*, vol. 222, no. 4, 1970, p. 58.

LINNETT, J. W. Chemical bonds. *Science progress, Oxford*, no. 60, 1972, p. 1.

MALATESTA, L.; CENINI, S. Recenti aspetti della chimica dei composti di coordinazione. *Chimica e industria*, no. 53, 1971, p. 1047.

MALOZEMOFF, A. Chemistry by computer. *IBM research reports*, no. 4, 1968, p. 2.

MULLIKEN, R. S. Spectroscopy, molecular orbitals and chemical bonding. *Science*, no. 157, 1967, p. 13.

OLBY, R. The macromolecular concept and the origin of molecular biology. *Journal of chemical education*, no. 47, 1970, p. 168.

RANDALL, E. W. Carbon–13 nuclear magnetic resonance. *Chem. Brit.*, no. 7, 1971, p. 371.

REES, D. A. Shapely polysaccharides. *Biochemical journal*, no. 126, 1972, p. 257.

SCHILL, G. *Catenanea, rotaxanes and knots*. New York, Academic Press, 1971.

SLATER, J. C. The current state of solid state and molecular theory. *International journal of quantum chemistry*, no. 1, 1967, p. 37.

SMITH, B. H. *Bridged aromatic compounds*. New York, Academic Press, 1964.

SNYDER, J. P. *Non-benzenoid aromatics*. New York, Academic Press, 1969.

SWEIGART, D. A.; DAINTITH, J. Molecular photoelectron spectroscopy. *Science progress, Oxford*, no. 59, 1971, p. 325.

WAHL, A. C. Chemistry by computer. *Scientific American*, vol. 222, no. 4, 1970, p. 54.

WOODWARD, R. B.; HOFFMANN, R. Die Erhaltung der Orbitalsymmetric. Weinheim-Bergstrasse, Verlag Chemie, 1970.

17 Chemical dynamics

Reaction mechanisms

Early studies of chemical reactions were limited to identify the kind and quantity of the end-products. Actually few reactions are simple since reaction paths are usually complex, and proceed in a series of distinct steps. The elucidation of the nature of the transient species or intermediates formed during a reaction reflects the main effort in the field of the studies of reaction mechanisms. One of the most important recent advances was the application of the most sensitive analytical methods to the detection and identification of reaction intermediates, even in small amounts. Gas chromatography, mass spectrometry, NMR and ESR spectroscopy (see Chapter 15) played particularly useful roles in this work. Concentrations as low as a few parts per billion can now be easily detected and identified by such techniques. This fact, together with our knowledge of the dependence of reaction rates on experimental conditions, allowed the correct formulation of some reaction mechanisms (i.e. the detailed description of the reaction path), even in some very complex cases.

The use of ESR spectrometry, for instance, allows the detection of all species with one or more unpaired electrons. Such radical species participate as intermediates in many reactions of the kind of homolytic cleavage of the peroxide linkage, thermal decomposition of metal alkyls, and peroxi-catalyzed polymerization of olefins (these reactions are very important also in industrial production of synthetic rubber and plastic materials).

Much effort has been made, also, in theoretical research concerning the analysis, by means of quantum chemistry, of the energetic paths of some reactions. It is worthwhile to mention the studies on the so-called electro-cyclic reactions in which certain olefinic systems give highly stereospecific rearrangements initiated by thermal energy and photochemical energy. The theoretical analysis of these reactions brought about the formulation of some inherently simple rules of particular interest to organic chemists. Other studies concern the molecular orbital description of the non-classical ion in 1,2-rearrangements, allowing the elucidation of the main characteristics of such reactions.

The study of reaction mechanisms is very important from a theoretical standpoint, but in many cases leads to a better understanding and realization of industrial chemical processes.

Chain reactions

In many reactions the rate is much greater

161

than what is expected in terms of any simple collision theory. In these cases, the presence of an 'initiator' can prime a chain reaction mechanism in which a given species (e.g. a radical molecule or an ion) reacts and continuously reforms very quickly.

Recent developments in the field of chain reactions are related to studies on the thermal cracking of paraffins, polymerization of olefins, diolefins, etc. Particular attention is focused now on thermal cracking of paraffinic hydrocarbons, with the aim of obtaining a suitable reaction rate model—useful in the design of industrial pyrolytic reactors. Reactors of this type are employed extensively in the production of the lower olefins (ethylene, propylene, butenes) and aromatics (benzene, toluene, xylenes) from petroleum fractions.

Other important chain reactions are combustion reactions and flames. In combustion reactions the chain propagation may take place in steady or non-steady conditions and, since chain reactions may frequently be very fast, sometimes they give rise to an explosion. The phenomenon may be caused either by the quick rise in temperature of the reacting mixture (steady state reactions), when the heat evolved is not sufficiently quickly removed, or by the fast branching of reacting chains, when one chain carrier gives rise to more chain carriers (non-steady state reactions).

Special attention in this research is focused on the material and thermal balances of flames, in order to determine the form of kinetic equations of the very fast reactions occurring in these systems. Special methods (e.g. photographic, or equilibrium perturbation techniques) have been developed in recent years for such studies.

Kinetics and catalysis

There are very few reactions which are simple, in the sense that they involve the decomposition of one molecule or the collision of two molecules to give reaction products. In certain circumstances, by-products or intermediates may be isolated only in small amounts. In recent years, efforts in the field of chemical kinetics have been devoted essentially to the detection and estimation of these low-concentration intermediates and particularly to the elucidation of their role in the general reaction pattern of a given process. The techniques of chromatographic analysis have greatly facilitated such researches.

From a different point of view, the use of computers has allowed the development of kinetic models suitable for the description of complex reaction networks. One example is offered by the reaction of n-butane on chromia-alumina. The compound can be dehydrogenated to n-butenes and butadiene, useful raw materials for plastics and synthetic rubber. Together with the main dehydrogenation reactions, isomerization and cracking reactions also take place. Careful gas chromatographic analysis of the reaction products has permitted us to determine the mechanism of the reactions.

The determination of the kinetics of catalytic reactions, particularly of industrially important ones, has been the subject of a very great number of scientific publications in recent years. New experimental microcatalytic reactors have been developed, in which the complications resulting from mass and heat transfer are minimized. Some newer techniques (e.g. the use of tagged molecules and catalysis poisoning) were introduced in these kinetic studies.

Particular attention has also been devoted to the development of techniques useful in the elucidation of the physical structure of catalysts, their surface area and pore geometry, the physical and chemical characteristics of solid surfaces and the nature of adsorbed species, including contact potentials, spectra of adsorbed molecules, acidity of surfaces, electron and field-ion emission microscopy, low-energy electron diffraction, ultra-high vacuum techniques, and so on.

Chemical kinetics are also very important in applied chemistry, in connexion with industrial reactors and their design and optimization. Since in the formulation of the complete kinetic model of an industrial chemical process many different disciplines take part, a general logic scheme must be followed.

The goal of the studies in such a complex field would be the complete reproduction of the overall process by means of a suitable mathematical model, which should simulate the behaviour of the chemical reactor and the changes of its performance resulting from the change in the values of the physical and chemical parameters involved.

Electrochemistry

Recent developments in the field of research in electrochemistry aim to deepen an understanding of the mechanisms through which some fundamental processes occur to emphasize the interdisciplinary character of electrochemistry itself. This is particularly so in what concerns both its superposition with other branches of physics (as, for instance, solid state physics and biophysics) and the use of methodologies peculiar to other fields of physical chemistry, such as statistical thermodynamics and chemical kinetics.

Analysis of the electrolytic gas generation mechanism can be, for instance, done either on the basis of what is already known about the processes of gas chemiadsorption on solid surfaces or by the employment of models proposed to explain the activation processes of gases in adsorption on solid surfaces.

Similarly, the study of electroplating mechanisms has implied a growing knowledge of the discharge mechanisms of metallic ions in contact with a solid surface and of the diffusion mechanisms of other adsorbed atoms on a metallic surface.

The study of the conversion of chemical energy into electrical energy also plays an important role in electrochemistry. Even from a historical viewpoint, in fact, the study of galvanic cells has always represented a wide field for inquiry. The attention of researchers, however, has been focused more recently on fuel cells.

Fuel cells differ from their galvanic counterparts only because electrodic reagents are both fed and withdrawn continuously from the electrochemical system. So fuel cells are more similar, in the way they work, to a heat engine—in which the fuel and its oxidizer are continuously introduced.

Research in the fuel cells field is partially connected with the ones first mentioned, especially in what concerns the relations between the mechanism of surface catalytic reactions and gas generation at the electrodes. In this case, the study of oxidation catalytic reactions is of particular interest.

Finally we must mention bio-electrochemistry, which concerns mainly the application of electrochemical principles and methods to the solution of biological problems.

A fascinating and very interesting problem concerns, for instance, the diffusion of inorganic ions (particularly sodium and potassium) through biological membranes. This problem is closely related to that of impulse transmission in the nervous system.

Ions transport through biological membranes also involves, as one of its components, a so-called 'active process' in which ion flux occurs in a direction opposite to that corresponding to the concentration gradient. The driving force in the active transport processes comes from the energy delivered in the degradation of organic molecules of high energy content (ATP); such phenomena can be described using the methods of irreversible process thermodynamics.

Radiation chemistry

Research in the field of radiation chemistry can be divided in two classes: (a) fundamental processes, including the analysis of interactions between radiations and molecules; (b) the mechanisms of reaction stimulated by radiolytic processes.

The first category aims chiefly to probe the physical nature of the process associated with the collision of a molecule with radiation and, above all, the possibility of describing this in quantitative terms by applying quantum mechanics.

For what concerns reaction mechanisms, the second category, it is extremely important to determine whether they are controlled by physical factors (as, for example, scattering) or if the reaction goes on through a chain mechanism.

Inquiries carried out during these last years were designed particularly to study the mechanisms of processes taking place in aqueous solutions, largely because of the interest that such research has for what concerns the influence of radiation on biological systems.

One of the most important reducing species formed in water radiolysis is the solvate electron e_{aq}^-. These electrons are, indeed, present in holes of the solution and stabilized by the interactions with a certain number of polar water molecules. They are formed by the interaction of high energy radiation with water itself, through a series of stages involving a gradual energy loss of the primary electrons formed as a consequence of the radiation-molecule interaction.

The role played by the hydrated electron in the mechanism of many radiolytic processes justifies the efforts spent in determining its physical properties. In particular, it has been possible to determine the radius of its charge distribution, amounting to 2.5–3.0 Å, its scattering constant (4.9×10^{-5} cm^2 sec^{-1}), its mobility (1.98×10^{-3} cm^2 V^{-1} sec^{-1}), as well as its thermodynamic properties.

A fundamental reaction of hydrated electron is that corresponding to its natural decay, by interaction with a water molecule, as follows:

$$e_{aq}^- + H_2O \rightarrow H' + OH^-.$$

In this reaction, H' radicals are formed which may start radical processes as well as the formation of hydrogen molecules by recombination.

As already mentioned, a fundamental aspect of radiation chemistry concerns the study of reactions resulting from the interaction of a solvated electron with molecules of biological interest, for example, amino acids. The interaction between e_{aq}^- and alanine (which is present in solution as a hybrid ion) is probably associated with the breaking of the C-N bond through the following series of reactions:

$$e_{aq}^- + NH_3^+ CHRCOO^- \rightarrow NH_3 + C^0HRCOO^-$$
$$e_{aq}^- + NH_3^+ CHRCOO^- \rightarrow NH_2 + CH_2RCOO^-$$
$$NH_2^{\cdot} + NH_3^+ CHRCOO^-$$
$$\rightarrow NH_3 + NH_3^+ C'RCOO^-.$$

Finally, remarkable results have been obtained, too, in the study of gaseous phase processes. The mechanism of reactions between charged particles, or ions, and molecules has been thoroughly investigated by both studying from an empirical viewpoint the kinetics of such processes and formulating theoretical models capable of explaining their mechanisms.

These models aim principally (a) to point out how much the reaction rates depend on the ions' kinetic energy and (b) to clarify the nature of possible interaction complexes that may form on collision.

Hot-atom chemistry

The chemistry of hot atoms, that is of atoms endowed with a much higher kinetic energy than that of those usually taking part in chemical reactions, is extremely interesting because of the variety of new reactions to which they can give rise.

This field of research is particularly important for what concerns the study of processes occurring in extraterrestrial space, on the moon's surface and in the atmosphere of stars.

Nuclear reactions producing atoms with a very high amount of kinetic energy are the most common source of hot atoms. If, for instance, a ^{3}He nucleus [made up of one neutron (n) and two protons (p)] absorbs a slow neutron, its transforms into tritium, a radioactive isotope of hydrogen (^{3}H=T), made up of two neutrons and one proton:

$$^3He + n \rightarrow p + T.$$

The proton emission gives to the tritium atom a recoil kinetic energy amounting to about 200,000 eV. Actually, with so high an energy content, the tritium atom behaves as an α-particle (^{2}He nucleus) and, during its motion, it gives rise only to processes of breaking and ionizing of the molecules it meets on its path.

Only when, in consequence of such collisions, its kinetic energy was decreased to about 10 eV (that is, when it becomes comparable to that of a chemical bond) the atom may chemically combine with other substances.

Tritium is one of the reagents most often employed in hot-atom chemistry. If interacting with methane, it can give reactions of hydrogen substitution and extraction to yield HT. The substitution reaction may take place through two mechanisms, that is with inversion of configuration

$$T + H_3C-H \rightarrow T-CH_3 + H$$

or with conservation of configuration:

$$H-CH_3 \rightarrow T \overset{H}{\diagdown}-CH_3.$$

Likewise, the interaction of tritium and methane may form HT and a methyl radical (CH_3) as a consequence of extraction or stripping of a hydrogen atom from a methane molecule.

Other reactions occur because of the participation of hot halegens produced, for instance, by the capture of thermal neutrons, as in the reaction

$$n_X + \gamma \rightarrow (n-1)_X + n,$$

where X is a halogen atom.

Reactions involving atomic carbon are most interesting as they permit the emphasis of some characteristics peculiar to such an atom. Hot carbon may be obtained through the reaction

$$^{12}C + \gamma \, ^{11}C + n.$$

Using hot carbon opens new opportunities for research seeking to clarify within which limits a reaction mechanism can be affected by the electronic state of the carbon atom itself. An example is given in the following insertion reaction:

$$H_2C=CH_2 + {}^{11}C \rightarrow CH_2{\overset{}{\text{——}}}CH_2$$
$$\diagdown{}_{^{11}C}\diagup$$
$$\rightarrow H_2C={}^{11}C=CH_2.$$

Moreover, the kinetic results obtained with the use of hot atoms have a very interesting theoretical interpretation, comparable to a detailed analysis of molecular collision dynamics.

Following this trend, research is being conducted in which chemical reactions are described, taking into account the reacting particles' motion on the multidimensional surface which describes the potential energy of the system itself.

Further reading

BAKER, B. S. *Hydrocarbon fuel cell technology*. New York, Academic Press, 1965.

BURWELL JR., R. L. Catalysis. *Scientific American*, vol. 225, no. 6, 1971, p. 46.

CLAVERT, J.; PITTS JR., J. N. *Photochemistry*. New York, 1966.

DIXON LEWIS, G.; WILLIAMS, A. Methods of studying chemical kinetics in flames. *Quarterly review*, no. 17, 1963, p. 243.

DORFMAN, L. M.; FIRESTONE, R. F. Radiation chemistry. *Annual review of physical chemistry*, no. 18, 1967, p. 177.

ENRY, H.; HENSERSON, D.; JOST, W. *Electrochemistry*. Vol. IX of *Physical chemistry*. New York, Academic Press, 1970.

ERDEY-GRUZ, E. T. *Kinetics of electrode processes*. A. Hilger, 1972.

GERMAIN, G. E. *Catalytic conversion of hydrocarbons*. New York, Academic Press, 1969.

KONDRATIEV, V. N. Chemical kinetics of gas reactions. London, Pergamon Press, 1964.

LAIDLER, K. J. *Chemical kinetics*. New York, McGraw-Hill, 1965.

MAGEE, J. L. Radiation chemistry. *Annual review of physical chemistry*, no. 12, 1961, p. 389.

MOELWIN-HUGHES, E. A. *Chemical statics and kinetics of solutions*. New York, Academic Press, 1971.

NATA, P.; FESSENDER, R. W.; SCHULER, R. H. Evidence that $e_{aq}^- + H_3O_{aq}^+$ does give H atoms. *Nature (Physc. science)*, no. 237, 1972, p. 46.

NECKERS, D. C. Mechanicistic organic photochemistry., New York, 1967.

PETERSEN, E. E. Chemical reaction analysis. Englewood Cliffs, N.J., Prentice-Hall, 1965.

PHELAN, N. P.; JAFFE, H. H.; ORCHIN, M. A molecular orbital description of the nonclassical ion in 1,2 rearrangements. *Journal of chemical education*, no. 44, 1967, p. 626.

RITCHIE, M. Chemical kinetics in homogeneous systems. London, Oliver & Boyd, 1966.

SOCHET, L. R. La cinétique des réactions en chaînes. Paris, Dunod, 1971.

THOMAS, J. M.; THOMAS, W. J. *Introduction to the principles of heterogeneous catalysis*. New York, Academic Press, 1967.

THOMSON, S. J.; WEBB, C. Heterogeneous catalysis. London, Oliver & Boyd, 1968.

VOLLMER, J. J.: SERVIC, K. L. Woodward-Hoffman rule: Electrocyclic reactions. *Journal of chemical education*, no. 45, 1968, p. 214.

WALKER, D. C. The hydrated electron. *Quarterly review*, no. 21, 1967, p. 79.

WOLFGANG, R. Hot atoms chemistry. *Annual review of physical chemistry*, no. 16, 1965, p. 15.
. Chemistry at high velocities. *Scientific American*, vol. 114, no. 1, 1966, p. 82.

18

The structure of matter in bulk

Crystals

The understanding of the macroscopic properties (mechanical, thermal, electric, magnetic, etc.) of solid bodies in terms of their constituent atoms and molecules, and the laws of quantum mechanics, is the main objective of solid state physics, and has extremely important technical applications, such as the transistor and the laser. The greatest advances have so far been made in the field of crystalline solids, that is those solids whose atoms are arranged in a regular lattice of great symmetry. The symmetry properties of crystals are the key to the mathematical description of crystalline solids, making it possible to use relatively simple theories in spite of the huge number of microscopic bodies involved. The most important of these theories is known as the band theory, since the atomic levels of the atoms in the lattice group themselves together in bands which can be used to explain the thermal and electrical properties of the solid concerned: the phenomena of insulation, conduction and the all-important semi-conduction can all be explained on the basis of the band theory.

But this idealized picture of crystals has many shortcomings, since real crystals contain many defects (point defects and line defects or dislocations) which make them differ from an ideal, geometric lattice: they also contain impurities which are very important in the design and manufacture of electronic devices such as transistors. A recent development in the study of real crystals is the discovery that charged particles (radiations) do not travel through a crystal with equal ease in all directions, but have preferred directions, or channels, in which they travel farther: this phenomenon is called channelling, and has very important implications for the materials used in nuclear reactors, which are exposed to very high fluxes of radiation. Ion channelling can be used to study crystal structure in a manner analogous to that of the classical method of X-ray diffraction. A striking feature of this new research is that it is mostly done using computer models; crystalline structure can be fed into the memory of a computer as a mathematical model, and experiments with ions can be simulated by appropriate programmes. The results of these simulated experiments can then be checked by real experiments.

The use of computers has also made it possible to refine band theory to take into account the interactions of two or more electrons in the potential field of the crystalline lattice; classical band theory was limited to the description of one electron in the averaged field of the others. The situation is very similar to that which now exists in chemistry in the study of

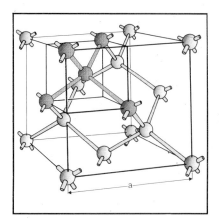

FIG. 33. The crystalline structure of semiconductors, such as silicon and germanium, can be illustrated as shown in the diagram. The crystal is cubic (dimension a) but each atom is bound only to four other atoms situated on the verticals of a regular tetrahedron, inscribed in a smaller cube. Germanium and silicon, the ideal semiconductors, have four valence electrons to each atom, and may have precise mathematical models in regard to crystalline structure. They are far less satisfactory from the chemical point of view, since the bond between the atoms cannot be assigned to an ideal category.

ence of a critical value for the ionicity parameter, and divides crystalline structures into two groups: in the first, every atom has four nearby neighbouring atoms, and the crystal is a semi-conductor; in the second, every atom has six nearby neighbouring atoms and the structure is an insulator (see Chapter 19). This is an example of how a microscopic property (the character of the chemical bond between two individual atoms) determines the macroscopic properties of the solid.

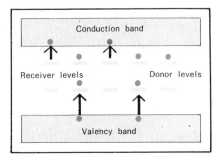

FIG. 34. In a semiconductor, the electrons of the 'impurity' atoms create energy levels inside the 'prohibited' space.

molecules: an idealized crystal can be thought of as a huge molecule, and problems related to the bonds existing between atoms in molecules, that is chemical bonds, are also very important in the study of crystals. There are two main types of chemical bonds, the ionic bond between two ions of opposite charge, such as the one existing between Na^+ and Cl^- in common salt (NaCl), and the covalent bond, which can be explained solely by quantum mechanics, which is very important in organic compounds. Bonds existing in real molecules, however, are always of a mixed character, partly ionic and partly covalent; the percentage of ionicity of the bond determines whether or not the compound is crystalline.

Recently American scientists have suggested a new scale for ionicity of bonds which is both simpler and more accurate than those previously used. This theory predicts the exist-

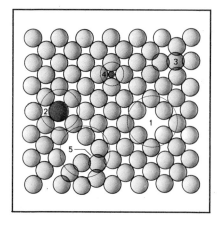

FIG. 35. Specific defects in a crystal:
1. vacancy (atom missing);
2. substitution of an atom of another kind; 3. interstitial atom; 4. interstitial atom of another kind; 5. Frenkel defect.

FIG. 36. The crystal lattice acts as a guide for particles in that it determines their trajectory through collisions related to target atoms, i.e. to the crystalline structure of the target and the location of the lattice in relation to the incident beam. This is the phenomenon known as channelling, or ion channelling, illustrated in the diagram. A reverse phenomenon also exists: a particle emitted by an atom in the direction of an atomic chain is defracted at a wide angle by the succeeding atom, yet a detector in the line of the crystallographic axis cannot record it. (From D. A. Marden, Chalk River Nuclear Laboratory.)

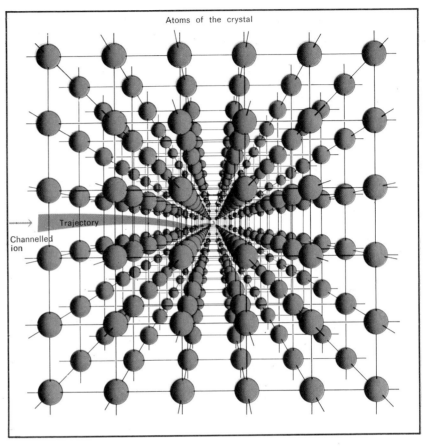

Atoms of the crystal

Trajectory

Channelled ion

Amorphous solids

Not all solids, of course, are crystalline. Glass, although very similar to diamond, will not produce the regular diffraction patterns of diamond when irradiated with X-rays. Solids like glass are said to be amorphous, and until recently very little was known of their structure, because they do not possess the symmetry properties of a lattice which so greatly simplify structural theories. Thus the problem is mathematically very complex and even given the chemical bonds between individual atoms, the gross properties of amorphous solids cannot be determined. Such solids were therefore little investigated until it was recently shown that they exibit semiconducting effects similar to those of crystalline solids (called Ovshinsky effects, after the name of their discoverer). Scientists then felt challenged to explain these properties, without the help of any previous theory such as band theory in the case of crystalline semiconductors. The only existing theoretical concept was that of a short-range order, as opposed to the long-range order in crystals. Amorphous solids are not completely disordered materials, but their atoms are arranged in an order which exists for a few atomic distances but is not propagated further. Experiments on very thin films of amorphous germanium have shown that its optical properties are similar to those of crystalline germanium, a material which is widely used in semiconductors. Since experiments with X-ray and electron diffraction show that the covalent bond between atoms of germanium, on which the optical properties depend, is found in both amorphous and crystalline germanium, the short-range order is probably due to the covalent bond. This is by no means true for all

169

amorphous materials, some of which are metallic, others alloys and compounds; however, amorphous solids in which the covalent bond is prominent are easier to describe, and can be said to represent ideal amorphous solids, for which analogies to the bond theory can perhaps be found. In this field, too, the use of computer simulation can be very useful. British scientists of the National Physical Laboratory at Teddington have used computer simulation to build a physical model of silicous glass, from which the observed diffraction patterns can be obtained. The study of amorphous solids and the explanation of their observed properties has thus just begun, and will represent a new and great challenge to solid state physicists, with important possible applications in electronics.

Surfaces and layers

The surfaces of solids are involved in many complex and important activities such as friction, catalysis and corrosion, to mention only a few. A better understanding of the microtopography of surfaces, especially crystalline surfaces, therefore has important technical implications.

The use of optical methods to study surfaces has now largely been replaced by methods which take advantage of the smaller wavelength of beams of electrons and other particles. The electron microscope is widely used in surface studies, both in its transmission and scanning mode. So are electron diffraction methods such as reflection of high energy electron diffraction (RHEED) and low energy electron diffraction (LEED) since the electrons, which are made to glance off the surface in the case of RHEED and to strike it at a larger angle in LEED, are reflected by the surface and not by the bulk of the material. Another new tool for the study of surfaces is the field ion microscope, which is even sensitive enough to produce images of individual atoms in the surface. These methods complement one another in the sense that they can be used to make observations of surfaces at different magnifications.

Chemically, the surface layer is studied by using techniques such as electron spectroscopy for chemical analysis (ESCA), in which atoms on the surface are excited by a beam of radiation and made to emit electrons, whose energy is then measured by instruments developed in nuclear physics. This method is very suitable for the study of chemical bonds on surfaces. A typical feature of all solid surface studies is the development of a very advanced technology: very pure and clean surfaces are needed, and a high vacuum is essential to avoid unwanted interactions between the surfaces and atoms of gas, such as adsorption and chemisorption.

A field that is closely related to the study of surfaces is that of layers or thin films of materials, especially semiconducting materials, which are widely used in microelectronics. Surface physics and chemistry methods are used in order to deposit layers of materials with the desired electrical properties on a 'wafer' of silicon or germanium, and to inject into them impurities to modify their properties: this last process is called doping, while the process of layer deposition, which is done in ovens in a selected atmosphere, is called epitaxial deposition.

Composites and reinforced materials

The mechanical properties of solids depend largely on their crystalline structure, which in turn depends on the nature of the chemical bonds prevailing in the solid. Covalent crystals, such as diamond, offer a high resistance to traction stresses, and this is true of many monocrystalline materials, that is single crystals. Most materials, however, are polycrystalline, and are made up from many, smaller crystals connected by weaker bonds; they therefore have less resistance to traction stresses.

For this reason monocrystalline fibres, called whiskers, have been artificially grown and found to have a tensile strength close to that predicted by the theory of crystalline structure. Unfortunately, such crystals are rather brittle and offer little resistance to shear stresses. To maintain the advantage of light weight and high tensile stress, and to minimize other disadvantages, these whiskers can be imbedded in a matrix of plastic material: the resulting composite material is very resilient and very light. Fibres used in this way can be made from boron, carbon (in the graphite form, of course, not in the diamond form), silicon carbide and others. They are mainly lightweight covalent fibres, and are not necessarily monocrystalline. They have been used in the aircraft industry, particularly carbon fibres, to form turbofans in jet engines and have found many uses in spacecraft where lightweight materials are essential. Other fibres have been embedded in metallic matrices of aluminium, copper and similar light metals.

The macroscopic properties of a solid such as polymer can also be modified at the microscopic level by irradiation. A polymer is a macromolecule obtained when a smaller molecule (its monomer) is joined together in a long chain and is conceptually very similar to a crystal. It has been shown that the irradiation of polymers with beams of electrons greatly improves their mechanical and thermal properties and their electrical resistance. The radiation ruptures many of the existing bonds and creates new ones across the long polymer chains, producing a material more similar to the fibres just mentioned. The bonds between the chains give the material a greater strength and stiffen it considerably. Irradiated polymers are used in food packaging and as insulating cables in the aircraft industry.

Liquids

The mathematical descriptions of the structure of liquids is yet another unresolved problem in the study of matter. Atoms or molecules in liquids are free to move in any plane, as in gases, and are not constricted by a regular or irregular framework, as in crystalline or amorphous solids. In contrast to gases, however, the molecules are close together and interact more with each other. Liquids thus pose very complicated structural problems because neither symmetry assumptions, as in crystals, nor statistical assumptions, as in gases, can be used to simplify the theory; liquids are similar to amorphous solids, but their atomic constituents are free to move in all the volume taken up by the liquid. All attempts to give any but the crudest mathematical picture of liquids have so far failed, and physicists have therefore recurred to physical models.

One of the simplest, and most successful, is the Bernal model of rigid spheres packed closely but randomly in a given volume. It turns out that if a sphere is surrounded by six other spheres, in each plane the arrangement is similar to that of a crystal lattice; if it has only five, this packing arrangement cannot completely fill the available space, and there remain some holes, so that the spheres are free to move from one point to another. This crude model suggests that symmetry considerations, which can be used quantitatively by the mathematical formalism of group theory, might be very important in the study of the liquid state

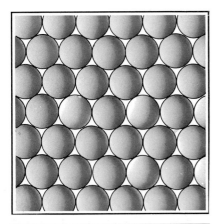

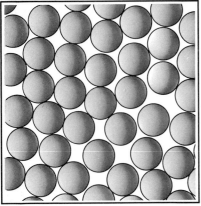

Fig. 37. Bidimensional representation of the structure of a solid (top) and that of a liquid (bottom): as it is not possible with five small spheres, to fill all the space available, they remain free to move.

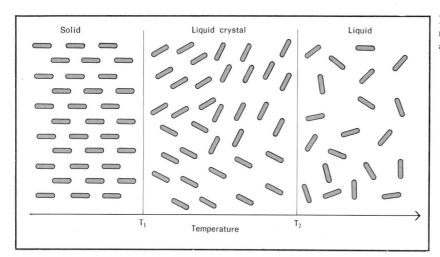

Fɪɢ. 38. Liquid crystalline, or mesomorphous, state, with the molecules arranged in partial order.

of matter. But advances in this area have thus far been slow, and more progress has been made by statistical studies, using computer simulation, on the possible distributions of random packings of the spheres. It must be pointed out, however, that molecules often have shapes which are far from that of a sphere, and that the Bernal model is therefore inadequate for all but the simplest molecules.

Other problems in the study of liquids are posed by liquid metals: how can the conductivity of these materials be explained when the band model is no longer applicable since there is no longer a crystalline lattice? Evidently, some properties of the crystalline state remain unchanged in the transition to the liquid state. In this area good progress has recently been made in the study of states of matter which are intermediate between the solid and the liquid state. Some organic materials, known as liquid crystals, are dense liquids but exhibit the typical optical anisotropies of crystals and a number of other interesting electrical and magnetic properties: for instance, their colour changes in an electric field of varying intensity. Their structure is known to consist of elongated molecules which tend to move in parallel directions. Their properties could find applications in electronics, either in television screens or in optical devices for displaying results of electronic computation, and their connexions with some biological molecules are also proving intriguing.

Gases and plasmas

There is no clear-cut demarcation line between a gas and a plasma: the latter is a gas at a temperature sufficiently high to ionize it partly or fully. Usually a gas is considered a plasma (the term was coined by I. Langmuir in the 1920s) if it is more than 50 per cent ionized, that is if more than 50 per cent of its atoms have lost their electrons. Gases have been described since the nineteenth century by the kinetic theory of gases which, however, applies only to ideal gases composed of point-like atoms. Real gases are made up from atoms and molecules which have a small but finite volume; therefore the basic equation describing an ideal gas, relating pressure, volume and temperature, must be modified in order to take account of the volume of its finite atomic constituents. The equation was modified in this way by Van der Waals in 1873, and subsequently other scientists have tried to refine it by the introduction of more empirical parameters; there exist five-parameter equations (like the Bettie-Bridgemann equation) and eight-parameter equations (like the Benedict-Webb-Rubin equation), all of which try to describe the behaviour of gases under particular experimental conditions. A recent trend in the study of collisions between gas atoms and molecules is the introduction of quantum mechanics, instead of classical mechanics, and of corresponding experimental apparatus capable of more precise measurements of relevant quantities, such as cross-sections for scattering of atomic and molecular beams.

In contrast to the theory of gases, the study of plasmas was begun in the 1930s by astrophysicists, since plasma is very common in the universe. Great progress in plasma physics, and especially in the problem of plasma confinement by a strong magnetic field, has been made recently by research on controlled thermonuclear reactions (see page 146) aimed at producing a nuclear fusion reactor for energy production. This goal has not yet been achieved, but great advances have recently been made. The chief problem is that of the diffusion, or leakage, of the plasma across the lines of the magnetic field. Diffusion can be estimated either from the classical diffusion law, in which the rate increases with the density of the plasma, is inversely proportional to the square of the magnetic field and to the square root of the temperature, or from Bohm's law, in which the diffusion rate is independent of the density, inversely proportional to the magnetic field and increases linearly with temperature. The first law is based on theoretical reasoning derived from astrophysics, the second on experiments done in radar research during the Second World War. If the classical law is at least partially valid a nuclear fusion reactor can be achieved; if Bohm's law is valid no such thing is possible. All nuclear fusion experimental devices built until the late 1960s had confinement times very close to those predicted by the Bohm theory, and were therefore insufficient for nuclear fusion. In 1968–69 Soviet scientists announced they had obtained, in their Tokamak device, plasma of sufficiently high density and confinement times some one hundred times greater than those predicted by Bohm's theory. The announcement was greeted with some scepticism, but the result was confirmed by a team of British scientists who flew to Moscow with their own measuring equipment, including a laser, to determine the characteristics of the plasma obtained by their Soviet colleagues. The results found by the British team were even better than those announced, and physicists throughout the world were greatly encouraged; the 'Bohm barrier' had been overcome.

Condensed matter

At the other extreme of the density scale is matter in a state of very high density. Such matter cannot be found on Earth, except perhaps deep in its interior, and its existence and unusual properties were foreseen in the 1930s as the final state of stellar evolution—dwarf stars and neutron stars (see Chapter 10).

It was not until the discovery of pulsars in 1968, however, that research in this field took on new importance because it is now generally accepted that pulsars can be identified with neutron stars, and their characteristics explained on the basis of the properties of this extremely dense matter. Although the densities and temperatures which can be obtained in the laboratory are much lower than those existing in the interior of celestial bodies, an increasing amount of work is being done in the field of high energy density physics, which purports to study such conditions of matter in the laboratory (see Chapter 16). Theoretically, the transformations undergone by matter beyond the limits of laboratory conditions are thought to occur as follows: molecules split into atoms; electronic shells of atoms are rearranged and filled more regularly, with the outer electrons being torn away; and the crystalline structure of all solids becomes the same (cubic and body centred). This is the 'universal' state of matter, which is achieved at a temperature of the order of 10 eV (electron volt) and a pressure of 100 Mbar. Any material which is solid under these conditions acquires metallic properties; gases or liquids become plasmas. As temperature and pressure become higher other electrons are stripped from the atomic shells and form a 'gas' of high energy electrons; the remaining innermost electrons behave like valence electrons in a metal. At still higher temperatures matter consists of a plasma of nuclei and electrons. But, if the pressure increases while the temperature remains constant, as in stars at the end of their evolution, nuclei crystallize in a superdense solid; at still higher pressure nuclei may break up into their constituents, and elementary particle interactions may play a very important role in the structure of superdense matter. Thus, elementary particle physics is the key to research on the structure of matter under extreme conditions of pressure and temperature, such as those encountered in modern astrophysics.

Further reading

BÉNARD, J. Apparence et réalité des surfaces. *Sciences*, no. 76, 1972, p. 14.

BILLARD, J. Les états mésomorphes de la matière. *Sciences*, no. 76, 1972, p. 3.

BRANDT, W. Positions as a probe of the solid state. *Scientific American*, vol. CCXXXIII, no. 1, 1975, p. 34.

CHISTYAKOV, I. G. Liquid crystals. *Soviet Physics Uspekhi*, vol. IX, 1967, p. 551.

COTTRELL, A. H. The nature of metals. *Scientific American*, vol. CCXVII, no. 3, 1967, p. 90.

DEARNELEY, G. La canalisation des ions. *Sciences*, no. 76, 1972, p. 39.

FINNEY, J. L'état liquide. *La recherche*, vol. I, no. 4, 1970, p. 336.

GUINIER, A. Ordre et désordre dans la matière. *La recherche*, vol. II, no. 17, 1971, p. 941.

KELLY, A. Les matériaux composites fibreux. *La recherche*, vol. II, no. 14, 1971, p. 631.

KIRZHNITS, D. Extreme states of matter. *Soviet science review*, vol. III, no. 4, 1972, p. 199.

MASSEY, H. Collisions between gas atoms. *Endeavour*, vol. XXVII, no. 102, 1968, p. 114.

MOTT, N. The solid state. *Scientific American*, vol. CCXVII, no. 3, 1967, p. 80.

ORSAY, Groupe des cristaux liquides. Les cristaux liquides. *La recherche*, vol. II, no. 12, 1971, p. 43.

PHILLIPS, J. C. La liaison chimique dans les cristaux. *La recherche*, vol. II, no. 14, 1971, p. 616.

ROSE, D. J. Controlled nuclear fusion: status and outlook. *Science*, vol. CLXXII, no. 3985, 1971, p. 797.

TUCK, J. L. L'énergie de fusion. *La recherche*, vol. III, no. 27, 1972, p. 857.

WOLF, R. La pentacoordination. *La recherche*, vol. VI, no. 60, 1975, p. 818.

YOUNG, R. D. Surface microtopography. *Physics today*, vol. XXIV, no. 11, 1971, p. 5.

19   The behaviour of matter in bulk

Imperfections and dislocations in solids

Perfect crystals are the exception, not the rule, in nature. Materials such as metals are made up from a great many smaller and often imperfect crystals, in such a way that relating the macroscopic properties of the metal, such as plasticity, brittleness, tensile strength and creep, to the microscopic properties of the crystalline lattice is an almost hopeless task. Of course, dislocations are known to be the cause of plasticity in metals, if they can move through the crystals, and of brittleness, if they are pinned down to particular sites in the crystals. But to bridge the gap between the physical theory of dislocations and the empirical properties of metals studied by metallurgists, empirical equations have recently been proposed which are not immediately related to basic physical laws.

Imperfections in the crystal lattice, such as the presence of foreign atoms, have important electrical effects on solids, even in very small proportion; when their number is increased, however, they tend to form a lattice within the lattice and their distribution is no longer at random. In fact, this is the basis of the old technique of producing metal alloys with differing strength, plasticity and brittleness. Only recently, however, have the properties of alloys been studied in relation to defects within the crystals, and the orderly way in which such defects occur, particularly in alloys such as Au-Cu and Si-C. Another kind of ordering process, in which groups of defects behave very similarly to molecules in a liquid and hence maintain an average separation, has been verified in Al-Ag and Al-Cu alloys which are important in the aircraft industry. Research in this field is concentrated on very strong but light alloys; high strength tends, however, to be accompanied by unwanted brittleness. For this reason, there is a trend towards the use of composite materials, where strength can be combined with elasticity (see Chapter 18). Composite materials so far studied are made from such fibres as metal wires, glass fibres, crystalline whiskers, tungsten wires coated with boron, carbon filaments and polycrystalline ceramic fibres embedded in plastic or metal matrices.

Imperfections and dislocations in crystals play an important role in surface processes such as crystal growth, oxidation and corrosion; these processes severely limit the utility of many materials, and many attempts are being made to minimize their occurrence. The microscopic structure of solids also plays an important role in determining the effects of high temperatures, high radiation fluxes, particularly in materials used in nuclear reactors, prolonged vibrations, as in aircraft materials, and very low temperatures in materials used in cryogenic engineering. To answer these and other problems, many new materials are being developed, including pure crystals (of Si and Ge), polycrystalline metals, alloys and ceramics (metal oxides such as SiC and SiN) and amorphous solids (glasses). Whenever possible attempts are made to relate the properties of these materials to their microscopic structure. Recently, some success has been achieved in making computer models of solid state structures which can be used to predict the behaviour of the material under any given set of conditions.

Metals, insulators and semiconductors

The difference in the electrical properties of metals, insulators and semiconductors is easily understood in terms of the electron band theory and the energy gap between the valence band and the conduction band: if the conduction band is empty of electrons the material is an insulator; if it is empty but the energy gap is small it is a semiconductor; if it is not empty it is a metal. In a perfect insulator each atom is surrounded by six other near neighbours; in a perfect semiconductor only by four; and in a perfect metal by from eight to thirteen. These situations correspond respectively to an ionic bond, a covalent bond, and a metallic bond.

Real, as opposed to ideal, crystals often share some of these properties because of the mixed nature of the chemical bond. Recent advances in the theory of the chemical bond (the spectroscopic theory of the chemical bond, so-called because its parameters are obtained through optical measurements) have made it possible to predict, among other things, the colour of the light emitted by a new semiconductor, GaN, during electroluminescence; to predict the magnitude of an electric dipole moment in a new piezoelectric crystal and some non-linear laser effects (frequency doubling) in a crystal irradiated with laser light. The field of semiconductors has seen spectacular advances recently, both in the understanding of physical phenomena and in their technological applications. The transistor

itself, the basic solid state amplifying device, has been almost completely replaced by the integrated circuit, in which hundreds of transistors, resistors, capacitors and other electronic components are formed on a single chip of silicon. This revolution has resulted in an enormous advance in the miniaturization of electronic circuits. Electronic computers, to mention just one application, have been greatly reduced in size and can now be carried aboard both spacecraft and heavy aircraft, for navigation and guidance purposes. Solid state circuits have all but replaced vacuum tubes in all applications, except where very high power is needed as in the generation of microwaves for radar and particle accelerators. In 1963, however, an effect was found in GaAs semiconductor (the Gunn effect) which makes it possible to generate microwaves with sufficient energy density at least for short range applications, for instance in a radar mounted on a car to avoid collisions in foggy weather.

Another important area concerns the temperature-dependence of these properties. This is particularly evident in the so-called Mott insulators, for which the classical one electron band theory does not work, and the interactions between two or more electrons have to be considered, which adds great mathematical complexity to the problem. In one of these materials, V_2O_3, which is an insulator at low temperatures, the conductivity jumps by a factor of 10 million within a temperature range of 1 degree; this is almost a case of superconductivity in reverse.

Recently, renewed interest has been shown in solid electrolytes, in which conductivity depends on the ions and not on the electrons, and which are used in batteries, fuel cells and other devices. Compounds which are being studied for these purposes include sodium aluminate and calcia-stabilized zirconia (CSZ). Their conductivity is explained as the effect of the presence of a number of vacancies in the ion lattice.

One of the greatest scientific surprises of the 1960s was the discovery of a semiconduction effect in amorphous materials (doped glasses), sometimes called the Ovshinsky effect from the name of its discoverer. This effect had not been foreseen because of the lack of a suitable theory of amorphous materials analogous to the band theory of crystalline materials, and for this reason the announcement of the Ovshinsky effect was greeted with some scepticism. Further research has shown it to be a real effect, reproducible under controlled conditions, and capable of being exploited for the construction of switching devices of the on-off type in logic circuits. The discovery led to more research on the amorphous state of matter, and the effect is now understood in terms of local disorder-order transitions in amorphous solids. So far the effect has not found widespread applications, but it holds considerable promise for the future, especially for radiation-resistant semiconducting devices.

Magnetic materials

The theory of magnetic materials is by now well established and research is mainly aimed at the discovery of new materials for special applications. For very strong magnetic fields, the emphasis has shifted to superconductor magnets, but soft magnetic materials are still widely used in modern electronics, for instance in the memory cores of computers. Ferrites, that is compounds of the general $MOFe_2O_3$ formula, where M is a metal, are used to this end. They are essentially ionic compounds and good insulators, and are prepared by sintering from powders. Ferrites can be used if the frequency of the external applied field is lower than 100 MHz; for applications in which the frequency is higher, such as telecommunication equipment, the rare earth-iron garnets, whose chemical formula is $M_3F_5O_{12}$, must be used. The most important of the garnets is yttrium iron garnet (YIG).

In the field of hard magnets, very high permanent magnetic fields have been obtained with finely divided iron powder mixed with a suitable binder; a development of this method makes use of elongated particles of FeCo which are each a single magnetic domain (the domain is the basic 'building block' of magnetization in materials: it is a region in which the magnetization is everywhere constant in magnitude and direction). New materials of the FeX type, where X is a compound containing a rare earth metal, such as $SmCo_5$, appear to have extremely promising permanent magnetic properties.

A very recent development in the investigation of the magnetic properties of thin films of materials, as opposed to the properties of magnetic materials in bulk, has produced an idea which looks extremely promising for the construction of high capacity and high-speed magnetic memories for computers. Small magnetic domains of cylindrical shape, called bubbles, can move freely within thin films of ferrite or garnet; suitably driven by externally applied fields, their presence or absence in a given location can be interpreted in terms of a 0 or a 1, thus giving rise to the typical binary digit system memory of electronic computers.

Superconductivity and superfluidity

Research on low temperature phenomena has been concentrated on two areas. First, the search for materials which are superconducting at relatively high temperatures, partly because they would find applications in advanced technology; and second, an attempt to understand superconducting and superfluid behaviour in terms of quantum mechanics.

In the first area, the properties of many compounds and alloys have been investigated; while Nb_3Sn, which is superconducting up to 18° K, is by now widely used in superconducting magnet windings, in 1973 a new record was reached with Nb_3Ge which was found to have a transition temperature of 22.3 K (−251° C). A variety of stratified compounds formed by layers of metals and organic materials was also found to be superconducting: this fostered speculation that some kind of organic material, because of its high degree of ordering, might be superconducting at room temperature. No evidence has been found, however, in favour of this very attractive idea.

On the fundamental side, an extremely interesting tunnelling effect in what is today called a Josephson junction was predicted from theory in 1962, and experimentally verified the following year. A Josephson junction is formed by a very thin insulator sandwiched between two superconductors; a current is able to flow through the insulator without any resistance and, if it is greater than a certain critical value, an alternating current appears whose frequency depends on the ratio of two fundamental constants, the electron's charge e and Planck's constant, h. Thus a measure of the current's frequency gives a very precise determination of the ratio of two fundamental constants of atomic, nuclear and elementary particle physics. Combined with new, more precise measurements of the speed of light c by piezoelectric methods, a whole new set of measurements for the fundamental constants of physics e, h and c was obtained in 1969.

A conceptually similar phenomenon in superfluid liquid helium (the continuous Josephson effect) was experimentally verified by French workers in 1970; like the alternating Josephson effect, it depends on the presence of pairs of particles (electrons or atoms) whose motions are correlated with each other. This is typical of superfluid and superconducting systems, and is the basic explanation of their strange properties. Among one of these is the presence, in superfluid helium, of quantized vortex rings, observable through a common microscope, whose angular momentum is quantized, that is, is a multiple of the Planck constant divided by 2π, the typical atomic and nuclear unit of angular momentum. This appearance of a typical subatomic unit in a macroscopic measurement is again due to the high degree of ordering. In 1971, superfluidity was found to exist also in the lighter isotope of helium, helium 3, below a few thousandths of a degree above absolute zero.

An important technological application of the properties of superconductors could be the transmission of electric power over long distances by superconducting cables; other uses may be found in the generation of very high magnetic fields for plasma containment in nuclear fusion reactors, levitation of high-speed vehicles and scientific instruments where low friction is essential (gyroscopes for spacecraft, gravitational wave detectors, etc.). Finally, the attainment of very low temperatures, below that of the lambda point of liquid helium (2.17° K), is in itself a scientific problem of great difficulty, requiring very advanced techniques such as magnetization, cooling and subsequent adiabatic demagnetization of a

paramagnetic salt. Recent advances in this field include the adiabatic depolarization of a paraelectric material, which could be used to cool a material even in the presence of a magnetic field.

Molecular acoustics, ultra- and infrasonics

The acoustic spectrum can be divided into four regions: the infrasonic (from 10^{-2} to 10 Hz), audible (from 10 Hz to 14 kHz), ultrasonic (from 14 kHz to 10^9 Hz) and hypersonic (above 10^9 Hz) regions. Infrasonics and audible sound are not very important for the study of the structure of matter (but are relevant in other sciences, such as geophysics, biology and medicine: infrasound was recently found to have damaging effects on the human body, for instance, and has been considered as a possible future weapon). The science of molecular acoustics, however, which is concerned with the effects of ultrasound and hypersound on molecules, has achieved some notable results. This is particularly true in the case of fluids (gases and liquids), where internal equilibria are particularly dependent upon temperature and pressure, both of which are affected by acoustic waves. In solids, acoustic waves may interact with the vibrational modes of the crystal lattice, producing what in terms of quantum mechanics is called a phonon-phonon interaction. Acoustic waves in solids may also interact with atomic and nuclear momenta and with free electrons in metals and semiconductors. A recent and interesting field of research is the study of acoustic surface waves (or Rayleigh waves) in solids; by using a piezoelectric crystal it is thus possible to couple an acoustic wave to an electromagnetic wave, and vice versa. In this way a hybrid acoustoelectric device, which combines the attractive features of both electronic and acoustic apparatus, can be made. This has given birth to the new science of microsonics, in which electric signals are converted to acoustic signals, and vice versa. The chief advantage of using acoustic waves lies in the fact that their speed is 10^5 slower than that of electromagnetic waves; this means that, at a given frequency, their wavelength is 10^5 smaller than

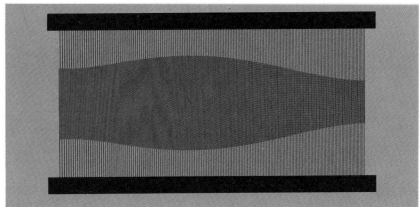

Fig. 39. Electronic devices based on the propagation of acoustic waves are used mainly in the construction of small delay-lines or filters with an assigned frequency response; this is because acoustic waves are propagated at a velocity approximately 10^5 times less than that of electromagnetic waves, and suffer much less attenuation. The upper illustration shows the structure of the code filter for surface acoustic waves in lithium niobate, for detection and generation of code signals. Right, transducer for acoustic waves, with suppression of the lateral lobes of the pass band. (*Source*: Lincoln Laboratories.)

that of electromagnetic waves of the same frequency. This effect is called demagnification, and allows for a reduction in the size of devices which are critically dependent on wavelength to make them comparable with those of typical integrated electronic circuits. The technology needed to make acoustic devices dependent on surface waves is similar to and compatible with that used in manufacturing integrated circuits.

Luminescence and photochromism

Luminescence is a process in which electrons are excited to higher energy levels and subsequently fall back to a lower level emitting a quantum of light (a photon). Recently, attention has been focused on electroluminescent devices, in which the excitation of energy is provided by an electric current. This has been achieved with good efficiency (30 per

cent) in the light-emitting semiconductor diode, such as the GaAs diode, but this emits infra-red light and is not suitable for applications where visible light is essential. The infra-red light must thus be converted into visible light by some frequency raising process, with corresponding loss of efficiency. Another semiconducting junction used is a GaP, which emits in the red region of the spectrum, but generates more heat in the process and is likely to overheat if used at high power. An alloy of the two compounds has been investigated which provides a red light without much overheating. Other compounds which should be capable of emitting in the green or even in the blue are InP-GaP alloys and GaN, if they can be prepared in a very pure form. Further, if the diode junction is built in such a way that most of the light emitted first undergoes multiple reflections on two parallel planes, a process of stimulated emission of radiation will

179

follow. In other words, the diode will emit coherent light, or laser light: this is the basic principle of the semiconducting laser.

Renewed interest has recently been shown in the phenomenon of photochromism, the process in which materials temporarily change colour when irradiated by light. This phenomenon is a rather complex one, which cannot be described by simple band theory but requires more refined descriptions which take into account the transfer of electronic charge from one ion to the other (the impurity ions are responsible for the colour). Research in this area promises to provide a better understanding of the structure of real crystals and could lead to possible applications of photochromic materials in three-dimensional colour television screens and optical memory devices for computers.

Lasers and quantum optics

The laser was one of the great scientific discoveries of the 1960s. The first laser was developed in 1960, after preliminary work had been done in the 1950s on the maser, which is its direct antecedent (laser stands for *l*ight *a*mplification by *s*timulated *e*mission of *r*adiation). Both are direct applications of quantum mechanics to the interactions between matter and electromagnetic waves. Their main feature is that of emitting coherent radiation, that is radiation in which individual waves are all in phase with one another; the waves are microwaves (radio) in the case of the maser, infrared, visible and ultra-violet waves in the case of the laser.

The laser (or maser) action is due essentially to a quantum effect; the stimulated emission of radiation occurring in a quantum system in which a higher energy level is more populated (that is, more atoms happen to be in that energy state) than a lower energy level. This is called population inversion, and is the key to the 'lasing' process. The population inversion is achieved by a 'pumping' process which can be optical, electrical or chemical. Laser action has been achieved in solids, liquids and gases (the first laser made use of a ruby crystal, but doped glass is widely used today). Another important feature of the emitted light is that

it is strictly monochromatic, although the frequency of the light emitted by a particular laser may vary somewhat. The light may be emitted either in pulses or continuously; the pulses are generally of very short duration, and therefore carry very high energy per unit of time.

Almost all the peculiar properties of laser light are now being used in some scientific or technological application: the phenomena of coherence and monochromaticity have made holography possible; the very short duration of laser pulses is used to measure events on the subatomic scale; and the very high energy densities are being exploited in attempts to create the conditions necessary for controlled nuclear fusion. Lasers are being used to machine metals, to treat skin tumours, to measure the distance between the Moon and the Earth, to study hyperfine interactions in atoms, to communicate at long distance and to implode frozen deuterium-tritium pellets to high density for thermonuclear fusion experiments; the number of possible applications seems to be almost countless.

With the advent of laser light a new branch of optics was opened up, the optics of coherent light, or quantum optics. A whole range of new optical phenomena thus began to be investigated in the 1960s, including non-linear optical effects such as the ability to make infra-red laser light visible by doubling its frequency. Other scientific measurements made possible by quantum optics are new measurements of the speed of light; the measurements of strains in the Earth's crust for the purpose of predicting earthquakes; new measurements of far greater precision in Raman spectroscopy and atomic spectroscopy in general (see Chapter 14). The laser is one of the most versatile scientific and technological tools ever created by man, but it must be recorded that military applications, including laser-guided bombing, were among the first widespread uses of this new invention.

High-pressure effects

As materials as subjected to increasingly high pressures, their structures tend to become more and more similar. Semiconductors tend to become first semi-metals, then metals which are superconductors at low temperatures (it has been calculated that one-third of the elements known to become superconducting do so only at high pressure). Such materials are difficult to study because of the extreme and conflicting conditions of very high pressures and very low temperatures which must be maintained; some superconducting compounds, however, synthesized at high pressures and then quenched to room temperature, have been found to be superconducting at atmospheric pressure. In all cases the appearance of superconductivity is accompanied by a change in crystal structure, which may lead to a better understanding of superconductivity. Even the magnetic properties of some elements, such as chromium, are very sensitive to pressure; the electronic band structure of others, such as caesium and cerium, is seriously affected by high pressure. Thus the study of the effects of pressure often lead to a more complete picture of the structure of materials and their structure-dependent properties.

High pressures are obtained in materials either by hydrostatic compression or by dynamic methods which employ shock waves. The pressures so far achieved are of the order of 500 kbar (1 bar=0.966 atm) for static pressures and of 10 Mbar with the use of chemical explosives. Explosives are also used to create very high magnetic fields by imploding steel tubes surrounded by a layer of explosive around the lines of force of an existing magnetic field: the imploding tube compresses the magnetic field to a very high energy density. This is one aspect of research on the physics of very high energy densities, which also includes the study of detonation, of plasmas created by

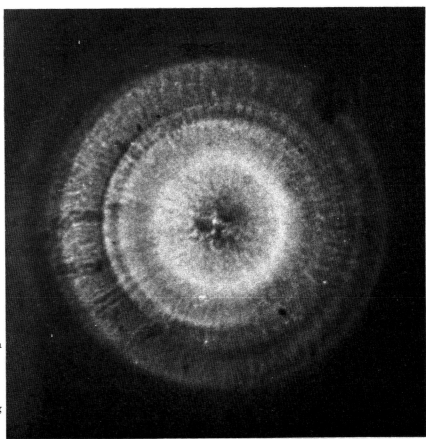

Fig. 40. By focusing a laser beam inside a nitrobenzene (lattice) cell, this series of rings was obtained on a shield perpendicular to the direction of propagation: the cell causes the rays to diverge at different angles corresponding to different wave lengths. (*Source*: Ford Motor Company.)

electric discharges in gases and which merges on one hand with astrophysics (Chapter 10) and on the other with nuclear research (see Chapter 14).

Further reading

BENIÈRE, F. Les superconducteurs ioniques. *La recherche*, vol. VI, no. 52, 1975, p. 36.

BLOCH, E.; JOUBERT, J. C. Les solides sous pression. *La recherche*, vol. III, no. 26, 1972, p. 742.

COHEN, M. H. Theory of amorphous semiconductors. *Physics today*, vol. XXIV, no. 5, 1971, p. 26.

DAVIDGE, R. W. Les céramiques. *La recherche*, vol. I, no. 1, 1970, p. 35.

ELECCION, M. The family of lasers: a survey. *IEEE spectrum*, vol. IX, no. 3, 1972, p. 26.

FRITSCHE, H. Les semiconducteurs amorphes. *La recherche*, vol. I, no. 6, 1970, p. 527.

KIRTON, J.; WHITE, A. M. Electroluminescent devices. *Endeavour*, vol. XXX, no. 109, 1971, p. 34.

KISS, Z. K. Photocromics. *Physics today*, vol. XXIII, no. 1, 1970, p. 42.

LEE, E. W. Magnetic materials. *Science progress*, vol. LVII, no. 227, 1969, p. 393.

McWHAN, D. B. The pressure variable in materials research. *Science*, vol. CLXXVI, no. 4036, 1972, 751.

MARAVIGLIA, B. L'hélium superfluide. *La recherche*, vol. II, no. 9, 1971, p. 142.

PETHRICK, R. A. Molecular acoustics: the study of sound waves to study molecular dynamics. *Science progress*, vol. LVIII, no. 232, 1970, p. 563.

SCHMITT, R. W. Solid state physicists and metallurgy. *Physics today*, vol. XXIV, no. 1, 1971, p. 44.

SUFFCZYNSKI, M. L'électroluminescence. *Atomes*, vol. XXIV, no. 269, 1969, p. 577.

THOMAS, J. M. The chemistry of deformed and imperfect crystals. *Endeavour*, vol. XXIX, no. 108, 1970, p. 149.

VOLLMER, J.; GANDOLFO, D. Microsonics. *Science*, vol. CLXXV, no. 4018, 1972, p. 129.

ZIMAN, J. The thermal properties of metals. *Scientific American*, vol. CCXVII, no. 3, 1967, p. 180.

Life

20 Molecular biology

Introduction: the genetic code

Deoxyribonuclei acid (DNA) is the genetic material of all cells and therefore of all unicellular and multicellular organisms (whether prokaryotes or eukaryotes). While DNA's structure had been already discovered in 1953 by Watson and Crick, its significance for life processes became only clear when the 'genetic code' was broken: the first breakthrough was made by Marshall Nirenberg in 1961 with the discovery that a succession of three thymine bases along the DNA strand codes for the amino acid phenylalanine; by 1966 the full correspondence between the possible sixty-four nucleotide triplets and the twenty essential amino acids was made clear. The code universal is the same for all organisms.

The basic structure of DNA has the following main characteristics: (a) each helix within the double-stranded molecule consists of a common backbone—P (phosphoric acid) and S (sugar: deoxyribose) (P–S–P)—to which the pirimidine bases, thymine (T) and cytosine (C), and the purine bases, adenine (A) and guanine (G), are linked; (b) the pairing between bases is strictly specific, i.e. A pairs always only with T, and G with C; (c) two DNA molecules, and two genes, may differ for the length of the chain of bases and for the different sequences of base pairs along the double helix: $A=T$ and $G=C$.

DNA's functions, which correspond to those previously attributed to the gene before its chemical nature was discovered, are two: (a) self-reproduction and (b) transmission of the genetic information to the chemical machinery of the cell. To achieve self-reproduction, both helyces of DNA participate: along each strand, through the action of DNA-polymerizing enzymes, a new complementary strand is formed, so that, at the end of the process, two double-stranded molecules will result, identical to the original one. Only one helix and always the same, on the other hand, is involved in the transfer of the genetic information present in each gene: molecules of messenger RNA (ribonucleic acid) are synthesized along the coding strand through a process of complementary formation of the RNA molecule, in which thymine is replaced by another pirimidine: uracil (U).

The so-called 'central dogma' of molecular biology (Crick) states that DNA determines RNA and this, in turn, protein: DNA→RNA →protein. The problem to be solved was that of finding what correspondence there could exist between sequences of four different bases, along the DNA double helix, and the innumerable combinations of the twenty amino acids present in protein molecules.

Together with nucleic acids (DNA and RNA), proteins are the most important molecules for

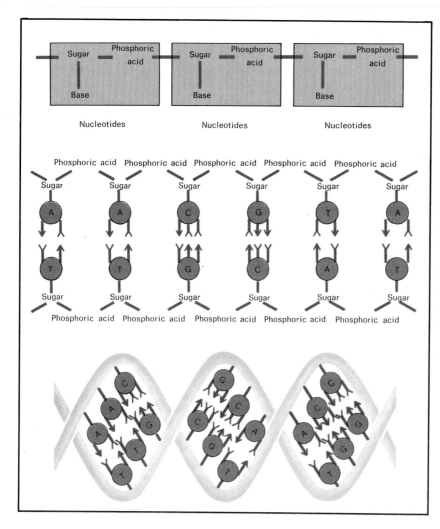

FIG. 41. The nucleic acids, DNA and RNA, are chains of subunits, the nucleotides, each of which is composed of a base, a sugar and one molecule of phosphoric acid. These chains may be of varying lengths: some of them contain up to 200,000 nucleotides. Each DNA molecule is composed of two chains of nucleotides, interconnected by hydrogen bonds which link the bases of one chain to the bases of the next, and which are wrapped round one another to form in space the characteristic double-helix structure.

life. Enzymes, which regulate all chemical cellular reactions, are indeed proteins. Not only from the biologist's viewpoint but from that of the chemist as well, they are extraordinary molecules. They consist of several tens of thousands of atoms, belonging to a relatively small number of elements. The relative positions of oxygen, hydrogen, carbon, nitrogen, sulphur, and occasionally of phosphorus and a few other elements' atoms confer to each protein molecule exceptional, and to not a minor extent, yet unexplained properties. Within the animal and vegetable kingdoms, each species differs from every other one for the presence of typical proteins; such extreme chemical differentiation extends much beyond the species level, to that of the individual; each one of us

human beings can be distinguished for a unique protein pattern.

Proteins are large molecules consisting of folded chains of amino acids, which contain an amino group (NH_2) and a carboxyl group $\overset{O}{\overset{\|}{C}}-O-H$. The most common amino acids present in living matter are twenty in number. They are built on the model:

$$H-\overset{H}{\underset{H}{N}}-\overset{\overset{H}{|}}{\underset{\overset{|}{R}}{C}}-\overset{O}{\underset{O}{C}}\quad H$$

where R stands for an atom or a group of

184

atoms. When R is just a hydrogen atom, we have the simplest amino acid: glicine. The remaining nineteen amino acids, in order of increasing complexity, are: alanine (ala); valine (val); isoleucine (ileu); leucine (leu); serine (ser); threonine (thr); proline (pro); cysteine (cys); methionine (met); lysine (lys); arginine (arg); aspartic acid (asp); asparagine (asn); glutamic acid (glu); glutamine (gln); phenylalanine (phe); tyrosine (tyr); tryptophan (try); histidine (his). Series of amino acids linked

$$\underset{\underset{-C-N-}{|\quad|}}{O\quad H}$$

through peptide bonds –C–N– are known as polypeptides. The primary structure of proteins depends on the particular sequence of amino acids in the polypeptide chain characteristics for each protein. To give an idea of the immense number of theoretically possible different proteins, let us assume that a given protein molecule were formed by a sequence of only twenty amino acids, in which each of the twenty previously listed were represented once. If any one were the first in the sequence, any other of the remaining nineteen could occupy the second position; each one of the remaining eighteen could be in the third position, etc. The end result shows that there could exist 2,432,902,008,176,640 different protein molecules: more than 2 million billion! But in reality things are even more complex: if we consider a protein consisting of a sequence of 150 amino acids, similar to that of the haemoglobin molecule in our blood, and if all the possible sequences and frequencies of the twenty amino acids could be realized, the number of potentially different protein molecules would equal 1 followed by 215 zeros.

Although DNA holds the primary instructions for protein synthesis, it does not itself directly participate in the process. In eukaryotic cells most of DNA is locked within the nucleus, while protein synthesis occurs in the cytoplasm, through the co-operation of small bodies: the ribosomes. The ribosomes were found to consist of a mixture of some twenty different proteins and several forms of ribosomic acid (rRNA). The main differences between RNA and DNA are: (a) the sugar component in the nucleotides is ribose in the former and deoxyribose in the latter, (b) most

RNA is single stranded rather than double stranded. The messages from DNA to ribosomes consist of one more type of single-stranded RNA: messenger RNA (mRNA).

Assuming some kind of co-linearity of the mRNA and the protein to be synthesized, it must be the base sequence of the former that specifies the amino acid sequence of the latter. Since we are dealing with twenty different amino acids and only with four letters in the DNA or RNA 'alphabet' (A, G, T, C in the former, and A, G, U, C in the latter) there cannot be a one to one correspondence between such four letters and the twenty amino acids, nor would two-letter words formed by the four DNA or RNA letters be sufficient, since one can form only twelve such words with four letters. It follows that the smallest number of letters that would suffice is three per 'word', i.e. the 'codon' which specifies precisely which amino acid is to be incorporated into a growing polypetide chain. Since with a four-letter alphabet one can form sixty-four three-letter words, the problem was to discover which relationship could exist between the four-letter alphabet of the nucleic acids and the twenty-letter alphabet of the proteins.

As previously mentioned, by 1966 the genetic code had yielded to the concentric attacks of molecular biologists. The main conclusions to be drawn from the prodigious successes of the 1960s can be summarized as follows.

There exists a perfect co-linearity between the base sequence in DNA and the amino-acid sequence in proteins, along a definite 'reading' direction.

The unit of the genetic code is the codon, corresponding to a 'triplet', i.e. the sequence of three successive bases along the DNA strand. Each amino acid in a protein is specified by a nucleotide triplet.

The property of having several codons for the same amino acid is described as the 'degeneracy' of the code. Out of the possible sixty-four, sixty-three triplets make sense, in that they correspond to one amino acid. Only two amino acids (methionine and tryptophan) are codified by a single triplet, while the others are specified by two, three, four or six different codons.

There are no 'punctuation marks' in the code,

inasmuch as there are no structures to indicate where a triplet begins and where it ends. At the beginning of the base sequence of a gene there must be something, much sought for but not yet identified, the so-called 'initiator' which starts the 'reading' of the message. There are no mechanisms that 'proof read' the primary structure of a protein. If the message present in the DNA is right, a functioning protein will be formed; but if there are mistakes (mutations) the resulting protein may not function. Protein synthesis, as guided by the DNA code, goes on until a 'chain terminator' is encountered. Protein synthesis occurs in two stages.

The first is 'transcription': a one to one transfer of the genetic information from the DNA to the mRNA. The second is 'translation', i.e. the shift from the polynucleotide 'language' to the polypetidic one.

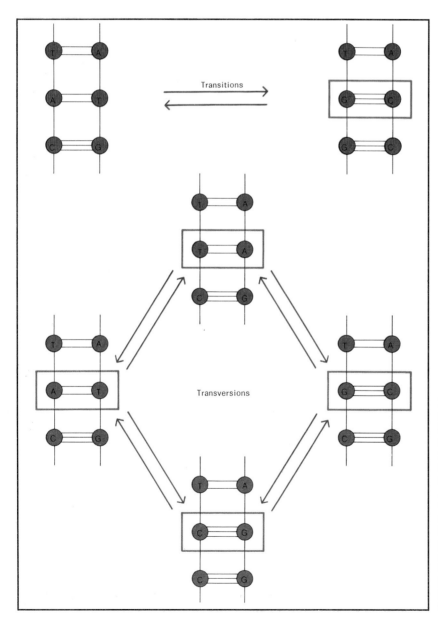

FIG. 42. The specificity of a DNA molecule (and hence of a gene, which is composed basically of DNA) lies in the sequence of the bases: thus genic mutation may consist of one variation only of this sequence, e.g. of substitution of the bases, as shown in the diagram. At top, transitions: replacement of one pyrimidin by another pyrimidin or of one purine by another purine. At bottom, transversions: replacement of a pyrimidin by a purine or vice versa (the diagram shows the eight possible transversions). In fact, there is no need for the mutagenic event to affect both bases (as in the diagram); indeed this never happens.

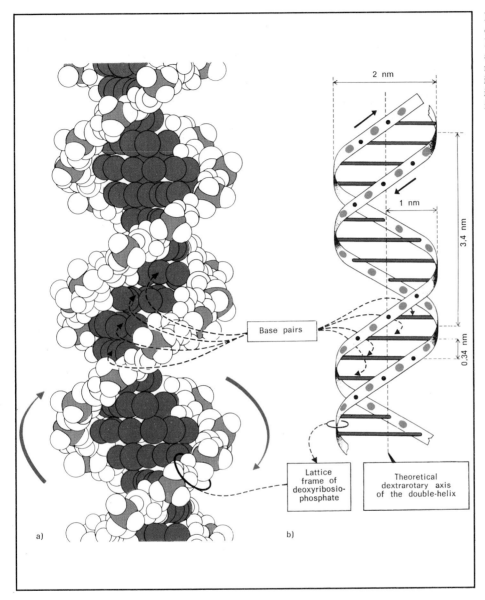

2 nm

1 nm

3.4 nm

0.34 nm

Base pairs

Lattice
frame of
deoxyribosio-
phosphate

Theoretical
dextrarotary axis
of the double-helix

a)

b)

FIG. 43. The molecular structure
of the double-helix of DNA
has been determined with the
aid of X-ray diffraction images,
making it possible to ascertain
its dimensions (which are
indicated on this simplified
illustration).

The *minutiae* of protein synthesis can vary from species to species. There are, for example, differences among various organisms with respect to the longevity of RNA-molecules. In some bacteria, mRNA lives only one or two minutes because their needs for one or another protein can be drastically changed by changes in the environment; in the homeostatic conditions of the vertebrate cells that are producing, e.g. haemoglobin, mRNA can be very long-lived.

Just as the genetic code provides for heredity, so it also provides for the production of variation. A genetic mutation can be defined as a change in the message encoded in the polynucleotide sequence, or, in chemical terms, as a permanent variation in the specificity of a DNA molecule. Since the specificity of a gene depends on the sequence of its nucleotides, a mutation should consist in a change in such sequence. Such an event could be due to either (a) a substitution of one or more bases

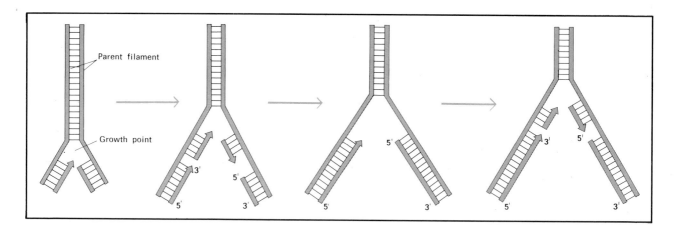

FIG. 44. The process of duplication of DNA: the molecule splits longitudinally at Y and, on the sections which grow gradually out of the parent filament, there appear short lengths of complementary filament (in the direction 5′–3′) which are subsequently bound together by a specific enzyme.

by others, without ensuing change in the length of the polynucletide chain; or (b) insertion or deletion of one or more bases, accompanied by a change in the chain length.

Two types of base substitutions (BS) have been distinguished: 'transitions', when a pirimidine is replaced by a different pirimidine and a purine by a different purine, and 'transversions', when a pirimidine is replaced by a purine or vice versa (Fig. 42).

The described types of changes in DNA's base sequences will result in changes in the meaning of the message. Once the change has occurred, it will be remembered and transmitted to the following generations of cells or of organisms, unless the distortion is so severe as to prove lethal before any descendants are produced.

The severity of the effects of a mutation depends on many factors. If, for example, the DNA codon AAA mutates to AAG, no observable effects will ensue, since both code for the same amino acid, phenylalanine. Slight changes involving only a few amino acids in a polypeptide may have only a small effect. Other self-correcting errors, for instance the deletion of a whole codon, such as appears to account for

one of the abnormal haemoglobins, may have trivial genetic consequences.

But some mutations may produce proteins that will have no biological activity, or that cause significant abnormalities. And others, those in which the stop codon appears in the wrong place in the instruction may prevent the cell from producing a necessary enzyme or protein, thereby creating a metabolic bloc.

Synthesis of macromolecules

The basic macromolecules of living things are nucleic acids (DNA and RNA) and proteins. In addition to an improved understanding of their structure, during the 1960s attention was paid mainly to the way they synthesize within the cell.

In 1953 the unravelling of the DNA structure by Crick and Watson allowed us to deal with and partly solve the problem of how genes, which are elements of DNA, duplicate themselves during cell division. The fact that the two complementary nucleotide chains of DNA coil in opposite directions, 5′→3′ and 3′→5′—i.e. one chain coils from the end in which the glucose residue has its free carbon in position 5′ to the end in which the same residue has its free carbon in position 3′, and vice versa—makes it difficult to understand in which way the enzyme or enzymes participating in DNA replication operates. In 1967, DNA-polymerase (an enzyme which induces the replication of DNA) was isolated but shown to be active only in one direction. Such behaviour has raised doubts concerning the action of this enzyme, and some hypotheses have been put forward. Among other things it has been supposed that, in the course of DNA replication, DNA-polymerase reverses several times the direction in which it operates. This would suggest that at least one thread of DNA is made up initially of short fragments which are linked afterward by another enzyme, a ligase. The hypothesis has been confirmed by the discovery by Japanese investigators of a linking enzyme, who have also demonstrated that DNA is at first made up of short fragments.

These results have emphasized the doubts concerning the real function of DNA-polymerase. These doubts are strengthened by the observation, made in 1969, that the separation of the two threads of DNA occurs all the same in organisms deprived of such an enzyme; the process seemed to be catalysed by a complex enzyme which is localized at the level of the cell membrane. This complex enzyme has been called DNA-polymerase II. DNA-polymerase I seems, on the contrary, to catalyse the repair reactions at the level of damages caused to DNA by external factors.

In 1972 another DNA-polymerase, DNA-polymerase III was identified. It is necessary now to establish whether II or III is the enzyme which catalyses DNA replication. Experimental work on mutants of *Escherichia coli* would have demonstrated that DNA-polymerase III is the enzyme of this replication. It is necessary to emphasize that none of the three DNA-polymerases is able to initiate the replication process. A short fragment of DNA is usually used *in vitro* as a primer; it was thought that the process would be initiated *in vivo* in the same way. But in 1972 the primer's role *in vivo* was assumed to be taken not by DNA but by an RNA which, once its function accomplished, is destroyed by a suitable enzyme. Japanese investigators have confirmed this hypothesis by demonstrating that each fragment of DNA presents during its replication, at one end a sequence of RNA made up of 50 to 100 units.

DNA, storing genetic information of living organisms, indirectly controls protein synthesis. It transcribes information on intermediate templates, which are molecules of messenger RNA. This process if catalysed by the DNA-dependent RNA-polymerase enzyme. Each RNA thread shows a direction of growth which is defined by the orientation of the sugar-phosphate backbone; this growing direction is believed to be $5' \rightarrow 3'$.

The enzyme binds to DNA, and opens its double-stranded structure in specific sites. The enzyme moves along the DNA, by using only one thread as the template to polymerize the four ribonucleosidophosphates in the complementary molecule of RNA.

While it has been demonstrated that DNA-polymerase is made up of only one polypeptide chain, DNA-dependent RNA-polymerase has a very complex structure. It is made up of β and β' sub-units, two equal α sub-units and a sigma (σ) sub-unity. Altogether these sub-units ($\alpha_2\beta'\beta\sigma$) constitute oloenzyme, of which two functional parts may be distinguished. These are the central part (the core enzyme $\alpha_2\beta\beta'$), which can catalyse the synthesis of RNA but is unable to initiate the process, and a factor, which catalyses the initiation of RNA but is unable to initiate the process, and a factor in determining the specificity of the bond and in enhancing the initiation of the process are thoroughly investigated.

The notion of σ factor as a primer in the transcription process dates back to 1969. It is now associated with the question of the positive control of gene expression, too.

Whereas the occurrence of the sigma factor seems not to arise in *Escherichia coli*, its presence is doubtful in the phages studied until the present time.

Some investigators have supposed the existence of other protein factors, which would participate in the transcription process (ρ, ψ and ppGpp factors), but the absence of experimental evidence questions the validity of this.

Messenger RNAs, which are synthesized in the transcription process, translate the genetic information to the protein. This translation process occurs on very complex subcellular structures, the ribosomes present in the cytoplasm (see Chapter 21).

The 1960s investigations on the modes of protein synthesis outlined this mechanism. By the beginning of the 1970s, experimental work in this field concentrated on the definition of its details and confirmation of the universality of some conclusion (which for the moment seem to be valid) at an experimental level, but only for relatively simple organisms such as bacteria.

The initiation end of the polypeptide chain constitutes a good example of the extension to eucaryotes of the validity of what is known on bacteria. In bacteria the first amino acid of every polypeptide chain is methionine, whose NH_2 group is formylated (i.e. the NH_2 group is blocked by the HCO group); when protein synthesis occurs, the formyl group is removed, often with the methionine residue. In bacterial cells, only one kind of transport RNA for methionine has been found: not only is it able to recognize the initiation sequences of messenger RNA, it also allows the formylation of the methionine residue, thus priming the synthesis.

Eucaryotic cells have been shown to contain two kinds of methionine transport RNA and, though deprived of transformylase, one of these two transport RNA may be formylated by bacterial enzymes. It has been demonstrated that in a system of cultured cells this transport RNA acts as a primer on a synthetic messenger RNA, whereas the other transport RNA incorporates methionine inside the polypeptide chain. It is necessary now to demonstrate that methionine is the universal primer, *in vivo*, also.

The universality of the genetic code would have been further confirmed by research on the termination of polypeptide chains during protein synthesis; the same triplets in bacteria and eucaryotes would function as full stop.

The identification, in the nuclei of different types of mammalian cells, of metabolically unstable nucleic acids, and particularly of an RNA of variable size, made up from 5,000 to 20,000 nucleotides and defined as heterogeneous nuclear RNA (Hn-RNA), brings forth the hypothesis of different regulatory paths in protein synthesis of bacteria and eucaryotes. The study of the synthesis of biological macromolecules has confirmed the validity of the central dogma of biology (DNA→RNA→protein), although some results obtained in 1970 on the possibility that oncogenic viruses and developing cells synthesize DNA from RNA by means of an inverse transcriptase have put into question this universal validity.

The protein structure is a direct consequence of the DNA structure. All that is known of protein structure has been acquired in the last twenty years. As late as 1950, little was known about their spatial interrelationships, although it had been established that proteins are made by amino acids. The complete sequence of the fifty-one insulin amino acids was first determined in 1955 and by 1966 its synthesis *in vitro* accomplished. This was the first protein synthesis *in vitro*, followed by that of ribonuclease (124 amino-acid residues) in 1969 and of the growth hormone (188 residues) in 1970.

The amino-acid sequence of a protein constitutes its primary structure. The chain is often coiled into a helix, and the arrangement represents its secondary structure. It may then fold itself in different ways in order to form its tertiary structure. Finally, the aggregation of several globular sub-units makes up the quaternary structure. The only method which can give satisfactory results in the analysis of the three-dimensional structure of proteins is X-ray diffraction. By using this technique both fibrous proteins, which are essentially structural proteins (they occur in the ligaments, tendons and hairs), and globular proteins, which are mostly functional proteins (enzymes, oxygen-carriers, antibodies), have been analysed.

The first protein whose spatial configuration was established is myoglobin (1960), followed immediately by haemoglobin. In 1965, 1967 and 1969 the spatial arrangement of lisozyme, ribonuclease and insulin, respectively, also was defined.

The correlation between structure and function is being steadily investigated in the muscular proteins. An effort is being made to find the correlation between muscle contraction and the structure of the proteins which are constituents of the muscle fibres.

The sarcomere in a muscle fibre is a segment between two Z lines, and is considered to be the contracting unit. It consists of threads of various diameters. By means of specific extractive substances, the thick filaments have been shown to be constituted of hundreds of myosin molecules, and the thin ones by another protein (called actin).

These two protein molecules have been analysed by physicochemical techniques and the electron microscope. Morphological investigations, such as those done through the optical and electron microscopes, are always associated with biochemical and biophysical investigations. Each of the two means of investigation does not supply exhaustive results. Formal translation of electron micrographs is not purely descriptive; it is necessary as interpretation. This can be done on the basis of histochemical analyses, biochemical and biophysical examination of the isolated and purified structures, and by molecular disaggregation.

Myosin occurs as an elongated and slender molecule with a bulbous head. By proteolytic methods it can be split in two: light meromyosin (LMM) and heavy meromyosin (HMM). LMM is elongated and slender and it has a prevailing α-helix configuration, while HMM is compact and globular. LMM consists of four sub-units, and HMM of two. Both the structure and function of myosin remain in question. Myosin has been supposed to function as an isoenzymatic complex, in which there exists a monomer-dimer equilibrium; this may be the cause of the difficulties encountered and the confusion generated in defining its characteristics. Interation would depend on the pH and ionic strength of the medium and be related to the formative mechanism of the thread. The myosin's structure also can be studied by immunochemical methods. The main research interest is focused above all on LMM and the four light chains (two DTNM, A–1 and A–2) of which it is made up.

Enzymes: structure and mode of action

Enzymes are globular proteins which catalyse chemical reactions in living things. This function can be accomplished thanks to an intimate association between enzymes and reacting substances (or substrates), and the formation of an 'enzyme-substrate complex' which dissociates itself at the end of a reaction,

delivering the enzyme as well as the products of the reaction.

The formation of the enzyme-substrate complex is thought to cause some alteration in the electronic structure of the substrate, activating it and making it able to react. The enzyme-substrate association also is thought to be specific; therefore each enzyme can catalyse only one or a few very similar reactions.

Investigations carried on by physicochemical methods, X-ray diffraction, and the use of competitive inhibitors (so-called because they can react with an active site and then can compete with the normal reagents in a chemical reaction) would have confirmed the existence, in the structure of lysozyme—an enzyme with a compact globular structure—of a groove which would be the active site. That is to say, this would be the region of the molecule which is responsible for the specific activity of that enzyme. The active site of ribonuclease has an analogous configuration. In both cases it would have a configuration like that of the depression in which the heme is fitted in the molecule of haemoproteins (haemoglobin, myoglobin). In this connexion, the results of the investigations on haemoglobin announced in 1970 can help us understand the way of action of enzymes. It has been stated that the changes in conformation which were observed in haemoglobin during the respiratory process are related to the oxidized and reduced state of the molecule. The fact that a little molecule like that of oxygen can upset the spatial configuration of a protein may usefully orient further research on the control of enzymal activity in relation to similar mechanisms.

The best-known enzymes are those studied by crystallographic methods. They are all constituted of only one polypeptide chain. The enzymes present in these cells are oligomers, i.e. they consist of several polypeptide chains or sub-units, which are defined as monomers. Research on oligomeric enzymes is designed to ascertain whether the monomers are catalytically active or if this activity occurs typically only in the whole enzyme. Investigations on aldolase, carried on with both the whole enzyme and the isolated sub-units, have shown a slight difference between catalytic properties

191

of the whole enzyme and those of the single sub-units. At the same time, they have also shown that the dissociated sub-units are more easily inactivated than the whole enzyme. Therefore, at least in the case of aldolase, the oligomeric state of the enzyme is important for its stability if not for its catalytic activity. Enzymes rarely operate individually. More often they are associated in a multienzymatic complex, which can be defined as an assembly of different functionally correlated enzymes bound together by non-covalent forces in a highly organized structure. An example is that of the multienzymatic system which is associated with the mitochondrial membranes and controls the oxidative phosphorylation.

All enzymes seen to be associated to particulate fractions of cells (lysosomes, mitochondria, ribosomes), although the kind and mode of this association are not yet completely understood.

One of the most striking properties of the cell's activity is its capacity to correlate all biochemical activities in order to maintain a steady equilibrium between the synthetic and catabolic processes. In the 1960s, more and more precise data on the regulatory processes were accumulated; a particular kind of enzyme, the allosteric enzyme whose activity is controlled by metabolic regulation, was identified. These enzymes have structural and kinetic properties, of which other enzymes are usually deprived: they may be activated or inactivated by metabolites (effectors) different from their usual substrates. Often effectors (activators and inhibitors) and substrates do not show structural analogy. The effector, which is not a steric analogue of the substrate (i.e. it does not show the same spatial properties), is defined as an allosteric analogue; the enzyme sites, which possess affinity for the effector, are called 'allosteric enzymes'. These do not correspond to active sites and, on the contrary, are situated in different regions of the enzyme's molecule. By extension, the enzymes which possess allosteric sites in addition to active sites are called allosteric enzymes.

These enzymes show a typical variation in their activity, which is related to substrate and effector concentrations; they may be desensitized by different substances: proteolytic agents, pH changes, or low and high temperatures. Some may be protected against the thermic inactivation by their allosteric effectors, and are subjected to reversible inactivation by cooling them to 0º C.

All the allosteric enzymes which have been studied thus far seem to be constituted of several sub-units. This characteristic could be important in bringing into focus the various theories on the mechanism of allosteric regulation. According to the models proposed to explain this mechanism, it is believed that the allosteric effector, by binding itself with the allosteric site, causes a spatial rearrangement of the active site, making the effector unable to react with its own substrate if the allosteric effector is not removed, thus blocking the activity of that enzyme.

Molecular regulation of genes

Genes are not continually and uniformly active; in relation to cellular requirements, there are some genetic control devices which regulate their activity.

The first model of genetic control was proposed in the early 1960s by Monod and Jacob who had done research on the bacterium *Escherichia coli*. This model made clear that bacteria produce some enzymes only when they are necessary. It explained particularly how a set of genes, whose code for enzymes is necessary in the production of lactose, function only when lactose is present in the medium.

The Monod-Jacob model hypothesizes that genes, which code in *E. coli* for three functional proteins, are associated in a lactose system. The system contains—besides the structural genes which induce the synthesis of the three proteins—a regulator gene (which controls the synthesis of a repressor), a promotor gene (which is the starting point for the synthesis of messenger RNA), and an operator gene. The complex of these genes is called operon and, in this particular case, lac-operon. Within the system, the operator gene is a kind of switch. It is switched off when a protein repressor is bound to it; when the repressor is removed, genes can be transcribed.

Initially the operator gene was thought also

to be the site on which the RNA-polymerase binds itself and begins transcription. Furthermore, research on bacterial mutants has demonstrated the existence of a separate binding site which has been called the promotor. In the absence of a repressor on the operator, RNA-polymerase can act on DNA at the level of the promotor, thus catalysing the transcription.

Further work has demonstrated that this model is also suitable for some viruses, e.g. for the lambda phage, whose genes are under analogous control when they infect *E. coli*.

Investigations on lambda phage have been most helpful in improving our knowledge of molecular biology, especially the understanding of the modes of gene control at molecular level as well as of some details of transcription.

During the reproduction process which follows the invasion of a bacterial cell, lambda phage inserts its own DNA in that of the host (lysogeny). Then a protein repressor inactivates the DNA, although at each division of *E. coli* it is copied and transmitted to bacteria of the succeeding generations. At a certain moment, the repressor may be removed, thus reintegrating the transcriptive power in the viral genes; this, in turn, can synthesize viral nucleic acids and proteins, thus producing new viral particles which cause the lysis of the bac-

terium and, once delivered, can infect new hosts.

The viral DNA trait which is inserted in the bacterial DNA has been called 'prophage'; it seems to be continuously localized near the *gal* locus, controlling the galactose metabolism.

Lambda phage is a temperate phage, i.e. a virus which may be in a lysogenic condition, a state of association with the host only rarely broken off by lysis. Two vital processes are typical of this virus cell association: prophage incorporation and lysogenic immunity, which are also defined as integration and prophage genetic regulatory processes.

It has been noted that the integration-excision problem (i.e. the problem concerning two of the most important events in the lambda phage-host interrelationships) depends on well-defined gene action. Whereas integration occurs with only one gene (*int*), excision requires the combined action of two genes (*xis* and *int*). The gene array on the lambda phage map corresponds to the successive steps in physiological and morphogenetical processes leading to the production of new viral particles. Up to the sequence of events related to development, a regulatory system has been found which is responsible for the choice between viral development and co-existence with the bacterial cell. Transcrip-

FIG. 45. If bacteria sensitive to the lambda phage (such as the bacteria of a specific stock of *Escherichia coli* which have survived strong ultraviolet radiation) are infected with this phage, there are two possible responses. In some cells, the phage immediately enters into a vegetative phase, and follows a lytic cycle: new lambda phages result. In others, the process is quite different: the bacteria are able to free lambda phages after induction, and become immune to these phages. In other words, they have become lysogens. Thus the process may be regarded as a race between the synthesis of precocious proteins and that of the repressor: if the former wins the race, there is lytic response, if the latter, lysogen response. In the first case, the genetic material of the phage remains autonomous and is replicated; in the second, it is inserted again in the bacterial material.

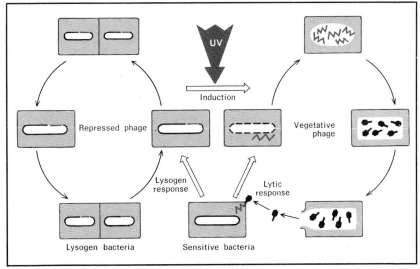

tion of the initial genes *O* and *P* causes the replication of phage DNA, while *N* transcription determines the production of a positive regulator. The latter enhances considerably the transcription of lambda phage and activates *O* gene. The product of *O* has, in turn, a positive regulatory effect on the expression of genes arrayed successively from *A* to *F* and responsible for the phage heads morphogenesis, and of genes arrayed successively from *Z* to *H* and responsible for the tail's morphogenesis, phage assembly and antigenic properties.

The isolation of a protein repressor from lambda phage (1967) and the demonstration of its interaction with specific sites in phage DNA

(1968) confirmed the operon model of Monod and Jacob. The repressor has helped, during a certain period, the understanding of how lysogeny is maintained. Further results (1968, 1972) have hypothesized that the regulatory system may be a little more complicated than it was at first believed.

The most important new acquisition is the existence of two regulatory phases which depend on the expression of two genes. One of these is the determinant of the classical repressor, while the other seems to be a different regulator element, with an action antagonistic against the first (the antirepressor, 1970, 1971). Investigations on its mode of action have shown that the antirepressor has a mode of action very similar to that of the repressor. Data (of 1970, 1971, 1972) indicate that the antirepressor acts negatively on both the synthesis of the repressor and the transcription of N-CIII and CII-O-P operons.

Future work should explain the meaning of the surprisingly similar modes of action of the two correlated regulatory systems, as well as

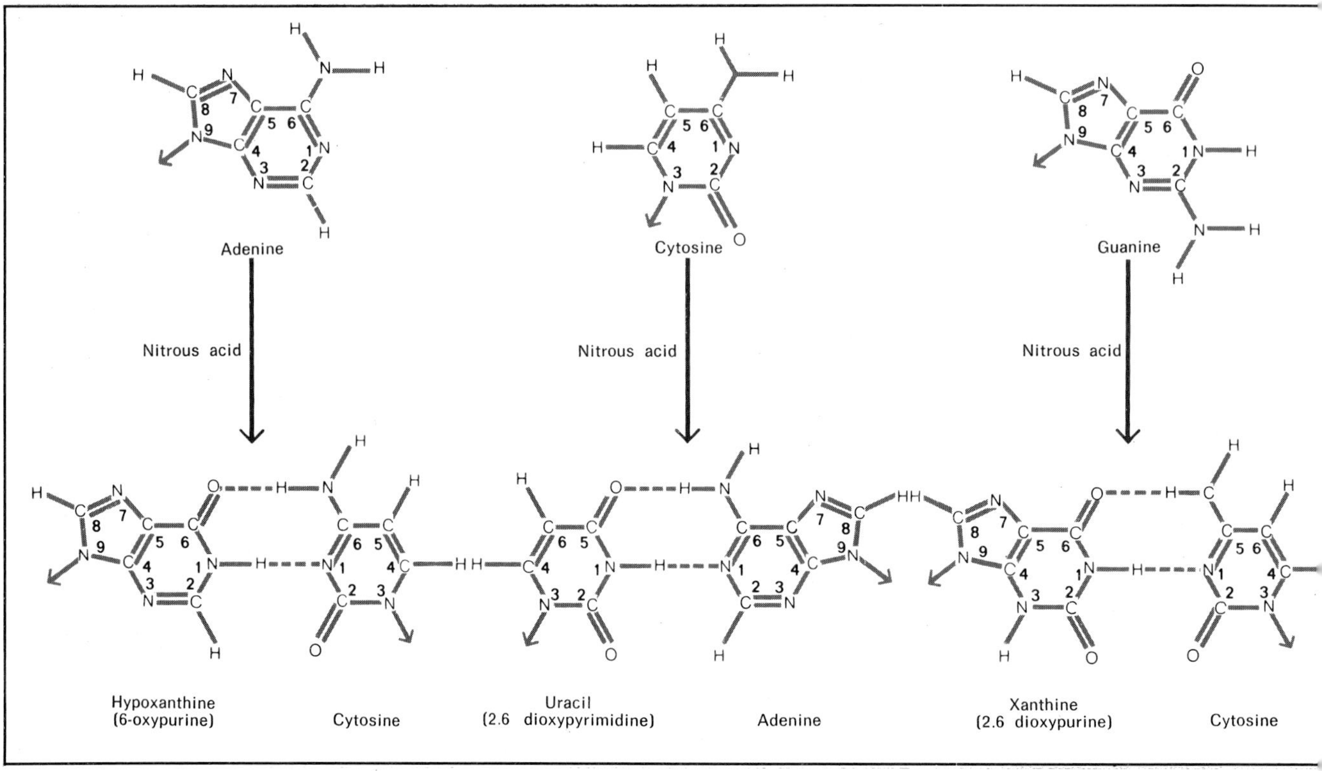

FIG. 46. Left, induced effects of nitric acid on DNA base. Right, the processes according to which induced transitions are made on DNA molecule. Diagram shows only the two phases in which mutagenous event takes place; the bold line corresponds to the original helix, whereas the broken lines represent new helixes formed in the process.

the meaning of the two states of regulation (the sensitive and the immune states) which have been observed in cells.

The cell possesses two kinds of control mechanism: the negative control mechanisms which inactivate certain genes, and the positive which activate those genes.

Production of L-isoleucine in *E. coli* is related to the amount of this amino acid present in the cultural medium. When the amount is exceedingly high, the bacterium ceases to produce it. This fact is explained by the existence of a feedback regulatory system. Indeed, the presence of an exceedingly large amount of L-isoleucine inhibits the activity of L-threonine deaminase, an enzyme which is necessary in the first synthetic reaction and blocks the production of all enzymes needed for the synthesis of L-isoleucine. The two control mechanisms are independent of each other. By means of mutant bacteria, it was possible to establish that the genes which control them are completely separate on the chromosome.

An instance of positive control is that of the storage of substances in the cell and of energy utilization. Animal cells store energy in the form of glycogen. Glycogen is synthesized from a precursor, glucose-6-phosphate, through three enzymatic steps. When a cell has a good energy supply, it produces large amounts of glucose-6-phosphate which functions as a signal to stimulate the synthesis of glycogen. When the energy supply diminishes below a certain level, it becomes necessary to activate an enzyme which splits glycogen (the enzyme being known as glycogenphosphorylase). The activation occurs thanks to AMP (adenosinmonophosphate), which behaves as a chemical signal. It is produced by the splitting of ATP, the main source of energy in the cell. ATP accumulation indicates, therefore, that the cell has consumed its own energy. Regulatory phenomena are known in most of the syntheses which occur in the cell: lipid, amino acid (protein precursor), purine, and pyrimidine (nucleic acid precursor) synthesis.

A particular kind of gene regulation is

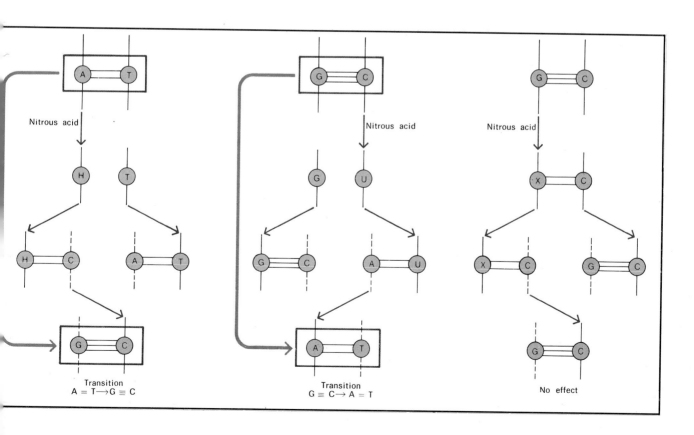

Transition
A = T → G ≡ C

Transition
G ≡ C → A = T

No effect

catabolite repression, which is determined by glucose or by biochemically related compounds such as glucose-6-phosphate, fructose or glycerol.

Although catabolite repression has been observed since the beginning of this century, our understanding of the mechanism has been acquired only recently. For a long while it was hindered by the impossibility of determining which effects (among several glucose effects) were directly correlated with catabolite repression. The determining event in understanding the phenomenon was the demonstration (1965) that, in the presence of glucose, an abrupt decrease of the intracellular level of cyclic-AMP is observed. Afterwards, some investigators succeeded in inverting the repressive effect of glucose in enzymic synthesis by adding cyclic-AMP. Other investigators have demonstrated that modification or deletion of the *lac*-promotor not only eliminates the catabolic sensitivity of the *lac*-operon but also causes the loss of stimulation by cyclic-AMP. This result shows that the site of catabolic sensitivity is closely associated in the *lac*-operon to the promoter. As the initiation of the synthesis of messenger RNA is thought to occur at the locus of the operon promotor, cyclic-AMP (or a very similar derivative) is likely to act on this gene site.

Cyclic-AMP is believed to act together with a protein factor, the catabolite gene activator protein or CAP. Two different modes of action have been suggested: according to the first, cyclic-AMP may react with the protein factor CAP in order to form a complex which, in turn, interacts with the promotor and DNA-dependent RNA-polymerase to initiate transcription. The second mechanism suggests, on the contrary, that cyclic-AMP could react with CAP to catalyse directly or indirectly a change in DNA-dependent RNA-polymerase, such as phosphorylation or adenylation, so that it can react better with the promotor.

The second of these two mechanisms is the simplest and is likely to be more compatible with the fast kinetics and typical reversibility of the phenomena of catabolic repression. In both hypotheses, in order to react, the CAP must bind itself with cyclic-AMP.

Having attained, by experimental means,

quite good certainty of the fact that cyclic-AMP binds itself with CAP, it then becomes necessary to understand the interaction between CAP and macromolecules of biological interest. Attempts have been made to recognize the formation of an RNA-polymerase complex, but these have been unsuccessful; attempts to describe the interaction between CAP and DNA have been more encouraging. It has been possible, in fact, to demonstrate that cyclic-AMP stimulates the formation of a CAP-DNA complex. The mode of action of CAP seems to be similar to that of the sigma factor, and its functioning may require the presence of the sigma factor.

Chemical mutagenesis

Mutations can arise spontaneously under normal conditions during the life of every organism (spontaneous mutations), or can be induced experimentally by physical or chemical means. The origin of spontaneous mutations is, at the moment, little known and dubious, whereas there is a great deal of experimental data concerning induced mutations.

Nowadays, it is extremely easy to find physical conditions or chemical substances capable of inducing mutations. In 1927 H. J. Müller discovered the mutagenic action of ionizing radiations (X-rays and γ-rays) and shortly afterwards it was possible to demonstrate that the most potent mutagenic agent was ultra-violet radiation (UV). For about 20 years attention was concentrated on these physical methods until, during the Second World War, C. Auerbach found the first chemical mutagen, a derivative of mustard-gas, unfortunately notorious as a result of the 1914–18 conflict. From that moment the announcements of chemical compounds endowed with mutagenic properties proliferated impressively, so much so that it justifies some experts in maintaining that it is easier to show a compound to be a mutagen than to prove unequivocally its complete innocuousness towards the gene. It would not, therefore, be of any significance to provide the reader with a long catalogue of chemical formulae. Instead it seems more logical to take a few small groups of compounds, which are recognized to have a definite specificity, and to describe their known, or at least presumed, mode of action at the molecular level.

The problem of mutagenic specificity has always troubled geneticists. No one is able to visualize, given our knowledge of the structure of DNA, a substance capable of altering one specific gene and not others. Instead the dream, too often the delusion, of those who have applied themselves to this field of research has been to discover agents capable of inducing one particular type of mutation alone; for example an $A = T \rightarrow G \equiv C$ transition, or a specific transversion or deletion, etc. Among the hundreds studied so far, only two have shown a high degree of specificity, and a very few others have shown a satisfying tendency to induce preferentially one or the other of the known types of mutation. We will limit ourselves to discussing the action of these rare exceptions.

The base analogues are substances very similar to the normal purine and pyrimidine bases, from which they differ only by some altered or additional chemical substituent. Some of these are able to be incorporated into duplicating DNA in place of the normal base, and later cause errors in base pairing. A characteristic of the base analogues most used nowadays (5-Bromouracil and 2-Aminopurine) is that they act only during cell division; they are completely inactive on the cell in its resting phase precisely because the prerequisite of their mutagenic activity is their incorporation in the form of nucleotides into a DNA molecule, and this can only happen during the process of multiplication. Another important aspect of these mutagens is that they cause exclusively transition type substitution of the bases.

An example will clarify the mode of action of these base analogues. The normal base pairs $A = T$ and $G \equiv C$ are stabilized by hydrogen bonds between the opposed bases. Such bonds are able to form between the nitrogen atoms of the aromatic rings or between the substituent groups. But, in each case, always between two atoms, one of which is saturated, the other unsaturated. This explains the pairing of adenine with thymine stabilized by two hydro-

gen bonds, one between the two nitrogen atoms of the rings and another between the amine group and the ketone group substituents in their respective bases. It also explains the pairing of guanine and cytosine, with three hydrogen bonds, and shows clearly why guanine is unable to pair with thymine or adenine with cytosine.

5-Bromouracil (BU), the most commonly used mutagen in this category, is an analogue of thymine, from which it differs by the substitution of a bromine atom at position 5 in place of a methyl group. This molecule is able to exist in two tautomeric forms, one the normal keto isomer, the other the rarer enol form. This ability to exist and to change between two forms, one rarer than the other, is not a property unique to BU but is shared with the other pyrimidines and purines. The difference, and hence the mutagenic power, is due solely to the high electronegativity of the bromine atom, which increases the frequency of formation of the less stable form in BU. Bromouracil, in the normal ketone state, can be incorporated into DNA in place of thymine, making a pair with adenine, and leaving the polynucleotide undisturbed. In fact it is possible to produce DNA in which the thymine is quite significantly substituted by BU without any functional damage. But if at the moment of incorporation a molecule is in the enol form, it pairs itself with guanine (i.e. it commits an error of incorporation) and makes the base pair $BU \equiv G$. At the subsequent replication of the DNA it is highly improbable that the BU remains in the enol form, which is a rare, unstable and transient state. When the two DNA strands separate the BU returns to the ketone form, pairs itself with adenine in the regular way, and at the next division the adenine pairs with thymine. The result, after two cellular divisions, is a transition of $G \equiv C \rightarrow A = T$.

The error can occur also at another moment. If one imagines a BU molecule incorporated normally and bound to adenine; during replication the BU may pass to the enol state and pair with guanine, committing an error of replication. The guanine in its turn, at the subsequent division, pairs with cytosine and the result is a transition $A = T \rightarrow G \equiv C$.

Another important mutagen is 2-amino-purine, an analogue of adenine, which, in its normal state pairs with thymine, but, in the rare imino form pairs with cytosine, thus committing in its turn errors of incorporation and replication. Bromouracil gives a higher frequency of transitions $G \equiv C \rightarrow A = T$ than $A = T \rightarrow G \equiv C$, whereas 2-aminopurine shows the reverse tendency. Both, however, are highly specific for the induction of transitions.

The base reagents are substances capable of altering the chemical structure of the bases even during the resting phase of the DNA. This first event, however, has to be followed by a duplication before the mutation can be fixed in a stable way. In this group are found some of the best known and most effective mutagens.

Nitrous acid (HNO_2) oxidatively deaminates the bases with free amine groups (adenine, cytosine and guanine), with the result, shown in Figure 46, that adenine is transformed into hypoxanthine, which at the subsequent replication becomes paired with cytosine. The cytosine in its turn pairs with guanine and the result is a transition $A = T \rightarrow G \equiv C$.

If the nitrous acid reacts with a cytosine, it transforms it to uracil, which cannot pair with guanine and causes, therefore, at the second generation, the transition $G \equiv C \rightarrow A = T$. Nitrous acid converts guanine into a compound (xanthine) which normally pairs with cytosine and therefore does not cause any alteration in the nucleotide sequence.

The action of nitrous acid is, therefore, specific for transitions in the two directions, depending on the base which has reacted; but at a very much higher concentration the same mutagen can provoke more gross alterations in the double-helix structure with consequent breakage and unpredictable mutagenic results.

Hydroxylamine (NH_2OH) is perhaps the most important experimental agent to date. It acts on the pyrimidines, and at a weakly acid pH (about pH 6) its action is 90 per cent restricted to cytosine; this is probably transformed into N-6-hydroxycytosine (IC), and perhaps also into other compounds which, however, all pair with adenine. The sequence of events during the mutation can thus be written schematically:

$$G = C \xrightarrow{NH_2OH} G? = IC \rightarrow A = IC \rightarrow A = T.$$

This mutagen, therefore, if used at optimum conditions of pH and concentration (1M) is unique, at the moment, in its absolute molecular specificity, in that it causes a unidirectional transition $G \equiv C \rightarrow A = T$.

The acridines are mutagenic compounds of exceptional importance, both for their unique mechanism and for their specificity. Some of the acridines are also of great use in medicine (some types of anti-malarial agents) but it seems as though the medical authorities of various countries have paid little heed to the danger of their mutagenic effect. Noticed by geneticists for about twenty years, these substances entered the history of mutagenesis when their action in the absence of light was distinguished from that due to photochemical phenomena in the presence of a light source. In this discussion we are interested solely in the mutagenic action in the dark, which has been explained by the work of several independent research groups.

L. S. Lerman has demonstrated that the acridines form fairly stable complexes with DNA. On the basis of the physiochemical characteristics of these complexes, he formulated the hypothesis of intercalation. A molecule of acridine fits between two adjacent pairs of bases distorting the structure and widening the interstice by 0.34 nm, that is the space occupied by one pair of nucleotides.

The mutagenic action of the acridines in the dark takes place only during particular cycles of cell multiplication. They are completely inactive on bacteria, and equally have no effect on eucaryotic organisms, such as yeasts, during mitosis. They are instead highly mutagenic for bacteriophage and for yeast during meiosis. In other words the action of these compounds is linked to the reproductive cycle accompanied by active recombination of genetic material, consisting of an exchange of large fragments between the homologous DNA molecules.

S. Brenner *et al.* (1961) have demonstrated that mutations induced by the acridines are different from those obtained with any other mutagenic agent of another type. The acridines produce, in fact, almost exclusively insertions and deletions.

This is the result expected if two identical molecules of DNA pair for an exchange of genetic material and one of them carries an intercalated acridine. When the exchange takes place between only two of the four helices, or between the entire molecules, one of the two DNA molecules will have an insertion, the other a deletion.

Recently some new acridines have been synthesized, which are capable of inducing insertions and deletions even during bacterial development without genetic recombination, but their mechanism of action is completely unknown.

One has, therefore, in the acridines a unique mutagenic agent, highly specific for the production of insertions and deletions.

Energy transformations

Energy transformations occur in the cells of living things as a result of mechanisms which are very similar in the animal, plant and bacterial kingdoms. The energy cycle in living matter takes place according to the following series of processes. The first source of energy is the Sun; plants absorb solar radiation in the process known as chlorophyll-photosynthesis and make up carbohydrates which are then utilized and oxidized in metabolic processes in order to produce CO_2 and H_2O, with the release of a remarkable amount of energy. This energy is trapped in the chemical bonds of ATP (adenosintriphosphate) by oxidative phosphorylation at the level of the respiratory chain, in the inner mitochondrial membrane.

Since the process of oxidative phosphorylation was discovered during the 1930s, several theories have been put forward in order to explain its mechanism. In the 1950s the chemical coupling hypothesis was proposed: three sites for energy conversion by energy-rich non-phosphorylated intermediates would be present along the respiratory chain. The theory has not yet been confirmed experimentally, given the inability to define the chemical nature of the intermediates.

In 1961 a second, chemi-osmotic, hypothesis was made. It does not postulate the presence of energy-rich intermediates, but maintains that electron transfer produces a separation

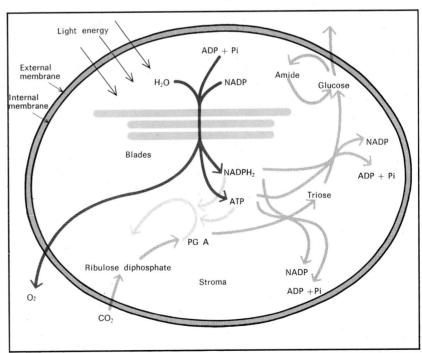

Fig. 47. When chloroplasts are illuminated by light of appropriate wavelength, they reduce the carbon dioxide to glucides, with production of oxygen. Photosynthesis occurs in two successive stages: a light reaction, located in the lamellae, and a dark reaction, occurring in the stroma or intercellular material. ADP, adenosine diphosphate; NADO, nicotinamide adenine dinucleotide phosphate; NADPH₂, NADP reduced; Pi, inorganic phosphorus; ATP, adenosine triphosphate; PGA, phosphoglyceric acid.

of charges on both sides of the inner mitochondrial membrane: the positive charges (H^+) outside, and the negative charges (OH^-) inside. The mitochondrial membrane would have several little channels through which hydrogen ions would flow, carrying the enzyme ATP-synthetase, which then catalyses the ATP synthesis.

In 1967 a third, conformation coupling, hypothesis was proposed. It is based on the fact that, during the steps of oxidative phosphorylation, a conformational change in the mitochondrial membranes is observable.

Most of the investigations which have been carried out so far—to determine the molecular mechanism which makes the energy of the electron flow to be stored in the chemical bond—emphasize that membranes constitute a major element in the process of energy coupling. Indeed, it has been observed that the process of oxidative phosphorylation needs good co-ordination to ensure that the sequence

of reactions will occur with the least loss of energy. The rhythm and efficiency of such a metabolic complex can be achieved because the enzymes and other elements involved are related chiefly to membranous structures.

Electron microscopy gave evidence (1970) of compact aggregates of particles of various sizes within the mitochondrial membranes. As these sizes correspond to those of membrane-bound enzymic molecules, the aggregates might be interpreted as being enzyme molecules with an average diameter of 7 nm and a width of 15 nm.

On the basis of these observations, a model has been designed to explain the behaviour of the enzymes in mitochondrial membranes. Enzyme molecules of the respiratory chain would form an aggregate of closely packed molecules and the enzymes which catalyse the successive reactions would be situated closely together. In this way, the sequence of reactions would reveal spatial correspondence in the arrange-

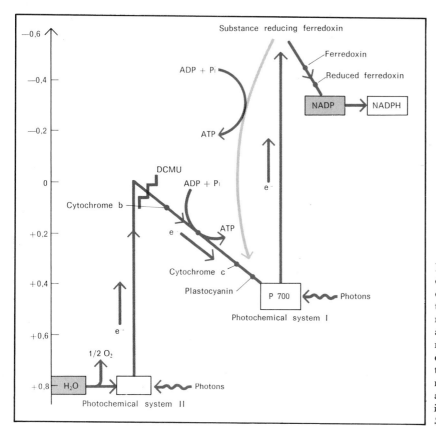

FIG. 48. Diagram of the transport chain of electrons in photosynthesis: the electrons subtracted from the water are transported along a series of donors and receptors until they reach the NADP and take part in its reduction. The energy required for pushing them against the electrochemical gradient is supplied by the photons which excite the chlorophyll molecules in photochemical systems I and II. DCMU is a substance which interrupts electron flow. (Copyright: *Scientific American.*)

ment of the enzymatic molecules. Accordingly, only the outer portion of the molecule would be involved effectively in the reactions and would represent the active site. The need for each component of the respiratory chain to enter into contact alternatively with the foregoing and the following components, in order to carry electrons, presupposes the possibility of positional changes in the molecules in order that active sites might come into contact with each other.

The process of oxidative phosphorylation is studied by means of different decoupling agents, which allow respiration to take place but protect the transformation of ADP into ATP during this process, with the subsequent accumulation of bond energy.

The first decoupling agent used was 2.4-dinitrophenol; later, there were discovered many other decoupling substances, including several antibiotics.

Oxidative phosphorylation also has been studied in phosphorylating particles, obtained by treating mitochondria with membrane-dispersing agents such as digitonine and by sonic irradiation. These particles evidently were formed by membrane fragments which did contain most of the enzymes essential to electron transfer and oxidative phosphorylation.

By means of these submitochondrial samples, it was possible to identify protein factors necessary in the process. In particular, two proteins have been identified: an ATP-ase with a high molecular weight (about 280,000), called F_1, and an F_0 protein.

F_1 factor has been purified and does not contain electron-carriers. It is likely to be the enzyme responsible for the phosphate transfer on ADP. Now it is necessary to define the action by which it achieves this transfer and how it receives energy from the reactions which are responsible for the electron transfer.

The action of some enzymes taking part in the breakdown of glycogen into pyruvic acid has thoroughly interested several investigators. The action of phosphorylases, the enzymes which catalyse glycogen transformation into glucose-1-phosphate, has been defined. Phosphorylases are known to occur in two interchangeable forms, phosphorylase a and b.

Phosphorylase b is not active in the absence of AMP, whereas its affinity to inorganic phosphate and glycogen increases according to AMP concentration; on the contrary, by increasing the concentration of one of the substrates, its affinity for AMP increases. These interactions result from conformational changes in the molecules of phosphorylase a, caused by phosphorylase-b-kynase, an enzyme which seems to be activated by hormonal substances such as adrenalin.

During the 1960s, very important achievements were made in our understanding of the distinctive steps of photosynthesis. Photosynthesis is performed by autotrophic organisms in order to transform simple inorganic substances, such as CO_2 or H_2O, into complex organic substances. This is accomplished by using solar energy, which constitutes the source of almost all biological energy.

The conversion of solar energy into chemical energy also involves a process of electron transfer and a coupling phosphorylation process similar to those occurring in respiration in heterotrophic organisms. Besides these fundamental analogies, other common features between photosynthesis and respiration exist. In fact, in higher organisms both processes occur in membrane-bound cellular organelles, such as chloroplasts and mitochondria (very similar in their ultra-structural and molecular organization). Both kinds of organelles possess their own DNA and show autoduplicative properties, which to a certain degree are independent from nuclear chromosomes.

In the photosynthetic organisms, the absorption of light energy is attributable to the presence of chlorophyll and other pigments (such as the carotenoids and phycobilins) localized in the chloroplasts. Light is absorbed as quanta of energy. Each photosynthetic pigment has a typical absorption spectrum range; it absorbs more or less light at different wavelengths according to its own molecular structure. According to some investigators, chlorophyll b, carotenoids and phycobilins do not participate directly in the photosynthetic processes but transfer excitational energy to chlorophyll a. Only the last is directly involved in subsequent reactions.

Water is the electron donor from which

oxygen is released, except in photosynthetic bacteria; these utilize H_2S, H_2 or organic compounds as electron donors. The main electron acceptor is CO_2.

The first series of reactions in photosynthesis is represented by the reduction of $NADP^+$ and the phosphorylation of ADP into ATP, made up of two light reactions. Afterwards, the NADPH and ATP thus formed are released in the non-membranous part of the chloroplast, where CO_2 transformation into carbohydrates occurs through the participation of a series of enzymes. CO_2 transformation reactions also may occur in the absence of light; they are known as 'dark reactions'.

It has been observed that in eucaryotic cells photosynthesis takes place in thylakoids, which are flattened vesicles inside the chloroplasts, packed in grana. The two light reactions of photosynthesis occur in two different photosystems. Photosystem 1, which contains most of the chlorophyll a, β-carotene and a molecule of P-700 (a form of chlorophyll a which has an absorption maximum at 700 nm), is responsible for the trapping of energy; it is activated by the longer light wavelengths. Photosystem 2, which contains most of the chlorophyll b and other accessory pigments, is activated by shorter light wavelengths and is responsible for the release of oxygen. The excitation of photosystem 1 causes the $NADP^+$ to be reduced through a series of carriers, which include ferredoxin and ferredoxin-NADP-oxydo-reductase. The two systems function in series in order to restore the loss of electrons. Electrons, necessary to restore the integrity of photosystem 1 after excitation, are supplied by the excited photosystem 2 through a series of electron carriers which include plastoquinone, cytochrome b_6, cytochrome f and plastocyanin. System 2 is regenerated by water through dehydrogenation. Efforts are being made now to discover the mechanisms operating in dehydrogenation.

Electron carriers yet to be identified are believed to be present in the two photosystems. Between cytochrome b and cytochrome c, at least one component should exist (still to be identified), as has been demonstrated by experiments on mutated strains of green unicellular algae.

In order to release 1 molecule of oxygen and to reduce 1 molecule of CO_2 to sugar, 8 light quanta at 700 nm are necessary. When chlorophyll or one of the other photosynthetic pigments absorbs photons, it passes from a lowest energy state, or ground state, to a higher energy state. The excited state is unstable and the pigment may return to its initial state in 10^{-9} seconds. If advantage is not taken to produce NADPH and ATP during this short period, energy is released as fluorescent light.

In 1965 and 1966 two research groups demonstrated the occurrence of a different pathway for CO_2 fixation in tropical plants, particularly in sugar cane. In this cycle, malic and aspartic acids (two four-carbon compounds) instead of phosphoglyceric acid (a three-carbon compound) occur as stable intermediates.

Further reading

BECKWITH, Zipser. *The lactose operon*. Cold Spring Harbor, 1970.

BELOZERSKIJ, A. N. *Nukleinovye kisloty i evoljucionnaja sistematika*. Moskva, Znanie, 1969.

BERNHARD, S. A. *The structure and function of enzymes*. New York and Amsterdam, W. A. Benjamin, 1968.

CASSELTON, L. A. Suppressor genes. *Science progress*, vol. 59, no. 234, summer 1971.

CHAPMAN, D.; LESLIE, R. B. *Molecular biophysics*. Edinburgh, Oliver & Boyd, 1967.

CHOKE, H. C. Genetical studies on active transport. *Science progress*, vol. 59, no. 233, spring 1971.

COULT, D. A. *Molecules and cells*. London, Longmans, 1966.

DUBININ, N. P. *Genetika i ee gorizonty*. Moskva, Znanie, 1965.

GREEN, D. E.; GOLDBERG, R. F. *Molecular insights into the living process* (2nd ed.). London, Academic Press, 1968.

HANDLER, P. *Biology and the future of man*. London, Oxford University Press, 1970.

HERSHEY, A. D. (ed.). *The bacteriophage lambda*. Cold Spring Harbor, 1971.

HOCH, F. L. *Energy transformations in mammals: regulatory mechanisms*. London, W. B. Saunders, 1971.

HOLLAND, I. B. DNA replication in bacteria. *Science progress*, vol. 58, no. 229, spring 1970.

KENDREW, J. *The thread of life*. Cambridge, Harvard University Press, 1966.

KOLLER, P. C. *Chromosomes and genes: the biological basis of heredity*. Edinburgh, Oliver & Boyd, 1968.

LAETSCH, W. M. Relationship between chlotoplast structure and photosythetic carbon-fixation pathways. *Science progress*, vol. 57, no. 227, autumn 1969.

MARTIN, D. T. M. The operon model for the regulation of enzyme synthesis. *Science progress*, vol. 57, no. 225, spring 1969.

MEZEL'SON, M. S. *Reduplikacija i rekombinacija genov*. Moskva, Znanie, 1967.

MOSBACK, K. Enzymes bound to artificial matrixes. *Scientific American*, vol. 224, no. 3, 1971, p. 26.

MOSS, D. W. *Enzymes*. Edinburgh, Oliver & Boyd, 1968.

NEEDHAM, J. (ed.). *The chemistry of life*. London, Cambridge University Press, 1970.

PTASHNE, M.; GILBERT, W. Genetic repressors. *Scientific American*, vol. 222, no. 6, 1970, p. 36.

RYZKOV, V. L. *Vzaimodejstvie jadra i citoplazmy v nasledstvennosti*. Moskva, Znanie, 1968.

SMITH, C. U. M. *Molecular biology*. London, Faber & Faber, 1968.

TEMIN, H. M. RNA-directed DAN synthesis. *Scientific American*, vol. 226, no. 1, 1972, p. 24.

TROŠIN, A. S.; BARENBOJM, G. M. Ot kletki k molekule i ot molekuly k kletke. *Budušee Nauki*. Moskva, Znanie, 1968.

VOLFIN, P. La mitochondrie, centrale énergétique de la cellule. *La recherche*, vol. 2, no. 15, 1971, p. 741.

VOL'KENSTEJN, M. V. Voda, belki, geny. *Nauka i Čelovečestvo*. Moskva, Znanie, 1968.

WILLS, C. Genetic load. *Scientific American*, vol. 222, no. 3, 1970, p. 98.

21 Cell biology

The nucleus

The nucleus is a cell organelle which is present in all animal and plant cells, except in bacteria, where the nuclear material is dispersed in the cytoplasm. Chromosomes and the nucleolus are contained in it and are immerged into a liquid matrix. Chromosomes are not visible to the naked eye in the nucleus at rest. The functional activities of nucleus are very homogeneous and are bound to the presence of nucleic acids, which are the starting point of all metabolic activities of the cell. Particularly, DNA replication and RNA transcription occur in the nucleus. By different methods (the use of tracers, autoradiographic techniques, nuclear transplantation and so on), it has been possible to ascertain that the synthesis of all the known RNA—messenger RNA, transfer RNA, ribosomial RNA—are present in the nucleus and afterwards they migrate in the cytoplasm through the pores of the nuclear membrane.

Most of the investigations which have been carried out on the nucleus during the 1960s were of molecular biology concern. It has been attempted through these to find a correspondence between the structures which are visible through the microscope and the knowledge of the structure of nucleic acids, acquired by other means. Electron microscopy has been one of the most useful means to define which are the persistent structures in the nucleus at rest and which events occur when chromosomes become visible in the dividing nucleus. According to some investigators, structures which will give rise to chromosomes are already present in the nucleus at rest as elementary unwound threads, which assemble at the moment of division in order to form chromosomes. According to other specialists, the chromosome formation at the beginning of cell division is due to molecular condensation. Molecular biological data which were acquired during the 1960s support the hypothesis of the persistence of the chromosomal structures in the nucleus at rest. In addition, electron microscopy has supplied structural data on the nuclear membrane and nucleolar structure.

It has been established, for instance, that the nuclear membrane is made up of two layers, the outer one in contact with the cytoplasm and the inner one in contact with nuclear material. Between these two layers is a space called perinuclear cistern. Its projections are continuous with the endoplasmic *reticulum cisternae* which are recognizable in the cytoplasm. In the nuclear membrane of all cells holes have been observed, which have been named membrane pores; they are delimited by a thickened rim, derived from the fusion of the two membrane layers. A possible direct communication between cytoplasm and nuclear material has been debated: investigations carried out by improved techniques have demonstrated the existence of a very thin septum. But it has been also demonstrated that different kinds of material, including some large macromolecules, such as ribonuclease, may pass through the membrane pores. Septa, or reinforced rings which exist at the level of the membrane pores, would behave in certain circumstances as regulatory devices, by controlling the substances entering the cytoplasm from the nucleus and vice versa. In addition, particular attention has been paid to research on the structure and function of nucleoli, which are present in various amounts in different kinds of nuclei. Some of them appear through the electron microscope as dense and spongy masses, while others have a long thread to which chromatin masses are associated. By cytochemical procedures it has been possible to define their chemical composition: about 40 per cent water associated with proteins, phospholipids, polysaccharides and RNA. Generally, a very small amount of DNA is also present. RNA storage has been demonstrated in the nucleolus by using radio-active precursors: the RNA is synthesized by that portion of chromosome which is associated with the nucleolus and in which the transcription of ribosomial RNAs occurs. These RNAs then migrate into the cytoplasm, where they assemble with proteins to form functioning ribosomes, ready for protein synthesis.

Chromosomes

Chromosomes are thread-like structures present in the cell nucleus and are characterized by a complex organization made up of DNA and basic proteins, called histones. Chromosomes become visible at the beginning of division of the cell. In the nucleus at rest only masses of chromatin are evident. According to our knowledge of molecular biology acquired in the last ten years, chromosomes and chromatin seem to be two states of the same material.

Chromosomes originate by duplication of pre-existing chromosomes. In 1958 (by Meselson and Stahl), replication of DNA in

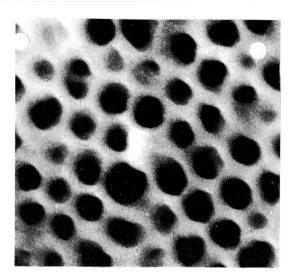

FIG. 49. The nuclear membrane is found, when examined by electron microscope, to be formed of two laminar strata separated by a perinuclear cistern. All the nuclei have a number of small openings, known as membranopores, but there does not appear to be direct communication between the karyoplasm and the cytoplasm because of the existence of a thin boundary sector. In this photograph the pores of the nuclear membrane of the protozoon are magnified 50,000 times. (*Source*: A. Bairati, G. Steinart.)

bacteria was shown to be semi-conservative: each DNA helix acts as a template for the synthesis of a complementary helix; two new helixes are then obtained, half new and half old, identical to the original one. A similar process seems to occur also in the duplication of the chromosomes of higher organisms. It has been observed that the DNA replication of certain chromosomes is initiated at different sites. In mammals, a starting point has been identified at every 100 μ, but the observation needs more experimental evidence. By means of autoradiographic techniques, it has been demonstrated that growth occurs in both directions from any single starting point.

At the end of the cell's division, some portions of chromosomes unwind and stain weakly (such as euchromatin), whereas others remain condensed and stain deeply (heterochromatin). According to some authors, heterochromatin would represent chromosomal regions in which genes are inactive.

The study of giant (polytenic) chromosomes of *Diptera* has shown that there is intermittent activity of chromosomal traits containing genes. Polytenic chromosomes result from successive division of single fibres, without their subsequent distribution in the daughter cells. A multistranded structure is thus formed, 100 times as thick as the normal chromosomes and 10 times as long.

In these chromosomes, active genes occur as 'puffs', i.e. as zones scattered along the chromosome, which appear as swellings due to the unwinding of the strands (or chromonemes) of the chromosome. By means of direct staining or autoradiography, a considerable amount of RNA has been identified at the level of these puffs. The RNA present in a puff is different from that in other puffs of the same chromosome. This property and that of high molecular weight suggest that the RNAs in puffs are messenger RNAs. The occurrence of a puff at the level of a certain chromosome band is not a constant fact: during the larval life, swelling and unswelling cycles alternate. By studying these cycles, time patterns of genic activity in different tissues of the developing larvae of insects have been constructed. By supplying hormones and other substances, it has been possible to initiate, to block and to prevent the activity of the various puffs, thus demonstrating a correlation between some physiological activities and the occurrence of certain puffs.

The study of 'puffing' has been utilized above all in the analysis of the mechanism by which hormones regulate genes. This has been carried out by observing changes in the pattern of puffs during the metamorphosis and, particularly, in relation to ecdysone hormone, from which the initiation of the process and its

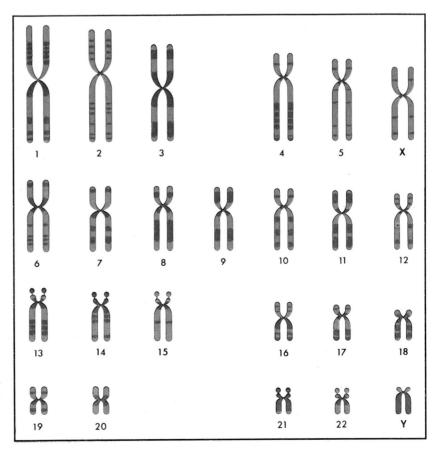

FIG. 50. Representation of a human karyotype deduced from observations made by Giemsa's technique, making it possible to identify all the pairs of homologous chromosomes and different regions within the arms of the chromosomes themselves. The bands of darker colour are formed, probably, by repetitive DNA.

normal development depend. These experiments have confirmed that ecdysone regulates the activity of single sites during mould, and that such an action manifests itself in the appearance and disappearance of various puffs at different times.

Although DNA is the genetic material in both procaryotes and eucaryotes and the genetic function is very similar, therefore, in these two groups, eucaryote chromosomes have been shown to differ from those of procaryotes for their chemical composition, molecular structure and probably their way of action.

The amount of DNA which is present in the chromosomes of higher organisms is thought to be much larger than the amount necessary for the production of the required m-RNAs, but it remains to be explained why much more DNA is present in a unicellular eucaryote (for instance, an alga or a protozoon) than in a bacterium with similar metabolic pathways. It has been hypothesized that the additional amount of DNA has to be correlated to the control of structures and to a division cycle which are more complex in the higher organisms.

For the first time, in 1964 (by Britten and Kohne) it was observed that most of the DNA in cells of mice is constituted of copies of identical or very similar base sequences which repeat themselves (repetitive DNA). Repetitive DNA has been identified in all the higher species which have been analysed. However, repetition does not occur exactly and the members of one 'family' of repetitive DNA are generally very similar but not identical. It must now be established whether the repetitive traits are genes, or portions of genes, or if they have a different function. According to some authors (Britten and Davidson), the DNA redundant sequences are control sites for RNA transcription. Some repetitive DNA families, after having been centrifuged, appear as separated bands and are called satellites. By hybridization and autoradiographic techniques, satellite DNA (which is present in mice) has been shown to be constituted of some 10^6 copies of a sequence of 300–400 pairs of bases and to be localized at the centromeric ends of chromosomes. A centromere is a constriction which divides the chromosome into two arms; at its level, during

cell division, a chromosome attaches itself to the spindle. All mouse chromosomes, except one of the shortest chromosomes (probably Y-chromosome) have a satellite DNA. A similar centromeric localization of repetitive sequences has been observed in the chromosomes of *Drosophila melanogaster*. The analysis of some satellite DNAs has determined that these cannot code for any protein; in the case of guinea pig satellite DNAs, among the six possible readings of the triplets contained in them, two have been observed to give alternatively nonsense or terminating codons. According to these investigations such a DNA is believed to have the role of maintaining the chromosome's integrity.

The chromosome-associated proteins have been studied above all in the chromosomes of the higher organisms, where they are present in large amounts. In contrast to what it was generally thought, these seem to be present also in the bacterial chromosomes where they would have a coadjuvant action with regard to the repressor.

The best studied chromosome proteins—for their peculiar chemical properties and their ability to be easily isolated—are the basic proteins called histones. Some students classify them in four groups on the basis of their chemical properties. According to investigations carried out in order to establish their function and to explain their ways of interaction with DNA, histones appear to have the role of diminishing the priming template capacity of DNA in DNA-dependent synthesis of RNA or in DNA synthesis, whereas the chromatin which has been deprived of such a protein is a much more efficient primer than the native chromatin. According to these results, one would tend to ascribe to chromosome proteins the role of repressors in the process of genetic regulation.

In the field of acquired notions, different specialists have attempted to build a model of the eucaryote chromosomes and of their structural and functional organization. In 1971 a model was proposed (by Crick), according to which most of the DNA in higher organisms does not code for the proteins, but serves as a control. In 1972 another model was proposed (by J. Paul) which partly modifies the preceding one. In others, an attempt has been made to consider also the presence of the repetitive sequences of DNA. These models offer the analogy between the polytenic chromosomes of some insects and plants (in which a side-to-side apposition of identical copies of the entire genome occurs) and the monotenic chromosomes of mammals and other higher organisms (in which an opposition, by chromatin packing, of similarly corresponding repetitive sequences in the same genoma occurs). Some evidence suggests an association between gene repression and chromatin packing in the higher organisms.

The cytoplasm

The cytoplasm is that portion of the cell, extending from the cell's membrane to its nucleus, which contains most cell organelles, imbedded in an amorphous matrix. In the 1960s electron microscopy, associated with cytochemical and cytophysical techniques, allowed a better understanding than before of many structural and functional aspects of cell organelles. A few in particular have been singled out for special attention. Ribosomes, for instance, are organelles on which protein synthesis occurs. They are made up of two

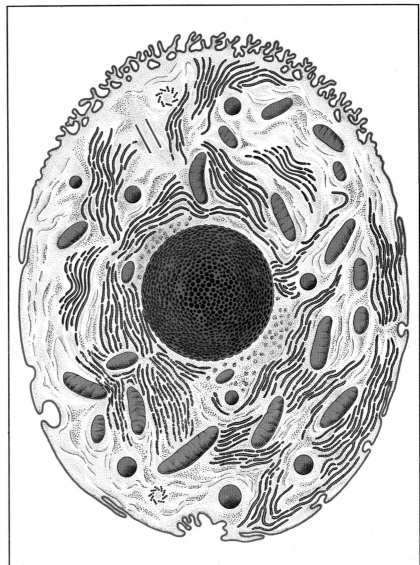

FIG. 51. The sketch shows a schematic representation of a typical cell: in the centre is the nucleus, enclosed by the porous nuclear membrane. Immersed in the cytoplasm are the mitochondria, large and elongated, biscuit-shaped; the lysosomes, which are roundish; the endoplasmic reticulum, small lamellae on which there are minute dot-shaped bodies, ribosomes; free ribosomes, which are seen scattered throughout the cytoplasm; the Golgi apparatus, flat vesicles piled up in a sinuous way; the centrioles, a series of small rods grouped together in series of three to four cylindrical structures (of which only two can be seen in the diagram). (*Source:* F. Albergoni; E. Matturri; A. C. Allison.)

sub-units called 30S and 50S, according to their sedimentation coefficient. It has been observed that the 30S sub-unit in *Escherichia coli* contains an RNA molecule (16S) and about twenty different protein molecules. The 50S sub-unit seems to be constituted by two RNA molecules (23S and 5S) and about thirty protein molecules.

Experimental data suggest a certain heterogeneity in the ribosome population, in which different protein groups are present according to the time and the kind of activity which they carry out. On the basis of this kind of observation, the very occurrence of 70S ribosomes as functional entities has been questioned. It is thought that a cycle of sub-unit association and dissociation occurs, according to which 30S and 50S sub-units would join and proceed altogether, as 70S ribosomes, along the messenger RNA, parting again at its end. They would present, as free 70S sub-units, in the cytoplasm only when excessive in relation to the cell's requirements. All steps of a sub-unit's association and protein synthesis seem to be mediated by soluble protein factors, which would link provisionally with the ribosome.

The discovery of the occurrence of specific ribosomal linking sites for the messenger RNA and the transfer RNA has confirmed that ribosome has an active role in protein synthesis and that it is not only an assembly bench.

The specification of the chemical composition of these organelles has been obtained by dissociation of their constituents and their subsequent reassociation, thus obtaining intact ribosomal particles with an excellent functional integrity. The reassociation *in vitro* has also demonstrated that the ribosome assembly is at least partly spontaneous and does not require an extrachromosomal template or the presence of specific enzymes. The assembly of the 30S sub-unit was obtained in 1967.

The cytoplasm of *Escherichia coli* has been shown to possess, *in vitro*, a rather high salt concentration: these salts may inhibit the non-specific aggregation between RNA and the proteins, whereas they promote the assembly reaction between RNA and specific proteins. In such a process the assembly is independent from the arrangement the proteins are given

when they join the RNA. Ribosome analysis by dissociation into components and their successive reassociation is one of the most spectacular achievements of molecular biology.

Given the high complexity of the interactions among the various constituents of ribosome, its overall activity is far from being simply the sum of the partial functions of its different components. Whereas the function of several ribosomal proteins has been specified by means of mutant strains (for instance, antibiotic resistance is due to a mutation in the structural genes which code for ribosomal proteins), the role of ribosomal RNAs is still unknown and no structural mutant has been isolated till now.

In the ribosome activity during protein synthesis, a key function is that of transfer RNA; in fact, the transfer RNA molecules have the role of interpreters of the genetic code. They are the intermediaries between the abstract language of the genetic code and the operative protein molecules which are specified by that code. A portion of the molecule is apt to recognize one code triplet and another one to recognize the amino acid which corresponds to that code triplet. Transfer RNAs, one for each amino acid, have been shown to act inside the ribosomes, at the level of the messenger RNA which bears the codon (the code unit composed of a triplet of nucleotidic bases), corresponding to the anticodon (the triplet of nucleotidic bases complementary to those of the codon) of transfer RNA.

In 1964, the nucleotidic sequence of a transfer RNA molecule, precisely the alanine transfer RNA, was first determined (by Holley). Several models have been proposed to define the three-dimensional structure of such a molecule, but a group at MIT (S. H. Kim and A. Rich) was able to specify its configuration only by X-ray crystallography. The molecule is made up of two main portions which are arranged in L-form. The anticodon which may recognize, on messenger RNA, the codon corresponding to a certain amino acid is situated at an end of one of the two arms of the L, whereas the receptor site is at the end of the other arm. The discovery of the three-dimensional structure of the transfer RNA, the first nucleic acid after DNA whose spatial

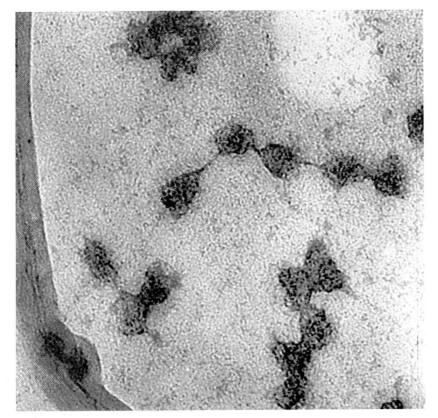

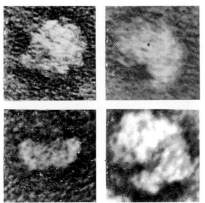

Fig. 52. Above, electron photomicrographs (800,000 ×) of subunit and complete ribosome of *Escherichia coli*. From left to right and from top to bottom: subunit 50 S, front view, same, side view; subunit 30 S; complete ribosome 70 S. Left, electron photomicrograph (1,200,000 ×) of several ribosomes linked by a filament of messenger RNA, forming aggregates known as polysomes; the number of ribosomes contained in polysomes may vary between a handful (4–5) and several dozen. (Courtesy M. Lubin, Dartmouth Medical School; A. Rich, MIT.)

configuration has been defined, now provides the opportunity to improve the actual knowledge on gene activity.

Lysosomes are sac-like cell organelles, often round-shaped and containing a large amount of enzymes, which may break down all the major constituents of living matter: proteins, carbohydrates, lipids, nucleic acids. Their action has been determined to occur both in normal and pathological processes, and has suggested the possibility of their specific role in the inducement of cancer.

Cytochemical studies have given evidence that a large range of lysosomal enzymes occur in the vacuoli of plant cells. In animal cells (in which lysosomes have been better studied), they are thought to digest the unwanted materials, thus determining the breakdown of intracellular organelles during differentiation, metamorphosis or organ regression. They are also helpful in removing injured organelles, after some sub-lethal damage has occurred in the cytoplasm. From a pathological point of view, lysosomal enzymes have been studied particularly in relation to immune processes and in the autoimmune diseases.

A certain lysosomal heterogeneity has been observed in Protozoa. Two lysosome classes have been put into evidence in *Tetrahymena*: light lysosomes, with a density of $1.14 \, g \times cm^{-3}$, with a high concentration of proteinases, ribonucleases and phosphatases, the main enzymes which are added to the ingested material to form the digestive vacuoli; and heavy lysosomes, with a density of about $1.24 \, g \times cm^{-3}$, which contain amylases, α- and β-N-acetyl-glucosamminidases, large amounts of which are secreted in the extracellular environment during normal growth and storage of material. Bacteria do not contain lysosomes in the forms which are recognized in the higher organisms, but they may release hydrolythic enzymes whose properties are very like those of the lysosomal enzymes.

In 1960 some investigators at the Rockefeller Institute demonstrated that leucocytes contain typical lysosomal enzymes. Digestion of bacteria by leucocytes is a fundamental kind of defence against diseases. Certain bacterial strains may defend themselves against lysosomes, thanks to a kind of resistance to their enzymes. Rare hereditary diseases, such as the Chédiak-Higashi syndrome, are known in which lysosomes are ineffective. In some instances, the inactivity of single lysosomal enzymes may provoke storage disease, because the metabolic product which should be broken down in the cells is not affected, and accumulates in pathological amounts.

Lysosomes may also have a role in the activity of certain hormones. A thyrotrophic hormone supply has been shown to increase the amount of colloid substances in the lysosomes of thyroid cells; the administration of glucagone causes striking changes in the lysosomes of liver cells. Some substances which are concentrated in the lysosomes cause foetal anomalies, whereas high doses of vitamin A reduce the stability of the lysosomal membrane, giving analogous effects. Finally, it has been noted that in cells ready to divide, lysosomes are relatively few and they are peripheral instead of being, as usually, perinuclear. This observation suggests that the lysosome rupture may be the primer of mytosis in mature cells.

Informosomes are ribonucleoproteic aggregates, first discovered in 1964 in the cytoplasm of fish embryos. Their analysis has demonstrated that the RNA present had a high molecular weight and was ribosomal in origin. According to the most common hypothesis, such ribonucleoproteic particles would transfer information (newly synthesized RNA) from nucleus to ribosomes. Hence their name. Subsequent research on the role of informosomes led to the hypothesis that these represent a kind of passive defence and 'm-RNA stabilization' against enzymic activity, or a kind of protein synthesis regulation at the level of messenger RNA translation.

The endoplasmic reticulum is an assembly of vescicles and communicating channels that fills almost entirely the cytoplasm. In some cells (especially in those which are more active in the protein synthesis), the prevailing vescicles have been shown to be thin and lamellar cisternae. They often possess parallel array. Through the electron microscope, this pattern in some cells seems to be made up by communicating channels which join to give a three-dimensional reticulum. All the vescicles are bounded by three-layered membranes which, besides dividing the cytoplasm into internal and external phases in relation to the reticulum, supply an extensive intracellular surface on which several functional loci are sited. Associations between endoplasmic reticulum membranes and other cell components have been noted; they are believed always to have a functional correlation.

In some instances, the endoplasmic reticulum behaves as a collector, i.e. it stores in its own cavities proteins which have been synthesized elsewhere. In addition, it has been ascertained that the role of the cavities of the endoplasmic reticulum is to transfer both extracellular and intracellular substances from one side to another side of the cell and to distribute them inside the cell.

In striated muscle cells, miofibrils are surrounded by an agranular endoplasmic reticulum (lacking ribosomes on its surface) in which cavities are variously shaped according to the level of the sarcomere to which they correspond. This reticulum, called sarcoplasmic reticulum, is intimately associated with the plasma membrane. According to a theory formulated in 1965 (by Porter and Franzini-Armstrong) the sarcoplasmic reticulum represents a calcium and ATP reservoir, allowing contraction to occur by rapid diffusion in the bulk of myofibrils. Finally, some experimental evidence suggests that the synthesis of vertebrate steroid hormones occurs at the level of the reticulum membranes.

Microtubules are thread-like organelles of a rather constant diameter (about 25 nm) and various lengths in different cells. Microtubules are the most recently discovered cell organelles; they have been detected in all nucleated cells. They have been shown to participate in the constitution of elaborate structures such as flagella, cilia, developing spermatozoa and mytotic spindles. When sectioned, they appear as unbranched tubular, stiff structures.

FIG. 53. Function of lysosomes, illustrated in the case of ingestion of a particle (a bacterium, for instance) by a macrophage. After the ingestion phase, three things may happen, according to the type of bacterium: the bacterium may multiply, remain unchanged, or be digested. In the last case, the lysosomes merge with the phagocyte vacuole to form a complex within which the phagocyte material comes into contact with the lithic enzymes contained in great quantities in the lysosomes, and are therefore absorbed: the lysosomes are then transformed into vesicular structures of variable content, which may either decompose or produce residues *(Source: La Recherche.)*

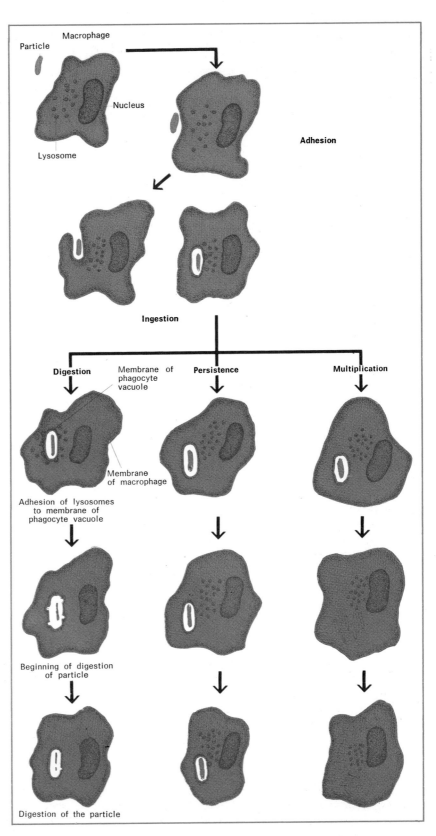

It has been postulated that microtubules are mainly responsible for a cell's shape and the maintenance of such a shape during the cell's activity. Electron microscopy associated with biophysical procedures suggests the existence of a complex of sub-units and that, at least in some systems of these sub-units, the balance between them (which is very unstable) is altered by a minimal change in environmental conditions. It has been ascertained that microtubules present in the flagella are much more stable than microtubules forming the mytotic spindle and assist chromosomal movement during cell division.

The variation in stability from one type of microtubules to another has been explained by admitting the existence of cross-bridge systems. Such cross-bridges would be responsible, on the one hand, for the maintenance of an order in the structural complex of microtubules and on the other hand, for the motility of the various species.

It has been explained that the autoassociation from which microtubules derive is not a random process, that there are some devices to determine the site and time of formation. Cell locations, such as microtubule organizing centres or nucleating sites, in which assembly occurs, are generally thought to exist. In dividing vegetal cells, an association of microtubules with the membrane has been noted during the destruction and reconstruction of these structures.

Membranes

Cell membranes are structures which delimit cell and cell organelles from the outside. They control the relationship between the cell and its environment by regulating exchanges of material and the mechanisms of impulse propagation.

Through the electron microscope plasma membrane appears as a three-layered thick structure, 7.5–9 nm. Plasma membrane is an asymmetrical membrane in which the intermediate layer, electron-transparent, is situated between two electron-opaque layers (of which the inner is wider than the outer). In some organelles, three-layered symmetrical membranes have been observed; the observation

has led to the concept of the 'unit membrane', according to which the fundamental organization of all the membranes would be identical. This concept, prevailing at the beginning of the 1960s, has been successively censured and abandoned. It was based above all on the model of H. Davson and J. F. Danielli (1935), according to which cell membrane is built up by two phospholipid layers with their hydrophobic regions confronted and the polar regions oriented toward the outside and covered by proteins. This model was reconsidered (1962) when phospholipids were shown to assume, in water, different configurations according to temperature and concentration. In these investigations, membranes different from the three-layered variety, particularly hexagonal or globular membranes, were detected.

Furthermore, other models have been proposed for the molecular organization of membranes. Above all, the possible artefacts which derive from the different fixation techniques necessary for the electron microscope observation have been thoroughly examined. In 1968, a model for a biological membrane was proposed (J. Lenard and S. J. Singer). In it, the hydrophilic regions, both of proteins and of phospholipids, would be oriented toward the watery interfaces; the continuity of the membrane would be assured by protein-protein interactions. According to another model (S. J. Singer and G. L. Nicolson) the membrane would constitute a lipid layer in which proteins float. The study of membranes in higher organisms is made difficult by the difficulty to isolate them; this was facilitated by setting up, in 1962, a model of artificial membrane which is similar to that of natural membranes.

Biological membranes regulate selectively the flow of substances and ions from the inside to the outside and vice versa. The hypothesis of the occurrence in membranes of special molecules called carriers, which facilitate the passage of metal ions and of tiny polarized molecules, was confirmed in 1965 (C. F. Fox, E. P. Kennedy) by identifying the mole cule of the lactose-transport carrier, a protein which is able to bind itself to lactose by a mechanism very similar to that of an enzyme. Since then, many other carriers have been identified. They have been demonstrated to be lipoproteins, and are called permeases. Their mechanism of action is still to be defined.

One of the peculiar features of membranes is their ability to discriminate between very similar ions and molecules (such as Na^+, K^+, Ca^{++}, Mg^{++}), alcohols and aldehydes. Understanding this peculiar property would help to explain active transport, passive diffusion, enzyme activation and electrical excitability—studied by means of membrane models. Artificial membrane models are built

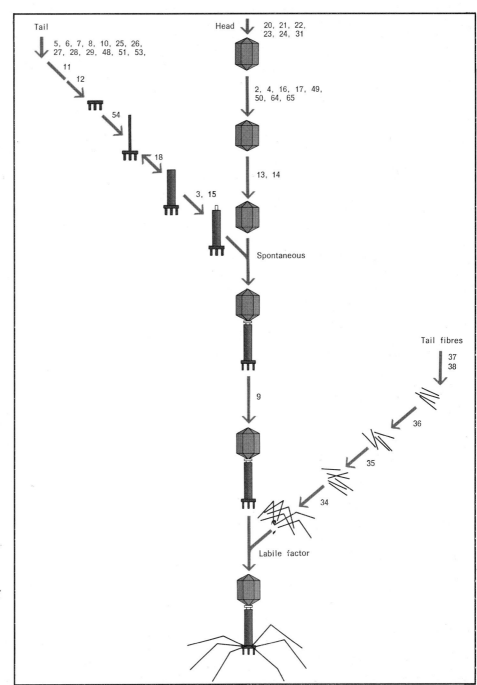

FIG. 54. Successive phases of the process of assembly of the bacteriophage T4. These phases can be induced with bacterial extract in DNA cultures of the phage: each phase involves the intervention of certain genes (those indicated by numbers alongside arrows marking the successive phases). There are three main 'paths' leading independently to the formation of the head, of the tail, and of the tail fibres. The primary structure of the head depends on the genes indicated by the numbers alongside the arrows between the various phases in the formation of the head; in the absence of any such genes, only the tail and the fibres are produced in the culture. The union of head and tail occurs spontaneously, without the need of genes; the union of fibres with the tail requires the intervention of the labile factor. (*Source*: Wood and Edge.)

of bimolecular phospholipid layers. Although it is known that natural membranes are not essentially constituted of bimolecular lipid layers, these layers certainly have a very important function. Two techniques are generally used to develop these models: the black membrane technique of Müller and Rudin (1962) and the liposome technique of Bangham (1965).

Another peculiar trait of cell membranes is their specificity. An example of this is given by the occurrence of antigens (agglutinogens) on surface of red cells, another by the ability to recognize foreign material (such as virus and bacteria) by macrophages. The cell membrane seems to have a very important task in the immune processes which occur in certain cells. According to the most modern immunological theories, lymphocytes bear membrane receptors which are very specific against all possible antigens. As every membrane receptor is specific for only one antigen, an extreme structural variability of the surface is supposed to exist in the lymphocyte population.

Cell surfaces are extremely important in organ and tissue transplantation. The occurrence of histocompatibility or transplant antigens is the property causing the rejection of transplanted organ or tissues. The knowledge of the number and distribution of histocompatibility antigens on a cell's surface would help to explain their mode of action and the setting up of preventive therapy in organ transplantation.

Modifications of the cellular surface have been observed in tumour cells. Particular investigations have been carried out on the action of some substances (such as cyclic AMP) on membranes, which succeed to block the uncontrolled growth of tumour elements and to restore contact inhibition. The action of some drugs which may modify the cell surface has also been investigated.

The recognition of certain substances by membranes is important throughout the animal kingdom. The specificity of action, which characterizes some phenomena as fertilization and the aggregation of like cells to form tissues, is based on this. Cell membranes often may be associated with enzymes. At the level of mito-chondrial membranes, molecular aggregates have been observed which are thought to be enzyme aggregates belonging to the respiratory chain; the DNA-polymerase II also may be a membrane-bound aggregate of enzymes. In addition, the occurrence of several cell functions in proximity of the cell's membranes suggest that this kind of association is more frequent than it has been experimentally demonstrated to be until now. DNA replication in bacteria seems also to be associated with the presence of membranes; some virus (lambda phage) replication also is surely associated with membranes. Thus membranes can be considered as complex biochemical systems.

The participation of acetylcholine has been postulated in nervous activity, and the action of such a substance has been observed to occur at the intermembrane level. It has been experimentally demonstrated that acetylcholine acts on a particular receptor protein, localized on membranes; a series of chemical reactions which increases the membrane's permeability to ions has its origin there. Using different techniques, comparative analyses have been made on the active site of the protein receptor of acetylcholine and of acetylcholinesterase, an enzyme which inactivates acetylcholine and restores the receptor protein to conditions of rest. Further work to isolate the receptor protein is needed to confirm diversity in the two proteins.

Viruses and the cell/virus interaction

Viruses are infective agents characterized by very small size, making them invisible through normal optical microscopes. They are also characterized by the absence of cellular organization and the ability to reproduce themselves only inside living cells. They are composed of a DNA or RNA core and a protein coat.

The question of the reproductive behaviour of viruses was cleared up during the 1960s, by observing how phage T4, having infected an *Escherichia coli* bacterium, gives rise to new phages identical to itself. DNA of the phage T4 is contained inside the protein coat forming the head of the virus. A springlike tail is joined to this head through a short neck; it is held in place by a contractile sheath which envelops a central core, in turn connected with an end-plate from which six short spikes and six long and slender fibres originate. The life cycle of phage T4 starts when it sticks, by its end-plate, to the surface of *Escherichia coli*. By contracting its sheath, the phage injects in the bacterium its own DNA, which diverts the cell machinery in a few minutes towards the production of viral proteins in order to restore the complete viral particles. It has been determined that there are, among the first synthesized proteins, the enzymes necessary for viral DNA replication. Afterwards, a new series of genes enters into action, coding for the production of the structural proteins of head and tail which are needed in the process of viral morphogenesis. When the bacterium is completely invaded by new phages, breakdown of the bacterial wall and release of the newly formed viral particles occur.

By using mutated T4 strains, it has been possible to trace the steps of the assembly process. Indeed, when a gene is mutated it codes for a different protein, which is often a nonfunctional one. In case of viral assembly, the mutation of a gene which codes for a certain protein makes the process stop when the presence of such a protein is requested. This has permitted a genic map of the virus to be prepared.

When mutations occur in genes which control the successive steps of the life cycle, an accumulation of nonassociated virus components is observed. By means of the electron microscope, it is possible to identify the structures lacking consequent to the mutation of a certain gene and to establish the normal function of the gene.

More than fifty genes are known to be responsible for the morphogenetic process, and probably there are more. Of all codes for proteins constituting the virus, this one must be much more complex than it looks now. Some genes are likely to function, however, only in the assembly process and not code for structural proteins.

It has been possible to carry out the assembly process *in vitro*. This experiment was very useful to interpret the specific task in the morphogenesis of many genic prod-

ucts, although the steps which have been reconstructed are only a part of those that occur *in vivo*.

The result has been achieved by associating genetic analyses with biochemical techniques, i.e. by observing, on the one hand, the changes produced by a well-defined mutation in the generations following that in which such a mutation has occurred and by identifying, on the other hand, the substances produced. The usefulness of such an experiment of assembly *in vitro* yields not only a deeper knowledge than before of the structure and function of the virus, but also a better explanation of how genes control the building of biological structures. Viral morphogenesis is a particular instance of molecular interaction; in the specific case considered here, it concerns an interaction between nucleic acids and proteins.

Infaction by phage T4, which is similar to that by other bacteriophages of T-series, has as a unique consequence the lysis of the bacterium. A different behaviour has been observed in the lambda bacteriophage, which infects an *Escherichia coli* strain without causing its lysis; it remains in a latent stage of profage by inserting its own DNA in the bacterial DNA. Another property which has been discovered in the virus and studied for its possible practical implications is its ability to transfer from one cell to another fractions of the host genetic material by means of a process which is called transduction.

This phenomenon was discovered in the early 1950s and has been particularly studied in the transducer lambda phage, usually associated to *Escherichia coli*. It may transfer galactose-operon from one bacterium to another. In 1971 this property was utilized to introduce into cells with metabolic errors those genes which are necessary to code the enzymes lacking. Three investigators (C. R. Merril, M. R. Geier, J. C. Petricciani) have succeeded in introducing a λ_{gal} virus, which is a transducer phage transferring the bacterial genes coding for the first three enzymes needed in the metabolism of galactose (galactokinase, UDP-galactotransferase, UDP-galactoepimerase), into mammalian cells lacking kinasic, transferasic and epimerasic activities. The metabolic defect known as

217

galactosemia is caused by a lack of transferasic activity. In galactosemic cells, after infection by λ_{gal} virus or its DNA, the appearance of transferasic activity has been observed.

Thanks to the electron microscope and biochemical and biophysical analyses, it is possible to state that virus/cell interactions are varied and their effects are of differing seriousness, ranging from an infective process to thorough neoplastic transformations. In the last instance, experimental research of the last ten years tends to confirm that connexions between virus and cancer have to be sought at the genic level. A cell, after having been infected by a virus, contains two groups of genes, its own and those carried by the virus. Given the ability of viral genes to overcome the control of cell genes, the alterations which they may provoke are remarkable. Cell transformation may be determined by various agents (a virus, chemical substances, radiations). It has been observed that all transformed cells show common properties, so that the study of transformations by virus, which can be observed in tissue cultures in relatively simple experimental conditions, is considered the more suitable experimental model for general study of cell transformations.

After the life cycle and the mode of reproduction of DNA virus inside the infected cell were explained, the existence of RNA virus opened the question of defining how those viruses could accomplish their life cycle in a cell in which the genetic material is DNA. Experiments carried out (by S. Spiegelman) with MS2 phage tended to establish a hypothetical transcription of viral RNA to bacterial DNA, but hybridization experiments did not confirm an inverse transcription of RNA to DNA. In 1965 (S. Spiegelman), by using RNA Qβ phage, it was possible to demonstrate the existence of RNA-dependent RNA-polymerase (replicase), which may give rise to identical copies of the viral RNA being introduced to the infected cell. A series of experiments demonstrated that the newly synthesized RNA is as completely active as the original viral RNA, that it is able to programme the synthesis of viral particles, and to serve as a template for the production of even more copies.

The synthetic replicase of Qβ, by producing unlimited amounts of viral RNA, has facilitated the observation of natural selection in action at the molecular level. By varying the selective pressure of the system, it has been possible to obtain mutated RNA molecules having extremely varied properties. These studies opened the way toward research on the RNA oncogenic virus.

In 1970 the discovery (by Temin, Baltimore) of RNA-dependent DNA-polymerase, i.e. an enzyme able to catalyse the synthesis of DNA from RNA in RNA oncogenic viruses, was announced. The process has been defined as an inverse transcription, as it acts in a direction which is opposite to what had been hypothesized in the central dogma of biology (DNA→RNA→proteins), according to which only the DNA transcription into RNA would occur. It had been believed that the RNA oncogenic viruses possess an RNA-dependent RNA-polymerase, which is analogous to that discovered in the RNA bacteriophages.

The inverse transcriptase (the RNA-dependent DNA-polymerase) has been discovered in the Rous sarcoma virus, in the Rauscher mouse leukaemia virus and, later, in other oncogenic viruses.

The template function of viral RNA has been confirmed by DNA-RNA molecular hybridization. The attention paid to this discovery is widely justified by the fact that explanation of this mechanism would help to clear up the pathogenesis of some kinds of human cancer. In fact, although some indirect evidence of the viral origin of some kinds of human cancer exists, no direct experimental confirmation exists.

And yet, discovery of an inverse transcriptase in RNA nononcogenic viruses has led us to assume that this is not specific to the oncogenic viruses. Further investigations have demonstrated, in normal cells, two enzymes which transcript hybrid synthetic DNA-RNA templates.

Inverse transcriptases of normal cells seem to be rather different than those of the oncogenic viruses; an antiserium prepared with inverse transcriptase, which had been purified from RNA virus, succeeded in inhibiting the viral enzyme in transformed cells without altering the enzymatic activity of normal cells.

Investigations on normal cells (Tocchini-Valentini, Crippa) have demonstrated not only that this enzyme exists but that it has a vital role, having an active part in the phenomenon of the gene amplification studied in amphibian eggs.

Cellular immunology

Cellular immunity is a kind of natural immunity which protects an organism from microbic infections. The cells accomplishing this function occur in three different localizations: in any part of the body in which an infection takes place, in the lymph node chain and in the blood system.

Molecular interaction phenomena are at the basis of the study of the immune processes in the organism. Although not definitively, investigations carried out since the early 1960s have largely clarified the part played by lymphocytes in the immune processes, besides revealing the extreme complexity of the overall immune system.

Most of the immunological research in this period has been concentrated on antibodies, protein molecules occurring in the serum and able to link closely with an antigenic determinant. Antibodies, the key immune molecules, are composed of a heterogeneous group of proteins, immunoglobulins, which are able to recognize the foreign molecules (antigens) in the organism. According to physical, chemical and antigenic properties—antibodies which have been inoculated in foreign organisms may also act as antigens—five classes of immunoglobulin have been established: IgG, IgA, IgM, IgD, IgE. Immunoglobulins G (IgG) are the most widely spread in man and other mammals.

In the last decade the structural analysis of antibodies has been carried out mainly by biochemical methods and aid of the electron microscope. In 1969 the primary structure of an immunoglobulin was defined by Edelman and his co-workers. On that basis, a new model has been designed (by Valentine) whose configuration is very similar to that seen through the electron microscope. According to this model, immunoglobulins are all made up of two kinds of polypeptide chains: light chains, constituting about 220 amino-acid residues and heavy chains, which have at least double the size. Each immunoglobulin molecule contains two identical light and two identical heavy chains. One of the most interesting features of the immunoglobulin molecule is the arrangement of the polypeptide chain into a variable region (V) and a constant region (C). The variable regions, with different amino-acidic sequences in different antibody molecules, are functionally correlated to the association with the antigen molecule. Several observations lead one to suppose that separate genes occur for the two C and V regions; on the other hand, it seems that genetic translation of the two variable and constant parts of the chain occurs thanks to only one messenger RNA. This fact would lead us to suppose that a linking mechanism of the two distinctive genes for V and C exists at the level of DNA.

The most interesting portion of the antibody molecule is the active or combining site, which is characterized by an extreme variability; at this level, an antigen-antibody interaction occurs. According to Edelman's model, the antibody molecule is symmetrical and each specular half contains a combining site which is identical to that in the opposite side. The portion with the antibody function is made up of the total light chain and part of the heavy chain. The antigen-antibody recognition is steric, i.e. one occurring according to the spatial similarities between molecules, by means of non-covalent bonds; this recognition is based on a complementarity between structures and is not absolute. Indeed, each antibody recognizes one antigen with a greater and a more specific affinity but other antigens with a lesser affinity.

It has been observed that a unique model for antibodies and antigens is exceptional: it occurs only in myelomas or in monoclonal antibodies, which are derived from a non-tumoral but rather unusual clone. Subsequent to these observations, the concept of a unique antibody has been substituted by that of a population of antibodies.

In the antigen-antibody interaction the antigenic determinant reacts with the active site

219

of the antibody. It is a molecule or a fraction of a molecule which may attach firmly to the combining site of the antibody. One or several antigenic determinants may occur on each antigen. Antigenic determinants have been separated (by Sela, 1967) in two classes: sequential determinants, which are constituted of a linear sequence of elements (amino acids in the case of a protein antigen) and conformational determinants, which are formed by the grouping of amino-acid residues belonging to different portions of the peptide chain and linked together by different interactions.

The size of the sequential determinants has been precisely defined: from four to seven monosaccharide residues in the polysaccharide antigens and from three to four amino acids in the synthetic polypeptides. The size of the conformational determinants is more difficult to define: it is supposed that several aminocidic residues are involved (1971).

It has been found that the sequential determinants are not affected by heating, while the conformational ones are lost as soon as protein is heated and denaturation occurs. A simple model has been designed (Sela) in order to study the conformational changes by specific antibodies which are likely to occur. The artificial antigenic system which has been utilized has been obtained by the linear polymerization of a tripeptide, tyrosine (T), + alanine (A) + glutamic acid (G), into chains of different lengths, by repeating the unit TAG from one to about 220 times.

In every organism numerous populations of specialized cells (lymphocytes) exist which can counteract the action of every antigen which is introduced into the organism. When a specific lymphocyte has been introduced by an antigen, it is transformed into a big cell (blastocell), which then rapidly divides itself into cells increasingly efficient in antibody synthesis. Finally a plasma cell is produced, which synthesizes antibodies at a growing speed and no longer divides itself. It has been confirmed that the production of antibodies is an asynchronous process, as lymphocytes are circulating cells: indeed the transformation into blastocells occurs at different times at which the cells of the specialized populations come into contact with a determined antigen.

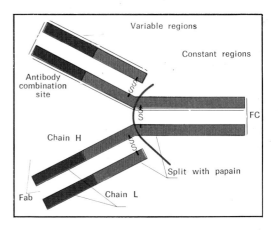

Fig. 55. Small diagram shows structural characteristics of an antibody molecule, as described according to the immunological research done by Edelman and Porter. This molecule possesses 1,300–1,400 amino acids, and may be divided into four polypeptide chains, two heavy (H) and two light (L) linked by disulphide bonds. Porter split the molecule with papain, obtaining two equal fragments (Fab) which are those determining the properties of the antibody, and a fragment FC, substrate of the biological properties of a general nature. Large diagram, constitution of the various immunoglobulins.

The problem of production of antibodies which may react to differing stimuli has induced several investigators to put forward some theories on their origin.

Two of the more debated genetic theories follow. The subcellular selection theory claims that every lymphocyte contains genes for the synthesis of any immunoglobulin. According to this idea, the antigen would function only as an inducer, by provoking the derepression of

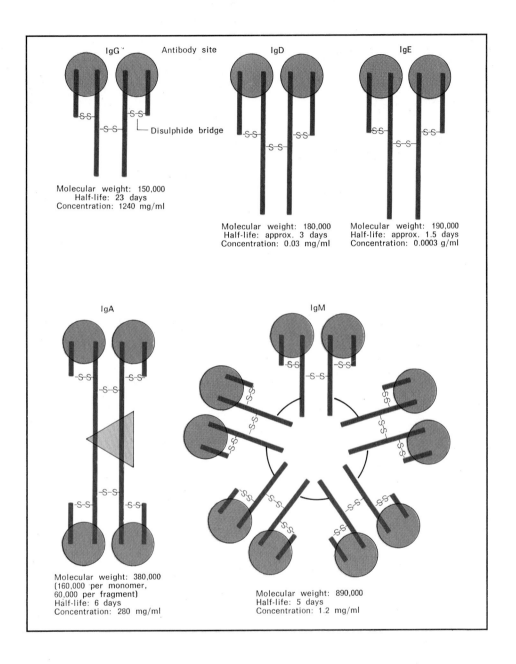

IgG   Antibody site   IgD   IgE

Disulphide bridge

Molecular weight: 150,000
Half-life: 23 days
Concentration: 1240 mg/ml

Molecular weight: 180,000
Half-life: approx. 3 days
Concentration: 0.03 mg/ml

Molecular weight: 190,000
Half-life: approx. 1.5 days
Concentration: 0.0003 g/ml

IgA   IgM

Molecular weight: 380,000
(160,000 per monomer,
60,000 per fragment)
Half-life: 6 days
Concentration: 280 mg/ml

Molecular weight: 890,000
Half-life: 5 days
Concentration: 1.2 mg/ml

the gene corresponding to the necessary anti-body. The clonal selection theory sustains that lymphocytes are a heterogeneous population of cells and that each lymphocyte is genetically predisposed to produce a certain type of anti-body. The antigen entering into the host or-ganism selects among many lymphocytes those that are forming the antibodies at a slow rate, thus stimulating an increase in such a speed of division and forming a clone from every selec-ted cell. According to this second theory, there is a very low number of cells which may react to certain antigens and each lymphocyte would form only one kind of antibody.

In the vertebrate immunological system, two immune (cellular and humoral) systems seem to coexist. They are mediated respect-ively by lymphocytes T, which are formed in the thymus and destroy directly foreign cells without intervention of soluble antibodies and

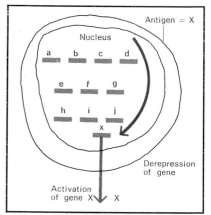

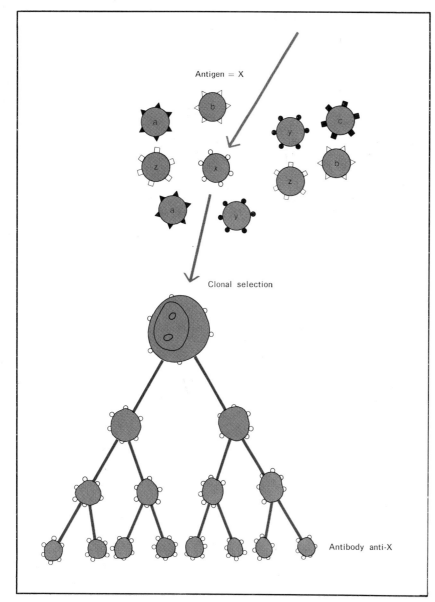

Fig. 56. Diagram of clonal selection in the formation of antibodies (left) and of the subcellular formation (above). In the first case the antigen selects, from among the lymphocytes, those which are forming the antibody at low velocity, and by stimulating an increase in the speed of division in these cells, clones from each of the selected cells are formed. In this case, each lymphocyte would only be able to form one type of antibody, and there would exit only a very few cells capable of responding to a certain antigen. In the second case, the antigen would function only as an inductor, by derepressing the gene corresponding to the requisite antibody.

lymphocytes B. These last are known to be derived from the bursa of Fabricius in birds, which produce soluble antibodies and which destroy foreign cells. The origin of lymphocytes B in mammals is still unknown.

A certain kind of co-operation seems to exist between these two types of cells (T and B): T cells would have an inducing function on most of the B cells, whose response would consist in the production of circulating antibodies. In addition, T cells appear to have the task of attacking and directly destroying foreign cells without the intervention of soluble antibodies. Rejection in grafts and coping with neoplastic aggression seem to be specific functions of the T-cell.

The antigen-antibody recognition in B cells occurs by means of a molecular receptor, made up of immunoglobulins. The mode of such recognition in T cells is controversial: rather convincing evidence seems to exclude the occurrence of a molecular receptor constituted of immunoglobulins. Other evidence indicates that T cells possess very little of one kind of immunoglobulin on their surface, while still other evidence supports cell-to-cell interaction.

Further reading

BAROJAN, O.; LEPIN, P. *Epidemiologiceskic aspekty sovremennoj immunologii.* Moskva, Medicina, 1972.

BERKALOFF, A.; BOURGUET, J.; FAVARD, P.; GUINNEBAULT, M. *Biologie et physiologie cellulaire.* Paris, Hermann, 1967.

BRITTEN, R. J.; KOHNE, D. E. Repeated segment of DNA. *Scientific American,* vol. 222, no. 4, 1970, p. 24.

COHEN, S. N. The manipulation of genes. *Scientific American,* vol. CCXXXIII, no. 7, 1975, p. 24.

COULT, D. A. *Molecules and cells.* London, Longmans, 1966.

CRUICKSHANK, R. (ed.). *Modern trends in immunology.* London, The Butterworth Group, 1967.

DE DUVE, C. Les lysosomes. *La recherche,* vol. V, no. 44, 1974, p. 815.

DE ROBERTIS, E. D. P.; NOWINSKI, W. W.; SAEZ, F. A. *Cell biology.* Philadelphia, London and Toronto, W. B. Saunders Company, 1965.

DURAND, M.; FAVARD, P. *La cellule. Structure et anatomie moléculaire.* Paris, Hermann, 1974.

FOX, C. F. The structure of cell membranes. *Scientific American,* vol. 226, no. 2, 1972, p. 30.

GIESE, A. C. *Cell physiology.* Philadelphia, London and Toronto, W. B. Saunders Company, 1968.

HAUROWITZ, F. *Immunochemistry and the biosynthesis of antibodies.* New York, London and Sydney, John Wiley & Sons, 1968.

LENTZ, T. L. *Cell fine structure.* Philadelphia, London and Toronto, W. B. Saunders Company, 1971.

LURIA, S. E.; DARNELL, J. *General virology.* New York, London and Sydney, John Wiley & Sons, 1967.

NÉEL, J. Les membranes artificielles. *La recherche,* vol. V, no. 41, 1974, p. 33.

NOMURA, M. Ribosomes. *Scientific American,* vol. 221, no. 4, 1969, p. 28.

PAUL, J. Cell biology. London, Heinemann Educational Books Ltd, 1965.

SALOMON, J.-C. La surveillance immunitaire. *La recherche,* vol. VI, no. 58, 1975, p. 640.

STERN, H.; NAUNEY, D. L. *The biology of cells.* New York, London and Sydney, John Wiley & Sons, 1965.

WALLACE, B. *Chromosomes, giant molecules and evolution.* Basingstoke, Hants., MacMillan, 1967.

22

From cell to organism

Fertility and implantation

Fertilization is the fusion of the egg and spermatazoon, the two germ cells, to form a zygote which, by successive cell divisions, will develop into an embryo. During the first steps of mammalian development, the zygote implants itself in the uterine, mucous membrane, where mechanisms to nourish the embryo will be acting until birth.

Electron microscopy of ultrathin sections, scanning microscopy and microcinematography have helped in analysing and tracing many phenomena occurring during the first stages of development; they have supplied a better knowledge of germ cells.

According to some investigations, it seems already ascertained that the morphology of the spermatozoon is related to the mode of fertilization. Organisms which release spermatozoa in water exhibit a primitive type of spermatozoon, whereas in organisms with internal fertilization or in which spermatozoa are released in proximity of the female genital opening, the spermatozoon's morphology differs in varying degrees from the primitive type. Spermatogenesis has been followed in Nemertina, Anellida and Mollusca by observing the changes to which the primitive cell has been subjected during such a transformation. Deviations from a primitive constitution are observed, above all, in the head of the spermatozoon as well as in the middle section. The typical arrangement of mitochondrial material in relation to the requirements of the spermatozoon's locomotion is particularly interesting: in external fertilization, a more intense motility of the spermatozoon is required and mitochondria, in which energy is transformed in such a way as to be utilized by the cell, are functionally related with the flagellum, which is the motile organelle.

Particularly aberrant spermatozoa have been observed in Platyhelminthes, Aschelminthes and Arthropoda; they may be related to an extremely modified and specialized biology of fertilization. Investigations carried out in Drosophila have confirmed that spermatogenesis is under the control of genetic factors which can be located on the Y-chromosome. Males in which the Y-chromosome is partially or wholly deleted exhibit aberrations in the differentiating spermatidia. Such abnormalities seem to be caused by the lack of organization of most structural elements of the spermatidion's organelles. A few animal groups show flagellate sperms with mitochondria irregularly scattered in the cytoplasm or flagellate sperms with mitochondria located in the nuclear area. In most animal groups, mitochondria are assembled behind the nucleus in the middle section. They may be simply as-

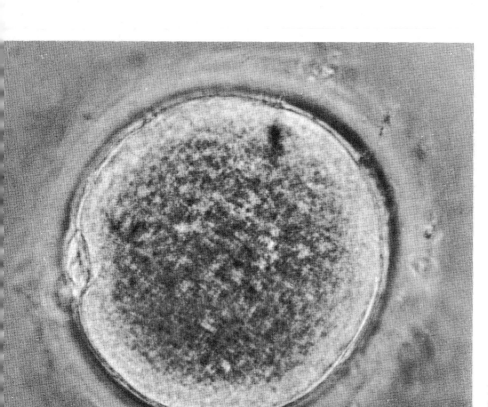

FIG. 57. The human ovule or egg-cell. The spermatic nucleus penetrates the cytoplasm of the egg, passing through the plasma membrane and the nuclear membrane, and blends with the nucleus of the egg itself to form the nucleus of the zygote. (*Source*: R. Edwards.)

sociated, without structural modifications, or fused and coiled, with more or less extensive modifications. The largest rearrangement is observed in insects and Gasteropoda.

The study of human spermatozoa is complicated by a pleomorphic cell pattern, in comparison with those in most other animal species. The normality range is very difficult to define clearly (Freund, 1966). Controversial hypotheses about the mechanism of energy conversion during the spermatozoon's metabolism have been put forward. The possible relationships among sites of ATPase activity (i.e. sites of the adenosintriphosphatase enzyme) and their possible implications in the motility of the spermatozoon have been considered. Co-ordination and propagation of the flagellar wave are essential conditions for the spermatozoon's propulsion: glycerinated models of mammalian spermatozoa may be reactivated by adding ATP. In this way, how-

ever, only repeated contractions along the flagellum and no progressions have been observed. As treatment by glycerine causes morphological alterations with loss of some substrates and enzymes, the spermatozoon function also is likely to be involved.

The study of correlations between structure and function of the spermatozoon as a peculiar kind of cell, variously improved in the living kingdom, is helpful in this respect, as is the comparative study of spermatozoa in the most modern of animal and vegetal systematics.

Electron microscopy has confirmed that, during spermatogenesis, male germ cells develop into clones, in which all cellular components from a single source are joined by intercellular bridges. A paper presented in 1971 (by Dym and Fawcett) stated that the number of germ cells which participate in a syncitium is beyond earlier expectation, and the connexions do not remain unvaried during spermatogenesis but form transversal septa. These isolate temporarily the cellular units inside the syncitium. According to some authors, the intercellular bridges among differentiating male germ cells help to synchronize the differentiation phenomena.

The study of spermatogenesis has reopened discussion of the role of centrioli during this process. Centrioli are considered to be autoduplicating organelles, with a morphological continuity from one cellular generation to another. Electron microscopy tends to confirm their duplication during mytotic division and their morphological continuity in daughter

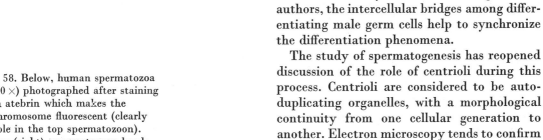

Fig. 58. Below, human spermatozoa (8500 ×) photographed after staining with atebrin which makes the Y-chromosome fluorescent (clearly visible in the top spermatozoon). Below (right), spermatozoon heads carrying, presumably, chromosomes X (above) and Y (below).

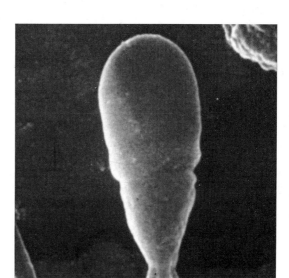

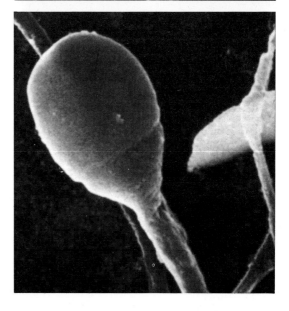

cells, but this duplication may not occur according to traditional theories. That is to say, centrioli would be subjected neither to transversal nor to longitudinal division, but the new centriole would originate in a specific terminal region of the pre-existing centriole. The matter from which the new centriole originates, i.e. the procentiole, is an annular condensation of dense fibrogranular substance, with the same diameter as the original centriole, but lacking initially microtubular sub-units in its wall. The newly formed centriole, by gradually elongating itself, tends to keep its organization perpendicular to that of the parental centriole. Accordingly, the centrioli appear to duplicate themselves neither by division nor by a template mechanism: the precursor centriole seems to act exclusively as an inducing site for an autoassociation process of the new centriole from precursors synthesized elsewhere in the cell.

During spermatogenesis it has been observed that centrioli have different fates. In the mammalian mature spermatozoa, for instance, the proximal centriole remains whereas the distal one disappears. The lack of a typical centriole at the basis of mammalian spermatozoon axonema has forced abandonment of the traditional concept of the basal body as an essential kinetic centre which functions as a site of initiation of the ciliar and flagellar beats. According to these studies the distal centriole would be essential to axonema formation, and it would be superfluous as a basal body in the mature spermatozoon. There is also question of the need of the presence of centrioli in the division spindle organization after fertilization. On the contrary, centrioli would participate during spermatogenesis in a kind of patterned differentiation in which they would function as sites of deposition for proteins of the axonema microtubles.

No other cell gives the opportunity for studying directly the correlations between structure and function as does the spermatozoon. In this way, the problem of explaining the role of the X-chromosome during spermatogenesis has been dealt with. According to apparently non-correlated cytological and genetic observations, it has been suggested that the single X-chromosome is inactivated during a critical stage of spermatogenesis in all heterogametic male organisms. As inactivation process is an essential control, any interference with the process could modify the course of development of the spermatocyte, leading to male sterility.

The study of female gametes is prevailingly oriented toward the fertilized egg and, therefore, toward problems which are more closely related to fertilization and the first stages of embryogenesis.

It has been observed that egg cells, somewhat less than spermatozoa, show morphological differences which may often be related to functional divergences. Studies on eggs are carried out mostly in relation to the foodstuffs they store and to the chemical mechanisms which are needed to attract a spermatozoon and to block the penetration of further spermatozoa.

In the last decade the study of gametes has been oriented prevailingly toward problems of molecular biology. Particular attention has been paid to the nucleo-cytoplasmic interactions which occur during development. Study of RNA synthesis and RNA transfer from the nucleus to the cytoplasm has helped explain the still poorly known aspects of the developmental process. An important contribution has been that of the study of ribosomal RNAs, in conjunction with the well-localized corresponding genes in DNA. In the eucaryotic cells at least two RNA-polymerase have been recognized; one, nucleole-bound and very likely to be specifically implicated in the synthesis of ribosomal RNA; the other, localized in the nucleoplasm and whose transcription product is DNA-like RNA. This last enzyme, RNA-polymerase I is known to be active mainly during oogenesis and developmental stages in which ribosomal RNA is actively transcribed (1970). In mature oocytes of *Xenopus* a protein factor has been discovered (1970) which interacts highly specifically with ribosomal cystrons and seems to depend on the requirements of a repressor gene.

In the pre-gastrular stages of amphibian development, when no RNA synthesis occurs, it has been observed that introduction in the system of an initiating factor of RNA-polymerase from *Escherichia coli* induces the synthesis of RNA molecules transcribed by an RNA-polymerase I (evidently in the inactive form). The observation suggests that the lack of RNA synthesis during well-defined stages of development is caused by the absence or the temporary inactivation of an initiating factor of the specific RNA-polymerase.

The phenomena of oogenesis and spermatogenesis make the gametes ready for fertilization with the formation of a zygote, whose nucleus contains the whole genetic complement from both parents. In order to fertilize the egg cell, the spermatozoon has to cross the envelopes which protect most eggs. According to some authors (I. H. Colwin and A. L. Colwin), while the spermatozoon interacts with the envelope's material, it undergoes a change called acrosomal reaction. The acrosome is a small body situated at the apical end of the spermatozoon; it is a vesicular structure, wrapped by its own acrosomal membrane. The acrosomal granule in this is thought to be constituted of enzymes which may dissolve the egg envelopes.

In the mammalian egg, which is protected by several envelopes, a particular enzyme occurs in the fertilization. This is called jaluronidase; it is probably contained in the sperm's acrosome, and detaches jaluronic acid from a residue of lysine. (The enzyme was first isolated in 1965 from bull and rabbit acrosomes.)

All the processes mentioned are possible only after the spermatozoon has been 'capacitated' in the female genital tract. It has not yet been possible to ascertain what capacitation is; however, it can be obtained also *in vitro*, by suspending mature spermatozoa in the tubarian secretion or in the follicle's fluid. Thanks to these achievements (in 1969), fertilization of mammalian eggs and also of human eggs have been obtained *in vitro*.

Fertilization causes some physico-chemical changes in egg membrane. One of the first and more important of these is the change by which the egg becomes refractory to interactions with other spermatozoa.

The activation process of the egg is a biochemical one mainly at the level of RNA and protein synthesis. The programme for embryonic development is being transcribed during oogenesis; fertilization triggers the mechanisms, which sets in action cell division and the whole embryonic development processes. Most ribosomes involved in the first developmental stages, as well as most messenger and transfer RNAs, are synthesized during oogenesis. In amphibian eggs, the synthesis of these substances seems to occur within a rather restricted period of time, when chromosomes become lampbrush (1966). During this period the egg cell will build up those ribosomes which will be required until gastrulation. Therefore a conspicuous synthesis occurs in a relatively short period of time. DNA meets an enormous requirement of ribosomal RNA by 'amplifying' the specific sites (cystrons) on which r-RNA transcription occurs. *Xenopus* oocytes have been shown to contain about 1,000 copies of genes which transcript ribosomal RNAs, each one present in its own nucleolus. Each normal cell generally contains, for every chromosome group, in nucleolus, which is derived from the nucleolar organizer, a region contained in one chromosome. The *Xenopus laevis* oocyte contains four chromosomal groups. Thus, gene amplification for ribosomal RNA has occurred in order to produce 1,000 occasional copies present in 1,000 occasional nucleoli. The scope of amplification is to produce a large amount of ribosomal RNA and, therefore, a large number of ribosomes in order to cope with the exceptional requirement of protein synthesis during the first stages of development. According to these requirements, genes for ribosomal RNAs transcript RNA molecules which, in turn, behave as templates for the synthesis of DNA by using an inverse transcriptase. Gene DNA for ribosomal RNAs is then replicated several times, through intermediate RNAs.

The egg cell has been demonstrated to possess enough m-RNA to direct protein synthesis until gastrulation. It has not yet been explained how m-RNA is stored. Two alternative hypotheses have been put forward. According to the first, maternal m-RNA would be bound to ribosomes and inactivated by a protein factor, which would be then removed at the moment of fertilization (1965); according to the second, in the unfertilized egg large ribosomal complexes are present together with messenger RNAs, which are available after fertilization (Stavy, Gross, 1967).

Study of the first stages of development has been attempted by tracing the development of eggs fertilized *in vitro*. It has been possible to obtain cleavage, *in vitro*, of fertilized eggs of the mouse and rabbit up to the point of blastula formation. In 1971 the development *in vitro* of a mouse blastocyst was obtained up to the stage in which the first heart beat occurs. Other investigators have cultured foeti *in vitro* in six days' time.

The above-mentioned research tends to establish, especially, the interaction between the embryo and uterine tissue, and to determine how widely this is essential to embryonic development. Attempts have been made to ascertain the hormonal requirement at the moment of egg implantation in the uterine mucus membrane, and if it is the trophoblast (tissue wrapping the fertilized egg) which digs niches in this membrane to implant itself, or if it is the uterine mucus membrane itself which is submitted to an autodigestive process. Hormonal action may be correlated also, with the transformation of the uterine mucus membrane at the moment of implantation of the fertilized egg; a normal mucus membrane, with dense and compact connective tissue, rich of glands, is transformed into a mucus, strongly oedematose, membrane without glands and but a few cells. Currently, investigation is being carried out in order to establish whether specific receptors for hormones of the progestinic type are present or not.

Blastokinins, a protein group which is probably secreted by womb glands, have been shown to occur in the uterine fluid only in the period of time immediately preceding egg implantation. These would be correlated with egg implantation in the uterine mucus membrane.

It has been observed that if a blastocyst, before being implanted in the uterine mucus membrane, is grafted in the maternal kidney or spleen, the transplant 'catches on' but only the trophoblast develops, becoming enormous. On the contrary, no embryo appears. The same occurs with cultured blastocysts. If a few drops of uterine fluid is added to the culture medium, however, a normal embryo develops.

Blastokynins have been isolated; they have a relatively high molecular weight.

Cell division and differentiation

Cell division is a process during which a mother cell gives rise to two daughter cells. According to some investigators, the control and regulation of the cell division, as well as the synchronism of the cell divisions in early embryos, would be attributable to specific regulatory substances.

According to some authors (such as J. D. Pitts, 1971) an animal may be considered as a population of cells which interact; the growth and function of each cell would vary according to the requirements of the whole population. Cell-to-cell interactions are of two kinds: direct ones, in which cells are in contact and signals are transmitted through intercellular junctions or cell membrane modifications at contact sites; indirect ones, which occur by means of the substances (hormones) transferred by extracellular fluids in order to reach the target cells.

Cell division involves a series of processes, which produce (on the one hand, by DNA replication) the exact qualitative and quantitative reproduction of genetic information and (on the other hand) the equal distribution of such information in the daughter cells. By autoradiographic analysis of the chromosomes, some investigators (Howard and Pelc; Hubermann and Riggs, 1968) have demonstrated that DNA synthesis occurs during the interphase, a period between two successive cell divisions.

Furthermore, it has been shown that, during cell division, the cell has a reduced metabolic activity; reduced respiration and protein synthesis, especially, are observed. During this period, the cell's metabolic behaviour is comparable to that of an anucleate cell.

Among the mechanisms to be observed which occur during cell division, there has been particular attention paid to the mytotic spindle. It has been established that this is constituted of microtubles, the occurrence of which is accompanied by cell movements allowing well-defined chromosomal distribution in the daughter cells. Since 1952 (D. Mazia and K. Dan) the fibrous structures of the mytotic spindle of sea urchin's eggs have been successfully isolated. Chemical analysis of the mytotic apparatus has revealed that it is composed of essentially protein containing sulphur. Further evidence learned through the electron microscope, that the mytotic apparatus is made up mainly of microtubules, leads to the conclusion that the microtubules possess not only sulphur-containing protein but, precisely, polymers of such protein. In these, single molecules are linked together by disulphide bonds.

During cytodieresis (the last step of the mytosis), when the division furrow appears and separation of the two daughter cells occurs, formation of the division cleft in the equatorial region involves an increase of about 30 per cent of the cell surface and therefore a synthesis of the plasma membrane. The division furrow formation seems to be attributable to contractile proteins which are present in that region. Since the position of the cleft would be determined by that of the mytotic spindle, the localization of the contractile proteins would depend upon the position of the spindle.

A peculiar mechanism is that of meiosis in the gametes, consisting of two successive cell divisions by means of which reduction occurs of the chromosome number from a diploid to an aploid condition, i.e. to one in which chromosomes number only one chromosome of each kind. During the 1960s, particular attention was paid to the phenomenon of the meiotic drive. The term had been coined in 1957 (by Sandler and Novitscki) to denote situations in which one or several events associated with meiotic divisions reveal an unequal occurrence of alleles (from a heterozygote) in functional gametes. In 1966 (Rhoades and Dempsey) it was possible to find a correlation between this unequal occurrence of alleles and an aberration of meiotic behaviour.

Sperm malfunction is one of the most common consequences of the process of the meiotic drive. A pair of homologous chromosomes are not distributed according to Mendelian laws; the mechanisms of unequal division are such that each homologous chromosome is contained in half of the meiotic division's products. These products—secondary spermatocytes or spermatidia—give rise to spermatozoa which do not function uniformly during fertilization.

In 1965 the first case of non-Mendelian segregation was described in which, according to unelaborated evidence, the sperm malfunction appeared as the consequence of the functioning of a well-defined mechanism. Hemizygous males for RD (recovery disrupter) gene on X-chromosome produce a progeny made of about 67 per cent of females. By observing the meiosis of such males (Erikson), a high percentage of fragmentation of X-chromosomes as well as other abnormal behaviour have been noted. It was thus concluded that spermatidia which bear a Y-chromosome and come from males with an RD gene are unable to complete normal spermatogenesis.

As an embryo gradually forms from the zygote by cell cleavage, areas differentiate and give rise to the formation of different tissues and organs. It has been ascertained that during differentiation genes do not undergo qualitative variations, whereas the extragenic portion of the nucleus will vary somewhat.

Differentiation occurs above all in the cytoplasm, and must be considered as a production of specific proteins; the main factors and mechanisms which are involved, therefore, must be those which control protein synthesis.

In the embryonic cells regulatory mechanisms seem to exist, similar to those which have been observed in the genetic regulation of enzyme synthesis in bacteria (F. Jacob, J. Monod, 1963). There would be groups of active or derepressed genes, and groups of genes which are inactive or repressed by the action of complex microenvironmental interactions. It can thus be explained why different cells inherit an identical genome but manifest themselves in different ways, giving rise to cell groups exhibiting different specializations. In such a mechanism the nucleocytoplasmic interaction would be fundamental and, according to some authors, it consists of a transfer from the cytoplasm into the nucleus of enzymes which derepress certain chromosomal traits.

In 1967 it was demonstrated (A. Tarkowski) that, as far as the 8-cell stage, each blastomere is able to produce a whole organism. In 1971 single cells from a blastocyst were shown to alter their mode of differentiation when transplanted into another blastocyst.

Embryogenesis

Embryogenesis includes the vast series of phenomena which range from egg fertilization to the formation of a complete individual. These phenomena have been investigated in order to understand the correlation between gene action and the different developmental stages, i.e. the forces which lead the cells to organize themselves into well-defined structures. It is unlikely that in the fertilized egg, ready to differentiate, there is a gene or group of genes for any variant of a given structure, while it is probable that the genetic information contained in the fertilized egg instructs the egg about the functions which build the adult.

Among the theories which have been proposed to explain embryogenetic phenomena, one of the most debated since the end of the 1960s holds that cells behave according to the position they occupy (Wolpert). The differentiation of a cell into a given organ would require the activation of a well-defined group of genes, but the cellular organization in that organ would be determined by a mechanism which is based on a positional information.

The most interesting studies on this kind of positional signal have been carried out mainly on *Hydra*, selected as a model organism, and proving to be helpful in tracing some of these phenomena.

When the cephalic end of *Hydra* is removed, a new cephalic end is formed on the cut surface. This regeneration does not require cell division, but occurs through a rearrangement of the neighbouring tissues. The occurrence of a supernumerary cephalic end inhibits the regeneration of the cut end. It could be imagined that the cephalic end produces a substance inhibiting the formation of another cephalic end and probably supplying positional information to the internal cells of the *Hydra*. When a cephalic end is removed, the gradient of the inhibiting substance diminishes and—as it has been postulated—the formation of a new cephalic end starts when the concentration of the inhibitor dips a certain threshold limit.

If these postulates will prove to be correct, *Hydra* should prove to be an ideal experimental system to study the signals made to control a developmental process. The results obtained so far (Wolpert *et al.*) show that the transmission times of the signals are rather long, so that they cannot be compared to those of a fast signal (such as the electrical impulse propagated from cell to cell). The signalling process, according to parameters concerning shape and the production of an inhibiting substance, has been simulated in a computer; the results achieved indicate that the diffusion of a relatively small molecule, as for instance a nucleotide, would account for the propagation of an inhibiting signal. Another feature, which looks valuable for this kind of investigation, is *Hydra*'s ability to recuperate an entire individual from dissociated cells. This feature has been useful in examining the development of polarity, positional information and successive control of a cell's motility as well as of differentiation.

It has been observed that *Hydra*'s dissociation by mechanical disrupture, into single cells or small cell groups, may be followed by reaggregation and regeneration. The composition of groups of cells or cellular constituents determines the polarity of the regenerating organism by establishing its main axis. According to some investigations (Gierer *et al.*, 1972), the dissociated cells remember their initial position; in order to clear up the essence of such memory, it would be necessary to find the biochemical treatment which could cancel it. The chemical system responsible for such memory is likely to be correlated to the inhibiting signal which has been hypothesized (Wolpert) in the regeneration process.

Some investigators have tried to explain embryogenetic phenomena by an immunological approach. The occurrence of a large number of different genetically controlled antigens, which may be recognized by the cell's surfaces, has allowed the construction of an ordered cellular surface mosaic. It has been observed, further, that each of these antigens is subjected to variations which are determined by environmental conditions.

It is unlikely that the most minor details of morphogenesis are programmed directly by the genetic material. It is probable that contacts between cells, in the early phases of development, are discriminant. A fascinating hypothesis holds that detailed morphogenesis needs a methodical diversification of the surface structures and that only a relatively small group of genes is required for this.

According to this hypothesis, as well as to evidence of the antigenic individuality of cellular surfaces and their genetic and epigenetic control, the morphogenetic process is controlled by some, not yet well-defined, mechanisms which concern recognition of the cell's surface.

The study of these mechanisms has been performed, as is often the case in embryology, by analysing extremely abnormal situations (from which it is easier to get information than in the normal condition).

Mouse mutants for the *T* locus have been analysed to this end. The *T* locus in the mouse occupies a region of association group IX. It is marked by a dominant gene *I*, which prevents the development and the maintenance of the notochord both in the heterozygote and in the homozygote conditions, thus causing in an early stage of development the scattering of the cells of this structure, which will be replaced in the adult by the backbone.

A series of different lethal or semi-lethal alleles (*t*-alleles) have been put into evidence in the *T* locus. These have been isolated from wild mice (which seem to be polymorphic for this locus) or have been obtained by mutation of previously known alleles. The recessive alleles may be distinguished one from another according to two of their features: by a cross test it is possible to identify the different types of alleles. These types of alleles may then be distinguished as each one produces a well-defined error during ontogenesis, thus interrupting their development in a very specific way and at a typical stage, ranging from morula to advanced neurula. More particularly, the anomalies in embryos carrying the *T* locus indicate that this locus influences unequivocally the chordomesoderm, the first inductive system in the embryo.

The immunological studies of the *T* locus have confirmed from many points of view that morphogenesis would be mediated by a phenotypical diversity in the structure of the cell's surface. The first step in support of this hypothesis is to prove that localized products in the cell's surface correspond to genes which are responsible for critical functions in the embryogenesis. In fact, if the cell's surface has a key role in the morphogenesis, it is likely that genes (which have a major part in such a process) are the producers of the components of this cellular surface. Experimental work thus far has only confirmed that *T*-allele codes for a component of the cell's surface.

The action of genes which are localized in the *T* locus and the action of the various alleles of these genes would be limited to well-determined cells and only for a certain period in ontogenesis, particularly during the first developmental stages of cells which are responsible for the structure and function of the primitive streak and the notochord.

Another investigation, to clear up the fertilized egg's potentialities and further differentiation in other organs, has been carried out on allophenic mice (B. Mintz, 1965). In order to obtain the allophenic mice, two embryos are drawn from different mothers at the blatocyst stage. They are released from the surrounding material and slightly pressed together. In the course of the following twenty-four hours, cells mix at random and rearrange themselves, thus forming only one larger blastocyst. The blastocyst is then implanted in the womb of a foster mother. The mice thus born would correspond partly or wholly to the transplanted fused embryos. That is, they result from the fusion of the traits of all four parents.

This study allows two kinds of speculation. First, at least until the blastocyst stage, no cell is oriented toward the production of a well-determined type of differentiated cell. Second, the lack of immune reaction in the blastocyst's cells to the cells of another blastocyst would confirm the hypothesis that the foreign cells, only recently put together, might undergo certain reciprocal control. This thesis would support the argument that during embryonic development some differentiation processes would be due to cell-to-cell interactions.

Cell cultures

Cell cultures are helpful in studying some complex biological processes in cells or groups of cells which have been isolated from the whole organism and in estimating interactions with both surrounding cells within the whole organism and with the outside.

Mammalian cell cultures have proved helpful in both the genetic analysis of somatic cells (through the isolation of temperature and drug-resistant mutants, and so on), and the study of mutagenesis in somatic cells. The results of cell culturing have proved particularly successful in the study of nucleic acids and in investigations of the mechanisms of protein synthesis. The synthesis processes are studied *in vitro* by using synthetic or viral messenger RNA. The different steps of the synthetic process have been followed by means of inhibiting substances that are similar, from the chemical point of view, to substances normally occurring in the synthetic process.

Cell cultures are also helpful in studying differentiation in plants. Plant cells tend to dedifferentiate *in vitro*. It is therefore possible to make them evolve in the desired way, in order to obtain a certain part of a plant; it is also possible to follow the various phases of the process and to control in which degree the environmental factors intervene. Within this trend one must consider particularly the study of the vegetal hormones and that of the modifications of enzymes and of proteins, in general, which are associated with the differentiation phenomena. During the 1960s the development of a plant of *Daucus carota* was first obtained *in vitro* from undifferentiated cells (F. C. Steward).

Liquid plant-cell cultures have been utilized chiefly to explain many unsolved questions concerning cell physiology, particularly that of the synchronization of cellular divisions. In seedlings obtained *in vitro* from somatic cells, it has been observed that they fail to exhibit the traits of the maternal plant. Especially in tissues maintained for a long time in the presence of auxins, tumoral traits and frequent endomytoses have been observed.

In the study of oncogenic viruses, cell cultures are helpful in making observable the deep transformations which occur in infected cells. For instance, in human embryonic skin treated *in vitro* with the human papilloma virus, proliferating phenomena and loss of the properties of contact inhibition occur.

In cells treated with the polyoma virus, typical transformations are observed such as the neoplastic potential (which is expressed by the cells' ability to produce tumours if implanted in suitable animals), the occurrence of viral genes, the increase of tetrahaploid cells in subcultures, the occurrence of abnormal chromosomes and, finally, the occurrence of virus-specific antigens in the transformed cells.

Investigation on cultured cells suggest that all kinds of transformation depend on a common factor, as all transformed cells exhibit similar properties. Besides the characteristics already mentioned, it has been observed that, although in normal cells the nutrient transfer from outside to inside is controlled in the cell by external factors, such as hormones, serum proteins, or contact with other cells, in transformed cells, such transfer seems to be independent of external factors. Furthermore, a much lower cyclic-AMP concentration is observed in the transformed cells. At any rate, the differences between normal cells and transformed cells (very obvious in dense cultures) tend to lessen in sparse cultures in which cells are growing actively. These observations have suggested that the fundamental difference consists of the ability of the transformed cells to pursue their growth when confronted with factors blocking the growth of normal cells.

The transformation of cells by viruses can be studied *in vitro* and, because of the uniformity of transformation traits, may supply a model to help us understand the neoplastic transformation of cells. This transformation is also provoked by other agents, such as chemical substances, radiations and so forth.

Transformation by virus has been investigated by observing the mutations of viral genes, chiefly those which are temperature-sensitive and those dependent upon transformation; by identifying viral proteins through the introduction of labeled amino acids in the system; and finally, by observing the functions of viral genes according to the functional changes occurring in cells after infection.

Transplantation and hybridization

Nucleocytoplasmic interactions and specific action of certain parts of the cell have been explained partly by two techniques that have been greatly improved during the 1960s: nuclear transplantation and somatic cell hybridization.

The first successful experiments in nuclear transplantation were carried out in 1952 (R. W. Briggs and T. J. King). *Xenopus laevis* was the experimental animal, also used in subsequent researches. One of the first objectives was to ascertain whether or not, during normal cell differentiation, genes responsible for the first stages of development are gradually lost or only inactivated (so that they can reactivate if necessary). These experiments were carried out by transplanting, into an unfertilized cell, the nucleus extracted from a cell of an older embryo. The first experiments with unfertilized eggs, in which nuclei from endothelian tissue of the digestive tract of advanced larvae were transplanted, demonstrated that such nuclei possess potentially all the genes required to control the first stages of development. Some embryos obtained from such transplantations attained the adult stage.

In a clone of embryos derived from one nuclear transplantation, it has been noted that many subjects often have similar anomalies—different from those of embryos derived from another nuclear transplantation which, on the contrary, are all identical among themselves. It has been established that these divergences occur as a result of nuclear transplantation, and are not observed in normal differentiation. The anomalies in embryos seem to be correlated with chromosomal aberrations not occurring in donor embryos. Several investigators have tended to interpret these aberrations as the result of an incompatibility between the low rate of division of differentiating cells (a division every two days, from which the nucleus derives) and the high rate of division of a fertilized egg at the beginning (from which the cytoplasm derives).

Once it was determined that cell specialization involves differential activity of the genes occurring in the cell and not a selective elimination of genes already exploited, nuclear transplantation was applied also to the study of how genes are activated or repressed during early embryonic development. It has been ascertained that signals to which genes respond are normal cytoplasmic constituents.

This was made possible by experiments in which a cell nucleus with well-defined activity was inserted in an anucleated cell, the cytoplasm of which has a well-defined role. By this technique, it has been possible to investigate also whether the different functions of any single gene—DNA synthesis, RNA synthesis, chromosome condensation when cell division approaches—are induced by cytoplasmic constituents. The results have been very indicative: transplanted nuclei have changed their function, after one or two hours, in order to conform to the typical function of the host cell's nucleus.

The influence of cytoplasmic components in the repression or activation of genes present in the nucleus has been studied by a group of investigators who remarked that no RNA synthesis can be observed in *Xenopus* during the first ten cellular divisions after fertilization. They noted also that only at blastula stage cells synthesize large RNA molecules, probably messenger RNA molecules, whereas it is only near the end of the blastula stage transfer RNA synthesis seems to occur.

The cytoplasmic influence on these events has been investigated by transplanting nuclei from neurular cells into anucleated eggs and by tracing nucleic acid precursors. By autoradiographic procedures neurular nuclei normally synthesizing any kind of RNA have been shown to stop synthesis when transplanted in anucleated eggs. These and other results suggest that egg cytoplasm contains constituents which are responsible for independent control of the activity in different classes of genes.

By the same technique, attempts are being made to obtain helpful information in order to identify the cytoplasmic components which influence gene activity and their mode of action.

Genetic analysis, one of the most powerful tools of the biologist, helps him to understand how traits are inherited from one generation to another, was limited to cross-study until the discovery of a revolutionary technique, that of the somatic cell hybrids, i.e. differentiated cells which may be fused to form hybrid cells which survive and are able to multiply.

This technique was elaborated in 1960 by investigators at the Institut Gustave Roussy in Paris (G. Barski, S. Sorieul, F. Cornefert). They mixed two different cancer-cell cultures from the mouse, and, after several months, obtained cells in which the single nucleus contained the chromosomes of both sets of parental cells.

The same procedure was later applied to other, different, cellular types; it can now be utilized for different purposes. The somatic hybrids show two fundamental traits: both chromosomal groups are functional, and the hybrid possesses hereditary traits of both parental cells. During their multiplication, some chromosomes are lost; some cells originate in subcultures which have different complements of parental genes.

The somatic cellular hybrids have been used to study several biological processes. One of these is the localization of genes in the chromosomes of higher animals, man included. In fact, analysis by hybridization, carried out for this purpose in lower animals and in plants, proceeds too slowly in mammals and it is not realizable in man.

In man-mouse cell hybrids, human chromosomes are best eliminated during multiplication. The conservation or loss of a human chromosome may thus be correlated with the conservation or loss of certain traits. It is then possible to localize on a well-defined chromosome the gene responsible for a certain characteristic. The most annoying difficulty, however, is that it is necessary to identify each of the twenty-three human chromosomes. This is now possible thanks to recent staining techniques (Cassperson, 1970) which allow us to identify any chromosome according to a series of typical chromosomal bands.

During the 1960s, increasingly reliable techniques were elaborated in order to produce somatic cell hybrids (J. W. Littlefield, R. L. Davidson, B. Ephrussi). At the beginning, somatic hybrids were obtained from among two cell lineages of the same species, usually mouse lineages. Afterwards, the hy-

bridization of cells from different species was attempted. The first of these interspecific hybridizations was a mouse-rat hybridization, then hamster-mouse and, finally, mouse-man cell hybrids were obtained. Most experimental work now is being carried out on interspecific hybrids. Cell hybrids have been shown to synthesize hybrid enzymes which function efficiently.

Hybridization between differentiated and undifferentiated cells, or between variously differentiated cells, can provide explanations of the regulatory processes occurring in cellular specialization. The activities of somatic cells can be divided into two categories: 'essential' functions, needed for the maintenance and the growth of the cell, and specialized or 'luxury' functions (such as muscle-fibre formation, hormonal secretion or the production of pigments) which are necessary for the survival of the organism rather than that of the isolated cell. Hybridization experiments on 'luxury' cells (Ephrussi, Davidson, Yamamoto), and specifically with melanin-producing cells from different cell lineages, have yielded cells which did not produce melanin. It is obvious that, in the hybridization between melanin-forming cells and nonpigmented cells, synthesis of the enzyme which is essential to the synthesis of melanin is blocked by a substance present in normal cells. It remains to be determined at which step the blocking of enzyme synthesis occurs.

In 1972 some investigators (Nirenberg and co-workers) fused neuroblasts from mouse cerebral cancer and mouse fibroblast. The hybrids contained chromosomes of both parental lineages and, rather surprisingly, maintained several neuroblast luxury fonctions: production of the important cerebral enzyme acetylcholinesterase, as well as the presence of electrically excitable membranes and thin membrane digitations.

Most interestingly, it has become possible to isolate cell-hybrid subcultures which have lost one or more such functions. These cells contained even fewer chromosomes than the already mentioned cells, thus suggesting that defective chromosomes might be involved in defective functions. The somatic cell-hybrid technique has been shown to be ex-

Fig. 59. Hybridization of somatic cells may gain increasing significance for the development of strains resistant to disease and insects. The diagram shows the process of substituting a gene by means of somatic hybridization. The gene substituted may, for instance, be responsible for resistance to a certain disease. By repeated backcrossing of the somatic hybrid with the species A, it is possible to produce plants with the chromosome content of A to which a chromosome of B has been added. Treatment with X-rays could bring about a redistribution of the chromosome set of such plants whereby a gene of this B chromosome is transferred, as shown on the right in the diagram, to a chromosome of A. This form of hybridization involves the transfer of the characteristics of one plant species to another, without affecting the taxonomic integrity but increasing the original variability.

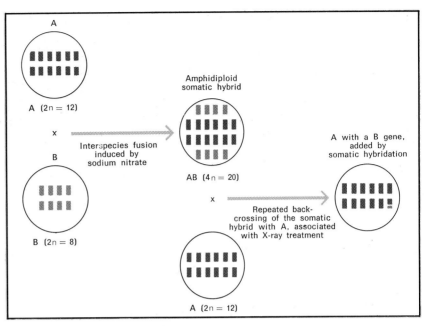

tremely helpful in the study of cancer. Experiments carried out by a group (H. Harris and co-workers) at Oxford, in collaboration with a research unit (Kein and co-workers) in Stockholm, have suggested that the loss of genetic material from a cell may be an important device to obtain a cell having rapid growth traits. The Oxford and Stockholm experiments involved cell lineages, from mice, called A9 strains. These lack a particular enzyme and are not grown on the cultural medium known as HAT. A9 cells are only weakly oncogenic when injected in mice. The research teams have obtained from an A9 lineage of the type a much more powerful A9HT cell lineage, one which is oncogenic in the most inoculated animals. The A9HT lineage has fewer chromosomes than the A9 lineage, suggesting that the loss of genetic material is an important source of malignancy. A9HT cells do not thrive in the cultural medium HAT, whereas tumour cells (produced by inoculating with A9HT) survive and grow rapidly when they are inoculated in the HAT cultural medium. These HAT-resistant cells look rather different than the A9HT cells: they contain a much higher number of chromosomes. It is most likely that they have acquired the enzyme which is defective in the A9 lineage. This can be interpreted to mean that the HAT-resistant cells stem from the fusion of A9HT cells with normal cells from the host animal. According to these results, complementation between malignant cells and normal cells would represent a way of recovering normal characteristics.

Further reading

Brochet, J. *Introduction to molecular embryology*. London, 1976.

de Grouchy, J. La carte chromosomique de l'homme. *La recherche*, vol. II, no. 14, 1971, p. 621.

Dollander, A.; Fenard, R. *Eléments d'embryologie*. Paris, Flammarion, 1971.

Ebert, J. D.; Sussex, J. M. *Interacting systems in development*. New York, Holt, Rinehart & Winston, 1970.

Houillon, C. *Embryologie*. Paris, Hermann, 1967.

——. *Sexualité*, Paris, Hermann, 1967.

McKusick, V. A. The mapping of human chromosomes. *Scientific American*, vol. 224, no. 4, 1971, p. 104.

POWER, J. B.; COCKING, E. C. Fusion of plant protoplasms. *Science progress*, vol. 59, no. 234, summer 1971.

SCHILDE-RENTSCHLER, L. Les protoplastes. *La recherche*, vol. VI, no. 56, p. 430.

SUNDERLAND, N. Anther culture: a progress report. *Science progress*, vol. 59, no. 236, winter 1971.

23 Tissues and organs

Muscles and nerves

Muscles are contractile organs, the function of which implies complex anatomical structures, the processing of chemical reactions, and fine control by the nervous system (see below). Contractile proteins have been shown to be of fibrous type and, through the electron microscope, appear as threads 5–10 nm in diameter, axially oriented. Such proteins have been subdivided as follows: myosin 54 per cent, actin 21 per cent, tropomyosin B 15 per cent. The remaining amount of 10 per cent is represented by proteins not yet identified. Myosin constitutes 95 per cent or more of the H-band or thick filaments; actin and tropomyosin together represent almost all the proteins of the I-band, or thin filaments. The actin molecules (molecular weight 70,000) form long threads which in turn constitute the two coils of a double helix. Myosin (molecular weight 428,000) may be subdivided into two components, light meromyosin (molecular weight 96,000) and heavy meromyosin (molecular weight 236,000). When actin and myosin solutions are mixed together, they react to form actomyosin. Myosin is an enzyme (ATP-ase), which catalyses the reaction $ATP \rightarrow ADP + P$. Actomyosin also is an ATP-ase which is activated by Ca^{++} and Mg^{++}, whereas the myosin ATP-ase is inhibited by Mg^{++}. It has been proved that through these enzimic activities part of the energy released in the ATP splitting is utilized by actomyosin when this one contracts. When actomyosin ATP-ase is active, the addition of ATP makes actomyosin contract; when it is inhibited, the addition of ATP makes it relax. Concerning the processes related to muscle contraction, available data have shown that ATP is directly involved in the formation of bridges between the thin filaments (I-bands) and the thick filaments (H-bands). It may be supposed that a tension would develop when, by action of Ca^{++}, ATP on the side of a thick filament attaches itself to a thin filament. Ca^{++} ions should neutralize some of the negative charges on ATP, thus inducing the shortening of the bridge projecting from the thick toward the thin filament. In turn, this should bring ATP near ATP-ase, thus causing the splitting of ATP and the breakdown of the bridge. Afterwards, phosphocreatine (PC) would regenerate, and another bridge is likely to be formed with a close point of the thin filament—if the muscle is contracted.

It seems that, when the muscle is relaxing, ATP is attached only on the side of the bridge where the heavy meromyosin is present, near the ATP-ase, thus inhibiting the combination of actin with myosin and making the filaments slide on each other.

It has been observed that the link between excitation and contraction is provided by a special system of conduction inside the muscle fibre; it is made up of a branched system of tubules, or endoplasmic reticulum. The transverse tubules (T-system) appear on the outside, on the fibre surface, as holes. They run inward, passing sometimes between pairs of external vesicles, the role of which would be that of pumping Ca^{++} ions from the sarcoplasm, thus lowering their concentration under the level that enables ATP to split. In the muscle at rest, a lot of Ca^{++} is present: when an action potential spreads on the fibre surface, an electrical signal goes along the transverse tubules and causes the Ca^{++} to be released in the sarcoplasm by the external vesicles, thus starting the ATP-splitting and fibre contraction. If no further action potential spreads along the fibre, Ca^{++} ions are pumped again into the vesicles, and the muscle relaxes. The energy sources of muscle contraction are provided by muscle proteins (which constitute the machinery of the fibre cell) and an enzyme system (which catalyses the chemical reactions), producing and splitting ATP, in order to transform part of its energy into mechanical work. Phosphocreatine, by reacting with ADP through the action of phosphocreatinphosphokinase, produces ATP and creatine. As soon as ATP has been split, it is synthesized again at phosphocreatine's expense. In turn, the phosphocreatine stores are reintegrated during the resting phase in which they gain energy from oxidation of glycogen stored in the muscles. If oxidation cannot occur, as in exhausted muscle, more energy is available from glycogen by hydrolizing it into lactic acid.

Muscle fibres contract only when an order comes along the axons from the central nervous system; the contact between the axon and the muscle fibre occurs at the level of the neuromuscular junction. The neuromuscular junctions are made up of nonmyelinated nervous endings, sited on the surface of the striated muscle as in a cleft, with an extracellular space between the nerve and the muscular membrane containing amorphous material.

By means of electron microscopy, the presynaptic ending has been shown to contain, in addition to the synaptic vesicles containing the transmitter, several mitochondria involved in the relevant metabolic activity of the junction. After the stimulus has passed across the synaptic junction, acetylcholine is eliminated through the action of a local enzymic system with acetylcholinesterase catalysing the hydrolytic splitting of the acetylcholine.

The kidneys

The kidneys are a pair of bean-shaped symmetrical organs, which lie on each side of the backbone, nestled against the psoas major. They carry out functions necessary to life maintenance; in fact, their removal causes death in a few days. Yet the removal of only one kidney is generally consistent with survival: the hyperfunctioning of the remaining kidney is sufficient to compensate the loss of the other. The vital functions of the kidney are: (a) maintenance of the volume of organic water; (b) maintenance of the overall concentration of solute in the organic fluids (osmotic pressure); (c) maintenance of the concentration of single substances and, above all, of the ratio between some ions such as Na^+, K^+, Ca^{++}, Mg^{++}; (d) regulation of the elimination of certain substances (vitamins and hormones); (e) maintenance of an acid-base balance; (f) removal of wastes produced by lipid, carbohydrate and protein metabolisms; (g) neutralization of toxic substances; and (h) regulation of the arterial pressure.

The study of factors influencing renal plasma flow (RPF) has demonstrated that such an index is higher in a lying-down position than in sitting or upright positions, as kidney vasoconstriction keep constant the pressure values against gravity. RPF is also lowered by intense physical exercise, pain, emotional state, and when a relevant blood volume is displaced from kidneys towards other body regions. RPF decrease also occurs in water and salt depletion and when a reduction in blood volume takes place. In pregnancy, RPF is increased as a consequence of the increase of blood and plasma volume. In haemorrhagia, shock and asphyxia, a marked decrease of RPF is observed, partly owing to increased vasoconstrictor activity of the sympathetic nervous system.

It has been demonstrated also that the blood flow rate through the cortex is five or more times higher than that through the medulla. In spite of the overall constancy of the glomerular filtration rate (GFR), the flow of the medullar blood decreases when no diuresis occurs and increases when diuresis occurs as demonstrated by means of photocells localized in the cortex and the medulla, permitting control of the appearance and disappearance of Evans blue dye injected in the renal artery. The dye rapidly disappears from the cortex, but much more slowly from the medulla.

The difference in the flow rates is probably due to: (a) the rich capillar anastomosis round the tubules in the cortex, whereas the long and straight vessels in the medulla offer stronger resistance; (b) the action of antidiuretic hormone (ADH) to which the medullary vessels are probably more sensitive. The kidney vessels are also under the control of the vasoconstrictor sympathetic nerves and are sensitive to adrenaline, noradrenaline and angiotensin action. The angiotensin, when injected, induces a generalized vasoconstrictor reaction. The adrenaline's and noradrenaline's effects depend on the amount which has been administered.

The study of the glomerular membrane permeability has demonstrated that it is not constant and would be influenced by ADH. The ways of action of this hormone have been elucidated by observing that its addition to the fluid which washes the internal surface of some amphibian membranes increase their permeability to water molecules and speed up the water flow across them as a response to an osmotic gradient. Experiments of the same kind have also demonstrated that ADH increases the net sodium transfer across a membrane, and that such an effect is associated with an increase in oxygen consumption by the membrane itself. According to the majority of the investigators, the increased permeability to water which has been observed in these experiments would represent the basis of the antidiuretic action exerted by the hormone on mammalian kidneys.

Two theories have been formulated to explain this action in molecular terms. According to the first, interaction would occur between the disulphide bond of the cystine residue of ADH and the sulphydryl groups of the membrane proteins in order to induce an increase in membrane permeability. According to the second, ADH is shown to increase the production of $3'$-$5'$-c-AMP in the tubular renal cells, probably by activating adenylcyclase. Moreover, $3'$-$5'$-c-AMP itself modifies membrane permeability to water through a mechanism which still needs to be elucidated. ADH secretion depends on ECF osmoticity. High osmoticity induces an increase in hormonal secretion, with removal of small volumes of concentrated urine, whereas a low osmoticity inhibits ADH release and causes an increase in diuresis.

Knowledge of renal physiology has enabled us to design artificial kidneys. An artificial kidney can substitute temporarily the function of the injured natural kidneys until they recover. It is essentially made up of a semipermeable membrane which separates the blood to be purified from a dialysis solution, which has a chemical composition nearly resembling that of blood. The hemodialyser operates according to the principle of equilibration (or near equilibration); equilibration is achieved by recirculating the patient's blood through the artificial kidney. Available data show that, in order to design a really efficient artificial kidney, one must attempt to satisfy the following conditions: (a) the dialyser must be efficient in the removal of the nitrogenous and toxic products of metabolism; (b) the dialyser must remove water from a patient who is excreting insufficient urine; (c) the dialyser should have a low internal volume; (d) the dialysing membrane should be presterilized, simple to assemble, and inexpensive enough to be disposable; and (e) an artificial kidney should be easily portable so that dialysis can be done continuously.

The heart

The function of the heart is that of making the blood circulate and to enable exchanges between capillary blood and alveolar air or tissues

241

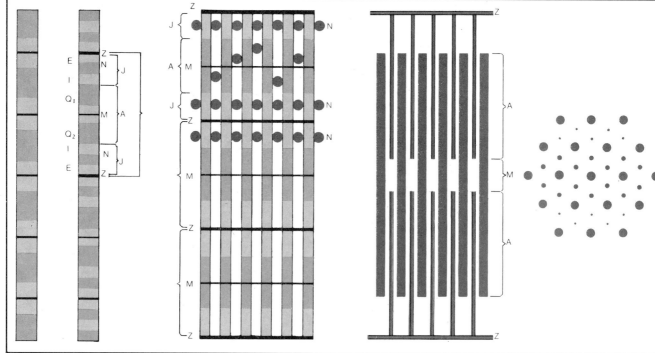
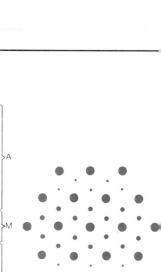

FIG. 60. A schematic representation of the banded structure of muscle fibres (technically known as sarcomeres), compared with a photograph by electron microscope (70,000 ×) showing myofibrils and sarcomeres in mammalian striated muscular fibres. Left to right: simple structure, and more complex one with sarcomeres (letters indicate different bands within sarcomeres); sarcomere structure, whose position is shown as pattern of dark spots, with current nomenclature; pattern of myofibrils, stretching without interruption from one stria to next; complex structure characterized by hexagonal symmetry; H. E. Huxley's representation of the twofold nature of myofibrils, differing both in their order of magnitude and in kind of protein. Thin filaments are constituted of actin; the thick ones, on the contrary, contain myosin. (From L. Amstrom and K. Porter.)

in order to provide an adequate metabolism to the various organs. According to physical and thermodynamic parameters, the heart and circulatory system may be considered as a volumetric pump connected with its own fluid flow. Every cycle of this pump has been shown to start with an atrial systole, in which simultaneous contraction of the two auricles occurs; pressure increases both in the auricle and in the ventricle, the atrioventricular valves being open and the two chambers communicating. When the ventricle begins to contract, the atrioventricular valves close and the endoventricular pressure increases rapidly, the semilunar valves being closed. At the moment in which the endoventricular pressure equals the pressure in the main aorta or pulmonary artery, the semilunar valves open and the blood is rapidly expelled from the ventricles. Then ventricular contraction gradually stops and pressure lowers both in the ventricle and artery (more rapidly in the first than in the second), causing the semilunar valves to close. After the decrease in endoventricular pressure, the atrioventricular valves open and the blood

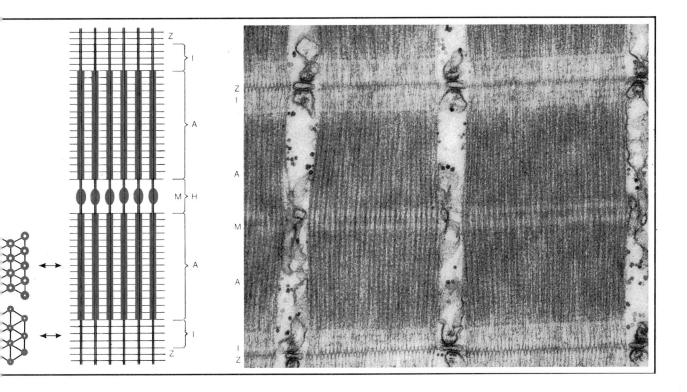

flows in the ventricle. The myocardium has been shown to receive its energy almost exclusively from carbohydrate, lipid and protid catabolism: in basal conditions, about 60 per cent of the energy expended by the heart is supplied by the catabolism of the free fat acids, and about 35 per cent by the carbohydrate catabolism (glucose, lactic acid, pyruvic acid). The remaining 5 per cent of energy metabolism is carried out as utilization of the ketone bodies and amino acids.

The rate of oxygen consumption (VO_2) in basal conditions has been observed to be 9 ml/min for each 100 g of tissue. When the heart beats without accomplishing any work, oxygen consumption increases at about 1 ml/min for each 100 beats/min and attains 5 ml/min per 100 g of tissue during ventricular fibrillation. The more interesting research concerning the myocardial structure has been carried out by means of electron microscopes and biochemical procedures. Although many aspects still need to be elucidated, the most important observation has been that myocardic fibres do not form syncitium, as it

was long claimed. In fact, the intercalated discs are tick cell boundaries which strongly bind the myocardial cells. Myofibrils are fixed at their internal faces. It has been demonstrated that the myocardium is formed by distinctive contractile cells, strongly bound by means of intercalated discs and arranged to form a long-mesh plexus. The same investigations also demonstrated that the conduction system is made up of myocardial fibres with very peculiar features: they appear to be very rich in sarcoplasmic material with a high glycogen content, while the myofibrils show a more irregular arrangement. Large vesicular cells are observed in nodes. Morphological-functional investigations have confirmed the validity of the 'law of the heart' according to which the contraction strength which may be attained in the ventricular chamber increases, within a certain range, the relaxing degree of the ventricles. Beyond this limit, the tension developed by cardiac contractions decreases progressively until it reaches a certain distension. Here cardiac contraction does not provoke an increase in the chamber's internal

243

pressure. The lax of the heart has been applied mainly in studies concerning the adaptive ability of the heart, especially in relation to sports medicine.

There is evidence that the cardiac activity (frequency and strength of the heart) is under nervous and hormonal control. Sympathetic and parasympathetic fibres are present in the cardiac nervous plexus, which surrounds the origin of the large vessels at the basis of the heart. The sympathetic supply is derived from T_2–T_4 tract and travels via the middle cervical and cervico-thoracic ganglia and the first four thoracic sympathetic ganglia. The postgangliar fibres, which branch out in the cardiac plexus, innervate the SA node and then whole cardiac muscle. The neuronal transmitter is mainly represented by noradrenaline. The cardiac effect of the sympathetic stimulation may be blocked by the beta-adrenergic blocking drugs. The vagus supplies the parasympathetic innervation: its fibres branch out in the cardiac plexus and then innervate the SA node and both auricles. No clear-cut data have been collected until now about the parasympathetic innervation of the ventricular muscle. The neuronal transmitter is acetylcholine and the effects of vagal stimulation may be readily reversed by atropine. Concerning the humoral control, investigations are designed chiefly to elucidate the feedback mechanism (realized through catecholoamines as mediators), the renin-angiotensin mechanism, and the aldosterone-controlled mechanism.

The study of cardiac activity has led to the design of pacemakers. A pacemaker is an apparatus (now reduced to the size of a pocket cigarette lighter) which supplies to ventricular myocardium rhythmic electrical impulses, the frequency and the amplitude of which may be regulated, when the conduction system has been injured or the very stimulus has not been formed. Stimulation may occur through both external and internal electrodes. The external electrodes are implanted on the thoracic surface and require rather high voltages (50–150 V) and amperages (50–150 mA). The internal electrodes are made up of wires which are introduced in the myocardium through the skin or in the right ventricle through the venous circulation. The excitation voltage is 3–15 V and the amperage 1–15 mA. The optimum length of each stimulus is 2 ms. The newest models use either nuclear energy or cadmium batteries developed for use on board space satellites.

Investigations in designing artificial hearts (apparatus which supply blood circulation in cardiac injury) are especially oriented towards the designing of (a) electrical hearts which have their control units on the outside; (b) artificial hearts which are driven by nuclear energy; (c) artificial hearts which are driven by biological energy, supplied by the combustion of tissue components (mainly glucose of the blood). The development of bionics has enabled the design of models to study the circulatory system. Investigations now in progress aim at considering; (a) conservation of mass; (b) equations of motion for linear fluids; (c) flows in the axial plasmatic gaps of capillaries; (d) laminar vortices; (e) flow through a bifurcation; (f) pulsatile flows; (g) pulsatile flow in deformabile tube; (h) turbulent flows; (i) models of the cerebral fluid system.

Nervous systems

The nervous cell of neuron is characterized by a peculiar morphology. From the cell body, tree-like projections protrude which show a more and more thin branching which finally forms a specific communication network.

The cell body's shape depends on the number and the way in which the projections originate; the shape can be stellar, ovoidal or conoid. Dendrites originate in a large root, and branch out at a short distance from the cell's body. A neurite is generally a unique projection, originating in the cytoplasm as a short cone; subsequently, it attains a size which remains constant as far as the terminal branching is concerned. The paraphytes, or accessory projections, are present only in some groups of large neurons usually deprived of dendrites.

The dimensional degree of neurons can be hardly assessed. For some typical neurons (such as spinal motoneurons), schematic representations have enabled us to visualize a standard form: the motoneuron appears as a sphere (with an average diameter of 40 mμ)

FIG. 61. Each nerve cell is composed of a cellular body, fibres (called dendrites), and another prolongation, generally of considerable length and in some cases also having ramifications, called an axon, along which neural impulses are propagated. The ends of the axon branch out, making synaptic contact with the dendritic fibres or with the cellular bodies of other neurons: through these synapses, the neuron enters into contact with another, and the message passes to the dendrites and then to the second neuron, and so on.

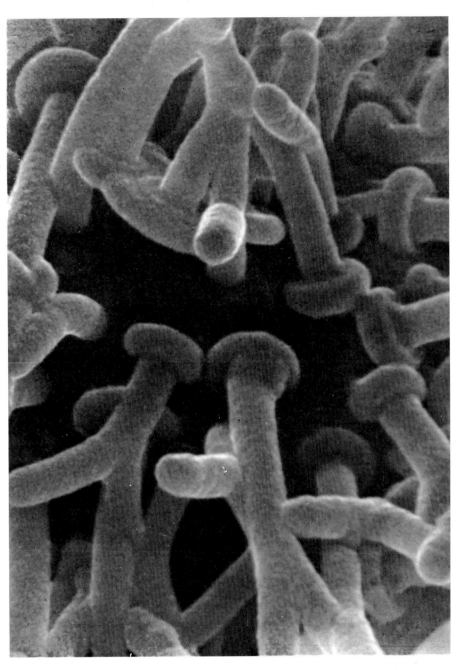

connected by five 40 mμ-long, 10 mμ-wide cylinders.

The cell body contains a rather large vesicular nucleus. Inside the nucleus a large nucleoulus appears, the extension of which seems to be related to continuous RNA synthesis. The cell body is made up by hyaloplasm delimited by the cell membrane, the properties of which are similar to those of the boundary membranes of other cells. The hyaloplasm contain several organites: (a) several mitochondria; (b) a well-developed ergastoplasm; (c) well-developed Golgi areas; (d) fibrous structures (such as microtubules and neurofilaments); and (e) different kinds of paraplasmic materials, especially glycogen granules. Dendrites are cytoplasmic projections containing a plentiful amount of cytoplasmic matrix, elongated mitochondria and several fibrous structures which are represented mainly by microtubles; neurofilaments are also present. The neurite possesses a thin boundary membrane containing the hyaloplasm in which neurofilaments (arranged parallel to the length of the fibre) and much elongated mitochondria are scattered.

The nervous fibre's structure depends on the unidirectional propagation of high-speed impulses. It has been demonstrated that the electrochemical gradient which occurs through the neuron's semipermeable membrane is created by the different ionic composition existing between the inside and the outside of the neuron, a composition which is asymmetrical but constant as concerns sodium and potassium ions.

The constancy of a 1 : 10 ratio for sodium ion concentration between the inside and the outside of the neuron is assured by a specific mechanism; this involves a cation active transport system, working against a chemical concentration gradient. Such a mechanism, the so-called sodium or potassium pump—by which sodium ions are actively pumped from the neuron's inside and potassium ions are actively pumped from the neuron's outside—is closely related to active metabolic systems. The latter are energy consuming and responsible for the maintenance of the potentials at rest, the conduction of action potentials, and the restoring of altered resting potentials.

At the moment in which the depolarization potential starts, it would seem that the lipid and protein molecules of the membrane orient themselves temporarily again in order to open some channels, to enhance the ion flow, or to release some sodium-ion carriers which are actively transferred inside the cell. The function of sulphatides and of the adenosintriphosphatase (ATP-ase) in the transfer of sodium ions across the membrane has been carefully studied. The correlation between ATP-splitting and signal transfer in the nervous fibres is still under investigation, mainly in relation to the point of the metabolic cycle in which the energy metabolism would release the requested amount of transfer energy. The great majority of available data

seems to indicate that stimulation energy starts impulse-generation. Successively, the impulses propagate in the form of explosion and autosensitization, using the energy of the electrochemical resting gradient of the nervous fibre as the leading force for such a propagation.

A special device, the synapse, is formed when nerve cells come into contact: a synapse represents the site of interneuronal transmission. The electron microscope reveals that it consists of an expanded presynaptic terminal fibre, the synaptic button, which closely surrounds a portion of postsynaptic neuron, the so-called subsynaptic membrane. The cleft between these two boundary membranes, 15–25 nm large, is called the synaptic cleft. The number of synaptic buttons may be very high for each neuron. It has been calculated that in spinal motor neurons as many as 10,000 synaptic buttons are present. This figure may be as much as 200,000 in the enormous dendritic branchings, as are those which occur in the Purkinje cells of the cerebellum. A gross calculation shows that the overall contact surface made up by the synaptic buttons in the neuron may be more than 30 per cent of the overall neuronal surface. The content of the synaptic buttons is made up of mitochondria, of plentiful vesicle-like structures (synaptic vesicles) and seldom microfilaments. In the majority of the synaptic buttons of the central nervous system, the rather homogeneous vesicles have a diameter of about 30 nm and their content is electrotransparent. In other buttons, especially in the peripheral vegetative system, considerably larger vesicles are present, which are characterized by a strongly electron-opaque granule inside themselves. Finally, the vesicles may be tubular or shaped as a flattened disc. Recent data from electron microscopy have demonstrated that the interneuronal contacts are not strictly limited to buttons that have somatic and dendritic contacts. This has been shown as follows:

1. Junction sites may form between neurons or their projections, as a consequence of more or less close contact between membranes, the structure of which is very different from that of the synaptic buttons.

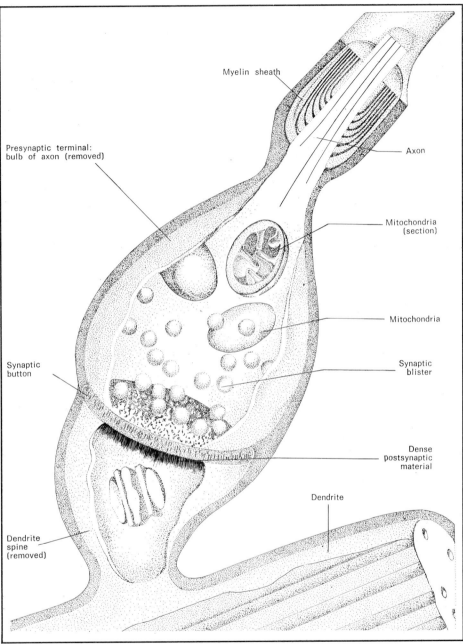

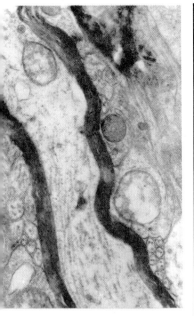

. 62. Diagram of synapsis showing
main structural elements, and
tomicrograph of the central axon.
urce: University of California,
R. Lewis.)

There are neither structural asymmetry nor
specific structures. These junctions are included among the general models of intercellular junctions. Desmosomic, pentalami-
nar and septated structures have been
described.

2. Junctions among dendrites exist, which are
called dendrodendritic synapses.

3. Axo-axonic junctions also exist, the best-
known examples being those of the junctions occurring at the level of the cone of
origin of the neurite and the close reciprocal
contacts between synaptic buttons of different fibres.

4. The occurrence of junctions with a general
morphology analogous to that of synaptic

247

buttons has been demonstrated, but this kind of junction has no vesicular content and no transmitter; it is considered as an electrotonic junction.

Biochemical investigations systematically carried out on synaptic buttons and the subsynaptic membrane have been made possible by the development of fractioning techniques applied to synaptic buttons in homogenates of the central nervous system. The analyses carried out have demonstrated that transmitters are contained in the vesicles, the number present in one vesicle being considered a 'quantum'. The occurrence of acetylcholine and of catecholamines (noradrenaline) has been clearly demonstrated, whereas the synaptic vesicular localization of serotonin, dopamine, gamma-aminobutyrric acid and of some amino acids is still to be ascertained. In the isolated synaptic button preparations, the occurrence of acetylcholinesterasic activity, bound with the presence of an enzyme (perhaps in the subsynaptic structure) which breaks down acetylcholine, has been also demonstrated.

In vitro experiments have taken into account the nerve trophic function. For this purpose, organ cultures have been used in order to determine the effects of tissue explants and nerve homogenates on muscular cholinesterasic activity of the newt *Triturus*. Cultures obtained with sensory ganglia, ganglia which have been separated from the muscles via millipora filter, spinal chord, liver and nerve homogenates have increased the activity of muscle cholinesterase, in contrast with non-treated muscles under the same conditions of culture and during the same period of time. Boiled ganglia, tubae, kidney and spleen homogenates have been shown to be ineffective. The method has proved to be a perfect bioassay to study the neurotrophic effect; it shows that the trophic effect is mediated by a diffusible chemical substance produced by nerves.

The hormone system

Hormones have been shown to be generally secreted in the circulation in extremely reduced amounts and they are very likely to be recognized only by specific tissues which react in a characteristic fashion. The plasma concentration of steroid and thyroid hormones ranges between 10^{-6} and 10^{-9} M values, whereas peptide hormones range between 10^{-10} and 10^{-12} M values. Moreover, in the target organs the cellular response to the hormone stimulation is generally limited: the trophic hormones induce the production of secondary hormones such as thyroxin or cortisol, which have different effects on the metabolism of other tissues. By contrast, the gonad steroids exert their effect on several responsive tissues. The collected results have demonstrated that the term 'hormone' may not be limited to substances which are secreted by a well-defined encodrine organ. Some hormones, for instance angiotensin II, are formed from circulating precursors; others are formed in tissues by conversion of a precursor such as testosterone in women. Therefore all the effects, as well as the causes of blood-borne stimuli, should be classed as hormones. Biological materials with an endocrine function have been proved to present different steps in their organization. There are complex organs, organelles or corpuscles and in addition also single cells which are localized in organs having other functions. The pineal body, the pituitary, thyroid, parathyroid and adrenal glands are true endocrine organs. The Langerhans islets in the pancreas and the corpora lutea in the female gonad are organelles or corpuscles. The interstitial cells in the testis, the theca cells in the ovary, the endocrine cells in the gut's mucus lining are all examples of endocrine tissues or cell groupings. In the endocrine organs, the glandular cells are arranged into nests or strings among which a rich vasal network occurs. Endocrine tissues occur as clusters or nests of cells in the connective tissue of other organs. All the embryonic layers have been demonstrated to be involved in the development of endocrine materials: thyroid and parathyroids and the gastroduodenal secreting cells are derived from the endoderm; the pituitary, the pineal body, neurosecreting elements, and aminogenic tissues are derived from the ectoderm: finally, some elements which produce lipid hormones are derived from the mesoderm.

Various endocrine glands have been shown to be able to adjust their activity to bal-

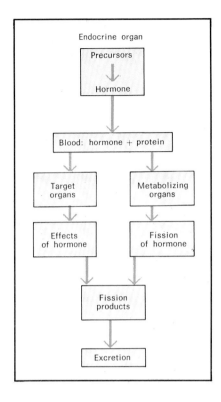

Endocrine organ

Precursors

↓

Hormone

Blood: hormone + protein

Target organs

Metabolizing organs

Effects of hormone

Fission of hormone

Fission products

Excretion

Fig. 63. Diagram of the path followed by the hormones from the endocrine glands, where they are secreted, through the target organs, or the metabolizing ones, where they are broken down until they are secreted.

ance between secretion and functional requirements. The most recent investigations have attempted to elucidate the relationships between the nervous system and hormones. The regulation of these interrelations would occur at the level of the hypothalamus in which an accord and harmonization occur between the pituitary neurochormones and the hypothalamus tracts descending to the brainstem and spinal chord. Along these tracts, and through the sympathetic nervous system, nervous impulses attain different organs.

The hypothalamus is in turn under the action of integrating nervous mechanisms which are localized in the telencephalic cortex, particularly in those areas in which occur the phenomena of emotional-instinctive life. Coenestesic modifications and continuous variations of the emotional states may thus induce,

through the hypothalamus, general modifications of endocrine activity and body homeostasis (see below). Hormones influencing sexual life act both in males and females, and are under pituitary control. In males these hormones are aimed at inducing androgen secretion in testes. They also control secondary sexual characteristics. Two pituitary hormones, FSH (follicle-stimulating hormone) and ICSH (interstitial cell-stimulating hormone) are called upon for the maturation and the production of sperm cells. In females, at the beginning of the menstrual cycle, the pituitary gland secretes FSH which acts on the follicle. Under FSH influence, the follicle begins to develop and in turn secretes other hormones: oestrogens. The main purpose of oestrogens is to prepare the uterus to receive the fertilized egg. Under oestrogen's influence, the uterine walls thicken and become well supplied with blood vessels.

Therefore an increase is observed in the production of FSH by the pituitary gland. As the oestrogen concentration increases, FSH secretion stops but, in turn, the pituitary gland begins to secrete another hormone, the luteinizing hormone (LH) corresponding to the male ICSH.

The luteinizing hormone speeds up follicle growth and prepares the egg separation. At this very moment the follicle transforms itself into the corpus luteum, which secretes progesterone, stimulating the uterus to supply foodstuffs to the free-swimming embryo, to enhance implantation and formation of the placenta.

The placenta continues to secrete progesterone, which controls and stops the further production of pituitary hormones, thus terminating the cycle. Many kinds of birth control pills are based on the presence of oestrogens and progesterone. Hormones thus taken reach the pituitary gland which is programmed in a way that, if a high concentration of these two hormones occurs, it stops the production of FSH and ICSH. These hormones are necessary to ripen the follicle and to start its transformation into the corpus luteum, which in turn is the precursor of the placenta's formation. A fertilized egg thus cannot implant itself nor initiate its own development. There is evidence

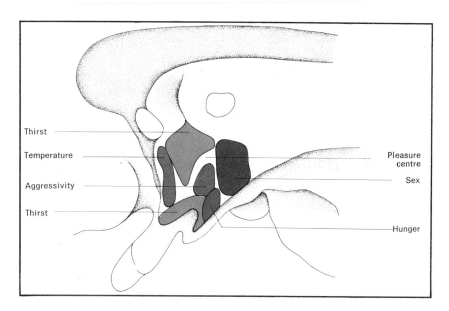

FIG. 64. Zones of localization of the main functional activities of the hypothalamus: it must be pointed out that the localization of the hypothalamic centres, and particularly their dividing lines, are at the moment extremely uncertain.

that the female's cyclical behaviour is determined very early and that a child would be able (in theory) to fulfil its sexual role if the process would not be under hypothalamic control. The existence of a biochemical sexual apparatus in the new-born, although genetically determined in its fundamental lines, seems to demonstrate that a further imprinting is requested, which is a kind of biochemical brainwashing in order to make it fully effective. It has been suggested that all people start off as female and only a successive hormonal imprinting determines the biochemical pattern of sexuality.

The primary event of hormone action is constituted of specific interaction between hormones and the responsive cells. Many cells do not recognize some hormones, proving that the receptor sites of cell membranes are highly specific. In 1964, the majority of the known hormones had been shown to regulate the protein and RNA synthesis. The collected data suggest that the steroid hormones act by (a) combining themselves with specific nuclear receptors and (b) modifying the gene transcription process, leading to RNA and protein synthesis. In the case of the peptide hormones, c-AMP seems to be an intracellular transmitter, transmitting specific messages which are received and decoded through interaction between the hormone and the cell membrane. A link between these two systems has been established through the observation that progesterone enhances the formation of c-AMP; this may act on the transcription of genes and the translation of their corresponding messages. Prostaglandins also could be involved; these would mediate the hormonal inhibition either directly or by blocking the formation of c-AMP.

Immunological systems

The immunity concept has more and more specified itself in recent years, by taking into account the phenomena which are responsible for the antoimmune manifestations, the manifestations which are related to rejection phenomena and those related to antitumoral immunity. The organism has been shown to be able to distinguish 'self' natural components, which are necessarily tolerated from an immunological point of view, from the 'nonself' components which are rejected. Immunological tolerance is considered to be a complex of homeostatic mechanisms by which the normal inviolability of the body's components is ensured by elements which are able to set up immunological normality. It has been demonstrated, however, that a lymphocyte may escape the control of the thymus and form eventually a 'forbidden' clone of cells. Another homeostatic level is likely to occur besides this first one, so that more mutations must occur in order to produce a 'forbidden' clone of cells. The development of such a clone involves creation by the organism of autoantibodies intended to destroy it. Two kinds of auto-antibodies have been shown to exist: the pathogenic autoantibodies (which cause auto-immune disease) and physiogenic autoanti-bodies, which would be correlated with the function of immunological surveillance, this in turn connected with the defence mechanisms against tumours.

Several diseases exist in which autoanti-bodies against DNA, mitochondria, micro-somes, the gamma-globulin Gm and Inv gen-etical traits, and so on, are shown to occur. Clinical and experimental data which have been collected show that, in order to be defined as autoimmune, a disease must fulfil the fol-lowing requirements: (a) the isolated antigen must be able to induce the same lesions if injected into an animal; (b) autoantibodies must always be present whenever the disease occurs; (c) in inbred animals the disease must be transferred from one to another by means of antibodies contained in serum or by means of immunocompetent cells.

The organ transplants which characterized the surgery of the 1960s put into evidence rejection phenomena; these are related to the host's reactions toward transplanted tissues or organs and have demonstrated that this be-haviour has an immunological basis and de-pends on certain tissue antigens (defined as histocompatibility antigens). Tissue typing has shown that antigens belonging to the ABO sys-tem (i.e. related to blood antigens A, B, O) and certain antigens of the HL-A system (i.e. antigens found in the leucocytes)—HL-A, HL-A2 (or Mac), HL-A3, HL-A5, HL-A7, HL-A8—are transplant antigens. As HL-A antigens may easily be individualized in the leucocytes, the leucocyte classification can be considered as an equivalent of tissue-typing. The existence of many antigens in loci differ-ent from the HL-A locus has been suggested, but the majority of the HL-A antigens is not yet known; this could involve serious clinical consequences.

An important series of investigations car-ried out in the last few years has been de-voted to the study of the action exerted by two lymphocyte families, the T-lymphocytes (thymus-dependent) and the B-lymphocytes (bone marrow-dependent) in immune phenom-ena. T-lymphocytes are mediators of cellular immune responses, whereas B-lymphocytes are involved in humoral immunity. T-lymphocytes seem to carry out another extremely impor-tant reaction. Through a mechanism not yet elucidated, they assist B-cells in initiating antibody production. When an interaction oc-curs between T- and B-cells, the T-cell reacts to an antigenic determinant, assisting the B-cell in reacting to another antigenic deter-minant situated on the same molecule. Ac-tivated T-cells seem to release a special kind of antibody against the carrier which connects itself with the carrier's fraction of the antigenic molecule, thus transferring it to a proper en-vironment favouring immunity induction (per-haps a macrophage surface). The aptene frac-tion, which is not covered by the antibody of T-cells, would be free to stimulate its own specific B-cell.

The results of investigations carried out with T- and B-cells have led to analysis of the thymus' role in the immune processes. Operatively, thymus turns out to be requested for the fulfilment of cellular immunity and many antibody humoral responses. It influ-ences these systems by stimulating inside itself the differentiation of haemocytoblasts into T-cells.

Induction of this kind is probably mediated by hormonal secretions from epithelial thymus cells. Several thymus hormones also would be likely to control a series of different steps. In fact it has been postulated that not only T-cells differentiate in the thymus but that autoreactive cells are eliminated therein.

According to investigations carried out dur-ing the 1960s in the immunological field, a belief has gradually consolidated that neoplas-tic transformation of the cell involves the ac-quisition of a new and specific antigenicity: i.e. in oncogenic cells, in addition to neoplas-tic antigens, species-, individual- and organ-specific antigens occur. On the contrary, from the antigenic point of view, two categories of tumour would exist: those caused by biological agents (the oncogenic virus) and those caused by chemical and physical agents, with peculiar antigens for each of them.

The occurrence of specific tumoral antigens has been clearly demonstrated in the Burkitt lymphoma, in the leukaemic cells, and in the colon adenocarcinoma (which is called carcino-embryonic adenocarcinoma). Specific tumoral antibodies have been observed in melanomes, by using immunofluorescence and in patients affected by sarcoma and acute leukaemia. In the last few years, the occurrence of immune reactions in cancer has led some to postulate interventions based on the immunotherapy. Active, adoptive and passive immunothera-peutical procedures have been assayed. Active immunotherapy has been applied with some success in the malignant melanomes and sar-comes of the soft tissues and in the acute lymphoblastic leukaemias. Adoptive immuno-therapy (normalization of the immune powers of an organism through the introduction in the organism of immune cellular elements) has been applied in the treatment of advanced cancers. Introduced into the organism were lymphocytes obtained from allogenic spleens which had been kept at –80° C. The passive immunotherapy (using heterologous immune sera obtained from sheep, goats, pigs and cancer-affected humans) have been applied, frequently with objectionable results.

Overall control

Investigations intended to study the homeo-static mechanisms by which the organism maintains dynamic equilibrium through a complex system of organic adaptation mech-anisms, have considered mainly: (a) the blood composition's constancy; (b) mechan-isms controlling metabolism; (c) constancy of the body temperature; (d) regulation of the thyroid function; (e) mechanisms of the muscle tonus.

It has been demonstrated that the main-tenance of pH in a range of almost constant values occurs with such precision that it re-mains practically unvaried too in cases of serious cell or tissue alteration, with marked variations of the acid-base equilibrium and pathological accumulation of basic substances.

Mechanisms which maintain the constancy of the blood's composition are very complex; they are related above all to its buffeting properties, to its peculiar protein composition, to the haemoglobin properties, to the in-terrelations between ions, and so on. The respiratory processes (intra- and extracellu-lar processes), those related to excretion, to metabolism, to the circulation rate, and so on, are also fundamentally important. The main-tenance of normal osmotic pressure occurs through a co-ordinated system of regulating mechanisms of water metabolism and ionic composition. The raising of osmotic pressure puts in action mechanisms which enhance the retention of water by tissues and salt dis-charge by the organism. On the contrary, the lowering of osmotic pressure enhances water removal (by an increase of diuresis and pulmonar hyperventilation) as well as salt retention.

In the maintenance of a normal hydrosaline balance, the blood protein's composition, cap-illary and cell membrane permeability, the connective tissue's condition, the endocrine apparatus' activity (secretion of antidiuretic hormone, the role of parathyroids in the regu-lation of calcium metabolism, and so on) are most important.

Observable modifications in metabolite con-centration have been demonstrated to contrast with metabolic alterations through neurohor-monal reactions. In the achievement of the physiological process, factors influencing its subsequent autoregulation appear. The kind of response of a certain physiological system to the action of different stimuli depends on these trends in the physiological process.

Nervous component impulses coming from enteroceptors, metabolic regulation by en-cephalic centres, reflex cortical activity, and so on—as well as endocrine organs and inter-mediate metabolism products (metabolites), formed during organ and tissue activity—these all participate in the achievement of the adapt-ive reactions. Some nitrogen metabolites have been shown to represent possibly adequate regulatory agents of proteolytic processes, oc-curring in the liver and kidneys, with a subsequent increase of azotemia. Lipid metab-olites, in turn, may regulate the lipolytic and chetogenetic processes, by controlling the action of regulatory mechanisms in lipid metabolism, particularly hormonal incretion.

In the hyper- and hypoglycemic states, homeostatic compensatory processes develop, contrasting with a further increase or decrease in the level of hematic glucose and tending to reintegrate normal glycemia. In hypoglycemic states, the compensatory reaction is achieved in three steps: (a) through the glucogenolytic action of adrenaline; (b) through the action of adrenal glycocorticoids, which activate gluconeogenesis from proteins and normalize the glycogen level in the liver; (c) through the somatotrope hormone, which inhibits gluconeogenesis and glucose oxidation whereas it enhances the protein synthesis.

Thermic balance occurs with the participation of cerebral nervous centres, situated in the hypothalamic region and are strictly connected with cortical activity. These centres regulate several chemical and physical processes, the aim of which is the maintenance of the constancy of body temperature. Some cephalic injuries have been shown to induce lasting damage in thermoregulation.

Concerning thyroid regulation, synthesis and secretion of the thyroid hormone have been shown to be stimulated by the hormone TSH (for thyroid stimulating hormone) produced by adenohypophysis. In turn, TSH production is stimulated by another factor, TRF (thyroid release factor) produced by diencephalic nervous nuclei. This complex system is regulated by at least three feedback mechanisms with negative retroaction. The first mechanism is produced by the intensity of the increase in oxidation which retroacts on the diencephalum and subsequently on TRF production. The second and the third feedback systems are produced by the level of thyroid hormone circulating in the blood,

FIG. 65. Sequence of control circuits showing the integration of the endocrine system and the central nervous system. The latter acts as integrator, in order to stabilize the system in stationary position.

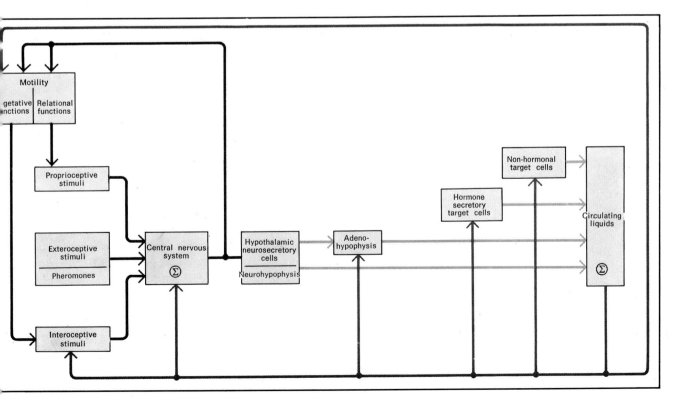

which retroacts respectively on the diencephalum (and therefore on TRF production) and on hypophysis (and consequently, on the production).

The muscle tonus has been analysed. During rest, muscles are in a state of balanced contraction as a consequence of homeostatic action which ensures equilibrium between the excitatory action of the fibres coming from the muscular spindles and the inhibitory action which is due partly to the postsynaptic inhibition (carried out by the Ranshow cells) and partly to the presynaptic inhibition carried out by internucial neurones.

The stretching of the muscular and spindle fibres sends a series of impulses to the alpha-motor neurons, which, in turn, discharge a series of impulses on the motor plate, thus making the muscle contract (the myotactic reflex). Muscle contraction, by shortening the fibres, contrasts with the elongation determined by the stretching of the muscle spindle; consequently, the reduction or abolition of the impulses coming from the spindle occurs. At the same time, muscular contraction causes the stretching of the tendineous organ, which will send inhibitory impulses to the alpha-motor neurons of the antagonist muscle, which will relax, thus permitting the contraction of the agonistic muscle.

24

Brain
and behaviour

Interpretation and processing
of sense data

The interpretation and processing of sense data which make up the coding process are concerned with the transformation of physical stimuli into nervous impulses. Such a transformation (which is called transduction) occurs at the level of several sense receptors localized in the eye retina (more exactly, in the rods and cones of the retina), in the ciliated cells of the ear cochlea, and in the nervous endings of the somatic sensitivity system.

The study of the mechanisms of coding the visual and acoustic stimuli, as well as the stimuli of the somatic sensitivity, allows us to trace the major lines of the integration of sense data. D. Hubel, T. Wiesel and H. H. Balow (1963, 1970) were the first to carry out and to interpret experiments by which it has been possible to differentiate the behaviour of the different receptors of the retina, demonstrating that some of them react selectively to lines and edges, i.e. to the contrasting of light and shadow, whereas others react to light moving through a visual field. These elements have been shown to react to both left-and-right and up-and-down motions; they can determine the direction of a moving line.

Further investigations have considered the second integration centre, which is on the path of the optical tract and consists of the lateral geniculated body, sited on the white matter which connects diencephale to telencephale. The most important result of these investigations has been that of demonstrating that neurons inside the retina reproduce a 'map' of the retina. Although the neuron arrangement of the lateral geniculate body is different from that of the retina, spatial correspondence is maintained which is rigidly exact between these elements. The role of lateral geniculate body in the process of visual information is not yet understood: in some experimental animals, it seems to play a precise role in the process of colour perception.

The modes of the cortical coding of information coming from the retina have been explained mainly by Hubel and Wiesel. In the cortex, as well as in the lateral geniculate body, there is a map of the retina which is made up (in this case) by a much larger number of cells. Some investigations would have demonstrated that such a map permits the three-dimensional, binocular vision which is typical of man.

It has been ascertained that the neurons of the cortical map are arranged in columns containing cellular elements which, according to the most recent investigations, are of three types: the simple ones, in which excitation may be induced on the retina by a ray of light; the complex ones, which are activated by a ray of light in a dark field or by a dark bar on a bright background or, finally, by a line of transition between light and darkness. For both these kinds of cells, it is necessary that the luminous lines move and be oriented at a particular angle in relation to the cell. A third kind of neuron is defined as upper-complex. It reacts selectively only to a line which shows a clear-cut beginning and end, in a particular way and at a well-defined angle.

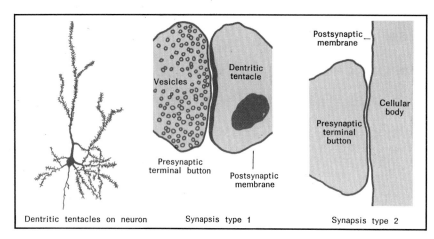

Dentritic tentacles on neuron   Synapsis type 1   Synapsis type 2

FIG. 66. The synapses, which represent the sole channel (as far as we know today) of communication between the neurons, are located mainly on the dendrites and cellular bodies. The first ones represent the cellular region, where information is summed up and sifted. The dendritic synapses with button, type 1 (shown in photomicrographic adaptations), are of the exciter type. The vesicles of the presynaptic button presumably contain the synaptic chemical mediator; the receptors of this mediator are supposedly located in the thickened fascia of the postsynaptic membrane. The type 2 synapsis, which is inhibitory and located on the cellular body, presents neither tentacles nor thickening of the membrane.

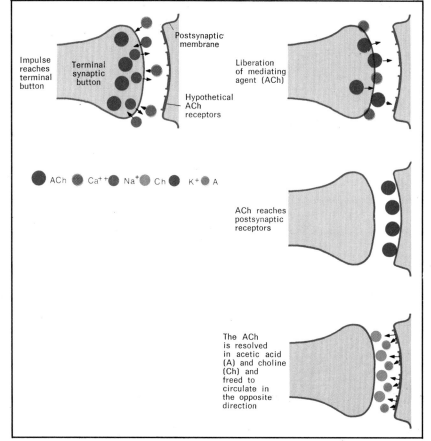

FIG. 67. Chemical phenomena occurring during synaptic transmission. The action potential reaches the presynaptic terminal, causing the entry of the sodium ions (Na^+) and the exit of the potassium ions (K^+), which in its turn causes the calcium ions (Ca^{++}) to move towards the membrane. This leads to the liberation of the mediator, in this case acetylcholine (ACh). When this reaches the receptor of the postsynaptic membrane, the postsynaptic electric response is generated. The mediator, acetylcholine, is split at the receptor into acetic acid (A) and choline (Ch), which can be reabsorbed by the presynaptic terminal and retransformed into acetylcholine.

The coding of acoustical stimuli has been studied especially by G. Békésy (1956, 1962) who demonstrated that a note of a certain frequency provokes the formation of a wave in the lymph of the cochlea, thus inducing a variation which happens to be maximal in a certain part of the basilar membrane. As a consequence of the differential rigidity of such a membrane, some mechanical effects of a rather complex kind would be produced: according to this, the high-pitched sounds produce deformations with a maximum near the base of the cochlea, the intermediate-pitch sounds deform the intermediate part of the cochlea and the low-pitched sounds would tend to deform all the membrane. The data which have been collected till now would show that in the coding process of the frequencies of a sound both the localization of the excitement and the frequency of the nervous response are most important.

It has been proved that a general map of sounds distributed according to their frequency is present on the auditory cortex. Furthermore, the cells of the auditory cortex code each for a different frequency and each neuron has a optimal frequency, which induces a maximum of reaction. The perception of complex sounds and the understanding of their meaning would be involved in other higher functions and in associative cortical areas.

The coding of the stimuli of the sematic sensitivity occurs through two different pathways. It has been demonstrated that the existence of an epicritical, recent and well-defined system, allows us to analyse the stimuli as touch stimuli, pressorial stimuli or stimuli determined by the position of the limbs, and of a more ancient system, which is specific for pain and thermic sensitivity. At the level of the somatic sensitivity cortex a columnar arrangement of the different cellular elements has been proved: every different category of the epicritical system is coded by separate columns of cells, making possible the cortical representation of different kinds of somatic sensitivity. In the last few years, it has been demonstrated that the brain controls, by means of systems of descending fibre, the sensorial peripheral, or afferent, information. Investigation now in progress would prove that the stimulation of several areas of the brain (especially that of the reticular formation of the brain stem) would alter the features of the afferent sensorial information. In this way, the amount and kind of information can be selected or even blocked at the sensorial level. These descending systems would seem to be involved in the processes of selective attention which depend on the power of concentration.

Theories of memory

Memory represents the long-lasting effect of a stimulation involving the activation of correlated electrophysiological, chemical, and neurophysiological processes.

It has been demonstrated that learning induces modifications of the electroencephalographic pattern. In fact, gross electrical alterations are noted at first at the level of the cerebral cortex. They then tend to normalize themselves as the learning process consolidates.

It has been noted that, in some experimental conditions, learning eventually induces constant alterations of the brain electrical activity. This would represent the electrophysiological equivalent of memory.

Some extensive and highly scrupulous investigations have demonstrated that the brain's gross electrical responses, which are evoked by a sensorial stimulation, are modified by experience. In man, the evoked responses would prove that the features of the potential wave are related to the meaning of the stimulus. Morell (1967) has shown that experience (or memory) can vary also the electrical reaction of single neurons. He has been able to prove, through electrophysiological investigation, that some cells may react simultaneously also to two or more kinds of stimulation and the electroencephalographic patterns derived therefrom seem to be related to learned responses.

The observation that some drugs (such as acetossicycloesymide and puromycin) block any protein synthesis for a few hours, if they are injected intrabrain has allowed improvement of investigations on the chemistry of memory. In fact, the protein structures which are synthesized in the cell are under the control of nuclear DNA, which stores the code for all the proteins that may be synthesized by this cell. The synthesis of particular proteins is organized by the molecules of RNA, which are replicas of certain parts of the DNA molecules; they move from the nucleus into the cytoplasm. It is indeed the property of the DNA molecules to carry information which supply the best means to form particular templates, the conformation of which agrees with the past experience. Through messenger RNA, this kind of coded information·could produce the cell components which would be renewed every time it is necessary to recall a certain item of memory.

The long-lasting excitation of neurons is likely to induce the formation of particular proteins, the number of which would be strictly determined by DNA. Some investigations seem to demonstrate that proteins of this type, formed during the learning process, move along the axon as far as to the synapses, modifying them in order to vary the efficiency and the strength of the junction with the post-synaptic neuron. According to other authors, proteins may reach the dendrites so as to alter the receptor sites and change the reactivity of cells to the stimuli arriving as far as that point. In both cases, however, permanent modification of the synaptic activity would result: a formerly active synapse may become inactive, and vice versa.

Investigations carried out through the electron microscope have shown that the variations induced by experience occur both in the size and in the number of the synapses, whereas the data of Krech, Bennett and Rozensweig have suggested a different crowding of the synapses in animals reared in stimuli-rich environments (having therefore a marked tendency to memorization) and in animals reared in stimuli-poor environments. The molecular synaptic mechanism involved in the maintenance of learned data has not yet been clearly explained: environments with a high level of sources of stimulation and facilities for moving would correspond to a higher cerebral acetylcholine (synaptic transmitter) content. Other chemical transmitters, equally involved in impulse transmission, seem to be noradrenaline, serotonin and the gamma-aminobutyric acid (GABA).

The occurrence of a chemical basis of memorization has also suggested a possible transfer of memory on a molecular basis. Investigations by Connel and Golub (1967, 1970) on experimental animals (planarias, rats, mice) have led to contrasting results, with which many authors do not agree. It would seem that a certain kind of transfer occurs, anyway. According to G. Ungar (1970), the transfer factor would not be a protein but a shorter chain of amino acids (about eight to fifteen amino acids), similar to the peptides (scotophobin). Such a factor would surpass the haematoencephalic barrier, where it would be absorbed selectively on particular synapses which would then code the learning only along certain pathways.

Psychological and psychophysiological investigations have demonstrated different kinds of memory: iconic memory, eidetic memory, short-term and long-term memory. It was demonstrated that the short-term memory may be lost in cases of concussion of the brain or as a consequence of electroshock. The results of these observations would put into evidence that the period in which short-term memory is active is less than thirty seconds and its localization would be in the hypocampus.

According to E. N. Sokolov and O. Vinogradova, functionally well-differentiated classes of cells would exist in such an area; they would seem to be involved in the processing of information and its short-term storage.

Research now in progress could demonstrate the possibility of improving long-term memory, by means of pharmacological treatment and electrical stimulation of certain areas of the brain.

Pathological approaches to brain study

The study of brain anomaly (inborn or acquired) is one of the major ways to localize well-defined cerebral functions. Experimental modifications are equally important.

The occurrence of a speech area has been demonstrated in which double polarity may

FIG. 68. R. W. Sperry's model of the split-brain. The two cerebral hemispheres, separated surgically, are capable of learning to perform two rival movements, and the animal in question behaves as though it had two brains. The diagram shows: 1. right-hand visual field; 2. left-hand visual field; 3. left eyeball; 4. right eyeball; 5. optic chiasma; 6. left hemisphere; 7. right hemisphere; 8. lateral geniculate body; 9. optic tract; 10. right-half visual field; 11. left-half visual field.
(From R. W. Sperry, *Scientific American*, January 1964.)

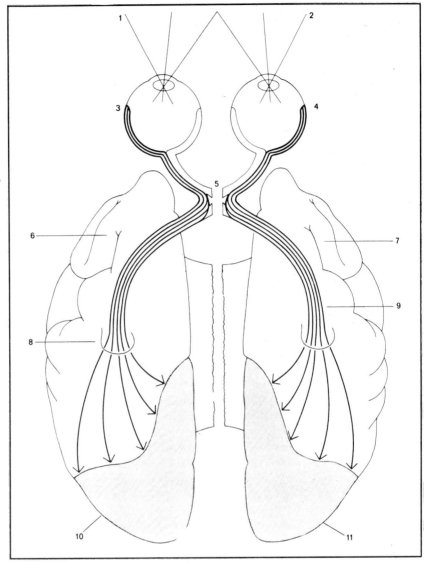

exist: frontal polarity (the damages in this part cause a loss in the expressive capacity of speech) and posterior temporal polarity (damages in this part cause deficiencies in the understanding of speech). When damage is localized toward the occipital region, trouble occurs in one's reading capacity. The observations which have been made have put into evidence, furthermore, the role of the amount of damage to brain tissue in the case of speech trouble.

Observations by neurosurgeons have permitted localization of brain functional areas. During the surgical treatment of epilepsy, it has been proved that the phenomena of vocalization, speech-blocking and aphasic aberration may be induced by stimulating some cortical areas. These aphasic reactions have been determined by stimulating several zones: the inferior frontal area (Broca's area), the parietal and temporal areas (which may be joined or separated), and the superior and internal frontal area (which is situated immediately before the supplementary motor area).

Analysis of these results and observation of the effects produced by cortical ablation carried out in order to eliminate epileptogenic scars have permitted us to establish a hierarchy in the brain areas related to the speech function. Thus it would seem that the most important area is the posterior temporo-parietal area, followed by that of Broca, whereas the least important is the additional or supplementary motor area.

According to investigations by H. Hecaen and R. Angelergues, the study of aphasia enables us to consider anatomoclinical aspects which may be outlined thus: expression aphasia would correspond to anterior damages, concerning not only Broca's area, but also the inferior part of the rolandic circumvolution; conduction aphasia corresponds to posterior parieto-temporal damage; sensory aphasia corresponds to damage to Wernicke's area; the intensity of reading trouble is dependent on the extension of damage in the direction of the occipital lobe; amnestic aphasia corresponds to different areas but, when it is isolated, it depends above all on posterior temporal damage.

In patients with a split brain, it has been demonstrated that each hemisphere has its own 'mind' perception, ideas, memories and experiences—each one independent of the corresponding experience in the other hemisphere. Although the patient in such cases may learn via both hemispheres, the hemispheres learn completely different data. The dominant hemisphere, which is usually the left one, is the only one appointed to the functions of speech and writing. Patients are thus able to relate in detail and to discuss only the information reaching the left hemisphere. Contrariwise, the right hemisphere is likely to function better in relation to the memory of tactile experience.

The best results in ablation have been achieved in the study of the cerebellum, where ablation of the flocculonodular node has been shown to induce serious trouble in the maintenance of static equilibria and of certain body postures, when the whole body or only parts of it are displaced. Ablation of the anterior lobe induces a tonus disorder which clearly modified the myotactical reflexes. (They appear abnormally emphasized.) Ablation of the posterior lobe causes hypotony with a general decrease of muscle tonus, and spontaneous quivering, nystagmus and ataxy.

Sleep and dreams

Sleep is a reversible, periodically recurrent state in which the motor and sensory activity of an organism with its environment diminish. The dream, on the contrary, is a biologically stabilizing function.

The complete-arousal-and-sleep cycle seems to be, at least partially, under the control of the pineal body, an endocrine gland situated behind the brain stem. It receives visual sensory stimuli and reacts to darkness by producing the hormone melatonin which influences the brain cells, using serotonin as a transmitter and representing the bulk of the active neurons during sleep.

It has been demonstrated that passing from arousal to sleep is not a continuous process, occurring in stages which are characterized by differently configured electroencephalographic patterns. In the first stage, the disappearance of the alpha rhythm is registered, electrical activity of high frequency and low voltage is prevailing. In the second, fast spindles and, in the third, long waves of small amplitude appear. The electroencephalogram of the fourth stage shows even longer and slower waves, corresponding to conditions of deep sleep.

The association of electroencephalographic and electroretinographic procedures has brought into evidence the existence of a particular kind of sleep, the REM (rapid eye movement) sleep or paradoxical sleep. This is associated with a decrease in the heart frequency of heartbeat and of arterial pressure, as well as with complete loss of the slight muscular tonus which is generally recognizable in normal sleep.

The main feature of REM sleep is a marked electroencephalographic desynchronization in an individual who, otherwise, seems to be in deep sleep. According to some investigators, this would prove a marked alertness towards the oneiric world. The collected data would also demonstrate that, in most cases, REM sleep is associated with dreams. It has been proved that the most serious physiological damage is induced by the deprivation of paradoxical sleep: it has been possible to show that this deprivation causes depressive symptoms, anxiety, psychical troubles, hypersexuality and sometimes death. It seems also that the deprivation of paradoxical sleep leads to a compensatory increase on subsequent nights; this sleep would attain, in some cases, as much as 60 per cent of the sleeping period. Some investigations are being carried out now in order to establish the existence of interrelations between some kinds of psychical pathology and disturbance during paradoxical sleep.

By means of fluorescence microscopy, it has been demonstrated that a different biochemical behaviour is manifested by the brain during slow-wave sleep and paradoxical sleep. An interaction would occur between two neurotransmitters, the serotonin which accumulates during slow-wave sleep, and the noradrenalin which is recognizable during paradoxical sleep. Each transmitter would activate a different series of neuronic tracts. It has been recognized that drugs which increase the amount of the existing noradrenalin also prolong the length of paradoxical sleep, whereas those which increase the amount of available serotonin prolong slow-wave sleep, thus reducing paradoxical sleep. On the other hand, a series of investigations would have demonstrated that reserpin, which decreases the amount of noradrenalin and serotonin in the brain, also decreases both the paradoxical and slow-wave sleep.

Investigations carried out in order to establish the localization of centres responsible for the appearance and disappearance of the sleeping condition have demonstrated that the more caudal regions of the reticular formation are mainly responsible for the deactivation or sleeping phenomena. The electrical stimulation of these regions may transform the electroencephalographic pattern from that typical of the arousal condition to that typical of the sleeping condition. The same effects would have been attained by stimulating a region out of the reticular formation and localized in the front of the hypothalamus; damage to both these regions induces insomnia and even death. According to some authors, two different areas of the reticular formation would control slow-wave and paradoxical sleep, under the control of a centre in the inferior part of the pons. The dissection of tracts which originate in this region and are connected to the cortex abolishes the paradoxical sleep. Collected data are still incomplete, however, and it cannot be excluded that an identical neuron reacts in a different way to different stimuli.

Parallel to these investigations on sleep, direct investigation has been carried out in order to ascertain the biological role of the dream which would accomplish a protective function with regard to the brain tissues.

The possibility to determine the exact period in which a dream occurs by registering REM sleep has allowed us to determine the length, the frequency and the physiological processes associated with the oniric period. The systematic interruption of dreams has induced anxiety, irritability and diurnal allucination in subjects thus treated. These data have to be confirmed.

One researcher recognized that, different from in normal subjects, the dream break does not cause trouble in scyzophrenic subjects, as if these subjects were compensated by diuranal allucinations. This would be the demonstration that the origin of mental diseases is associated with an alteration in the regulation mechanisms situated at the level of the brainstem.

The irregular cardiac and respiratory frequencies associated with sleep seem to prove the influence exerted by the emotional state. According to some investigators the consciousness-sleep-dream cycle is prolonged all day in recurrent periods, thus modifying (at the unconsciousness level) the subject's properties of alertness.

Biochemistry of behaviour

The study of affective behaviour from the normal point of view enables us to utilize data which have been already acquired in other fields (endocrinology, neurophysiology), to take into account the long-term function of the evolutionary processes in the interrelation between hormones and behaviour, and to utilize biochemical procedures to acquire

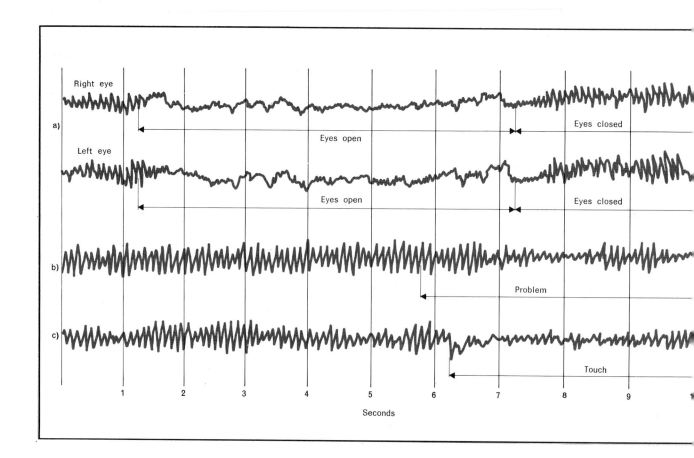

Fig. 69. The small diagram shows a typical night's sleep (electroencephalographic recordings made during the night) of a young adult. The main horizontal lines show periods of dreaming (s) characterized by periods of stage I and series of rapid eye movements. Above, changes in cerebral wave patterns according to a person's activities. When he is relaxed and lets his attention wander, there will be, as shown in the diagram at (a), an alpha rhythm: electric oscillations of approximately 0.00005 V and frequency of approximately 10 Hz. If he then sets out to solve a problem, or responds to a stimulus, the alpha waves are replaced by beta waves, with oscillations of 20–25 Hz. The recordings in the third diagram show the brain wave patterns in various states of consciousness, from active wakefulness to deep sleep. (From Hartmann, 1967; J. C. Eccels, 'The Physiology of Imagination', *Scientific American*, September 1958; H. H. Jasper, 'Electro Encephalography', in W. Penfield and T. C. Erikson (eds.), *Epilepsy and Cerebral Localization*, Springfield, 1941.)

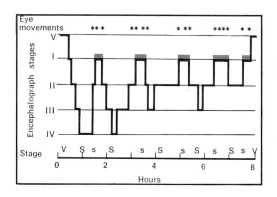

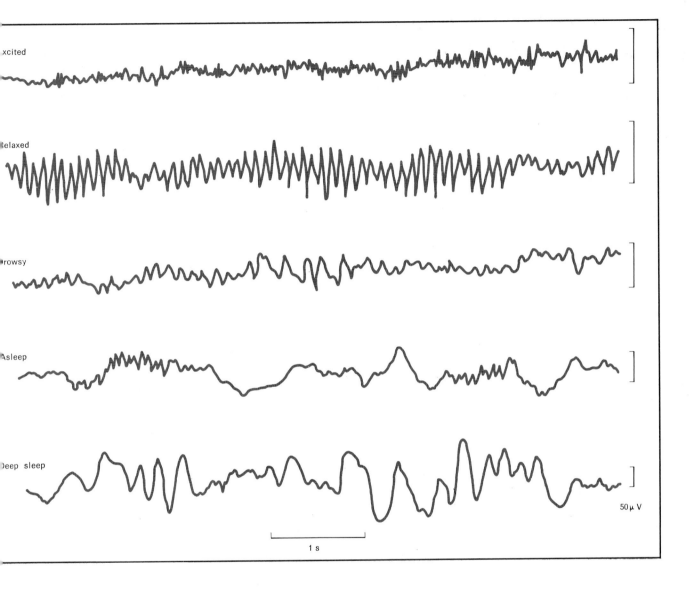

Excited

Relaxed

Drowsy

Asleep

Deep sleep

50 μ V

1 s

an objectively strict evaluation of behavioural response.

It has been demonstrated that, in the early stages of life, hormonal influences on the brain's organization will condition successive behaviour, particularly sexual and aggressive behaviour. Anatomical localization of the sites in which the brain's differentiation (according to sex) occurs, as well as the responsible mechanisms, has not yet been ascertained.

According to some investigators, testosterone provokes an increase of RNA-synthesis in some well-defined differentiating areas of the brain, which seem to be located in the fore-

hypotalamus and in the amygdaloid body. In these two areas, the hormonal effect would be limited to certain critical periods.

In the relationships between the foetus and its mother it has been observed that, if a correlation exists between androgen concentration and behaviour, it should not be interpreted as contemporary but evolutionary. A high level of androgens during early infancy appear to influence the brain's activity in order to predispose the subject to a certain type of behaviour.

The need to check the hormonal state during pregnancy and investigate the conditions which expose the brain to the action of testosterone during foetal life has been demonstrated. The evaluation of data collected after therapeutical miscarriage has shown that in pregnancy a fast transfer of steroids from foetus to mother occurs through the placenta. The foetus makes use of the urinary elimination path of its mother, therefore the occurrence of a certain hormonal condition of the foetus may be evaluated by determining the so-called urinary metabolites. Investigations are in progress, and the data already collected must be elaborated. The correlation between the sex hormonal function and adolescent and adult behaviour has been analysed in a series of biochemical investigations, which have considered dihydrotestosterone as the biologically active form of the testosterone.

The need has been demonstrated to obtain more exact data on androgens which are present in the adolescent plasma: their concentration would have to be studied in connexion with the cyclicity of the elimination of testosterone. Differences between the testosterone and the androstenodione levels have been demonstrated before puberty, but it seems that no increase in the values of these androgens occurs during the period between 3 and 9 years of age. From 9 on, males show a certain gradual increase in the levels of testosterone. The more interesting data concern boys from 10 to 15, in which a marked increase of testosterone is associated with reactivity variations towards a provoking stimulus, with reduction in the frustration threshold, high motor drive, a decrease in alertness, and so on.

Hormonal-behavioural interrelations have been studied in women during the menstrual period; the collected data are still fragmentary, but it seems that no strictly determined relationship can be established between serious behavioural troubles and the menstrual cycle.

According to some authors, the degree of anger in subjects participating in stressful conversation seems to be correlated in a significant and linear way with the hydrocortisonal concentration in their plasma. A relationship between genetics and aggressive behaviour has been observed in the Lesch-Nyhan syndrome. Trouble occurs in the purine metabolism, associated with serious behavioural problems

and unusual aggressive display. The purines involved are related, at a certain level, with caffeine and the teobromin-like purines, which, when administered in large doses to experimental animals, induce aggressivity and automutilation. The data are not yet complete, but they suggest the existence of an interrelation between the storage of excitants (like the purines mentioned above) and divergencies in aggressive human behaviour.

Imprinting and learning

It has been demonstrated that the nervous system is endowed with plasticity of response, which allows the organism to acquire an adaptive behaviour. The behaviour may be modified by a new kind of stimuli, by the effects of environmental events, by particular responses or by the association of exteroceptive and enteroceptive stimuli, based on their equality, in equality or contiguity. These behavioural modifications derive from gained experiences and are defined as learning. As learning needs the modification of future behaviour as a consequence of past experience, some neuronal mechanisms must be considered as involved in this process: (a) the neuronal mechanisms involved in the storage of a representation of an experience and (b) the neuronal mechanisms that are intermediaries in the retrieval of such stored information and its interaction with present experience.

The attempt made to explain the ways by which the learning mechanism functions has led to the formulation of some theories among which the following must be mentioned. First, the classical theory of the conditioned reflex according to which learning would be connected with the fact that an animal (including man) learns to react in a certain way to a stimulus to which it did not react before—or reacted in a different way. In such a case, a conditioned stimulus exciting the sensory receptors is systematically coupled with an unconditioned stimulus, which activates the muscular or glandular effectors. After a series of coupled presentations, conditioned stimulus alone may induce a conditioned response closely recalling the unconditioned stimulus. It would seem that the conditioned stimulus activates the neurons in the brain regions of the corresponding system of receptors, whereas the unconditioned stimulus would activate neurons situated in regions which mediate the specific functional reactions. The presentation of the coupled stimuli would seem to provoke the formation of connexions between the two neuronal regions.

Secondly, there is the instrumental conditioning theory. In this case too, the reflex responses are determined by the association of a certain formerly neutral stimulus with a given action. But it is the animal itself which will attempt to repeat, among all the possible actions, only that which is followed by reward or reinforcement. By positive reinforcement, one means the effect of an event which increases the probability that a certain response is given again. By negative reinforcement, one means an event which may decrease this probability. The frequency by which a certain response is given becomes a means to measure a certain type of learned behaviour.

Thirdly, there is Mowrer's two-factor learning theory. According to this, learning is partly associated with an acquired anxiety state; the responses by which this kind of anxiety state; the responses by which this kind of anxiety may be reduced are learned and stored responses.

Then there is the theory of learning to learn. According to this, animals subjected to prolonged training, in order to learn to discriminate different objects or stimuli, develop a learning disposition or tendency. That is, they become more and more able to solve discriminative learning problems.

Fifthly, comes the theory of learning by insight, according to which the solution of the problem to learn is not gradual, but occurs suddenly. The study of the learning mechanism has led Lorentz to formulate the imprinting theory, according to which the mode, time and object of learning are already programmed genetically in every animal. This learning process (which allows animals to recognize other animals of the same species) is extremely limited to time.

For instance, in the chick it may occur only in the period which lasts from five to twenty-four hours after egg hatching; it has an optimum between thirteen and sixteen hours. If the chick is kept in the darkness or it is not allowed to see anything moving during the first twenty-four hours after hatching, it will follow no moving object; it will not be able to recognize animals of its same species, avoiding even its own mother. Another kind of imprinting, connected with the former, is so-called sexual imprinting, by which (since birth) an animal accustoms itself to recognize its sexual partner. As in birds, this has been observed in fishes and animal. It is a very fast learning process which can occur only in a well-defined period, neither before nor afterward. This is called the sensitivity or critical period. The main features of the imprinting are its persistence during long periods of time and its irreversibility. The lack of a certain kind of information at the right moment can have very serious repercussions on the subsequent behaviour of the adult.

Further reading

Agranoff, B. W. Biological effects of antimetabolites used in behavioral studies. In: D. H. Efron et al. (eds.), *Psychopharmacology: a review of progress*. Washington, United States Government Printing Office, 1968.

Anohin, P. K. Novoe o robote mozga. *Nauka i Chelovechestvo*. Moskva, Znanie, 1967.

Best, C. H.; Taylor, N. B. *The physiological basis of medical practice*. St. Louis, C. V. Mosby Co., 1961.

Blesser, W. B. *A systems approach to biomedicine*, New York, McGraw-Hill, 1969.

Brown, J. H.; Jacobs, J. E.; Stark, L. *Biomedical engineering*. Philadelphia, F. A. Davis Company, 1971.

Carlson, D. F.; Wilkie, D. R. *Muscle physiology*. New York, N.Y., Prentice-Hall, 1974.

Catt, K. J. *An ABC of endocrinology*. Boston, Little, Brown & Company, 1971.

Chusid, J. G.; McDonald, J. J. *Correlative neuroanatomy and functional neurology*. Oxford, Blackwell Scientific Publications, 1968.

Dubois, J.; Hecaen, H. *La naissance de la neuropsychologie du langage*. Paris, Flammarion, 1971.

Evans, J. R. Structure and function of heart muscle. *Circulat. research*, vol. XV, suppl. II, 1964. (American Heart Association Monograph No. 9.)

GREENWALD, G. S. In: S. J. Behrman and R. W. Kistner (eds.). *Progress of infertility*, Boston, 1968.

HEIM, A. *Intelligence and personality*. London, Penguin Books, 1970.

HOCKMAN, C. H. *Limbic system mechanisms and autonomic function*. Springfield, Thomas, 1972.

HUNTER, I. M. L. *Memory*. London, Penguin Books, 1964.

LI, C. H. In: A. Pecile and E. Muller (eds.), *Growth hormone*. Amsterdam, 1968.

LUNDE, D. T.; HAMBURG, Da. Androgens and aggressive behavior. In: S. Levine (ed.), *Hormones and human behavior*. New York, 1971.

MARTINI, L.; GANONG, W. P. *Neuroendocrinology*. New York, 1966.

MILLER, G. *Psychology—the study of mental life*. Harmondsworth, Pelican, 1966.

MILSUM, J. H. *Biological control systems analysis*. New York, McGraw-Hill, 1966.

NOSSAL, G. J. V. *Antibodies and immunity*. New York, Basic Books, 1969.

NOSSAL, G. J. V.; ADA, G. L. *Antigens, lymphoid cells and the immune response*. New York, Academic Press, 1971.

RIETHMÜLLER, G. *Progress in immunology*. New York, Academic Press. (In press.)

SEGAL, B.; KILPATRICK, D. (eds.). *Engineering in the practice of medicine*. Maryland, Williams & Wilkins, 1967.

SOLLBERGER, A. *Biological rhythm research*. New York, Elsevier, 1965.

STARK, L. *Neurological control systems: studies in bioengineering*. New York, Plenum Press, 1968.

TALLAND, G. A. *Disorders of memory and learning*. Penguin Books, 1968.

TAYLOR, G. (ed.). *Immunology in medical practice*. London, W. B. Saunders, 1975.

TUTTLE, W. W.; SCHOTTELIUS, W.; BYRON, A. *Textbook of physiology*. St. Louis, C. V. Mosby Co., 1961.

WALSH, E. G. *Physiology of the nervous system*. London, Longmans, 1964.

YOUNG, J. Z. *A model of the brain*. London, Oxford University Press, 1964.

ZANGWILL, O. L.; WITTY, C. W. M. (eds.). *Amnesia*. Butterworth, 1966.

25 Evolutionary biology

The origin of life

The sequence of chemical events which led to the information of the first living things on the Earth has been the subject of speculation for many decades. Only in the past ten years, however, has a sustained research programme been mounted to simulate the conditions which must have existed on the Earth soon after its formation. The results have shown that a whole range of progressively complicated organic compounds can be formed, through the action of heat, electric discharge and shock waves, starting only with what are thought to have been the early constituents of the primitive atmosphere. More recently, these results have been given added significance by the discovery of many simple organic molecules in space (Chapter 10) and to a lesser extent by investigations on meteorites which have fallen on the Earth, some of which also appear to contain simple organic compounds. In this way, a picture has been built up of how many of the basic building blocks of life were formed on the Earth. The story is still far from complete, however, and in particular the evolutionary jump that must have taken place between these molecules and the formation of the first simple cells remains obscure: this is an area on which attention is likely to be focused in the near future.

One of the landmarks in the study of early chemical evolution was a series of studies carried out at the University of Chicago in 1953. There a mixture of methane, ammonia, hydrogen and water—all compounds thought to have existed in the primitive atmosphere—was exposed to the action of steam and electric discharge for a period of a week. Many new organic compounds were found to have been formed, including the amino acids glycine, alanine and glutanic acid as well as formic, acetic and propionic acid. These experiments have since been duplicated in many laboratories, particularly in the United States, the Soviet Union and the Federal Republic of Germany. Attempts have also been made to simulate specific features of the primitive Earth. For example, lightning has proved particularly effective in producing organic compounds which are the precursors of life; temperatures of 1,000° C have been used to simulate a flow of hot lava and have produced, from a mixture of simple gases, the compounds known as nitriles which are the precursors of amino acids; the electron beam from a linear accelerator and ultra-violet radiation have been used to simulate radiation effects which may have occurred early in the Earth's history; the base adenine, one of the five major constituents of DNA, has been formed simply by the action of heat on ammonium cyanide. And high velocity metal balls have been projected into mixtures of simple gases to simulate the effect of the shock wave formed, when a meteorite enters the Earth's atmosphere.

The main result of this work has been to prove that the organic building blocks of life—the chemicals which lie at the heart of all living systems—can be synthesized inorganically, simply by the action of such physical phenomena as molten lava and lightning on the mix of gases that was originally present in the Earth's atmosphere. The question of how the first living organisms were assembled from this chemical mixture—often referred to as the 'primordial soup', for the chemicals must have entered into solution in the oceans and rivers—remains obscure. Several experimenters have tried to produce primitive cells from mixtures of the gases concerned, and some success has been achieved in producing tiny spherical globules or microspheres of matter. If such a process led to the first living organisms on the Earth, then it must be supposed that these tiny cells were formed in such a way that the key ingredients of nucleic acids became trapped in them and the first rudimentary forms of the genetic code then came into operation. Early in the 1970s, however, there was very little idea of the steps that must have ensued between the stage of purely chemical evolution and the emergence of biological evolution.

Future investigations are likely to concentrate on this area and on the search for primitive forms of life on other planets. The problem of how life originated on the Earth is related to how it may have evolved in other areas of the universe and whether there is a possibility that living things elsewhere could be based not on carbon chemistry as is the case for the Earth but on alternative combinations of chemicals.

Natural selection

From the time when chemical evolution was replaced by biological evolution, the principle of natural selection—as outlined by Charles Darwin in the nineteenth century—began to operate. Recent research in this area has fo-

cused on a number of different aspects of the way in which living species are continually adapting to the environment in which they live. Such studies are concerned essentially with both the pressures leading to natural selection and the mechanisms by which it occurs. In the past decade or so, several new features have emerged. First, although Darwin's original ideas were often expressed in terms of the individual, it has been found more rewarding to investigate the concept of natural selection in terms of populations. This work has been stimulated recently by the arrival of the computer which now makes it possible to model the behaviour of large populations and to assess the effects on a population, of a given environmental pressure or mutation, over many generations in a very short space of time.

While such studies used to be largely confined to genetics, ethology and ecology now provide important information about the mechanisms of natural selection. This is particularly true in the study of animals with clearly defined social structures, including man. Computer studies are, of course, still less common than investigations in the field or laboratory where a renewed interest has been created by programmes such as the International Biological Programme and Man and the Biosphere. A key interest is the concept of genetic polymorphism, and the mechanisms which lead to changes in the differing frequencies with which various genetic forms of the same species are found to exist side to side.

From the ethological point of view, a special aspect of natural selection has assumed considerable importance: sexual selection, or competition between individuals of the same species for reproduction. Besides the traditional selective pressures, such as predation and competition for food and territory, an individual must also compete with individuals of the same species in order to be allowed to reproduce.

Experiments with animals such as doves and mice have shown that, contrary to previous ideas, sexual preferences are 'imprinted' on the individual at an early age. This is important because evolution would proceed in markedly different ways according to whether sexual preferences are learned or whether there is an innate tendency for individuals to mate with other individuals similar to themselves. The effect of these sexual preferences on the evolution of populations has been analysed by computer simulation.

Another aspect of sexual selection is competition between spermatozoa, which occurs in several animal groups. This phenomenon has been studied particularly in insects. It has been found that some phenomena may be advantageous for the species, although not for the individual; there are, for example, occurrences of morphological, physiological or ethological mechanisms which block fertilization by spermatozoa from a second mating. Equally, multiple insemination may occur although this process is advantageous only for the species and not for the individual.

Another recently studies aspect of natural selection is frequency-dependent selection, which has been observed in *Drosophila*. In this kind of selection, certain males which are rare in the population considered appear to have sexual advantages over more common forms. For instance, when male *Drosophila* which bear the 'white' mutation (males with white eyes) make up less than 40 per cent of the male population, they reproduce more often than normal males until the frequency of the latter has been surpassed; the advantage is then reversed in favour of the normal males. Mechanisms such as these help to explain the presence of commonly observed balanced polymorphisms.

Particular attention is now being paid to the genetic analysis of mimicry because some important aspects of natural selection can be studied experimentally in this way. For example, experiments carried out on different species of Lepidoptera (*Papilio dardanus* and *Papilio memnon*) have shown that some mimetic forms evolve gradually through natural selection and also that forms which are only partially mimetic are advantageous against predators. These results are in disagreement with conventional evolutionary theory which states that an evolutionary adaptation is of no use to the species if it is not a perfect one and that, consequently, adaptation cannot evolve gradually.

Finally, there has been increased interest recently in studying the system represented by a species and its environment as a closed loop First, a species adapts to the environment in which it finds itself. But that environment is never static and is liable to change over time. Hence the study of adaptation and evolution should include changes in the environment itself. Thirdly, those changes may be induced by the presence of the species in the environment, and its effects on the milieu. Thus arises the notion of cultural evolution in man in which evolutionary changes occur as a response to changes in the environment, themselves caused by the presence of man. The most remarkable example of this trend is seen in recent attempts to model the overall ecology and long-term future of the biosphere (see page 252 ff.).

Population dynamics

The way in which animal and plant populations increase, decrease and reach a stable situation is now being studied intensively for many reasons. First, the subject has its own intrinsic interest; second, it is closely related to studies on genetic improvement in plant and animal breeding programmes; and, third, it provides an important input to problems connected with the study of human populations. All species tend to modify their own environment, and the study of population dynamics is therefore closely related to the study of the dynamics of ecosystems as a whole. In the context, the way in which a new species adapts to an ecosystem, undergoes a rapid rise in numbers, and eventually reaches a situation of stability, either with or without a population 'crash', seems to have important parallels with the human population 'explosion'.

During the 1960s many censuses were taken of different animal and plant populations in different ecosystems, research which was stimulated in part by the International Biological Programme. This, in turn, enabled a number of studies to be made of how species react demographically to altering environmental conditions. The notion of a given ecosystem having a certain 'carrying capacity' was refined, and it was shown that a population which is either larger or smaller than this size is less productive than a population of inter-mediate size. Many of these studies were carried out to define the optimum size of a population in terms of their useful yields for man, and to elucidate mechanisms by which such optimal populations can be maintained. Work of this kind is proving particularly useful in conjunction with computer models of the population concerned; for instance, a model of the population of redwoods in California was used to deduce the optimal rate at which timber could be removed from redwood stands.

During the 1960s increased emphasis was placed on finding effective substitutes for pesticides and insecticides in the field of pest control. One interesting approach was the investigation of simple methods of manipulating the microenvironment of the pest concerned, by varying ploughing times, sowing seasons or the lighting régime, for instance, in such a way as to hinder the normal growth of the pest but to let development of the crop proceed as normal. The concept of biological control of pests, such as by the artificial introduction of predators of the pest concerned, was further advanced, particularly by studies of the interactions or predator populations with their prey. An alternative technique of biological control, which appeared very promising at the beginning of the 1970s, was the introduction into a pest population of a certain number of sterile or incompatible males. It has been found that each population has a certain minimal level of reproductive success, and if the level is made artificially to fall lower than this critical minimum, the population undergoes a 'crash'.

This concept of population crash was also further investigated. No population can undergo indefinite expansion without facing the risk of serious consequences, which in natural populations usually take the form of a sudden and drastic reduction in numbers; at this point the population either becomes extinct or population growth begins again, and is eventually followed by a further crash. Some populations, in fact, seem rarely to reach a stable level but to oscillate continually between the point of near-extinction and severe over-population.

For the first time, the implications of this

work were examined by computer models for the human population, although the latter differ from all other populations in that they can consciously gauge the effect of their own modifications of their environment, and react accordingly.

Population genetics

Population genetics is the study of genetic variations in a population of individuals which possesses a common genetic pool. A thorough knowledge of the type and the extent of genetic variations of polymorphism in populations is needed to understand evolutionary processes. It has been known for a long time that natural populations contain extensive genetic and chromosomal variations, in direct contradiction to the much earlier idea that genetic variations in populations were abnormal deviations. New techniques, such as electrophoresis and the molecular analysis of gene variations, together with the use of mathematical models and computers, have provided a great deal of information in this area and have made it possible to study new phenomena which were previously too difficult or too time-consuming.

The major research goals are to estimate the degree of polymorphism in different organisms; to understand the nature and extent of genetic differences between species; to find out whether some kinds of proteins are more polymorphic than others; to discover to what degree environmental differences influence population variability; and finally to elucidate the mechanisms by which genetic variability is maintained in populations.

One complex aspect of population genetics is concerned with reproduction. Organisms are called K-strategy organisms if they live in a stable environment and can therefore spend most of their energy in non-reproductive activities. R-strategy organisms, by contrast, live in very variable environments and devote much of their time to maintaining their demographic level. A given species is usually always either an R-strategy or a K-strategy group, although the choice is influenced by the environment and, probably, by the genetic structure of the population concerned.

The recent discovery of new polymorphisms,

FIG. 70. On the basis of the frequencies of five blood groups (AB, O, MN, Diego and Duffy) and four blood proteins, it has been possible, using computers, to make this diagram of the evolution of human ethnic groups. The analysis was made from the study of between two and four aboriginal groups per continent. Concentric circles represent periods of time, measured in tens of thousands of years. Modern man, divided into the different races indicated, and forming the groups now existing in the world, seems to have evolved in a period of not more than 50,000 or 60,000 years.

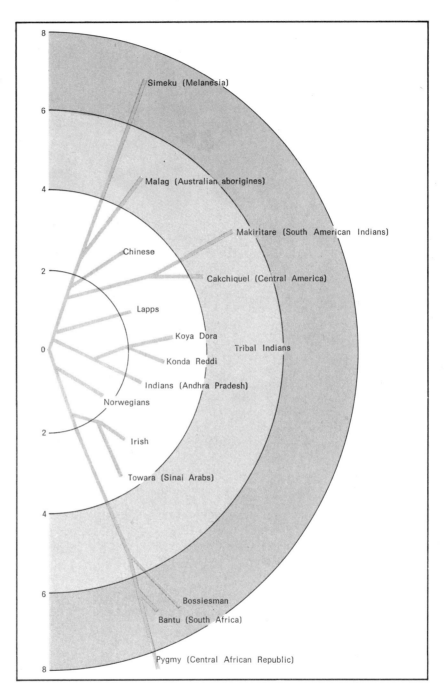

through serological and chemical analysis, has focused attention on the genetics of human populations. The study of isolated human populations should soon improve understanding of some basic evolutionary processes. In this connexion some South American Indian tribes were studied in the 1960s by multi-disciplinary teams from American universities using many of the new techniques for identifying genetic variations. A surprisingly high degree of genetic diversity was found, at least partially attributable to the social customs of the tribes concerned. The researchers contended that this led to a high evolutionary

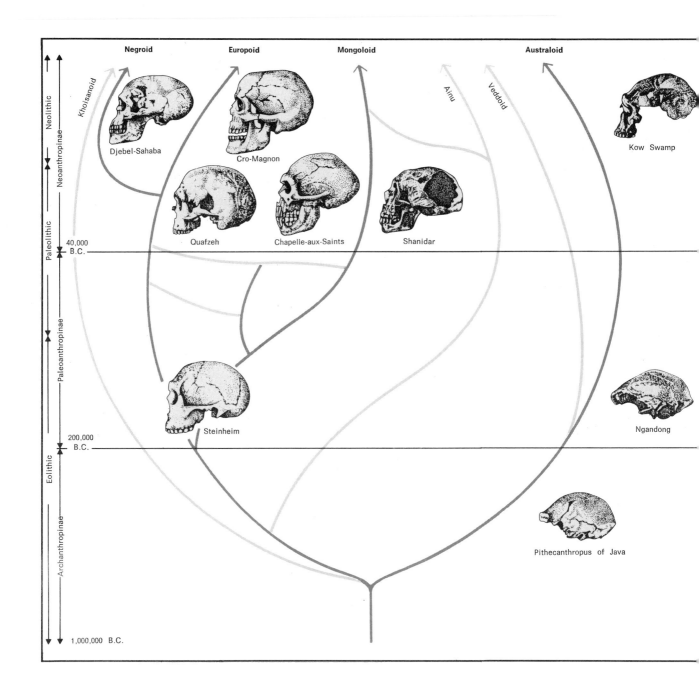

FIG. 71. Modern man, with his four main racial subdivisions, developed in three successive phases. The first differentiation occurred during the Archanthropinae era, giving rise to the Australoid stock; the second (Paleoanthropine era) produced the Mongoloid and Neanderthal stocks; the third (Paleoanthropinae), the Europoid and Negroid stocks. (Drawing by A. Thoma, University of Witwatersrand, Johannesburg; by courtesy of R. Leakey.)

rate and that, whether through accident or design, these primitive tribes were 'characterized by a genetic structure incorporating somewhat more wisdom and prudence than our own'.

The origin of species, including man

Recent investigations on the origin of species have been carried out mainly to demonstrate the microevolutionary mechanisms which act at the ecological and ethological level. This has been done by measuring the selective advantage of certain characteristics in well-defined natural populations, specially in mammals and birds. Special attention has been devoted to species which interact with man, i.e. invasive species whose ecological niche depends on the presence of man. This is the modern version of Darwin's research on domestic animals and important studies have been carried out on sparrows, a number of different animal species which are harmful to crops and finally on mice.

Many of these studies have a practical end. New information on how species originate has been provided by studies concerned with artificial selection, i.e. selection carried out by man to obtain hybrids and genetic variants with desirable traits. The objective of this work is to improve food production from domestic animals and cultivated plants, and to obtain other advantages from textile plants (such as hemp), animals (such as sheep and the silkworm), ornamental plants, animals used for traction, and plants and micro-organisms used in the production of drugs.

Crops and livestock can be genetically improved in many different ways: by either a limitation or an intensive exploitation of the reproductive ability of a certains species; by selection of the mode of contact between gametes by inbreeding, back-crossing and deliberately choosing in some plants to obtain seed either by autofertilization or cross-fertilization; and finally through recent techniques of creating new genetic variations (mutation breeding) or by producing new genotypes by physical or chemical methods. This procedure is widely used for plant species,

but among animals is still limited to the silkworm.

Parallel forms of artificial selection have been much discussed for man. The idea that human stock could be 'improved' either by taking steps to diminish the number of individuals with serious genetic defects (negative eugenics) or by adopting social measures to increase the frequency of 'superior' genetic varieties (positive eugenics) is far from new. However, during the 1960s the implications of eugenics were debated very extensively for a number of reasons. First, new techniques in molecular biology have brought the prospect of 'genetic engineering' a step closer. Second, there has been concern over the effect of medical advances which enable people with genetic defects to survive past reproductive age, whereas previously they would have died at birth or shortly after. And thirdly there has been concern over the effects of the rapid mixing of the human gene pool due to new means of rapid transport.

In relation to the origin of the human species, fossil remains have now traced tentatively the evolution of man from *Ramapithecus*, an early hominid which lived about 14 million years ago, through *Australopithecus*, *Homo habilis*, *Homo erectus*, and *Homo sapiens*. There is little doubt today that *Australopithecus* is an ancestor of man although its subsequent evolution is still hotly debated. Some specialists believe that only one species of *Australopithecus* has ever existed, whereas others claim there were two. Until recently, *Homo erectus* was considered the first 'true man', but this claim has recently been questioned. *Homo erectus* has been found in Java *(Pithecanthropus)*, in northern China (Pekin Man or Sinanthropus), in Africa, in Hungary and in Germany. Little is known about its evolution and its relationship to earlier and later fossil remains.

The discovery of a number of different fossils called *Homo habilis* early in the 1960s produced many new controversies about human evolution. According to some authorities this human ancestor, after the extinction of *Australopithecus* and *Paranthropus*, evolved directly into *Homo sapiens*, without going through the intermediate stage of *Homo erec-*

tus. Early in the 1970s the balance of opinion was still that *Homo erectus* was man's direct ancestor and that his ancestor was *Australopithecus*, of which *Homo habilis* was possibly only an incorrectly named variant. Only new fossil finds will answer these and other unsolved questions.

Macroevolution

Macroevolution—the evolution of groups of animals and plants comprising many species—was until recently studied exclusively from the fossil record. During the past twenty years or so, however, it has become clear that analysis of some of the principal molecules which are essential to many different species is a powerful means of retracing major evolutionary events in the past. Minor differences in these molecular structures can reveal evolutionary details which would otherwise remain hidden.

During the 1960s attention was focused particularly on the comparative study of the primary structure of some proteins (such as myoglobin, haemoglobin, cytochrome-c and ribonuclease). Now that the genetic code is well defined, such studies enable a comparison to be made of the structure of the corresponding genes. When similarities of composition and linear sequence of the amino acids in two or more proteins are such that they cannot be traced back to random events (as is the case for the different myoglobins and haemoglobins which have been analysed and extracted from different organisms), it can be safely concluded that the responsible genes are homologous. In other words, they originate from a common ancestral gene and any differences between them are the result of mutations.

The way in which such studies are done is well illustrated by work on cytochrome-c, a protein involved in photosynthesis. The complete amino-acid sequence of cytochrome-c from forty different species of eucaryota has been determined and a phylogenetic tree constructed from the collected data. The conclusions from these studies do not agree entirely with those of classical zoology, and the results obtained are therefore surprising. The explanation may simply be that as only one gene has been considered among the enormous number which constitute the genetic characteristics of a species, the results are misleading. Obviously, the choice of genes is critical in that some are more subject to mutation than others. The cytochromes-c of sheep, calf and pigs are identical and thus useless as a means of reconstructing the ancestral relationships between ruminants. On the other hand, this kind of protein would be of great interest in the study of the origin of vertebrates from invertebrates, of metazoa from protozoa, and of eucaryota from procaryota. In general, it is now possible to envisage that the comparative study of a number of the most important genes could lead to a greatly refined picture of the evolutionary relationships between species.

Similar techniques may also clear up another much debated problem of macroevolution: the origin of mitochondria and chloroplasts, and even of the cell itself. According to some workers, these organelles have a symbiotic origin. An ancestral procaryote would have parasitized an ancestral eucaryote and, in the course of evolution, they would have become symbiotic, each unable to exist without the other. This theory is now supported by the discovery that both these organelles contain their own specific nucleic acids which differ somewhat from the nucleic acids involved in protein synthesis.

A biochemical approach of macroevolution has also made it possible to measure the rate of evolution. In spite of the fact that paleontological records are very uncertain, it may soon be possible to extrapolate, with great precision, from one or two points in time which are known with accuracy from fossil evidence. In this way the evolutionary path of a species, and the times at which major changes were made, could be pinned down—once all the mutations to which a gene has been subjected were well defined—because it is thought that mutations which have accumulated in the genes have done so at a constant rate.

Another approach to the problems of macroevolution is that known as numerical taxonomy, which is now the best way to classify organisms in a manner which is not wholly evolutionary. This approach is based on mathematical models and on the use of computers. It is related to the biochemical approach, and produces results which tend to agree well with those obtained from comparative protein analyses.

The evolution of ecosystems

As concern with industrial pollution, the population explosion and the deteriorating environment of the industrialized nations mounted during the 1960s, a great many new studies and programmes—such as Unesco s Man and the Biosphere programme—were set in motion. Many of these studies analysed man's impact on the biosphere—the thin layer of air, water and soil which supports all living things on the Earth's surface. In general they showed that man's impact on his natural environment, in terms of many different phenomena such as thermal pollution, increased use of pesticides and the depletion of natural resources, for instance, had now reached substantial proportions.

From the scientific point of view, these studies were marked by a notable change of philosophy. Many of them, instead of analysing one detail of the ecosystem concerned in ever finer detail, were directed towards a greater understanding of that ecosystem as a closed system—one in which all the principal components such as energy or natural resources were considered finite, undergoing substantial changes due to modifications by man but all ending up in some altered form within the system. Ecology has always been concerned, of course, with the interdependence of species within such a system but the new approach which emerged during the 1960s did much to improve our understanding of the very large-scale cycles which comprise the flow of energy and materials on the Earth's surface. This approach was popularized in the now famous speech made by Adlai Stevenson to the United Nations Economic and Social Council in Geneva in 1965. The Earth, he said, was like a spaceship in that the survival of all forms of life on it depended on the presence of the resources to be found in the biosphere, which is essentially a closed system of finite size.

During the 1960s many studies were made of ecological systems in which the complete flow of either energy or of a specific resource such as carbon or water was traced through its entire cycle. One American anthropologist made a complete analysis of the energy cycle involved in the agricultural system of the Tsembaga, a tribe which lives in New Guinea. His analysis revealed the extreme complexity of their system, and the symbiotic relationship that existed between man and the plants on which he depended and which depended also on him. A similar study made of the energy flow within an Eskimo society revealed not only the relationship that now exists between the Eskimo and the Arctic ecosystem, but enabled the anthropologist to quantify the newer relationship that now exists between the Eskimo and industrialized society. A common feature of many of these studies was the ecological diversity inherent in systems which, although superficially simple, were in reality of great ecological complexity. During the 1960s the point was repeatedly made that diversity and complexity led to stability in ecosystems, in contrast to many of today's trends which are towards a drastic simplification of ecological systems through the use of advanced technology.

The most celebrated and all-embracing study of this kind was the one made late in the 1960s by scientists at the Massachusetts Institute of Technology under the auspices of the Club of Rome. Essentially these scientists made a computer simulation of the global ecosystem, starting from five basic parameters: population increase, agricultural production, depletion of non-renewable resources, industrial output and the generation of pollution. By linking each of these variables with each other through positive and negative feedback loops, an attempt was made to predict the outcome of continued population growth over the coming decades. The model was held not to be highly dependent on the initial assumptions, and it was shown that the present system might face a sudden collapse of the size of the human population within the next 200 years, unless radical changes were made.

The authors of the study urged that their model not be taken too literally, but for many scientists the approach simply confirmed what was intuitively obvious: continuous expansion

275

of the world's population at the standard of living set by the industrialized countries must soon lead to a situation in which demand for resources approached the finite limits existing on the Earth's (finite) surface area. Early in the 1970s this study gave rise to many similar efforts aimed at including yet more variables, refining the assumptions and producing models of the same kind on a regional scale. Although this exercise produced heated controversy in academic and industrial circles, particularly among those who maintained that any global model which specifically excluded social and political variables should not be taken seriously, it is likely that the study has produced a new wave of thinking among those responsible for global planning. The study had equal relevance to all those who in the 1960s had begun to study the relationship between science, technology and the future of human society. These issues are dealt with in detail in Part IV.

Further reading

BERNAL, J. *The origin of life.* London, Weindenfeld & Nicolson, 1967.

BOLOZORSKIJ, A. N. *Nukleinovye kisloty i evoljucionnaia sistematika.* Moskva, Znanie, 1969.

BOULTER, D. Protein structure in relationship to the evolution of higher plants. *Science progress,* vol. 60, no. 238, summer 1972.

CALVIN, M. *Chemical evolution.* New York, Oxford University Press, 1969.

——. L'origine de la vie. *La recherche,* vol. V, nº 41, p. 44.

CAVALLI SFORZA, L. L.; BODMER, W. F. The genetics of human populations. San Francisco, W. H. Freeman & Co., 1971.

CROWSON, R. A. *Classification and biology.* London, Heinemann Educational Books, 1970.

DOBZHANSKY, T. *L'hérédité et la nature humaine.* Paris, Flammarion, 1969.

DOBZHANSKY, T.; BOZSIGER, E. *Essais sur l'évolution.* Paris, Masson, 1968.

FLORKIN, M.; SCHEER, B. T. (eds.). *Chemical zoology.* New York and London, Academic Press, 1971.

FOX, S. W. (ed.). *The origins of prebiological systems.* New York, Academic Press, 1965.

JACQUARD, A. *Structures génétiques des populations.* Paris, Masson, 1970.

KEMP, William B. The flow of energy in a hunting society. *Scientific American,* vol. 224, no. 3, September 1971.

LEOUVRIER, C. L'origine de la vie, théories contemporaines. Paris, Laffont, 1970.

MAYR, E. *Animal species and evolution.* Cambridge, Harvard University Press, 1963.

MEADOWS, Donella H.; MEADOWS, Dennis L.; RANDERS, Jørgen; BEHVENS, William W. *The limits to growth* (2nd ed.). New York, Universe Books, 1974.

MESAROVIC, M.; PESTEL, E. The second report to the Club of Rome. New York, N.Y., E. P. Dutton & Co., 1974.

METTLER, L. E.; GREGG, T. G. *Population genetics and evolution.* Englewood Cliffs, N.J., Prentice-Hall, 1969.

NEEL, James V. Lessons from a 'primitive' people. *Science,* vol. 170, no. 3960, 20 November 1970, p. 815.

OPARIN, A. I. The chemical origin of life. Springfield, Ill., Thomas, 1964.

PETIT, C.; PRÉVOST, G. *Génétique et évolution.* Paris, Hermann, 1970.

PONNAMPERUMA, C. *Exobiology.* Amsterdam, North Holland, 1971.

RAPAPORT, Roy A. The flow of energy in an agricultural society. *Scientific American,* vol. 224, no. 3, September 1971.

UCKO, P. J.; DIMBLEBY, G. W. (eds.). *The domestication and exploitation of plants and animals.* London, Gerald Duckworth & Co., 1969.

WALLACE, B. *Topics in populations genetics.* New York, W. W. Norton & Co., 1968.

WATSON, A. (ed.). *Animal populations in relation to their food resources.* Oxford and Edinburgh, Blackwell Scientific Publications, 1970.

Part III

*The organization
of scientific research*

26

The scientific community

Resources: manpower and expenditure

Science is an expensive business. In the most scientifically advanced countries of the world, the amount spent on research and development (R & D) is over half that spent on all levels of education, and the world total, although difficult to calculate precisely, because of disparities in exchange rates, is roughly $60,000 million (in 1970).

Since the Second World War, there has been a dramatic increase in the resources devoted to science and technology. In the United States, for example, R & D expenditure has risen from $340 million in 1940 to $28,000 million in 1972. The United States spent more on space and military research in 1970 than the total spent on all areas of research and development by the United Kingdom, the Federal Republic of Germany, France and Japan combined.

Not only is science expensive, it also takes up a lot of human effort. At the end of 1972, there were almost 15 million persons in the world who had received a university degree or its professional equivalent in some area of science and technology. This represents an average of one for every 240 members of the population.

The distribution of graduates in science and engineering in different parts of the world, however, was far from even: almost 90 per cent were concentrated in North America and Europe, including the Soviet Union, although these areas account only for 25 per cent of the world's population. In terms of wealth these countries have approximately 90 per cent of the world GNP (Gross National Product): the distribution of scientific activities appears reasonably even with respect to this parameter. At the other end of the scale, Africa, with 10 per cent of the world's population, has less than 0.5 per cent of the world's graduates in scientific or technological fields, or about one for every 6,000 members of its population (see Table 25).

The difference between the advanced countries and the rest of the world are even more marked when we compare the number of qualified scientists and engineers involved in research and development (R & D). This category is used to define any creative and systematic activity undertaken to increase scientific and technical knowledge and to devise its applications. In many ways the number of scientists and engineers in R & D gives a better idea of the level of scientific activity in a particular country than either the total number of graduates in science and engineering, or the total amount of money spent on science and technology, comparisons of the

279

TABLE 25. World distribution of scientific and technological manpower[1]

	Percentage of world population	Percentage of world GNP	Total number of graduates in science and technology	As percentage of world's total	Scientists and engineers in R & D	As percentage of world's total
Africa	9	2.5	60,000	0.4	10,000	0.4
North America	6	42.3	2,000,000	13.0	550,000	23.7
Latin America	8	5.7	400,000	2.6	40,000	1.7
Asia (excluding People's Republic of China)	41	16.00	1,200,000	7.7	80,000	3.5
People's Republic of China	17		750,000	4.8	75,000	3.2
Europe (excluding U.S.S.R.)	12	32.00	3,000,000	19.4	450,000	19.5
Oceania	1	1.3	40,000	0.3	6,000	0.25
U.S.S.R.[2]	7		8,000,000	51.7	1,100,000	47.6
TOTAL			15,450,000		2,311,000	

1. These figures are very approximate, and are intended primarily as a rough guide. More detailed figures, where available, can be obtained from *Unesco Statistical Yearbook*.
2. Data for the U.S.S.R. include law, humanities, education and the arts or social sciences and humanities. Graduates employed in Soviet R & D in 1969 (excluding the social sciences) were probably 700,000. Qualified scientists and engineers (excluding social sciences) totalled about 2.7 million in 1965.

Source: Unesco Statistical Yearbook, 1973, Paris, Unesco, 1974.

latter being made difficult by differences in salary levels, cost of living and exchange rates. More than 94 per cent of the world's scientists and engineers involved in R & D were working in North America, Europe and the Soviet Union in 1972, i.e. an average of one for every 460 inhabitants; in Africa, this figure may well be as low as one for every 35,000. Again, scientific manpower is correlated to wealth.

Not only does the number of scientists and engineers vary widely from one country to another, but so too does the amount of money spent on R & D for every member of the population. The United States heads the list, spending about $130 a year for every American (of which at present about $50 goes on projects related to defence or space research) (see Table 26). It must be pointed out, however, that R & D is not more unevenly distributed than wealth.

The amount of money spent on R & D as a proportion of the gross national product, and the way that this varies over time, serves

as an important indicator of a country's activities in science and technology, and their relationship to other national priorities. In the most highly developed countries, between 1.5 and 3.5 per cent of GNP tends to be spent on R & D, as compared to between 4 and 8 per cent spent on all levels of education (and up to 10 per cent spent on defence). It must be added, however, that adding together expenses for R & D gives a somewhat distorted picture of reality. Indeed, in most cases the fraction assigned to development is much greater than that given to research, which is partly a reflection of industrial production expenses in high technology. Actually, most nations spend 0.7 to 1.0 per cent of GNP on research and the amount of development varies directly with aerospace and electronic production, mostly military or crypto-military. In developing countries, the situation varies widely. On the whole, however, the proportion is very much lower. It has become one of the main aims of programmes such as those outlined by the Advisory Committee on the Applications of Science and Technology (ACAST) of the United Nations to increase the effort of developing countries in this direction, suggesting that a target of 0.5 per cent be aimed at, as a proportion of GNP to be spent on R & D, by all developing countries by the end of the Second Development Decade (1971–80).

In recent years, there has been a notable falling off in the rate of growth of expenditure on research and development in the most highly developed countries. This has perhaps been most marked in the United States, where the proportion of the GNP spent on R & D fell from a maximum of 3.0 per cent in 1964 to its 1972 level estimated at 2.5 per cent.

Although these trends are in many ways a reflection of the economic and social problems which faced the United States at the end of the 1960s, similar trends in countries such as the United Kingdom and France seem to indicate that the explosive growth which characterized the development of science and technology during the 1950s and early 1960s has now at least temporarily come to an end.

The situation in developing countries is very different. Not only are they in an inadequate position to develop the scientific and technical potential required to combat their specific problems, but attempts to tackle these problems also represent a very small proportion of global R & D expenditure. It has been estimated that the amount of money spent on R & D in developing countries represents less than 1 per cent of the world total.

For the developing countries, the overall picture is perhaps best illustrated by a paragraph from a report by a group at the University of Sussex, prepared for the United

TABLE 26. Comparative expenditure on R & D[1]

	Total stock of scientists and engineers	Number engaged in R & D	Number engaged in R & D as percentage of total stock	Number engaged in R & D per 10,000 inhabitants	R & D expenditure as percentage of GNP	Annual per capita expenditure on R & D
U.S.S.R. (1970)	6,042,000	927,709	. . .	38.2	4.2	48.2 roubles
United States (1969)	1,694,300	535,500[2]	31.6	26.2	2.8	$129.3
Japan (1970)	. . .	286,439	. . .	27.7	1.8	10,405.1 yen
Federal Republic of Germany (1969)	. . .	76,332	. . .	13.0	1.8	DM. 185.1
France (1969)	992,000	59,020	5.9	11.7	2.0	282.4 F
Poland (1969)	569,000	54,500[2]	9.6	16.7	2.3	488.3 zloty
United Kingdom (1968)	211,231	43,588[3]	. . .	7.9[3]	2.4	£18.4
Czechoslovakia (1970)	277,148	36,927	13.3	25.5	3.6	825.5 koruna
Italy (1969)	. . .	25,214[2]	. . .	4.7	0.9	8,730.8 lire
Canada (1969)	. . .	21,052	18.8	10.0	1.3	$50.3
People's Republic of China (1970)	710,000	70,000	. . .	. . .	3.0	. . .
Argentina (1969)	240,000	4,452	1.9	1.9	0.2	652.0 pesos
Chile (1969)	. . .	4,904[2]	. . .	5.1	. . .	. . .
Nigeria (1969)	3,964	1,723	43.5	0.3	0.5	£0.2
Asia	864,480	65,288	7.5	0.05–1.2[4]	0.03–4[4]	$0.02–0.62[4]

1. Care should be exercised when using these statistics, which cannot be used for comparison between countries since criteria used in definition of scientists and engineers vary greatly.
2. Full-time equivalents.
3. Not including higher education data.
4. Ranges vary according to level of economic development of country surveyed.

Source: Unesco Statistical Yearbook, 1973, Paris, Unesco, 1974.

Nations World Plan of Action in 1969, which pointed out that

probably about $1,000 million p.a. are being devoted to R & D on synthetic materials (plastics, fibres and rubbers) in the chemical industries of the advanced countries. This is almost equivalent to the entire expenditure on research of all kinds in the developing countries, and is, of course, very largely devoted to new materials of primary interest to the advanced countries.[1]

When discussing the resources available for science, it is important to look at the supply of qualified manpower: two factors have influenced such supply over the past ten years. The first, commonly referred to as the 'brain drain', represents the migration of highly qualified manpower towards countries where they feel their skills will be most rewarded. In the mid-1960s, for example, it was calculated that over 6,000 qualified scientists and engineers were emigrating to the United States every year, of whom almost half came from Europe, and a third from various parts of Asia: in this phenomenon the chronic under-utilization of available manpower, even in developing countries, probably played a significant role. Although the flow seems to have slowed since then, and in some ways even reversed its direction, the tendency of scientists and engineers from developing countries to move to the developed world poses many problems to those who are attempting to build up an indigenous scientific potential in these countries.[2]

The second factor is that of increased unemployment and underemployment among scientifically trained manpower. Again this tendency has been most marked in the United States, where in mid-1971, 3.5 per cent of all R & D scientists and engineers were unemployed. This was due in particular to a 16 per cent decline in those employed in the aircraft and missile industries from 1970 to 1971. In addition, a survey carried out by the American Chemical Society (ACS) in 1972 found that 24 per cent of all chemists under 25 were unemployed, and that the overall unemployment and underemployment rate for ACS members was 7.2 per cent. Similar patterns are beginning to emerge in other scientifically advanced countries, partly due to a combination of the rapid increase in scientific and technological higher education during the 1960s, with the decrease in the rate of growth of R & D expenditure outlined above, and partly due to the saturation of the higher education system—which suddenly no longer needs to feed back 50–75 per cent of its output to become teachers. The scientific labour supply therefore increases by a factor of two or three quite rapidly. A number of developing countries also have a growing problem of unemployed and underemployed among qualified scientists and engineers.

The organization of the scientific community

Of the world's 15 million graduates in science and technology, less than 15 per cent are actively engaged in scientific research or associated fields of technological development. The rest tend to be engaged either in routine industrial production, in teaching, or for example in health and agriculture programmes which have a high scientific content, but would not be classified as R & D; in addition, a small but growing proportion of those with scientific and engineering qualifications are to be found in administrative jobs.

There are three main fields in which research workers are to be found. Traditionally, scientific research is held to be the responsibility of the academic community; and a large proportion of research scientists are indeed to be found in the higher education sector of society, which includes universities, technical colleges, and certain types of independent research institutes. However, as both government and industry have come to realize the potential and economic social benefits of science, so the number of research scientists and engineers employed by these two sectors have increased.

In the scientifically advanced countries, the number of scientists and engineers involved in government-sponsored R & D projects lies

1. Draft introductory statement for the 'World Plan of Action' prepared by a group at the University of Sussex. Published in *Science and Technology for Development*, New York, United Nations, 1970, Annex II, paragraph 45.
2. See, for example, *The Brain Drain*, edited by W. Adams, London, Macmillan, 1968.

broadly between 10 and 25 per cent (with the exception of 30 per cent in Canada). A much wider variation occurs in the distribution between the other two sectors; 47 per cent of Japanese research scientists and engineers, for example, are to be found in the higher education sector, as opposed to 12 per cent in the United States, and only 6 per cent in Czechoslovakia (see Table 27). In the market-economy countries, the distribution between those in productive enterprise and those in other sectors of the economy provides a rough reflection of the way responsibility for scientific research and development is divided between the public and the private sector. Considering the total number of qualified scientists and engineers active in R & D activities, one finds that the number of engineers exceeds that of scientists in Japan, the People's Republic of China and the Federal Republic of Germany, while there are more scientists than engineers in the United Kingdom, India, Italy and France. It must not be forgotten, however, that often the differences mentioned are due more to the definitions used and social organization than to function.

In the developing countries, it is even more difficult to generalize about the distribution of scientists and engineers between different sectors of the economy, since there are many other factors, such as the degree of foreign investment or the development grants assistance, that can distort the underlying picture. A rough indication, however, is provided by the three groups of Asian countries represented in Table 27 from which it will be seen that the distribution is broadly along the same lines as that in the scientifically advanced nations.

To understand the organization of the scientific community, however, it is necessary to go further than a broad description of the distribution of scientists and engineers. As a social activity, in the sense that the scientific community is a collection of individuals that have come together to pursue common goals, scientists find it necessary to formalize organizational structures in such a way that the community is able to pursue its task in what is held to be the most effective and efficient manner.

It is rare to find an individual scientist working in isolation, with little or no contact with others working in the same field. Indeed the importance, not only of the communication and exchange of information, but also of contact with a peer group that can exercise critical appreciation of developments in a par-

TABLE 27. Distribution of scientists and engineers engaged in R & D, by type of activity

	Total in survey	Productive enterprise	Higher education	General government
		%	%	%
U.S.S.R.				
United States	507,000	73	12	15
Japan	199,000	43	47	10
Federal Republic of Germany	63,000	64	19	17
France	51,000	54	34	12
Poland	45,000	66	17	17
United Kingdom	44,000	79	—	21
Czechoslovakia	366,000	79	6	15
Italy	20,000	46	40	14
Canada	21,000	36	34	30
Chile	4,800	7	74	19
Zambia	700	64	13	23
Asia	65,125	32–64[1,2]	19–20[2]	17–34[2]

1. Statistics represent 13–17 per cent agriculture, 15–33 per cent non-agriculture.
2. Ranges vary according to level of economic development of country surveyed.
Source: Unesco Statistical Yearbook, 1973, Paris, Unesco, 1974.

ticular field ensure that, on the whole, the scientist can function effectively only as an active member of the scientific community. It is possible, of course, to produce examples both of exceptional scientists who have worked successfully in relative isolation—such as Albert Einstein—and of those where the conservatism of the scientific community in accepting radically new ideas, such as Mendel's discoveries on the mechanism of heredity, has impeded the development of science. Such instances, however, tend to be the exception rather than the rule.[1]

The average scientist or engineer works as a member of a research team, usually with a number of colleagues and the assistance of one or more technicians. In the scientifically advanced countries, the ratio of scientists and engineers engaged in R & D to the number of available technicians can be as high as 1 to 2.4; in the developing countries, the situation is often the reverse, with as many as six or seven scientists having to share one technician and this alone can present problems in designing and carrying out programmes of research.

Within the academic community, the research team usually forms part of a larger unit, either an academic department with additional responsibilities for teaching, or as part of an institution specifically set up to undertake research. The individual scientist is usually free to publish the results of his research under his own name, and to share his discoveries with other colleagues in the community. Science, as we pointed out in Part I, is competitive on the individual, on the organizational and on the national level. This fact accounts to some extent for the 'publication explosion' we will examine later in this chapter.

In both government and industrial R & D departments, the situation is slightly different. Research is the responsibility of a research director, who is himself responsible to a higher authority for all the activities carried out under his direction. The scientist working in such a situation can publish the results of his work in scientific journals, but before doing so, must receive permission from the research director (or often the higher authority). Such permission can be withheld if it is felt that issues of national or industrial security are involved; most scientific research carried out under defence contracts, for example, is not published as part of the regular scientific literature, and neither is information which might be of value to rival competitors in private industry.

Within the scientific community, there are three types of organization to which scientists may belong. The first of these are the professional institutes or societies, usually referred to as the 'learned societies'. The role of the learned society is both to provide a channel of communication between its members, usually scientists in a particular specialized field, by arranging conferences and seminars, and publishing journals. They also provide professional status for their members, for example by creating fellows and associates. In addition, most countries have one or more 'super-learned' society, covering the whole of science rather than one particular field. Examples of these are the Soviet Academy of Sciences in Moscow, the National Academy of Science in Washington and two of the oldest, the Royal Society in London and the Accademia dei Lincei in Rome. Their main functions, as those of the learned societies, is to facilitate communication at both a national and an international level, and to confer social prestige; in the planned economy countries, the academies of sciences usually have scientific responsibility for the organization and conduct of all fundamental research (see Table 28).

The second group of organizations are those bodies which have been created by scientists to protect and further their professional interests. In the past, learned societies have attempted to embrace this function, but in recent years trade unions and other associations of scientists have sprung up in answer to the problems of what has been called the 'proletarianization' of science, by which the individual scientist finds himself further and further removed from the major decisions over the organization of his work. Sometimes these

1. See, for example, B. Barber, 'Resistance by Scientists to Scientific Discovery', *Science*, Vol. 134, 1961, p. 596–602.

bodies have been the result of situations—such as increasing unemployment—in which the learned societies have been unable or unwilling to take a firm stand. They are primarily concerned with issues such as the salaries and working conditions of scientists and engineers, very much along the same lines as the rest of the trade union movement. Their members are not usually confined to a single field of science.

The third group, somewhat smaller than the other two, but growing in importance in recent years, are those societies and associations of scientists formed around issues that are common to all fields of science. They are usually concerned, either with interdisciplinary, national and international issues, as well as with the dissemination of science in the popular domain (such as the American Association for the Advancement of Science), or with presenting a critique of the role of science in modern society (for example the British Society for Social Responsibility in Science). The membership of such associations is not usually confined to scientists, although they inevitably tend to make up the largest proportion of members.

Communication in the scientific community

More than 50,000 scientific and technical journals are published every year; between them they carry about 5 million articles, almost half of which are in bio-medicine alone, and these figures are thought to be doubling every ten to fifteen years.[1] Communication is vital to the development of science, yet faced with such a barrage of publications, the individual scientist can be excused if he occasionally feels himself awash in a sea of information which seems to threaten to overwhelm him at any moment.

In fact, the situation is not quite so bad as it might appear at first glance. To begin with, only a very small proportion of the total number will be relevant to a particular field, and of these, too, the quality and importance varies to such an extent that only a few need to be consulted regularly: for almost any purpose a tiny fraction of the journals will give a large proportion of the value, for example

half the significant articles of the world's 50,000 journals come from the biggest 250 only. A survey of British scientists, for example, found that a typical working scientist or technologist saw fairly regularly the issues of ten to twelve journals, and although the range was wide, few scanned less than five or more than twenty.[2] But the average scientist still has to spend five hours reading every week.

The 'information explosion', characterized by an increase by a factor of 1,000 in the number of scientific journals over the past 150 years, has caused headaches to both scientists and librarians. Various developments, including the publication of abstract journals, which give brief synopses of the articles contained in the main journals in a particular field, and the design of computer-based information systems, have helped to tackle this problem. Likewise, the weekly publication of *Current Contents* by the Institute for Scientific Information in Philadelphia offers the possibility of following the titles of papers published in some 2,000 journals throughout the world. As we shall see later, the *Science Citation Index*, prepared by the same institute, represents a novel and efficient method for literature classification and retrieval. On the whole, however, it is being realized that only a well-thought out and coherent national and international policy on scientific and technological information, in which each particular element has its own part to play, can meet the growing demands of the scientific community with the minimum of effort.

It is possible to divide broadly the channels of communication in science into two sectors; first there are the formal channels, including journals, magazines, monographs and books, as well as regular meetings of professional and learned societies. There are also the informal channels of communication—private letters, reprints and preprints exchange, individual acquaintances and small-scale seminars—which are felt by many

1. D. J. de Solla Price, *Little Science, Big Science*, New York, Columbia University Press, 1963.
2. R. H. Phelps and J. P. Herling, 'Alternatives to the Scientific Periodical', *Unesco Bulletin for Libraries,* Vol. XIV, 1960, p. 61-7.

TABLE 28. Number of learned societies, distributed by fields of interest, for selected countries

	Agri- culture and veterinary science	Natural sciences				Tech- nology	Medical sciences and medicine
		General	Biology	Math- ematics	Physical sciences		
Australia	3	7	14	2	9	8	12
Brazil	2	3	4	4	1	3	13
France	7	6	13	3	18	6	62
Federal Republic of Germany	7	6	15	7	26	30	36
United Kingdom	16	15	28	5	26	60	64
Hungary	3	1	3	1	8	12	1
India	4	8	—	4	1	6	6
Italy	4	8	4	3	14	12	24
Japan	29	1	17	2	36	26	55
South Africa	—	3	4	—	2	4	2
United States	12	13	22	5	28	36	66
U.S.S.R.	------------------------------------ 22 ------------------------------------						32

Source: World of Learning, London, Europa Publications Ltd, 1971–72.

scientists to be equally necessary, if not more so, to the healthy development of science.

In the last two decades, particularly in fast advancing fields of research, the distribution by the authors of pre-prints of articles accepted for publication by scientific journals has accelerated the rate of diffusion of information. One should also mention the important function of fast duplicating machines, which became available in the 1960s, in facilitating the access to literature for the research worker.

The most important of the formal channels are the scientific and technological journals. These are usually published weekly, monthly or quarterly, and contain the results of original research carried out in a particular specialized field. A survey of 12,000 United States life scientists found that on average each scientist published at least four articles per year of which at least two were full-length research articles, while he would write, on average, 0.8 abstracts of original research, 0.6 articles for in-house publications, 0.2 chapters in books, and 0.1—i.e. about one in every ten years—major reviews of his subject. Over 90 per cent of those who replied had written one or more research articles during the year in question.[1]

The publication of research findings in journals fulfils two main functions. The first is that it communicates information about a particular scientist's work to other members of the scientific community, ensuring the wide dissemination of scientific knowledge. The second is that the number of publications a scientist has to his name is often taken as an important indicator of professional ability.

This second function has undoubtedly been a factor in creating the current publications explosion (as, too, has the fact that publishing journals can prove to be a lucrative business for scientific publishers). But the most important scientific papers still tend to be limited to the few journals which enjoy a high prestige. One British study, for example, found that 95 per cent of important primary scientific work was to be found in only 9 per cent of the total number of journals (165 out of 1,900); extended to a world basis, this would imply that the most important science is confined to a 'core' number of about 4,500 out of 50,000 journals.[2]

1. Survey carried out by the National Academy of Science, Washington, D.C. on publications by 12,364 life scientists in 1966.
2. J. Martyn and A. Gilchrist, *An Evaluation of British Scientific Journals*, London, ASLIB, 1968.

The institutionalization of science, the rapid growth of the scientific enterprise, together with high status attributed today to a scientific career have prompted competition and even fierce rivalry within the scientific community. Aware that science is a highly stratified institution and that power resources are concentrated in the hands of a relatively small minority, scientists believe that it is necessary to publish significant work in order to obtain recognition. Hence the vernacular expression of the laboratory milieux: 'publish or perish'. The trend appears particularly clear in the so-called 'hot fields' which deal with 'hot subjects'. As R. K. Merton and R. Lewis have recently pointed out:[1]

'hot' fields are not only more active than 'cold' ones, but their results are taken to have implications that reach well beyond the borders of the specialty. They tend, at least for a time, to attract larger proportions of talented men who have an eye for the jugular, concerned to work on highly consequential problems rather than ones of less import.

It may well be that the impressive proliferation (or pollution, as some critics prefer to call it) of scientific information is more a consequence of the availability of funds than of the intrinsic dynamics of scientific advance. Such interpretation finds some support in a recent investigation on the quantity and quality of publications in physics,[2] which lead the authors to conclude that

science would probably not suffer from a reduction in number of new recruits and an increase in the resources available to the resulting smaller number of scientists.

In addition to journals, important scientific information is published both in research and technical reports, and in the patent literature. It is difficult to estimate how many reports are issued annually, since these are issued by a wide variety of bodies from small laboratories to government departments; but it is known, for example, that the United States Government publishes about 100,000 reports on science and technology every year. As for patents, it is estimated that 10 million out of the 100 million articles, papers and books ever published on science and technology have

been concerned with patenting a new process or technology.

Some 680,000 patent applications were filed in the world in 1968; 350,000 patents were granted in that same year, protecting approximately 100,000 inventions. Japan and the U.S.S.R. lead in the number of filed patent applications (see Table 29).

For the busy scientist who is unable to see all the journals relevant to his research, abstract journals provide a quick guide to current developments, and help him track down specific information in a quicker time than going to the original source. Several thousand such abstract journals are now printed, the

TABLE 29. Statistics on patents

	Applications for patents	Grants of patents	Patents in force
Canada	29,438	29,242	366,358
France	47,971	51,456	371,806
Federal Republic of Germany	65,756	18,149	118,676
Italy	30,826	17,500	—
Japan	105,785	36,447	217,845
U.S.S.R.[1]	124,021	33,631	4,966
United Kingdom	61,078	41,554	238,538
United States	104,729	78,316	931,416

1. Includes inventors' certificates.

Source: *Industrial Property*, Geneva, World Intellectual Property Organization, December 1972.

abstracts themselves ranging from half a dozen to two or three hundred words. As an example of the rapid growth of abstract journals (which, alone, provide a measure of the size of a particular field at any given time), *Chemical Abstracts*, first published in 1907, took thirty-two years to publish its first million abstracts. The second million was reached after a further eighteen years, the third after eight years, the fourth after four years and eight months, and the fifth after three years and four months

1. R. K. Merton and R. Lewis, 'The Competitive Pressures (I): The Race for Priority', *Impact of Science on Society*, Vol. 21, 1971, p. 151–61.
2. J. R. Cole and S. Cole, 'The Ortega Hypothesis', *Science*, Vol. 178, 1972, p. 368–75.

in 1972. At the present rate the 10 million mark will be passed in another ten years.[1]

The *Chemical Abstracts* staff surveys each issue of 12,000 science and technology journals published in fifty-six languages, as well as the patents issued by twenty-six nations. This requires a full-time staff of 1,000; in addition, 2,700 men and women in fifty-six countries write for *Chemical Abstracts* in their spare time.

A further recent addition to the scene has been services such as the *Science Citation Index*, which gives a guide to the work being done in all branches of science.

A citation index is an ordered list of cited articles, each accompanied by a list of citing articles. The citing article is identified as a source, the cited article as a reference. The *Science Citation Index* (SCI), published by the Institute for Scientific Information, is the only regularly issued citation index in science. It is prepared by computer and provides an index to the contents of every issue published during a calendar year of more than 2,000 selected journals. Journals covered by the index are chosen by advisory boards of experts in each of the topics represented and by large scale citation analyses.

The entry for a cited article (reference) contains the author's name and initials, the cited reference year, and the publication name, volume, and page number. Under the name of each cited author appears the source article citing this work. This line is arranged by citing author's name, publication, type of source item (article, abstract, editorial and so on), citing year, volume and page. The searcher starts with a reference or an author he has identified through a footnote, book, encyclopaedia, or conventional word or subject index. He then turns to the *Citation Index* section of the SCI and searches for that particular author's name. When he has located the name, he checks to see which of several possible references fits the particular one he is interested in. He then looks to see who has currently cited this particular work. After noting the bibliographic citations of the authors who are citing the work with which he started, the searcher then turns to the *Source Index* of the SCI to obtain the complete bibliographic data for the works which he has found.

After finding several source articles, the searcher can use the bibliographies of one or several of these as other entries into the citation index; this process is called 'cycling'. Since authors frequently write more than one closely related paper,

additional articles by the author of the starting reference can also be used as entry points to the index.

Basically, then, the SCI does two things. First, it tells what has been published. Each annual cumulation cites between 25 and 50 per cent of the 5 to 10 million papers and books estimated to have been published during the entire history of science. Second, because a citation indicates a relationship between a part or the whole of a cited paper and a part or the whole of the citing paper, the SCI tells how each brick in the edifice of science is linked to all the others. Because it performs these two fundamental functions so well, important applications for the SCI have been found in three major areas: library and information science, history of science, and the sociology of science. For example, the *Citation Index* has made it possible to discover that Mendel's work was actually cited by at least five different people before 1900.[2]

Books and monographs also provide a source of scientific information, although publishing delays mean that they are used primarily for reviews of particular areas of science, rather than as a means of presenting new work. The number of books published by a number of countries during 1972 are shown in Table 30.

In all published material there is the problem of language. A language breakdown of the 35,000 periodicals received by the British National Lending Library for Science and Technology correlates closely with the ranking of the most important scientific nations in Table 31. There is a greater imbalance in the major English-language abstracting journals. In *Chemical Abstracts*, for example, 50.3 per cent of the articles abstracted were in English, 23.4 per cent in Russian, 7.3 per cent in French, 6.4 per cent in German, 3.6 per cent in Japanese, and 9 per cent in another language. An even greater disproportion is to be found in a very recent study of the literature of synthetic chemistry, in which data on 1.2 million compounds based on more than 128,000 abstracts published in *Index Chemicus* have been analysed (see Table 32). In *Engineering Index*, over 82 per cent of the articles were in English,

1. *World of Learning*, London, Europa Publications Ltd, 1971–72.
2. E. Garfield, 'Citation Indexing for studying Science', *Nature*, Vol. 227, 1970, p. 669–71.

TABLE 30. Number of book titles published during 1972 on pure and applied science

	Pure science	Applied science
U.S.S.R.	7,706	33,887
United Kingdom	3,667	5,190
Federal Republic of Germany	2,619	4,315
United States	2,667	5,098
Japan	1,788	5,863
Netherlands	1,243	11,014
France	1,573	4,570
Poland	1,165	3,919
Spain	1,068	2,574
Czechoslovakia	638	2,793
Romania	467	2,330

Source: Unesco Statistical Yearbook, 1973, Paris, Unesco, 1974.

TABLE 31. Language breakdown of periodical literature received at the United Kingdom National Lending Library

	Percentage of literature received
English	46
Russian (and other Cyrillic)	14
German	10
French	9
Japanese	4
Spanish	3
Italian	3
Polish	2
Czech	2
Portuguese	2
Other	5

Source: D. N. Wood, 'The Foreign Language Problem Facing Scientists and Technologists', Journal of Documentation, Vol. 23, No. 2, June 1967. Reproduced with the permission of ASLIB, London.

leaving only 3.9 per cent in Russian, 8.6 per cent in German, and 5.5 per cent in another language. However, it seems likely that these figures represent as much a cultural bias in the selection and translating and publishing mechanisms as any indication of national strengths and weaknesses in particular areas of science.

The last category of formal channels of communication are the conferences and other large meetings at which original work is pre-sented and discussed. These are usually organized by national and international professional bodies, and the papers later published as a conference report. The study of American life scientists mentioned above found that in one year, 90 per cent of them had attended at least one meeting, 15 per cent had attended four or more and 10 per cent had participated in meetings outside the United States. It is only in Europe that over 400 international meetings on science, technology, agriculture and medicine are held every year, the top host cities being Paris, London, Geneva, Rome and Prague. On a national level, each professional association or learned body usually holds at least one major conference every year.

One of the major functions of conferences, in fact some might argue their most important function, is to stimulate informal contact between research workers in a particular field. The importance of the networks of colleagues and acquaintances to a professional scientist, sometimes referred to as an 'invisible college' in the sense of consisting of a group of mutual interacting individuals, has become increasingly realized during the past decade. Particularly in those areas of science where significant discoveries seem to occur at a high rate, the so-called 'hot fields' of research around which many brilliant minds concentrate their efforts, useful communications (useful in the sense that they contribute to hasten the pace of advance) occur almost entirely outside the official channels of journals and books. Within the invisible college, information is exchanged primarily through personal contacts, telephone calls, letters, gossip, informal parties and what not. A survey of mathematicians found that 51 per cent rated informal communication contacts as being very important. Diana Crane,[1] who studied this aspect of the sociology of modern science, concludes that

programmes for dealing with the 'information explosion' in science ought to be directed towards bringing isolated scientists in contact with scientists who are the foci of communication networks.

1. D. Crane, 'The Nature of Scientific Communication and Influence', International Social Science Journal, Vol. 22, 1970, p. 28–41.

TABLE 32. Language distribution 1960–69

	Language						
	English	German	Russian	French	Japanese	Italian	Others
1960							
Abstracts	49.9	16.9	14.7	7.5	7.0	2.4	1.6
Compounds	49.4	18.7	13.1	8.6	6.2	2.9	1.1
1961							
Abstracts	53.0	16.4	16.0	5.2	5.7	2.0	1.7
Compounds	53.3	18.8	12.6	6.1	5.3	2.6	1.3
1962							
Abstracts	52.1	16.2	16.4	5.8	6.5	2.0	1.9
Compounds	50.9	18.3	13.3	7.3	6.0	2.6	1.6
1963							
Abstracts	53.8	16.5	16.2	6.0	3.5	2.4	1.6
Compounds	52.5	19.1	13.4	7.2	3.2	3.4	1.2
1964							
Abstracts	55.5	15.9	16.8	4.8	3.1	2.1	1.8
Compounds	51.5	18.7	14.7	6.9	3.6	2.9	1.7
1965							
Abstracts	55.2	16.4	16.4	5.9	3.0	1.8	1.3
Compounds	54.6	17.4	14.4	7.1	3.1	2.3	1.2
1966							
Abstracts	56.0	15.3	16.4	5.5	2.9	1.7	1.9
Compounds	54.0	17.4	13.9	7.6	3.4	1.9	1.9
1967							
Abstracts	56.9	13.8	16.4	6.2	3.1	1.9	1.6
Compounds	55.5	15.4	14.0	8.3	3.4	2.3	1.3
1968							
Abstracts	61.3	11.8	15.4	5.7	2.3	1.6	1.9
Compounds	59.7	13.5	12.5	7.4	2.9	2.2	1.8
1969							
Abstracts	61.4	10.0	17.4	5.9	2.9	1.1	1.5
Compounds	60.2	12.6	13.8	7.2	3.0	1.6	1.2

Source: E. Garfield, G. S. Ravesz and J. H. Batzig, 'The Synthetic Chemical Literature from 1960 to 1969', *Nature*, Vol. 242, 1973, p. 307–9.

Such a programme, however, is easier said than done. On the more practical level, various ideas are being developed for attempting to tackle the information crisis that seems to be looming up. Perhaps the most far-reaching and ambitious of these is the proposed establishment of a World Science Information System (UNISIST) being developed by Unesco in conjunction with the International Council of Scientific Unions (ICSU). The aims of UNISIST are to make use of the latest advances in information storage and retrieval techniques including manual, mechanical and electronic techniques, to build a world-wide information system on science and technology, including the establishment of international standards and procedures for co-ordinating the use of such equipment. At the seventeenth session of the Unesco General Conference, held in Paris in 1972, it was agreed that almost $400,000 from the Organization's regular budget would be provided towards the development of UNISIST for the biennium 1973–74, with a further $400,000 entrusted to the Organization by the United Nations Development Programme (UNDP) to help with setting up national agencies responsible for the development of scientific and technical information resources and services, to initiate pilot projects to develop effective approaches for linking the developing countries with UNISIST, and setting up appropriate training facilities and developing their infrastructures for scientific and technical information services and networks.

Finally, the increasing mass and complexity of scientific and technical information is making the question of national policies in this area increasingly urgent. A survey of scientific and technical information services in Canada, carried out in 1971 by OECD,[1] emphasized the need for such a policy to coordinate and develop the following activities: the choice and acquisition of documents by libraries, the common indexing of documents for easy location, the holding of documents, the circulation and exchange of documents, the computerization of libraries, meeting the 'interest profiles' of users, setting up information analysis centres, and expanding information retrieval services to all parts of the country. Only by such planning—which in many ways represents a microcosm of the problems of science policy in general—is any country in the future likely to be able to provide its scientists and technologists with a comprehensive, efficient and reliable information service.

The education and training of scientists

There is little need to emphasize the crucial role of education in developing a country's scientific and technological potential. Perhaps this is illustrated by the case of Japan, where the total number of graduates in science and technology increased by two-and-a-half times between 1955 and 1965, and the number of doctorates alone almost tripled in the five years between 1960 and 1965.

Similar patterns can be seen in other advanced countries. In France and the U.S.S.R., in particular, similar expansions in the number of graduates have accompanied rises in scientific and technological activity. Even in the developing countries, it is noticeable that those who have achieved the greatest growth in science and technology, for example India or some of the countries in Latin America, are those with the highest production of qualified scientists and engineers.

TABLE 33. Science degrees per 100,000 population

	Number		Number
Africa	0.2	Europe	13
Arab States	3.0	Latin America	1.4
Asia	2	North America	37

Source: *World Plan of Action for the Application of Science and Development to Technology*, New York, United Nations, 1971.

Many problems, however, remain. Some have always been with us, in particular the gross distribution of higher education in the various parts of the world. The United Nations World Plan of Action has drawn attention to the great differences that occur in the number of science degrees per inhabitant in various parts of the world, again reflecting the disparity of resources described earlier (see Table 33).

Other problems are more recent in origin, and to a certain extent reflect the changing attitudes to science and technology that have occurred, primarily within the advanced countries, over the past decade. One problem is the over-production of pure scientists as opposed to applied scientists and technologists, a situation that is particularly crucial during the present period when pure science as a whole is experiencing a decrease in the rate of growth; the unemployment and underemployment problems this has created have been mentioned above, but it seems unlikely that the situation will improve for a number of years, as countries are obliged to make the structural changes in their educational system that the new situation demands. This itself is aggravated by a second aspect of the problem, the traditional teaching of science as a narrow academic discipline dedicated solely to the intellectual pursuit of knowledge; an inadequate awareness of the social dimensions in science teaching has left many scientists and technologists in a bad position to tackle the new problems which are confronting them.

Science education is commonly held to play two roles. These can be expressed as:

1. *Review of National Scientific and Technical Information Policy in Canada*, Paris, OECD, 1971.

291

1. To ensure that every person has such a grasp of science as to be ready to co-operate with understanding in the application of science to man's needs.
2. To ensure a sound foundation of the basic principles and facts of science in those who seek to make their careers and serve society as scientists or technologists.

It is the second of these that is the more important when considering the impact of education on scientific research, and although the role of science teaching at a primary and secondary level is of vital importance for both aspects of science education, it is in the higher education sector that the most important trends are to be found when considering the supply of qualified scientists.

Wide differences occur between the number of those leaving secondary schools with certificates in science, and those entering university courses in science and technology. At the end of the scale, 58 per cent of Norwegian secondary-school students leave with certificates in science, although only 23.3 per cent of those entering university do so to take courses in science and technology. On the other hand, in Switzerland the situation is reversed, with a higher percentage, 37.2, taking university courses in science and technology than those leaving school with qualifications in science. In the United Kingdom, for example, secondary education is treated as a systematic pre-university training, something that seems to be a direct result of the admission process adopted by universities. In countries such as the Federal Republic of Germany and Italy on the other hand, secondary-school graduates are offered much wider possibilities for study when they enter university, similar to the situation in the United States, where specialization at a secondary level is kept to a minimum. Both systems have their merits and disadvantages, and any differences are in a large part due to different pedagogical traditions. It appears that in France, for example, an appreciable fall in the number of first-year university students opting for pure science from 38.5 per cent in 1958 to 21.9 per cent in 1967 may be due to the larger number of preparatory classes for the *grandes écoles* of engineering and administration, whose compe-

tition with the university system is a characteristic of the French system.

By comparing the relative growth in the number of first degrees awarded in science and technology over the period 1955 to 1965, the much faster expansion of pure science can be seen (see Table 35). The situation with regard to technology is particularly emphasized by the fact that, during the period 1960–65, the proportion of students enrolled in technology fell from 16.9 to 13.5 per cent in the Federal Republic of Germany, from 13.6 to 8.8 per cent in Canada, from 9.5 to 7.1 per cent in the United States, and from 11.4 to 11.1 per cent in Italy, while rising from 18.5 to 19.2 per cent only in the United Kingdom.

Although this decline was more than compensated by an increase in the total number enrolled in education, it still reflects a move from technology to pure science that may leave national communities ill-equipped to tackle problems that seem likely to arise in the future. However, it should also be remembered that the growing importance of non-university type technical education has also made up to a certain extent for this situation.

Post-graduate education presents a slightly different picture. In the majority of OECD countries the pace of growth of higher degrees in science and technology has been much more rapid than that of first degrees (see Table 35). In the United Kingdom, for example, the average annual growth in the number of master's degrees and doctorates was 14.2 per cent over the period 1960–65, more than twice that of first degrees, while in the United States the relative rates of growth were 17.3 per cent for master's, 10.4 per cent for doctorates and 7.8 per cent for bachelor's degrees. Other important factors include the relative decline in the total doctorates awarded in the Federal Republic of Germany, the rapid expansion of higher education at all levels in France, and the spectacular increase in Japanese post-graduates referred to earlier.

Such a situation, however, does not give quite the cause for rejoicing that might have been predicted at the beginning of the 1960s. The stress on pure science rather than technology in universities has already been pointed out. The cost of the over-production of pure scientists is now being measured in the unemployment and underemployment figures described earlier.

Turning to the situation in developing countries, the existing disparity in the distribution of scientific degrees has already been referred to. Perhaps more importantly however, is the large proportion of students who opt for pure rather than applied science or technology. According to further figures published in the United Nations World Plan of Action, although the number of higher education students enrolled in natural sciences and engineering in Europe and North America represented 13.7 and 17.4 per cent respectively, in Africa these figures were 13.7 and 6.0 per cent, while the figures for Asia (excluding India) in 1964 were 30.4 and 6.8 per cent.

Thus in those parts of the world where the application of science would seem to be of far greater importance than the discovery of further scientific knowledge, not only is the proportion of engineers even lower than it is in the advanced nations, but the ratio between natural scientists and engineers is out of all proportion to the relative needs of the countries themselves.

This situation brings out the fact that developing the scientific and technological potential of these countries is not merely one of money, but equally importantly one of attitudes. It should be remembered that what we refer to as 'science' was developed against a Western cultural background very different from that of most developing countries. While science and technology fit naturally into patterns of Western education, considerable tensions can be felt when these are introduced into alien cultural patterns. Referring to the difficulties in getting scientific concepts across to a child from a developing country, one educationalist[1] has written:

A good deal more than translation and acquisition of new terms is involved. A new set of abstractions is needed, and the thinking process must accord itself more closely to a hypothetico-deductive model than to a holistic one, which both language habits and cosmology tend to favour.

In the light of the above uncertainties, many countries are now beginning to realize the importance of some form of manpower planning, by which an attempt is made to match the output of educational institutions with the demands of the surrounding economy. In the past, a mixture of chance and market factors has managed to keep this balance relatively stable; the migration of qualified scientists and engineers to the United States (for example) was at its peak when American R & D was going through its most rapid period of expansion, and the two factors were closely related. More recently, however, the growing number of unemployed graduates has given increasing cause for alarm, particularly in the developing countries where a large financial investment has gone into their education (the ratio of costs per pupil between higher education and primary schooling in Africa is often as high as 30 : 1, compared with 5 : 1 in Canada or Australia).

The United Kingdom is one country in which manpower planning has been the concern of central government for a number of years. The first Committee on Scientific Manpower was set up in 1950 to 'study the future needs of scientific and technological manpower for employment both at home and

TABLE 34. Years marking the beginning of a decline in the relative position of university studies in pure science and technology[1]

	Pure science	Technology
Federal Republic of Germany	1951–52	1958–59
Belgium	1964–65	1960–61
Denmark	+	1957–58
France	1964–65	…
Italy	1964–65	1959–60
United Kingdom	+	+
Sweden	1964–65	1959–60
United States	+	1959–60
Japan	+	+

1. + signifies that the relative growth was maintained, at least up to 1966–67.

Source: Development of Higher Education, 1950–67. A Statistical Survey, Paris, OECD, 1971.

1. S. Biesheuvel, 'The Ability of African Children to assimilate the Teaching of Science', in: P. G. S. Gilbert and M. N. Lovegrove (eds.), *Science and Education in Africa.*

TABLE 35. Evolution of the rates of first and higher degrees[1]

Country	First degrees in science and technology		Total number of doctorates		Doctorates in science and technology 1965/66	Intermediary degrees
	1955	1965	1960/61	1965/66		
Federal Republic of Germany	0.57	0.53	0.80	0.66	0.23	—
France	0.70[2]	1.70[2]	0.21	0.48	0.34	0.20
Italy	0.61	0.84	. . .	. . .	. . .	
United Kingdom	1.23	2.38	0.32	0.58	0.36	
Canada	1.10	2.14	0.12	0.29	0.22	(MA)0.56
United States	2.63	3.97	0.49	0.78	0.39	(MA)1.30
Japan	0.90	2.13	0.06	0.22	0.06	(MA)0.19
U.S.S.R.	2.07	5.72				

1. All figures expressed as percentage of single-year age group.
2. Estimates for technology degrees.
Source: Development of Higher Education, 1950–67. A Statistical Survey, Paris, OECD, 1971.

abroad', and triennial surveys have been made every three years since 1956. In the middle of the 1960s the concept of the 'stock' of qualified scientists and engineers as a natural resource was introduced. The committee was reconstituted in 1964 as the Committee on Manpower Resources for Science and Technology, and in its first report explained how it had interpreted the responsibility for advising on manpower resources

to include the effective use of the whole stock of scientific and technological manpower as well as the output of newly qualified individual in relation to employment opportunities. This is because we believe that the deployment within the economy of existing scientists and technologists, and their utilisation within any organisation, demands as careful and urgent consideration as the supply from the educational system.

Similar moves to introduce some form of programme planning and budgeting system have been introduced in the United States in an attempt to compare educational costs with output, essentially a modified form of cost-benefit analysis. In all centrally planned countries such as the U.S.S.R., educational planning has always formed an important part of economic policy, and many market economy countries are now following the same example.

Within this context, it seems inevitable that the nature of education will alter as greater demands are put on it by the economic and industrial system. At a conference on highly qualified manpower held in Paris in 1966, Seymour L. Wolfbein of the United States Department of Labour suggested that

the function of education can be viewed as the process which enables an individual to withstand the inevitable changes which will occur in the relationship between what he learns and what he will be called upon to do in the world of work.[1]

How traditional educational institutions will be able, or even prepared, to adapt themselves to this new role remains to be seen.

The sociology of science

The magnitude of the scientific enterprise in the world today, as briefly presented in the preceding pages, is the end product of a process of growth that has characterized industrial societies over the last few decades; the growth continues, even though one may perceive novel trends in the research system. (We will discuss this in Chapter 30.) As we all know, the advances in scientific knowledge and technological developments have had a profound impact on the world. In the process, the nature

1. *Policy Conference on Highly Qualified Manpower*, Paris, OECD, 1967.

and meaning of science have changed. No longer can these be considered as an individualistic and élitistic search for truth; science has become a major social institution. While this was happening, the ambitions, allegiances and behaviour of scientists underwent major changes. Momentous quantitative as well as qualitative events have opened new horizons to research by historians, philosophers, psychologists and in particular by sociologists.

Scientists first turned to a study of science itself during the 1920s and early 1930s. A number of reasons have been put forward why this should have happened. Some think that the value of science was being challenged as a result of the general disillusionment that surrounded most political and social life after the First World War, others that the need to understand science grew out of a general desire to apply it to the social problems arising from the Depression of the 1930s (the first attempts at technological prediction started at this time).

The sociology of science as a distinct discipline had a very slow start. Its birth can be placed at some time before the Second World War, with the appearance of works on the sociology of knowledge, of J. D. Bernal's *Social Function of Science*, of socio-economic interpretations of the history of science by the pen of left-wing scientists such as Joseph Needham and Lancelot Hogben, and by the early contributions of Robert K. Merton. In the 1940s and the early 1950s studies on the sociology of science were still relatively neglected; more recently there has occurred a flourishing of the field, so that the sociology of science has emerged as a distinct sociological specialty and has developed a professional and institutional identity.

While it would be impossible to render justice in a few pages to this very active area of study that some describe in terms of 'science as a science', it appears worth while to indicate some recent trends.

Derek J. de Solla Price wrote in 1966:[1]

Instead of the old feeling that writing on a favourite scientist or digressing on talk about science was a nice thing to occupy the retirement of someone excreted from the research front, there is something of a new feeling. The history of science is to be pursued because it is intellectually more stimulating, and perhaps more difficult than merely being a scientist, or pursuing those plainer parts of history that can be understood with the usual sort of Arts certificate of ignorance in the sciences.

And, indeed, in the course of the last few decades the number of professional science historians has grown at a remarkable rate. The data in Table 36 gives an idea of the present expansion of this field.

According to the prominent and articulate specialist just quoted, remarkable changes have occurred. He writes:

The most pronounced change is perhaps the way in which the Siamese twins of history and philosophy of science have now taken up virtually independent lives. Since Whewell they had been part and parcel of each other and the union is still preserved in the structure of several university departments. When both were rather amateur activities and it was easier to cultivate a universal Renaissance mind, it was possible; but now, for the greater number of specialist contributors, the width of field has become too great and the cross-fertilization between history and philosophy of science too small to be so strenuously sought. Philosophy of science, among its most eminent modern practitioners, has become a significant part of philosophy itself in a way that history of science has not entered history, except in the much less technical regions of history of ideas and economic history.

Debates on the relative importance of 'external' and 'internal' factors in the development of science have characterized the whole field since its beginnings. Early participants such as Bernal and J. G. Crowther sought to explain how science had grown in response to social needs. This was soon challenged by others, in particular by Michael Polanyi, who was one of the first to outline the nature of the scientific 'community'[2] and demonstrate the way in which it sets up its own rules and special norms that have little to do with outside influences. A prevailing influence of internalism is to be found in the scholarly works of those who attempt the analysis of exact

1. D. J. de Solla Price, 'Science as a Science', *Times* (London), Literary Supplement, 28 July 1966.
2. M. Polanyi, *Personal Knowledge*, London, Routledge, 1958.

TABLE 36. Total of state, higher education and private institutions involved in research and teaching on various aspects of science, distributed by main areas of interest[1]

	U.S.S.R.	United States	Federal Republic of Germany	France	Poland	United Kingdom	Czecho-slovakia	Italy	Canada
Philosophy/theory of science	6	25	5	12	3	5	3	5	4
Ethics of science	1	31	2	3	3	7	1	4	3
Sociology of science	9	37	4	10	4	17	4	8	4
Classification of fields and disciplines	4	9	3	4	3	4	3	2	1
Creativity/ psychology of research workers	6	22	1	13	4	10	4	3	4
History of the organization of science	5	37	4	5	3	14	2	3	5
Economics, etc., of R & D	13	48	11	26	3	28	4	8	8
Organization, administration and management of R & D	14	34	8	23	5	25	4	8	9
Statistics of science and technology	8	24	2	3	4	11	4	7	5
Planning of R & D	14	56	10	21	4	26	4	9	11
Technological forecasting and futurology	10	42	10	11	3	16	2	9	6
International aspects of science policy	5	33	4	12	4	8	4	5	7
Legislation on science and technology	3	39	1	4	2	6	–	5	7
Transfer of technology	—	8	1	1	–	4	–	2	2
Assessment of technology	2	26	1	2	–	5	–	1	2
Science and society/ popularization of science	1	24	1	2	–	2	–	1	—
Out of a total of	20	67	15	33	5	33	4	8	11

1. Figures should not be taken too literally, as they merely indicate institutions which have declared an interest in the field concerned, without specifying their degree of active participation.

Source: Science Policy Research and Teaching Units, Paris, Unesco, 1971.

sciences during Mesopotamian and classical antiquity as well as in mediaeval times. In the study of our century's science, on the contrary, externalist interpretations are to be found more frequently.

Indeed, the gap between the way of thinking of the externalist historian and the sociologist of science is not very wide. This is well illustrated by the personal evolution of one of the most respected and influential sociologists of science, Robert K. Merton. The recent publication in one volume of some of his papers on the sociology of science theretofore scattered in various journals, symposia and other books[1]—together with the introduction to the volume written by another distinguished sociologist and historian, Norman W. Storer—bears witness, as it were, to the necessary transformation of the historian into sociologist. In the early 1930s, Merton's interest was not primarily in the sociology of knowledge but rather in the history of science. But as the years went by, he realized more and more that the development of science cannot be explained without considering the ways in which scientists proceed and behave and how their institutions operate. It has been observed[2]

By 1945, Merton had laid down an approach which identified science as a social institution with a characteristic ethos, and subjected it to functional analysis. This was for a long period the only theoretical approach available to sociologists in this area, and it remains productive and influential today.

By now, the field of the sociology of science has expanded vigorously to embrace a vast variety of scholarly endeavours which extend much beyond the sociological interpretation of scientific developments *sensu strictu*. These studies include, for example, the influence of ideologies on science; competition within the scientific community; motivations in research activities; influence of early environment on the development of the scientist; models of scientific growth; problems in the social organization of science; communication among scientists and technologists; fashion and style in the selection of scientific research topics; scientific creativity and discovery; influence of the academic milieu and of the research

TABLE 37. Institutions concerned with science policy research and education in Europe and North America[1]

	State sector	Higher education	Private sector
U.S.S.R.	------ 20 ------		–
United States	5	56	6
Federal Republic of Germany	2	10	3
France	5	19	9
Poland	------ 5 ------		
United Kingdom	4	20	9
Czechoslovakia	3	1	–
Italy	3	1	4
Canada	4	7	–

1. Total number of units in Europe and North America, 343; total number of individuals involved in science policy research and teaching, 1,772.
Source: Science Policy Research and Teaching Units, Paris, Unesco, 1971.

laboratory on the scientist and his production; science and ethics; the social responsibility of scientists; science and ideology; and many more.

As it happens, the evolution of trends of thought in a novel direction is reflected, with a sort of feedback mechanism, on those very problems from which the early stimuli to explore new grounds had come. Thus, after the separation of the philosophy from the history of science and the filiation of sociology in respect to the latter, the original approaches of thinkers such as Thomas S. Kuhn[3] and John Ziman[4] show the significance of sociological factors on the process of discovery itself. As we shall discuss in Part IV, the developments mentioned acquire special significance when we attempt to evaluate the human implications of scientific advance.

It should be added that, side by side with the developments cited in the history, philosophy and sociology of science, an imposing number

1. R. K. Merton, *The Sociology of Science—Theoretical and Empirical Investigations*, University of Chicago Press, 1973.
2. *Sociology of Science: Selected Readings*, edited by B. Barnes, London, Penguin Books, 1972.
3. T. S. Kuhn, *The Structure of Scientific Revolutions*, Chicago University Press, 1962.
4. J. Ziman, *Public Knowledge—The Social Dimensions of Science*, Cambridge University Press, 1968.

of activities has characterized the related field of science policy in which we include the quantitative assessment of the scientific effort, the planning of goals, priorities and the allocation of resources, the organization, administration and management of R & D, and so forth. The current situation in this area is best illustrated by the results of a recent Unesco survey of science policy research and teaching units (see Table 37). Most institutions concerned with these topics are concentrated in the academic fields, although both the state and private sectors are showing growing interest in the particular aspects that are important for their own activities. Such matters will be further discussed in Chapter 28.

27

Sponsors and performers

The overall system

In almost all countries, the government is the biggest—and usually the most important—sponsor of scientific and technological research and development. In France, for example, the government provides 63 per cent of the funds for R & D, and in the United States, in Canada and the United Kingdom it provides 56 and 51 per cent respectively. The situation is even more marked in the developing countries. Here almost all research, except for that carried out by subsidiaries of foreign firms, is government financed and is carried out either in government research institutions or in the universities.

In some countries, however, the majority of funds for R & D are provided by the business sector. This is particularly true for Japan and the Federal Republic of Germany, where this sector accounts for 66 and 55 per cent of R & D funding respectively, two countries which have been characterized in the past two decades by high rates of industrial expansion but without the burden of heavy government spending on space or defence. This follows logically from the fact that so much of R & D is development, which is almost entirely in the business sector and more accurately thought of as an overhead cost of manufacture. Indeed Freeman's data[1]

indicate that the rate of this overhead is proportional in all countries to the cube of the growth rate of the particular industry. This is very high for aerospace and electronics, lower for fine chemicals and pharmaceuticals, low for the 'low' technologies.

Whereas government may be the main source of finance for R & D, most of the performance is carried out in the business enterprise sector (see Table 38). The figures in italics refer to the results of a survey among member countries carried out by OECD in 1969. Differences in definition of the exact categories used—in particular, the exclusion from the table of the OECD figures for the private non-profit sector—lead to certain discrepancies between the two sets of figures. The general picture is that government spreads its funds fairly evenly between the different sectors; industry carries out most of its own R & D (usually between 80 and 95 per cent), but contracts a certain amount to the universities, as well as to other non-profit institutions such as research associations; universities rely equally on government support and on their own funds, although this part of

1. C. Freeman, 'Research and Development: A Comparison Between British and American Industry', *National Institute Economic Review*, Vol. 20, May 1962, p. 21–39.

TABLE 38. Distribution of total R & D activity by sector of performance and source of funds (percentages)[1]

	Sector of performance								Source of funds			
	Productive enterprise		Higher education		General government		Government		Special	Productive enterprise		Foreign	Other
United States	66	69	16	...	18	14	56	58	—	38	37	—	5
Japan	62	67	27	19	11	12	27	27	—	62	66	0	11
Federal Republic of Germany	63	68	19	17	17	5	47	39	—	53	59	—	—
France	58	55	16	14	26	28	63	62	—	30	32	4	2
Poland	79		12		9		30		36	34		—	—
United Kingdom	67	65	8	8	25	24	51	51	—	45	41	—	1
Czechoslovakia	84		3		13		45		—	55		—	—
Italy	55	55	24	20	21	25	49	41	—	49	50	1	—
Canada	36	37	27	29	36	35	54	54	—	30	29	2	14

1. The discrepancies between the two sets of figures (OECD figures in italics) are partly due to differences in definition. For the Unesco figures, the general government sector of performance is taken to include all R & D activities not covered by the first two categories, while the productive enterprise sector includes private or non-profit institutions mainly or exclusively serving productive enterprises. The OECD figures, however, do not include the private non-profit sector; this is particularly noticeable in the Federal Republic of Germany, where this sector, which includes the Max Planck Gesellschaft, is particularly strong.

Sources: Unesco Statistical Yearbook, 1973, Paris, Unesco, 1974; International Survey of the Resources devoted to R & D in 1971 by OECD Member Countries, Statistical Tables and Notes, Vol. 5, Paris, OECD, 1973.

the picture differs widely between different countries; finally, the non-profit institutions get most of their support from government, provide a fair amount themselves (see below), and obtain the rest from industry (see Table 39).

The different patterns of funding and performance reflect a number of economic and political issues, the latter often of an ideological nature; a government trying to stimulate rapid scientific and technological growth can do this either by financing R & D activities itself, as in France, or by stimulating the private sector to make this investment and buy licences from elsewhere, as in Japan, with its own efforts concentrated on other areas. In the planned economy countries, R & D is held primarily to be the responsibility of research institutions (academies of science) and the 'design institutes' of the productive enterprise sector rather than that of the higher education institutions. This follows from an ideology that sees the value of science and technology being in their contribution to production rather than in their cultural value as academic knowledge. It must be added, however, that for many years in the social-

ist countries a rapprochement between academies and universities has taken place, thus proceeding according to the Western trends. This is brought out clearly in the U.S.S.R., although a lack of comparable definitions that might be used to measure R & D activities make evaluations difficult. Within the Soviet Union, there are clearly defined links between academic institutes, industrial research institutions, design bureaux and factories. As an OECD study of science policy inside the U.S.S.R. showed,[1] however, the original pattern, which worked well in the initial stage of widespread industrial development, is now beginning to create problems, particularly administrative barriers between the R & D system and industrial production, which are in their turn reinforced by the Soviet system of planning.

With regard to the situation in developing countries, it is difficult to make anything more than broad generalizations. The United Nations World Plan of Action points out that

1. Science Policy in the U.S.S.R., Paris, OECD, 1969.

TABLE 39. Intersectoral transfer of funds used in R & D in the United States, 1972 (estimates)

Source of funds	Performers					
	Federal government	Industry	Universities and colleges	Associated FFRDCs[1]	Others	Total (%)
Federal government	4,000	8,050	1,750	775	635	54.3
Industry	...	11,150[2]	65	...	105	40.4
Universities and colleges	...	...	1,060[2]	...	...	3.8
Other non-profit institutions	...	...	175	...	235[2]	1.5
Percentage distribution of performers	14.3	68.5	10.9	2.8	3.5	

1. FFRDCs (Federally Funded Research and Development Centres) are organizations exclusively or substantially financed by the federal government to meet a particular requirement or to provide major facilities for research and training purposes.
2. Includes State and local government funds.

Source: Technological Innovation; its Environment and Management, Washington, D.C., United States Panel on Invention and Innovation, Department of Commerce, 1967.

a country without an indigenous scientific and technological capacity has no means of being aware of its own needs, nor of the opportunities existing in science and technology elsewhere, nor of the suitability of what is available for its own needs.

It has consequently been accepted as part of the 'international development strategy' that the developing countries themselves should spend 0.5 per cent of their gross national product on research and development (the current figure in advanced countries is about 2–3 per cent). The developing countries need to 'acclimate' innovations to their own social, economic and physical conditions; there cannot be effective technology transfer without some such process of acclimation.

Developing scientific and technological potential, however, is not sufficient by itself to ensure economic growth. In the developed countries, for example, R & D has been estimated to account for only 5 to 10 per cent of the costs involved in the successful introduction of a new product or technique; by far the largest portion of costs goes into production processes such as tooling and general engineering. To quote the report of a Unesco conference on science and technology in Asia,[1]

measurements of expenditures do not provide any information about the exploitation of the results of scientific work by the economy through the process of technological innovation. It is reasonable to suppose that, as expenditure reaches size-able proportions, national authorities will see to it that proper use is made of these funds. This, however, has to be assessed, and not assumed.

A further problem is reflected in the paradoxical fact that in developing countries where the R & D expenditure is already very low (on average, about 0.2 per cent of GNP), there often appears a bias towards fundamental as opposed to applied research. A study of science policy in Latin America, for example, showed that 74 per cent of the research being carried out in Venezuela in 1963 was fundamental, 22 per cent applied and only 4 per cent experimental development, while in Mexico almost all research was devoted to the pursuit of pure knowledge. This has been characterized by Professor Varsavsky in terms of 'outward' looking scientific communities in the developing countries. Other students of this problem, however, do not share this view; they think that what is happening is that the developing country performs a fairly reasonable fraction of the research it should have (say, one-half or one-third) but only a tiny amount of the development (about 1 per cent) since it has little command over any manufacturing industry.

A similar bias can be seen in the choices of students in universities, as has already been discussed in the section on education. While not denying the right of developing countries

1. *Science and Technology in Asian Development*, Paris, Unesco, 1969.

to play their part in the world's scientific community—examples of the success that can be achieved here are: the International Centre for Maize and Wheat Improvement in Mexico; the International Rice Research Institute, Los Baños (Philippines); the International Centre for Insect Physiology and Ecology, Nairobi (Kenya); the International Centre for Theoretical Physics, Trieste (Italy) (this is situated in an advanced country but is playing a very significant role in favour of the scientific communities of developing countries)—it is important to stress that the status achieved by this is unlikely, by itself, to improve the economic conditions in the developing countries. It may indeed give the false impression that the problems faced by all countries, both developed and underdeveloped, are essentially the same, and that the same technological tools can therefore be used to deal with both.

The same problem of maintaining the appropriate balance between pure research, and that carried out with short- or long-term applications in view, also affects the developing countries. On average, fundamental research in the advanced countries absorbs between 10 and 20 per cent of the budget, applied research between 20 and 40 per cent, and experimental development between 40 and 65 per cent. In almost all countries applied research is about two times as extensive as basic research, in terms of expenditure: this is largely an artefact

TABLE 40. Distribution of R & D expenditure by type of activity, (percentages)

	Funda-mental research	Applied research	Devel-opment
United States	15	22	63
Japan	38	22	40
Federal Republic of Germany	31	...	...
France	19	31	49
Poland	12	20	68
United Kingdom	(11)	(26)	(63)
Czechoslovakia	15	65	20
Italy	21	41	38
Canada	21	39	40

Source: Unesco Statistical Yearbook, 1973, Paris, Unesco, 1974.

of the definition and of the fact that all countries have about the same balance between all fields of science: for example, high energy physics (basic) and electronics (applied). The amount of development, on the other hand, depends not on the size of the population and wealth, but only on the size of the high technology manufacturing factor (see Table 40).

An important factor in preserving equilibrium is that although, as was pointed out earlier, the growth of R & D expenditure in most advanced countries has begun to level off in recent years, the proportion of this devoted to fundamental research continues to grow. In France, for example, the proportion of the gross expenditure that went to fundamental research increased from 16.5 to 19.7 per cent between 1963 and 1967; in the Federal Republic of Germany, the proportion rose from 22.5 to 23.5 per cent over the same period, and in the United States, from 10.4 to 14.0 between 1963 and 1966. In these countries, the average annual expenditure on fundamental research therefore rose at a considerably higher rate than that of the total R & D budget (20.0 as opposed to 18.5 for France, 18.0 to 13.2 for the Federal Republic of Germany, and 11.2 to 7.2 for the United States). Only in the United Kingdom, where the expenditure on fundamental research dropped from 10.9 to 9.1 per cent of all R & D expenditure between 1964 and 1967, was this situation reversed.

This privileged growth of fundamental research is still growing, apparently more slowly than in the past, but still faster than that of the gross R & D expenditure or the GNP. Two reasons for this have been put forward by Georges Ferné;[1] the first is that

it is difficult to translate intention into fact, and that the desire to give priority to applied research and development has not always met as much response from industrial and allied circles as the authorities would have wished.

He continues:

It must also be recognised that the theoretical desire to limit the growth of fundamental research

1. G. Ferné, 'In search of a policy', *The Research System*, Vol. I, p. 23–55, Paris, OECD, 1972.

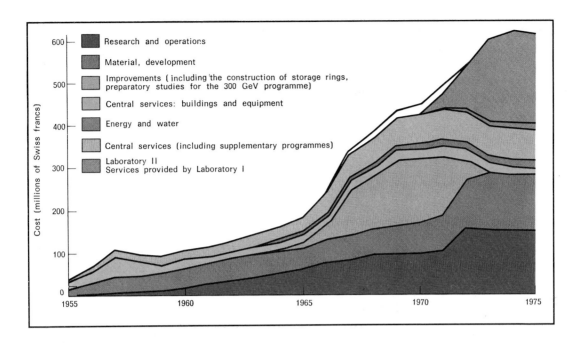

Research and operations

Material, development

Improvements (including the construction of storage rings, preparatory studies for the 300 GeV programme)

Central services: buildings and equipment

Energy and water

Central services (including supplementary programmes)

Laboratory II
Services provided by Laboratory I

FIG. 72. In 1954, several European countries agreed to set up a laboratory designed to work in what was at that time the most promising and most costly field of physics research—the study of elementary particles. Two years later, the laboratory went into operation. Since that date CERN has been identified with the rebirth of European science and the emergence of a new group of European scientists. Twelve countries are members of CERN, and three others are observers. The area occupied by CERN, on the border between France and Switzerland, is half in one country and half in the other. The graph shows CERN's budget, by main items.

runs counter to other science policy options; defence, the exploitation of nuclear energy, the conquest of space or the oceans and the mastery of computer science, are all objectives which entail highly advanced research programmes with a view to gaining fundamental knowledge. The development of higher education, moreover, lends in itself to an increase in the volume of university research.

Another indication of the strength of fundamental research is shown by its increasing importance within the private sector of R & D.

Spending on fundamental research as a proportion of the R & D budget in American industry has increased from less than 4 per cent in 1957 to its current value of 8 per cent, while non-federal spending on R & D in general has increased from about 60 per cent to 85 per cent of the level of federal spending over this period. This and similar observations have led some commentators to remark that certain sections of the business community seem to be taking over the research role traditionally left to the universities; as Jean Jacques Salomon, for example, has pointed out,

the first enzyme synthesis was achieved simultaneously in 1969 in a university laboratory (Rockefeller University) and in an industrial laboratory (Merck). Applied in the one case and fundamental in another, this research can no longer be distinguished except by the institution which houses it.[1]

One aspect not included in previous figures is the proportion of a country's expenditure on research and development which is devoted to international activities. These are particularly important in an age of 'big science' when often the only opportunity offered to small countries

1. J. J. Salomon, 'General introduction', *The Research System*, Vol. I, p. 11–22, Paris, OECD, 1972.

to keep up in the scientific race with their larger rivals is to pool resources on joint projects. A particular case of considerable importance is the European Organization for Nuclear Research (CERN) in Geneva, which was set up to provide European nuclear physicists with the equipment (initially a 600 MeV synchrocyclotron and a 28 GeV proton synchroton) on which they could keep up with the United States (at Brookhaven) and the U.S.S.R. (Dubna), but which no one country could afford to build on its own.

For the smaller countries, international commitments such as CERN and EURATOM can eat up a large part of the R & D budget allocations. Participation in international organization accounts for 15 to 20 per cent of national R & D expenditure on nuclear research, in Switzerland and Sweden, for example, about 30 per cent in Norway and the Netherlands and 50 per cent in Belgium. Even in the larger countries, such contributions are by no means insignificant. Almost half of the United Kingdom's expenditure of £34 million on space programmes for 1968–69 went towards the international programmes being carried out by ELDO, ESRO and INTELSAT, and its annual subscription to CERN alone costs £8.5 million or 2.3 per cent of the total R & D budget.

Another international aspect which must be taken into account is the transfer of technology and, perhaps more importantly, technological knowledge from one country to another. For in many cases it may appear economically worthwhile for a country to buy patents and ideas from outside rather than attempting to build up its own research potential. Japan presents a prime example of how this can be successfully achieved; the buying up of patents, particularly from the United States and Western Europe, enabled it to carry through a programme of economic expansion before the rapid expansion of its scientific and technological potential, which has really only occurred in the past ten years.

The situation itself, however, can pose problems, particularly to those countries which find the buying of patents creates an imbalance on foreign exchange. Calculations made by the Science Policy Research Unit in Sussex,

for example, in 1967 showed that the United States accounted for 12 per cent of the world's payments for technical know-how, licences and payments in 1964, but received 57 per cent of the receipts. But this flow into the United States is diminishing so rapidly as to become a major cause of the inflation and general economic problem of that country. On the other hand, the world's developing countries made 8 per cent of payments, although receipts amounted to only 1 per cent of the total (for Japan the figures were 13 and 1, while for the United Kingdom the technological payments almost levelled out, with

TABLE 41. Conduct of research and development, United States (in millions of dollars)

Department or agency	Obligations		
	1973 actual	1974 estimate	1975 estimate
Defence—Military functions	8,382	8,573	9,581
National Aeronautics and Space Administration	3,085	3,309	3,122
Health, Education, and Welfare	1,844	2,332	2,228
Atomic Energy Commission	1,361	1,429	1,702
National Science Foundation	480	530	654
Transportation	311	358	396
Agriculture	371	393	412
Interior	254	287	510
Commerce	191	210	266
Environmental Protection Agency	181	174	336
Veterans Administration	74	85	94
Housing and Urban Development	58	65	70
Justice	35	52	56
All other	176	132	128
TOTAL (conduct of research and development)	16,802	17,930	19,556
TOTAL (conduct of research)	6,478	7,287	7,607
TOTAL (conduct of development)	10,324	10,643	11,950

Source: J. Walsh, R. Gillette, B. J. Culliton, N. Wade and D. Shapley, 'R & D Budget: The Total is up 10 per cent, but . . .', *Science*, Vol. 183, 15 February 1974, p. 635–41. Copyright 1974 by the American Association for the Advance of Science.

11 per cent of the payments and 12 per cent of the receipts).

The World Plan of Action has suggested that developed countries and international organizations should contribute $1,250 million annually in financial aid, equipment, experts and related technical assistance, to help overcome this problem.

The government sector

Over three-quarters of the research and development programmes carried out by the United States Government in the last few years (see Table 41) were concerned with space or defence, and although this is a decline from the early 1960s, when the figure was nearer 85 per cent, it still represents more than 40 per cent of the total R & D budget. At $16,000 million, it is more than the total spent on all fields of R & D in the Federal Republic of Germany, France, the United Kingdom and Japan put together (see Table 42). A similar situation prevails in the U.S.S.R., where, according to United States estimates, the total Soviet R & D commitment was of about $21,000 million in 1970, out of which between $16,000 million and $17,000 million were spent on defence and space research.

Table 42 indicates the trends in the governmental financing of R & D as percentage of total budget; Figure 73 (p. 309) shows the percentages of R & D governmental financing, classified according to main groups of goals.

These figures give a slightly distorted impression of the amount of effort that is invested into these various fields, since they are based on financial allocations, and both space and defence are notoriously expensive businesses. But they give some idea of how governments interpret their R & D priorities, and indicate some of the broader objectives which are being attained.

The governments of different countries have slightly different reasons for supporting science to the extent that they do. In the U.S.S.R., for example, science and technology have played a central role in government policy since the October Revolution of 1917 and Lenin's Draft Plan of Scientific and Technical Work, drawn up in 1918, in which he de-

fined the basic aim of research in the U.S.S.R. as being to harness science and technology to the economic transformation of the country. In the West, government support for R & D only reached a significant level after scientists had proved their value in the Second World War. In the mid-60s, over half of the United States research scientists and engineers were involved in projects related to military ends; and at the same time, the United Kingdom's new Labour Government of 1964 was proclaiming its faith that the future was to be forged in 'the white heat' of scientific revolution,[1] underlining a general belief in science and technology as mainsprings of social and economic progress. In the developing countries, science and technology are often looked on as expensive luxuries when compared to the massive problems of underdevelopment, and government support, although increasing, remains small compared to that in the developed countries.

Although government is a major source of funds, only in a few countries does it carry out the majority of the R & D it finances, either within ministries or departments or in government-sponsored research institutes. In general, government performance of R & D ranges between 25 and 60 per cent of the R & D that it pays for. Exceptions to this are in the Federal Republic of Germany, where the highly decentralized system of research—in particular through the *Länder* governments and the Max Planck Society—as well as the relative lack of activities in the space and military fields, result in a low level of government performance at 12 per cent. On the other hand, in countries with a low industrial investment such as Greece or Ireland, or with a highly formalized division of responsibilities between different levels of R & D such as Sweden and Canada, the government sector tends to perform most of the activities which the government finances (see Table 43).

There are several reasons to suggest why governments accept responsibility for the performance of R & D, as opposed to the higher education or business enterprise sectors. It is

1. Speech by Harold Wilson at Labour Party Conference, 1964, London.

TABLE 42. Public financing of R & D, expressed as percentage of total State expenses[1]

Country	1961	1962	1963	1964	1965	1966	1967	1968	1969	1970	1971
France	7.0	7.6	8.3	9.7	11.2	11.8	12.5	11.6	10.6	9.6	9.1
United States[2]	10.5	11.2	13.1	12.6	12.7	11.5	10.1	8.9	8.0	7.8	7.5
United Kingdom[2]	8.5	8.1	8.0	8.4	7.9	7.7	7.2	7.2	7.3	7.0[3]	7.1[3]
Japan[2]	5.3	5.5	5.3	5.4	5.6	5.8	5.9	5.9	6.1	..	..
Federal Republic of Germany	3.7	4.3	4.4	5.1	5.3	5.5	5.9	5.9	5.9	6.0	6.0
Belgium	..	..	..	..	3.8	4.6	4.6	4.5	5.1	5.4	5.9
Netherlands	3.9	4.1	4.9	4.9	5.3	5.6	5.9	5.9	5.6	5.7	5.3
Sweden[2]	4.9	5.0	5.2	5.6	5.4	5.3	5.1	4.9	4.2	..	..
Canada[2]	3.4	3.1	3.4	3.6	4.1	3.8	4.0	4.0	3.9	3.5	3.4
Norway	3.0	2.8	3.0	3.1	3.6	3.3	3.6	3.7	3.7	3.5	..
Italy	..	..	1.4	1.8	2.2	2.4	3.1	3.0	3.1	3.7	3.3
Spain	..	..	1.5	1.6	1.8	1.7	1.6	1.8	1.9	..	..

1. According to the OECD National Accounts.
2. Data for the fiscal year 1961/62 calculated as percentage of total current expenses for calendar year 1961, and so on.
3. Figures revised since the completion of Chapter 27.
Source: Changing Priorities for Government R & D, Paris, OECD, 1974.

often in a stronger position to determine and work towards long-term social and economic objectives in applied research—what has been called 'strategic or goal-oriented' research—in a way that industry, with short-term economic objectives in mind, or universities, with their bias towards non-oriented research, are unable to do. Government is also sometimes in a better position to tackle multidisciplinary programmes where such co-operation in the scientific community is often made difficult by the highly specialist divisions and traditions.

The concentration of government activities in the applied research field is clearly shown in Table 44. Here it will be seen that for most OECD countries, between 40 and 60 per cent of expenditure on work carried out in the government sector is allocated to applied research, although as we saw earlier this only accounts for between 20 and 40 per cent of overall spending.

This point is particularly important when we look at current trends. In the United States, applied research has been expanding at a considerably faster rate than either fundamental research or experimental development; according to estimates made by the National Science Foundation, it increased by almost 14 per cent between 1969 and 1972, as opposed to 9 and 4 per cent respectively. The British Government, in its 1972 White Paper[1]

on the reorganization of the British research councils, affirmed its intention 'of extending [the customer/contractor principle] to all its applied research and development', by which a particular piece of applied research will only be carried out if a government department, or other such body, is prepared to 'contract' it to a research institute.

Turning to a more detailed look at the way governments distribute their research expenditure, this can be broadly divided into three main areas: those concerns which have traditionally been the responsibilities of governments, such as agriculture, medicine and defence; those which are aimed at increasing industrial efficiency; and those that include the 'new fields' of nuclear research, space and (most recently) oceanology.

For the traditional fields, each country seems to have its own slightly different way of allocating funds and determining resources. In agriculture, for example, research in the United Kingdom divides into two main blocks, the more applied research being carried out by the Ministry of Agriculture, and the more fundamental by the Agricultural Research Council, which comes under the responsibility of the Department of Education and Science.

1. Framework for Government Research and Development, London, HMSO, July 1972 (Cmnd 5046).

TABLE 43. Government as source of funds and performer of R & D

	Gross national expenditure on R & D ($ million)	Percentage		
		Government-financed (1)	Government-performed (2)	(2) of (1)
Austria[1]	77.6	41.5	7.6	18
Belgium	363.7	46.7	13.1	28
Canada	1,145.4	53.6	35.2	66
Denmark	143.2	49.5	24.6	50
Finland	90.9	40.1	19.3	48
France	2,920.4	59.5	28.1	47
Federal Republic of Germany	4,499.1	43.7	10.4[2]	24
Greece	17.9	71.6	56.9	79
Iceland	2.7	89.2	78.9	88
Ireland	32.8	51.7	44.7	86
Italy	929.0	41.2	21.7	53
Japan	4,041.0	27.9	13.5	48
Netherlands	783.8	40.3	17.6[2]	44
Norway[1]	111.6	56.0	20.3	36
Portugal	23.9	63.4	49.8	79
Spain[1]	77.9	49.2	54.0	110
Sweden	538.3	41.2	12.2	30
Switzerland	473.8	17.4	6.3	36
United Kingdom[1]	2,596.5	51.7	24.3	47
United States	27,527.6	55.2	15.9	29

1. Figures are for 1970.
2. Change of coverage of this sector.

Source: 'International Survey of the Resources devoted to R & D in 1971 by OECD Member Countries', *Statistical Tables and Notes*, Vol. 5, Paris, OECD, 1973.

In France, on the other hand, the National Institute for Agronomic Research (INRA), which comes under the authority of the Minister of Agriculture, has an almost complete monopoly on agricultural research, with about 850 scientists out of its staff of 5,000. As a third example, the organization of agricultural research in the Federal Republic of Germany is much more diffuse, responsibility being shared by the Deutsche Forschungsgemeinschaft, six institutes of the Max Planck Society, the Federal Ministry of Agriculture and the Ministries of Agriculture of the six *Länder*.

A detailed breakdown of government intramural expenditure on research and development reveals the dominance of military research mentioned earlier. Most classified research is carried out under the direction of government institutes of the ministry or department responsible for defence. Private industry and the universities play a considerable part. As a recent OECD survey pointed out,

a great deal of research is done by private firms which produce entirely or partly for defence; in (France and the U.K.) the military research institutions finance a great deal of fundamental university research whose potential application is very hypothetical; the importance of this source of funds for university research should not be underestimated.

Government financing of industrial research follows a similar pattern to defence research, although on a more modest scale. In many countries, governments help to finance research associations whose aim is to enable firms to carry out co-operative research, in a particular field, which they would not have had the resources to undertake separately.

The notable exception to this, however, has been the growth in certain new fields of science, referred to as 'big science'. These include nuclear research, space research and, to a certain extent, oceanology; large funds are assigned to a specialized agency which has

TABLE 44. Current expenditure on R & D in government sector by type of activity, (percentages)

	Basic research	Applied research	Experimental development
Austria[1]	24	58	18
Belgium	21	58	21
Canada	14	56	30
Denmark[1]	21	49	30
Finland	24	50	26
France	15	42	43
Federal Republic of Germany	25	----- 75 -----	
Greece	24	54	22
Ireland	1	57	42
Italy	30	51	19
Japan	20	33	47
Norway[1]	17	44	39
Portugal	9	41	50
Spain[1]	27	39	34
Sweden	3	49	48
United Kingdom[1]	18	32	50
United States	13	36	51

1. Figures are for 1970.

Source: 'International Survey of the Resources devoted to R & D in 1971 by OECD Member Countries', *Statistical Tables and Notes*, Vol. 5, Paris, OECD, 1973.

responsibility for all sectors of the field. Indeed a significant trend in many countries during the last three decades has been the organization of large governmental research institutions oriented towards well-defined goals, particularly in the 'new fields'. 'Big science' was born in the United States with the Manhattan project, and several European countries followed that model, the efficiency of which was clear. What are the major characteristics of the American model? Gabriel Drilhon replies in a recent study as follows:[1]

At its head there is a government agency endowed with ample finance in which scientific advisers meet politicians in order to decide on policy. When a new field is developing, the creation of a single agency for this field may be preceded by the establishment of liaison committees grouping representatives from the various bodies concerned (such was the case with space and oceanology). Research is carried out not so much in Federal laboratories as in university laboratories, in industrial firms and in non-profit institutions. The relationship between the agency and the laboratories is based on contracts either for research or for supplies

which include a certain percentage for research. To borrow a very eloquent image, the scientific infrastructure has been mobilised, but mobilised on the spot.

And the author later comments:

To a very great extent research agencies in the government sector have done what was expected from them. The advantages of locating research in the government sector are, however, accompanied by certain disadvantages. First comes the risk of setting fundamental research in an administrative and political framework whose characteristics conflict with the special requirements of fundamental research; turning scientists into civil servants, inflexible financing and management methods, excessive weight of cyclical factors. The attribution of legal autonomy has often not been enough to avoid this danger and to give research agencies all the genuine autonomy they need. The major problem is that it seems practically impossible to discontinue these agencies once their initial assignment has become obsolete or once other institutions have become more suitable to undertake the assignment.

This pattern is repeated in other countries, particularly those with a large stake in the above fields. France, for example, has an Atomic Energy Commission (CEA), a National Centre for Space Research (CNES), and a National Centre for the Exploitation of the Oceans (CNEXO), each with almost complete responsibility for its own field of research; 60 per cent of the whole French research effort in high energy physics is still located in CEA laboratories, which were originally designed partly to help France back on to her scientific feet after the Second World War.

The situation is slightly different in the United Kingdom. The Atomic Energy Authority plays very much the same role as the American AEC and the French CEA, but the United Kingdom's expenditure on space research is divided equally between the different ministries with an interest in possible pay-offs in the area.

In 'big science' fields, countries such as Belgium, Norway and Switzerland, with a smaller R & D budget, find themselves at a

1. G. Drilhon, 'Fundamental Research and the Agencies in the Government Sector', *The Research System*, Vol. I, p. 145–94, Paris, OECD, 1972.

FIG. 73. Triangular diagram of government expenditure on R & D, by main sectors, in some of the OECD countries. (From OECD, *Changing Priorities for Government R & D*, Paris, 1974.)

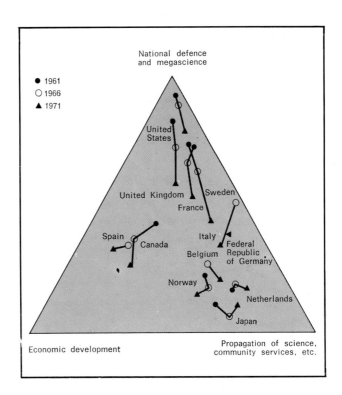

considerable disadvantage. Although trends and problems in the traditional fields are usually the same whatever the size of the country, there are marked differences in the new fields, and usually the newer the field, the greater the difference. Thus the scale of international co-operation in the nuclear field is much larger in the smaller European countries than in their bigger neighbours.

Planned economy countries, in which all research and development comes, by definition, under the government (the State), present a slightly different picture. In the Soviet Union, for example, R & D is mainly concentrated in the large number of 'research institutes' which come under the control of the U.S.S.R. Academy of Science (or under the Academies of the various autonomous republics), which are directly linked to productive enterprise organizations. Soviet science policy will be discussed more fully below.

The situation in the developing countries should also be mentioned. It has already been pointed out that in these countries practically all science is directly or indirectly financed by the government sector, and that there seems to be an emphasis on fundamental research rather than on the applied or developmental aspects. This is reflected in the distribution of government spending. In India, for example, considerably more money is spent on nuclear research than on medical or agricultural problems, representing 33, 7 and 11 per cent of the government R & D budget respectively in 1965–66.

Perhaps not too surprisingly, the scientific community has devised various ways of obtaining support from governments while at the same time getting round some of these problems. Typical methods are to keep a small proportion of the research budget in reserve for unforeseen circumstances, or to seek additional sources of external funding such as from foundations or industry to help ensure continuity and stability in research programmes. The crucial problem, however, has been expressed as 'the integration of fundamental research into the traditional framework which is ill-suited to a system which requires continual revitalization by new approaches and ideas'. In their attempts to attract new talent governments often

309

find themselves in direct competition with large corporations that are usually able to offer considerably better salaries, working conditions and opportunities; delays in introducing computers into health care in some countries, for example, are said to have been due largely to the fact that governments have been unable or unwilling to offer sufficiently high salaries to attract the quality of computer personnel required. As an OECD Advisory Group on Fundamental Research observed in 1966:[1]

If Governments are to attract to their own laboratories fundamental research scientists of quality of the level of those entering university or industrial research, pay and conditions of work must be at a competitive level, and superannuation and other schemes devised in a flexible manner to encourage a substantial and easy flow of research workers between laboratories, universities and industry.

The alternative, and one that some say already exists in various countries, is that government will end up as a safety net for second-rate scientists. One research director has been quoted as saying 'unless radical measures are taken, government research establishments will inevitably drift towards scientific mediocrity; in many areas they seem to have already run down the way of intellectual decline'. This may well be an overstatement of the case, but it appears to reflect certain widespread anxieties among both scientists and politicians.

Industry as a sponsor and performer

'The basic objective of industrial research', according to U. Colombo of the Italian chemical company Montecatini-Edison, 'is the transformation of an idea or intuition into a new, improved, or less costly product'.[2] More than half of all R & D scientists and engineers in the advanced market-economy countries work in the business enterprise sector, and in some countries such as the Federal Republic of Germany, the total is as high as 70 per cent. Montecatini-Edison itself accounted for almost 10 per cent of Italy's total budget for R & D in 1970, and 20 per cent of that spent in the private sector. In the Federal Republic of Germany the firm of Siemens AG spends more

on R & D than the total spent by Norway, Finland and Austria combined, and the spending of large United States firms is very much ·greater; in the United Kingdom, the chemical firm ICI spends almost twice as much on R & D as the British Medical Research Council (about $100 million in 1969–70).

These figures give some idea of the extent of industrial involvement in R & D. It has been estimated that between a third and a half of the R & D of advanced industrialized countries is 'economically motivated', and although industry is the main source of funds in only a few countries, it is by far the most important sector of performance, usually carrying out between 40 and 70 per cent of a country's R & D (see Tables 45 and 46).

The *raison d'être* of industrial R & D, in terms of developing new products, has been outlined as 'to match competitors, to stay ahead of changing customer requirements, and to keep pace with fast-moving technology'. As one business executive replied to a survey conducted in the United States in 1970: 'The problem will be to obsolete our products before someone or something else does.'

Economic factors, however, are not the only ones that are important. Often industry is used by government to help carry out its own programmes; the doubling of industrial R & D performance between 1957 and 1967 in the United States, for example, was largely due to a concentration of funds in the aerospace and telecommunications industries. These were both deeply involved in defence and space explorations, and are thought to have accounted for $12.7 billion of R & D expenditure in 1970.

It is less obvious why industry should want to support fundamental research, as opposed to applied research or experimental development. One reason is that the problems presented by modern complex technology often outstrip available knowledge, and require basic research to be carried out for their solution. A second is that various areas of research often

---

1. *Fundamental Research and the Policies of Governments*, Paris, OECD, 1966.
2. U. Colombo, *In Research lies our Future*, Milano, Montecatini-Edison, 1971.

TABLE 45. Total government intramural expenditure on R & D institutions (percentages)[1]

	Canada	France	Federal Republic of Germany	Italy	Japan	United Kingdom	United States
Central government institutes	79	...	53	27	40	83	...
Research councils and similar institutes	13	...	—	—	22	16	...
Provincial and local government institutes	8	...	47	73	38	0	...

1. These figures represent the total work carried out by these institutions, which is usually slightly higher than that sponsored by government; ... = no data available.

Source: 'International Survey of the Resources devoted to R & D in 1971 by OECD Member Countries', *Statistical Tables and Notes*, Vol. 5, Paris, OECD, 1973.

have fairly immediate economic applications, and conversely innovation in such fields may be impossible without an adequately high level of fundamental research; the plastics and the pharmaceutical fields are two prime examples of this.

Advanced market-economy countries can be divided broadly into two groups, those in which a large amount of industrial R & D is financed from government funds (particularly the United States, the United Kingdom and France) and those such as the Federal Republic of Germany, Italy and especially Japan, where government involvement is relatively low. This is largely a reflection of the relative amount of effort put into the aerospace and military fields. In the United States, for example, over half of the government funds are absorbed into the aircraft industry, and the same is true for both France and the United Kingdom, for which the Anglo-French Concorde project must take a large part of the blame. Other areas where government support is strong in these countries are electronic components and, in the case of France, motor vehicles.

In Japan, government assistance is concentrated predominantly on electronic and other types of machinery, and again in the Federal Republic of Germany and Italy it is electronic equipment (predominantly computers) which receives the most funds.

A further point which emerges from Table 46 is the importance of the industrial sector in smaller countries such as the Netherlands,

Sweden and Switzerland. This may be due to the relatively more flexible academic and institutional traditions in these countries, as well as the greater need to develop entrepreneurial skills in an economic world largely dominated by their bigger relations. The large number of multinational companies based in each of these countries is perhaps an indication of the importance of this last point.

Problems can sometimes result from conflicts between scientific procedures and the demands of industry, particularly when the latter fears that publication might be putting important information into the hands of its competitors. The quality of research in the laboratories of the British firm ICI, for example, is widely recognized in the chemical profession, but much of it is never published. One important area where this secrecy can actually hinder research is in the study of the potential, human and environmental, harmful effects of synthetic products and new industrial processes.

Some firms are beginning to realize, however, that the benefits to be gained from secrecy may not be as great as they had thought. According to Salomon Wald,[1]

some industrialists are already convinced that they have more to gain from first-class research to which their competitors may have access, than from the rigid persistence of a secrecy principle

1. S. Wald, 'Fundamental Research and Technological Application', *The Research System*, Vol. 1, p. 195–258, Paris, OECD, 1972.

TABLE 46. Funding and performance of R & D in the business enterprise sector,

	Percentage of total R & D funding	Percentage of total R & D performance	Expenditure by source of funds (%)			
			Own funds	Government funds	Other[1] funds	Abroad
Austria[2]	57.4	62.0		8.1	91.3	0.7
Belgium	50.4	56.7				
Canada	30.2	37.6	71.8	15.3	7.2	5.8
Denmark[2]	47.2	47.7	97.4	1.3	0.8	0.5
Finland	57.4	59.9	88.1	5.1	5.7	1.1
France	35.8	56.0	59.2	28.8	6.9	5.2
Federal Republic of Germany	54.9	67.4	80.3	18.2	0.2	1.3
Greece	26.0	25.9	100			
Iceland	8.0	1.1	100			
Ireland	42.6	40.8	97.8	1.1	0.1	1.1
Italy	56.8	60.8	81.5	4.5	11.1	3.0
Japan	72.1	66.5	96.8	2.0	1.2	0.1
Netherlands	56.0	60.3	90.0	5.7	0	4.2
Norway[2]	41.4	49.8	67.2	18.3	13.0	1.4
Portugal	30.9	27.6	88.5	2.0	7.9	1.6
Spain[2]	45.1	42.7	94.7	3.1	0.7	1.5
Sweden	55.8	64.7	81.8	14.3	3.2	0.7
Switzerland	81.2	80.4	99.7	0.3		
United Kingdom[2]	42.7	64.5	62.7	32.7	0	4.7
United States	39.3	66.5	58.1	41.9	0	0

1. The column 'Other funds' refers to other enterprises and national funds.
2. Figures are for 1970.

Source: 'International Survey of the Resources devoted to R & D in 1971 by OECD Member Countries', *Statistical Tables and Notes*, Vol. 5, Paris, OECD, 1973.

which would prevent this research from being carried out at all,

particularly when they are trying to get this research carried out by a university department. Machine building companies in the Federal Republic of Germany, for example, now accept that the Machine Tool Faculty at the University of Aachen publishes all research which it is carrying out for them.

Companies differ in the extent to which they believe in concentrating all research in a small number of research centres—these often taking on the air of 'mini-universities'— or whether they prefer to distribute research between different parts of the company. In either case, the company will usually rely on a single research director to co-ordinate all aspects of its research policy, and with a budget that usually runs at about 10 per cent of the firm's annual turnover, his responsibilities are large.

Not all R & D is carried out by individual firms. Table 47 shows the distribution of expenditure between different types of institutions for a number of OECD countries, from which the importance of co-operative and other research institutes, particularly in France, can be seen.

Developing countries have a particular need for industrial research services, since one of the basic problems is that industrial enterprises in these countries are often not large enough or sufficiently developed to create their own internal research activity, and therefore have to rely on 'external' services. A report to the Economic and Social Council of the United Nations[1] pointed out that

1. *Factors affecting the Effectiveness of Existing Industrial Research Organisations in Developing Countries*, New York, United Nations Economic and Social Council, 1971 (E/4960).

TABLE 47. Total intramural expenditure in business enterprise sector by type of institution in which carried out (percentages)

	Canada	France	Federal Republic of Germany	Italy	Japan	United Kingdom	United States
Private enterprises	96	61	96	97	96	92	...
Public enterprises	4	26	—	3	4	6	...
Co-operative and other research institutes	—	13	4	—	0	2	...

Source: 'International Survey of the Resources devoted to R & D in 1971 by OECD Member Countries', *Statistical Tables and Notes*, Vol. 5, Paris, OECD, 1973.

just as careful preparation should go into the founding of a research organisation as into an industrial enterprise. Perhaps even mòre so, for what is involved is not a tangible product with well-defined markets and an assured income, but a somewhat abstract commodity: knowledge.

The World Plan of Action[1] suggests that

the efforts of the developing countries to establish systematic industrial research networks on the national level should be encouraged. In particular, the aim should be for each developing country to participate in the operation of at least one first-class industrial research organisation.

The current situation in developing countries is reflected in the fact that a Unesco survey carried out in 1967 found that less than 10 per cent of R & D scientists and engineers in Africa were involved in industrial research.[2]

A second major contractor for industrial R & D is the higher education sector, in particular, the universities. A large number of university departments carry out research projects for industry; in the United States and the United Kingdom, for example, 6 and 9 per cent respectively of university research were financed from this source. A further indication of the increasing integration between the industrial and the academic community is shown by the fact that in the United States, 70–80 per cent of all university trustees are industrialists.

A recent phenomenon has been the growth of the 'technological' university relying heavily on industrial, rather than government, support. The University of Strathclyde in Scotland finances half of its research income

of $2.4 million from industrial contracts, and in France, the city of Grenoble has managed to combine an internationally recognized centre of scientific excellence based on the university with the establishment of many closely linked science-based companies.

The problems of university research will be discussed in the next section. Salomon Wald[3] has indicated, however, various obstacles that often arise in trying to establish links between universities and industrial research departments. From the university side, these include obstacles created by the higher prestige that has traditionally been attached to pure rather than applied research, the incompatibility of rules and procedures (such as those governing secrecy and publication) between industry and the universities, the psychological 'walls' that often seemed to exist between those working inside and those outside universities, and the strong objections of those who felt the role of the university to be to provide an objective critique of society rather than to act as an extension of the industrial sector.

The obstacles presented by industry include the type of demands for secrecy often included in research contracts, the desire to deal with individuals rather than institutions, and a certain ignorance of the workings of university science.

1. *World Plan of Action for the Application of Science and Technology to Development*, New York, United Nations, 1971.
2. *Survey on the Scientific and Technical Potential of the Countries of Africa*, Paris, Unesco, 1970.
3. Wald, op. cit

Research in the universities

In most countries of the world, universities are looked upon as the home of fundamental research. In the United States, 55 per cent of all fundamental research is carried out in universities, and in France as much as 65 per cent; conversely, 69 per cent of current expenditure on research carried out in American universities is for fundamental research, 80 per cent in France, and 50 per cent in the United Kingdom. This high concentration of basic research in universities is also true of the developing countries; in India, for example, 70 per cent of all fundamental research is carried out in universities. Only in the planned economy countries, where a major part of all research is carried out in research institutes, do universities play a less important role.

To talk about universities in general, however, implies a false degree of homogeneity. Patterns of education have developed in different ways in different parts of the world. Europe has been the traditional centre of learning in the Western world since the Middle Ages, and traditions established many centuries ago still flourish. University establishments in the Federal Republic of Germany and Italy, for example, are widely distributed throughout each country, reflecting a past lack of political unity. In the United Kingdom, however, the twin centres of Oxford and Cambridge had dominated the scene for many centuries and today still enjoy a relatively higher status above other British universities. In France, the situation of Paris as a mediaeval seat of learning was later reinforced by strong centralism of the Napoleonic system and the city's many universities together with the development of its *grandes écoles*, means that it still occupies the supreme position in French education.

Other countries have taken up various aspects of the European model, but have adapted it to their own conditions. In the United States, for example, strong anti-centralist feelings and the traditional Jeffersonian distrust of federal control led, more than a century ago, to the establishment of universities on a State basis; federal support for science and technology was concentrated almost entirely, up to the end of the last century, on the agriculturally based land grant colleges, established by the Morrill Act of 1862. In Russia, the traditional European pattern of universities was completely redesigned after the October Revolution of 1917, being replaced by a system of 'higher education institutes' and parallel 'secondary specialized institutes'.

The idea that a university should be not only a centre of learning, but one of research too, first became established in the Federal Republic of Germany at the beginning of the nineteenth century. The great prestige and influence enjoyed by German universities at this time has been credited to the ideas of Wilhelm von Humboldt. In a memorandum written some time between the autumn of 1809 and the autumn of 1810, and which has been referred to as the 'charter of the modern German university',[1] von Humboldt wrote that the essence of intellectual institutions (*wissenschaftliche Anstalten*)

manifested in the individual, consists of the combination of objective scientific and scholarly knowledge with the development of the person; in institutional terms, this essence lies in the articulation of the mastery of transmitted knowledge at the school stage with the first stages of independent inquiry. In other words, the task of these intellectual institutions is to effect the transition from the former to the latter.

The success of this formula in the development of educational techniques was indicated by the number of universities both in Europe and elsewhere that soon took up the same pattern. Whether this success was due to the correctness of the formula itself, or whether, as some have suggested, it was due to the economic and social pressures of the time, is open to some dispute. According to the sociologists J. Ben-David and A. Zloczower, the success of these educational techniques was a result of 'the unintended mechanism of competition which exploited rapidly all the possibilities for intellectual development open to the universities . . .'. They add:

1. W. von Humboldt, 'Ueber die innere und äussere Organization den Höeren wissenschaftliche Anstalte zu Berlin', *Gesammelte Schriften*, Berlin, B. Behr's Verlag, 1903. Translated in *Minerva*, Vol. 8, 1970, p. 242–51.

the freedom and prestige of the German universities seemed to be safest when the university was kept isolated from the different classes and practical activities of society; it pursued therefore a policy aimed at the preservation of an esoteric and sacred image of itself.

Whatever the true situation, Humboldt's idea of the importance of combining teaching and research in one institution still forms the backbone philosophy of most university research departments. The rapid expansion of higher education since the Second World War, however, and the increasing responsibilities for teaching that this has placed on the universities, has resulted in growing tensions within the traditional university structure.

In some smaller countries, for example Sweden, Norway and the Netherlands, these tensions were originally encountered in the establishment of technical universities. These have a strong bias towards industry and the practical application of technical knowledge; unlike the *grandes écoles* of France, whose aims are primarily to prepare students for high posts in education and the military and civil services, the technical universities have a strong research tradition. In Sweden, for example, 20 per cent of all R & D scientists and engineers come from technical universities, as opposed to only 4 per cent from traditional universities. The ability of the technical universities to meet the requirements of modern industry, both in research and in the production of suitably qualified graduates, has occasionally left the traditional universities in these countries with outmoded ideas and institutions ill-equipped to meet the new requirements of society.

In the Federal Republic of Germany, France, Italy and the United Kingdom, the traditional universities have successfully adopted von Humboldt's dictum, and have remained the main centres of both research and teaching. This has been achieved, however, only by setting up peripheral systems concerned solely with research. In the United Kingdom, for example, the research council

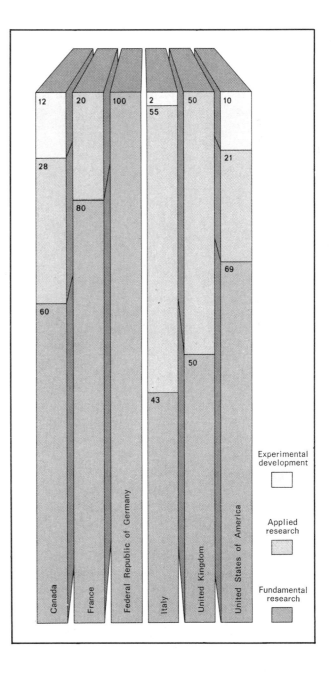

Fig. 74. Percentages of current expenditure on R & D activities in the higher education sector in some of the OECD countries. It is significant that, in the Federal Republic of Germany, 100 per cent is spent on basic research. (From OECD, *International Survey of the Resources Devoted to R & D in 1971 by OECD Member Countries.*)

TABLE 48. Current expenditure on R & D within the higher education sector by type of activity, (percentages)

	Canada	France	Federal Republic of Germany	Italy	Japan	United Kingdom	United States
Basic research	60	80	100	43	...	50	69
Applied research	28	20	—	55	...	50	21
Experimental development	12	0	—	2	...	0	10

Source: 'International Survey of the Resources devoted to R & D in 1971 by OECD Member Countries', Statistical Tables and Notes, Vol. 5, Paris, OECD, 1973.

system was established by the setting up of the Medical Research Council in 1920 and the Agricultural Research Council in 1931.

There are now five research councils, the original two having been joined by the Natural Environment Research Council, the Social Science Research Council, and the Science Research Council in 1965, following the dismembering of the Department of Scientific and Industrial Research.

In France, the National Scientific Research Centre (CNRS) was set up in October 1939, being given in 1945 the responsibility to develop, orient and co-ordinate all French science. With the rapid increase in the number of ministries supporting scientific research during the post-war period, this became an increasingly complex assignment, and in 1958, soon after the formation of the de Gaulle government, the General Delegation for Scientific and Technical Research (DGRST) was set up and made responsible to the Prime Minister and the Minister for Science for the staff work associated with national science policy. The CNRS and the DGRST are important sponsors of research outside the universities in France; the former, like the British research councils, also carries out a large amount of its own research, at present employing 218 research directors with the equivalent status of university professor and 3,100 attachés out of a total research staff of 5,689. The authorized budget for CNRS in 1972 was $50 million.

The Max Planck Society (MPG) in the Federal Republic of Germany, although a private foundation, exists almost entirely on state (Länd) and federal funds, and occupies a position similar to that of the CNRS, em-

ploying 160 institute directors, 200 researchers and 1,800 assistants in 1969. The MPG was established in 1949, when it was given the task of carrying on the work of the Kaiser Wilhelm Gesellschaft, which had been founded in 1911 in an attempt to make up for some of the shortcomings of university research being experienced in Germany at that time. The main function of the MPG has been the establishing and financing of research institutes of which there are now over fifty. The MPG and the German Research Association (DFG) share the responsibilities of the five British research councils, the latter being primarily responsible for aid to universities. Its annual budget in 1967 was $42.1 million.

The original aims of all three peripheral systems was to provide academic scientists with an opportunity to carry out research without the burden of teaching and other academic responsibilities. Although supplementary to and integrated into the university system, they have occasionally performed the additional task of protecting scientists from the political upheavals to which a number of universities have been subject in the past few years.

The system of university research is more complex in the United States. Only since the last war has the United States Government been prepared to support fundamental science to any great extent, and federal support for university research comes mainly in the form of research contracts given by particular government agencies, as the AEC, NASA and some departments, such as the Department of Health Education and Welfare and the Department of Defense, to individual universities. The National Science Foundation was established in 1950 with the responsibility of protecting and supporting both fundamental research and higher education. In 1970 the foundation paid $126 million for R & D in the universities, 14 per cent of the total from all government agencies; 19 per cent of that total came from the Department of Defense.

Fundamental research still forms the most important part of American university research. Between 30 and 70 per cent of research carried out at universities is fundamental research, although overall this only accounts for between 10 and 25 per cent of a nation's total R & D spending. The low figure of 22.9 for the United States again reflects the fact that most research here is carried out in the private sector.

The pattern of the different disciplines in a country is often a result of university organization. The tradition in France or the Federal Republic of Germany, for example, has been that universities have been divided into faculties under the control of a single professor, who occupies what is known as the 'chair'. These chairs have often been established for centuries, and reflect earlier and often outdated divisions of the academic field. In the United Kingdom and the United States, there has been a stronger tendency to organize universities around departments, each of which will have a number of professors representing different branches of the field (a physics department, for example, might be divided into theoretical physics, particle physics, plasma physics, etc.). This model is now being adopted by other countries, in particular by France since the disturbances of May 1968; in the latter case it is still too early to tell how successfully deeply engrained traditions can be overcome.

Recent trends in the pattern of scientific research are posing new problems in university organization. Factors that are becoming increasingly important include the desire to orient research more directly towards social and economic goals, the need for a multidisciplinary approach to certain types of problems for example, problems in the sciences of the brain and a need for a high level of specialization in particular institutions. Some of these require a flexibility in the university system if it is to meet the new demands being placed upon it.

Funds for university research received from the 'business enterprise and non-profit' sector have raised a certain amount of controversy in recent years. These include research founded by charitable foundations, and university research funded for various reasons by business interests. In the United States, universities received $65 million in 1970 from industry to carry out a wide range of research projects, from the most fundamental to the development and analysis of direct applications of technology; in the United Kingdom, 20 per cent of all funds for post-graduate research and 12 per cent of funds for post-graduate training came from industry in 1969.

The advantages gained by industry from close links with the universities are considerable. A former president of Bell Telephone Laboratories has written that 'enlarged, well-supported and effectively performed basic research in our academic institutions is of vital importance to the immediate interests of the nation's industries' (although he added that

TABLE 49. Performance of R & D in the higher education sector by source of funds as a percentage

	General government	Business enterprise and private non-profit	Own funds and higher education	Abroad
Austria	97	3	0	0
Canada[1]	78	8	14	0
France	97	2	1	0
Federal Republic of Germany	96	4	—	—
Italy	100	—	—	—
Japan	99	1	—	—
Netherlands	99	0.5	—	0.5
Norway	95	3	2	—
United Kingdom	82	9	8	1
United States	64	7	29	—

1. Together with PNP sector.

Source: 'International Survey of the Resources devoted to R & D in 1971 by OECD Member Countries', Statistical Tables and Notes, Vol. 5, Paris, Unesco, 1973.

'at least one half of the nation's basic research should be done by industry, which . . . is performing somewhat less than 30 per cent').[1]

Military and industrial demands are but two examples of the way in which traditional concepts of the university are being put under increasing pressure. An equally important one is the conflicts that are beginning to emerge between teaching and research responsibilities, in particular because the rapid increase in university education that has occurred since the Second World War has created a demand for teaching on a scale previously unknown.

Finally, it is important to stress again the different set of problems faced by universities in the developing countries, although in some cases these problems can be traced back to the same root causes (in particular the isolation of the university from the surrounding society) as those which are now affecting universities in the developed countries.

In many cases, the older traditional type of European university has been used as the model for setting up universities in developing countries. Although these have often proved helpful in raising overall standards of education, and made countries more aware of general patterns of world culture of which they sometimes tend to stimulate the growth of an educated élite, out of touch with community needs.

This is particularly true of science, where a research scientist, brought up in the traditions of his Western counterparts, can find himself torn between the values and requirements of the scientific community, and those of his native country.

Under these circumstances, faculty staff engaged in research, often in relative isolation, tend to follow the dictates and fashions of the world's scientific community, which is overwhelmingly of the developed world. Research work is thus frequently done which is more relevant to the problems of the developed world than to local needs. Overseas training may exacerbate these tendencies.[2]

To help resolve this problem, the United Nations World Plan of Action has recommended that an important goal for the Second Development Decade (1971–80) should be to increase the number of scientists and engineers from both developed and developing countries doing research and development work within the less developed countries. It adds that general policies towards research in these countries 'should emphasise practical problems of production and application rather than basic or fundamental research'.[3]

The words of President Nyerere of the United Republic of Tanzania seem equally applicable to developing and developed countries alike:

What we expect from our university is both a complete objectivity in the search for truth, and also a commitment to our society—a desire to serve it. We expect these two things equally. And I do not believe that this dual responsibility—to objectivity and to service—is impossible of fulfilment.

Foundations and other non-profit institutions

Although the financial contribution of charitable foundations to any country's R & D effort is small compared to other sources of funds, seldom exceeding 2 or 3 per cent of the total R & D budget, their influence on the development of science has been out of all proportion to this figure. Most of the largest are to be found in the United States; the annual budget of the Ford Foundation, at almost $300 million, is larger than the annual budgets of Unesco and the United Nations put together, and a considerable proportion of this goes to support scientific research. Foundations are equally important in Europe. It is unlikely that the European Molecular Biology Organization would have got started without substantial financial support from the German Volkswagen-Stiftung (foundation) and in the United Kingdom it has been estimated that over 10 per cent of all money for medical research comes from charitable foundations.

The success of foundations in promoting scientific research is partly due to their independence, which means that they have been in

1. *Handbook of Industrial Research Management*, edited by C. Reinhold Hehedel, New York, 1968.
2. G. Jones, *The Role of Science and Technology in Developing Countries*, Oxford University Press, 1971.
3. *World Plan of Action . . .*, op. cit.

a position to support new ideas when all conventional sources have been committed to keep existing programmes going. Proof of this is to be seen in the fact that quite a number of Nobel Prize winners have been working on foundation money and that the most important discoveries in genetics and biochemistry were made in laboratories supported by private foundations, especially the Rockefeller Foundation. On the other hand, most foundations rely heavily on the interest from invested capital, which can cause problems during times of economic austerity and lend a certain insecurity to funding procedures.

Foundations are not the only bodies with responsibilities for research and development that do not fall into the previous three categories of government, industry and higher education. Other types of institutions include independent research institutes relying to some extent on government funds for support, but responsible for their own research policies,

such as the Pasteur Institute in France, which receives not quite $3 million (20 per cent of its budget) a year from the French Government.

A foundation is a legal body set up according to aims expressed in a charter or deed of trust, and administered by a group of directors or trustees. If its aims are considered to be sufficiently 'charitable', usually defined as being to serve the 'public good' in some way, it may qualify for tax relief and other benefits. Most foundations rely on a sum of invested capital, often the bequest of a wealthy businessman or industrialist, which (as in the case of Volkswagen or Ford) can include a large block of company shares.

There are between 150,000 and 200,000 charitable foundations in the United Kingdom, over twice as many as the rest of the world put together, but the vast majority of these have long ago ceased to function in any viable form; only a small percentage, possibly 10,000 to 20,000 could be considered active. The

TABLE 50. Foundations: fields of activity by country (percentages)[1]

Country	Fields of activity									
	Arts and humanities	Education	Medicine	Natural sciences	Physical sciences	Social sciences	Social welfare	Other	General	Total
Austria	23	8	15	8		8		8	30	100
Belgium		18	42			18		12	10	100
Denmark	15	8	8				8		61	100
Finland	25	5	10	5	10	5			40	100
France	23	19	14	4	9	14	4	4	9	100
Federal Republic of Germany	7	20	3		13	10		7	40	100
Greece					33				67	100
Italy	17	4	17		12	42			8	100
Netherlands	11	5	17	5	5		17	17	23	100
Norway	12.5		12.5	12.5		25			37.5	100
Portugal	33	33							33	99
Spain	10	10	10			30	10		30	100
Sweden	3	3	23	6	9	3		9	44	100
Switzerland	10	3	35	3	7	13		7	22	100
Turkey		33		17			17		33	100
United Kingdom	7	7	29	2	7	11	4	4	29	100
Total percentage	11	9	19	4	7	13	2	5	30	100

Within the single areas, Belgium is the highest supporter of medicine (42 per cent of the foundations), followed by Switzerland (35 per cent), the United Kingdom (29 per cent) and Sweden (23 per cent). In the arts and humanities we have Finland in the first place (25 per cent of the foundations), followed by Austria and France (23 per cent in each). The Federal Republic of Germany has the highest percentage of strictly educational foundations (20 per cent), and Italy has by far the highest percentage of foundations strictly limited to the social sciences (42 per cent).

Source: Guide to European Foundations, Milan, Franco Angeli Editore, 1973.

TABLE 51. Major United States foundations and their activities in the field of science

Name and date established	Founded by	Capital ($ million)	Expenditure[1] ($ million)	Major activities in the field of science and scientific research
Ford Foundation (1936)	Henry Ford Edsel Ford	2,901.5	282.6[2]	Development problems, especially population studies for training and research in reproductive biology, and development and testing of contraceptives
Lilly Endowment (1937)	J. K. Lilly Sr, Eli Lilly, J. K. Lilly	777.8	8.0	(Mainly local activities in Indiana)
Rockefeller Foundation (1913)	John D. Rockefeller	757.1	38.9	Development problems, especially hunger, population and environment
Duke Endowment (1924)	James B. Duke	509.8	22.5	(Mainly local in North Carolina, including 32 per cent to Duke University)
Kresge Foundation (1924)	Sebastian S. Kresge	432.6	9.0	Mainly capital grants for building. No support for research programmes
Kellogg (W. K.) Foundation (1930)	W. K. Kellogg	392.6	16.8[3]	Application of knowledge. Education, health and agriculture, especially in North America, Latin America, Western Europe and Australia
Mott (C. S.) Foundation (1926)	Charles S. Mott and family	371.5	14.8	(Mainly local in Michigan area)
Pew Memorial Trust (1957)	Members of Pew family	367.0	6.9	Emphasis on hospitals and medical research
Sloan (Alfred P.) Foundation (1934)	Alfred P. Sloan	302.9	17.0	Interests in science, technology and related problems in society. Special programme in neuroscience
Carnegie Corporation of New York (1911)	Andrew Carnegie	283.4	14.0[2]	Broad interests in education. General for improvement of teaching and research

1. Total expenditure by United States foundations (1969) was $1,844 million; this represents about 6.5 per cent of assets of 5,346 foundations. Average grants were $20,000 for large foundations and $5,000 overall. Total grants for science of $10,000 or more (1970), $93 million; this represents 12 per cent of the grants made of $10,000 or more.
2. Financial year ending 30 September 1970.
3. Financial year ending 31 August 1970.
Source: *The Foundations Directory, Edition 4*, New York, The Foundation Center, 1971.

United States has about 26,000 foundations, the Netherlands has the same number, and Switzerland has 16,700. There are 3,800 foundations in the Federal Republic of Germany; and France, possibly as a result of a post-revolutionary atmosphere that distrusted private activity that claimed to be for the welfare of the people, comes at the bottom of the list with 250. In Japan, where the tradition of the foundation is less marked, the Japan Society for the Promotion of Science is specifically concerned with the furtherment of research.

The Japan Foundation and the International House of Japan are actively involved in exchange abroad, although not specifically in science and technology.

A large number of foundations contribute funds for scientific purposes, particularly research. A survey carried out in 1972 of 296 major European foundations expressing an interest in cultural affairs found that almost 30 per cent of the foundations were interested in at least one area of science, more than twice as many as were interested in any other single category; in Switzerland and Sweden, two countries with a relatively underdeveloped peripheral research system, the proportion was even higher (see Table 50). The United States reveals a slightly different pattern, with the majority of funds going to education (a far greater proportion of American students rely on scholarships and bursaries to get them through university than their European counterparts). However, two separate surveys have shown that the proportion of grants given for scientific projects is still high, being over twice that given to all branches of the humanities.

Most foundations finance research being carried out in universities and other research institutes, as already noted, rather than doing it themselves (although this is also done, for example, by the Agnelli Foundation in Italy, which conducts studies on the problems of industrial society). In the United States almost 6 per cent of all R & D is financed by 'non-profit institutions', nearly three times the amount financed by private industry. This is estimated to have come to $175 million in 1972. Besides direct research grants, foundations also support science by helping to finance science education, by promoting travel fellowships and other such awards, and by building research facilities and supplying equipment. The Alexander von Humboldt foundation in the Federal Republic of Germany, for example, awards about 30 per cent of its research fellowships to applicants from eastern and southeastern Europe.

Table 51 lists the ten major United States foundations and their interests and activities in the field of science. Some of these, such as the Lilly and Duke Endowments, are mainly restricted to local activities, in these cases to the two states of Indiana and North Carolina respectively. Others have world-wide interests, and some, such as Ford and Rockefeller, operate large programmes on an international scale. These last two, for example, set up the International Rice Research Institute in the Philippines and the International Centre for Maize and Wheat Improvement in Mexico in the early 1960s; 10 per cent of all the wheat and rice acreage in developing countries has already been planted with new varieties de-

veloped at one of these two centres, and the 'green revolution' for which the agronomist Norman Borlaug won the Nobel Peace Prize in 1970, was based on work carried out in Mexico since the 1940s with money from the Rockefeller Foundation.

Another large domain in which the Ford Foundation plays an important role, in line with its declared aims of advancing 'the public welfare by trying to identify and contribute to the solution of problems of national and international importance', is that of population research and family planning. In 1969, the foundation spent $8 million on population studies alone, and over $10 million on training and research in the problems of human reproduction, and the development and testing of contraceptives. Much of this money was spent on field trials to evaluate the effectiveness of birth control campaigns in developing areas, especially Puerto Rico and elsewhere in Latin America.

Other funds similarly have selected their own specific areas of interest. The Alfred P. Sloan Foundation, for example, is one of the main sources of funds for social science research in the United States, while the Carnegie Corporation concentrates on the development of teaching and education.

The proportion of European foundations which support activities in science, particularly research, is considerably higher than in the United States partly due to stronger traditions of Government support for welfare areas such as health and education (see Table 52).

There are many advantages of obtaining support from a foundation for scientific research. Most of these spring from the financial and scientific independence of the foundations. The first of these means that foundations are only accountable to themselves in the way that money is distributed (provided it lies within the legal framework of the foundation's charter or trust deed); with no government or industrial interests breathing heavily down their backs, foundations are less interested in an immediate return, whether social or economic, from the money they invest in science. Scientific independence, i.e. the fact that most foundations are free to decide which

TABLE 52. Major European foundations and their activities in the field of science

Name and date established	Country	Assets ($ million)	Annual expenditure ($ million)	Main activities in the field of science and technology
C. Gulbenkian Foundation (1953)	Portugal	303.7	15.8	General educational and scientific purposes (in 1967, 45 per cent spent on education and 14 per cent on science)
Volkswagen Foundation (1961)	Federal Republic of Germany	268.5[1]	27.0	Development of science and technology through promotion of research and higher learning. Projects 'to rationalize and modernize the sciences and education' (16 per cent of expenditure on scientific research)
Krupp Foundation (1967)	Federal Republic of Germany	125.0[1]	—	Promotes research, science, technology, education, etc., especially television in German schools and research at Dortmund University
Nuffield Foundation (1943)	United Kingdom	82.2	5.3	Supports medical and scientific research. Also major programmes in science education in the United Kingdom
Bosch Foundation (1921)	Federal Republic of Germany	64.7[1]	1.6	—
Carl Zeiss Foundation (1889)	Federal Republic of Germany	50.0	3.7	Support of natural sciences and mathematics
Carlsberg Foundation (1876)	Denmark	42.3	1.1	To contribute to the growth of science in Denmark. Support for laboratories and research
Juan March Foundation (1955)	Spain	28.9	0.4	Promotion of education in Spain
Fritz Thyssen Foundation (1959)	Federal Republic of Germany	25.0[1]	2.0	Promotion of scientific research and training
V. S. Foundation (1953)	Federal Republic of Germany	18.7	0.7	General charitable purposes
Nobel Foundation (1900)	Sweden	18.1	0.7	Award of annual prizes. Organizes 'Nobel Symposia'
Isaac Wolfson Foundation	United Kingdom	—	3.0[2]	Supports primarily scientific and technological education
Bank of Sweden Tercentenary	Sweden	—	3.3	Promotion and support of research in Sweden
Alexander von Humboldt Foundation	Federal Republic of Germany	—	2.7	Awards scholarships for research
Wellcome Trust	United Kingdom	—	2.4	Supports research in human and animal medicine and the history of medicine
Giovanni Agnelli	Italy	8.3	0.2	Promotes research towards economic, scientific and cultural development of Italy

1. Plus legal claims to dividends on 36 per cent of the stocks of the Volkswagenwerk A.G.
2. One-third of total of $8.9 million for years 1966–68.
Source: The Foundations Directory, Edition 4, New York, The Foundation Center, 1971.

branches of science they will support, means that grants can be directed to the areas in which they are most needed with far greater flexibility than either of the other two sources of support.

These two factors mean that foundations can play an important part in supporting new ideas which sometimes contain a high level of uncertainty. This power to innovate and to take certain types of changes has paid off again and again; as mentioned above, the 'green revolution' was made possible almost entirely by foundations' funds, while major innovations have occurred in science education through the activities of the Nuffield Foundation in the United Kingdom. Many foundations are now turning their attention to environmental problems; the Limits to Growth study, for example, was financed by the Volkswagen foundation on behalf of the Club of Rome, and a number of foundations are considering ways by which they might execute some of the recommendations made at the United Nations conference on the environment held in Stockholm in June 1972.

It should also be pointed out, however, that foundations have a number of disadvantages. The first is the purely practical one that when a foundation rests almost entirely for its income on the interest from invested capital, it is particularly susceptible to variable trends and perturbation of the stock market. The economic problems faced by the United States at the end of the 1960s, for example, had a considerable effect on the availability of foundation grants for research; at the same time, a general cutback of R & D funding elsewhere led to a general increase in demand for foundation funds. Total grants of 10,000 or more for scientific projects made by foundations in the United States fell from $114 million in 1969 to $93 million in 1970. This problem implies certain long-term insecurity for grant-aided research, and most foundations are unwilling to commit themselves to support projects for more than two or three years into the future.

The second disadvantage of foundation support springs directly from their independence and lack of public accountability. From one side, foundations attempting to support progressive social developments are often accused of having 'political' or 'subversive' intentions, and hence infringing on the privileges of their charitable position; from another side, foundations are often seen as a reflection of the ideological attitudes of those who set up and run them, these individuals often being enmeshed in a country's industrial and commercial superstructure. Both these criticisms (which are in many ways the same criticism voiced by right and left) have an element of truth; one of the main functions of any foundation is to fill the holes left by government policies (for example, early research on cancer or in mental health), and these 'holes' are inevitably subject to ideological definition.

Despite the above criticisms, however, it seems that foundations will continue to play a large and important role in financing the development of science. When one realizes that a large number of the advances that have occurred in biology, genetics and human behaviour (the Kinsey Report) have been supported since the last war primarily by foundations' grants, or that the social sciences might never have 'got off the ground' in the United States without massive support from the Alfred P. Sloan Foundation, the importance of the role of foundations becomes clear.

The second group of institutions that make up the 'private non-profit' sector are those which carry out independent research, but which rely on funds received from elsewhere, primarily government and industry.

An example of this is the Pasteur Institute, named after the founder of modern microbiology, and classed as an 'establishment of public benefit'. The institute was founded in 1836 in line with Pasteur's own words that 'it is desirable to create an establishment for vaccination against rabies', and almost all its work is concerned with pure and applied microbiology. Its annual budget of about $15 million is financed partly by the sales of vaccines and antisera, and partly by direct grants from the French Government. It has produced five Nobel Prize winners altogether and it has a world-wide reputation for the quality of its research, with subsidiary institutes in seventeen countries. The Paris Pasteur Institute employs 800 scientists and technicians in about 100 laboratories.

In the United Kingdom, the core of the 'private non-profit sector' is made up of a number of independent research institutes and laboratories that receive grants from one of the Research Councils. These include fifteen research institutes financed by the Agricultural Research Council, nine freshwater research laboratories supported by the National Environmental Research Council, and five or six institutions, mainly the big charity-aided cancer institutes such as the laboratories of the Imperial Cancer Research Fund in London and the Royal Society of Edinburgh.

An important type of institution in the United States is the independent type of research institute known as the 'think-tank'. These are institutes which undertake research, usually on a short-term basis, under contract from either government or industry. 'Think-tanks' have been called the élite of R & D and exercise a considerable influence on American political life. According to one estimate, almost 10 per cent of the $16,000 million spent by the federal government on R & D in 1972 went to institutes of this type.

The 'think-tanks' set up after the last war were modelled on the Princeton Institute of Advanced Studies, a brilliant invention by Abraham Flexner, and at first they worked primarily on military problems. The Rand Corporation, for example, one of the biggest and most important of the 'think-tanks', was set up in 1948 as an independent organization following the signing of a contract between the United States Air Force and Douglas Aircraft for a study of the general implications of intercontinental warfare. Throughout the 1950s and early 1960s, a large amount of military work of this nature was carried out by a growing number of 'think-tanks', which slipped terminology such as 'strategic deterrence' and 'second-strike capability' into our language. They had a considerable effect on the moulding of America's military policy, such as the regrouping of American forces abroad and the proposed deployment of anti-ballistic missile systems.

There are now about seventy 'think-tanks' which rely on the Department of Defense for more than half of their budget, and twelve of these have contracts guaranteeing annual support. One example is the Institute for Defense Analysis (IDA), directly financed by the Pentagon. As an example of the way in which many scientists are involved in this type of work, IDA has a special advisory committee known as the Jason Committee composed of forty-three academic scientists, thirty-three of whom are physicists. The committee meets for seven weeks every year for an intensive study of significant technical problems related to the national interest. Its fields of interest include oceanology and undersea warfare, strategic missile survivability and penetration, re-entry physics, tactical warfare systems, and laser research and applications.

In recent years, however, the stock of such 'think-tanks' has been falling considerably. A general reaction among the American public, and especially within the scientific community, against the military activities of the United States Government has led to difficulties in obtaining sufficient numbers of top-rate recruits. Similarly, political disenchantment with the effectiveness of the 'think-tanks'—combined with and added to the convulsions that followed the disclosures of the secret 'Pentagon papers' by two former employees of the Rand Corporation—has led to demands in Congress that the federal government should reduce its payments to 'think-tanks' by 25 per cent.

In the light of the uncertainty that they now face, many 'think-tanks', which have in the past relied almost entirely on military contracts, are now joining the hundreds of other contract research institutes, of which one of the most well known is the Arthur D. Little Inc., in Cambridge, Massachusetts, to turn to industrial and social problems. The Rand Corporation, for example, whose total budget in 1972 was $25 million, now spends 35 per cent of this on civilian research, both for government and private industry, the other 65 per cent being provided by its contracts with the Air Force and other military departments. In 1969, the Rand Institute of New York was set up, jointly financed by the main Rand Institute in California, the Ford Foundation and the City of New York. The aim of this new institute is to concentrate on urban problems, and those already looked at include police communications, the need for municipal psychiatric hospitals, and the frequency of fires in the city.

In a similar manner, the Hudson Institute, founded in 1961 by Herman Kahn, whose studies of the effects of nuclear war had led him to 'think the unthinkable', has now turned to environmental problems, and is carrying out a four-year study on the future of the collective environment.

Whether America's socio-economic problems are appropriate to the techniques previously used to decide military strategy remains to be seen.

As one science writer has written,

there are a number of sceptics who think that sophisticated mathematical techniques, such as games theory, or analytical methods such as the Delphi technique, which gave good results when the experts used them to predict the number of rifles, vehicles, missiles and boats in the possession of each side in a conflict simulation, will no longer be as valuable if one decides to apply them to events which are not quantifiable in advance, such as the number of criminals, the arms and the potential victims in a given block of flats.[1]

In 1972 the United States created an Office of Technology Assessment after having passed, in 1969, the National Environmental Policy Act. This law requires an assessment and full public disclosure of the environmental impacts of government projects; the assessment must accompany any proposal submitted through the decision-making process. There has been, concurrently, a general realignment of science policy in a trend away from large-scale 'prestige' projects (such as the exploration of space) and a trend more directly toward social and economic goals. In view of these developments; it seems likely that the future of most 'think-tanks'—as, indeed, of contract research institutes in general—will be centred more and more about social and environmental problems.

1. D. Shapley, 'La crise des "think-tanks"', *La Recherche*, Vol. 3, p. 821–7, Paris, 1972.

28　Science policies

In the previous two chapters we have discussed the resources, both in financial and manpower terms, that are available to science. We have also looked at the bodies and institutions that pay for science, and those that carry it out. In this chapter, we shall outline how the resources and institutions are brought together within a national framework, and how resources are channelled to and divided between major fields of science and technology.

There are two types of mechanisms that it is necessary to discuss. The first are the planning mechanisms, the responsibility of which is to try to ensure that plans for science fit in with overall economic and social plans or programmes; the tasks of such bodies can include compiling inventories of available resources and co-ordinating the scientific and technological dimensions of the plans of various government ministries and departments. The second set of mechanisms are those where responsibility lies for taking the major decisions about science, such as how much is to be spent and by whom, and sometimes for ensuring that any decisions made are carried out satisfactorily.

Institutions

In the planned economy countries such as the U.S.S.R., science and technology have always played an important role in planning, in line with Lenin's belief that communist society could only be built on the basis of scientific and technological achievements.

In some market economy countries, such as France (and to a considerable extent Japan), planning for science and technology is strongly linked with plans for overall development.

In other market economy countries—for example, the United States, the United Kingdom or the Federal Republic of Germany—no overall plan is prepared for science, as it is believed to be difficult to solve the fundamental problem of research policy of how to make the best possible use of resources of government and industry through rigid planning schemes. In such countries, each ministry, government department, or, in the case of the United States, agency, draws up its own plans for science, which are usually submitted independently to the major decision-making body, although coordination is achieved by interdepartmental discussions; centralized planning efforts tend to be confined to collection of data on resources and the prediction of future trends and requirements. This is also true for many smaller countries, such as Belgium or the Netherlands, not necessarily because it suits their economic system, but because in these countries the economy tends to be governed by the fluctuation of international markets and forecasts of market trends. In some cases, such as the European Economic Community, or the Andean Pact/Andres Belló Convention covering five countries in South America, attempts have been made to overcome this problem through regional economic grouping and the co-ordination of the R & D systems of the participating countries.[1]

Planning for science and technology in development has been undertaken in some developing countries. Both the Republic of Korea and the Philippines have announced five-year science and technology plans. More recently, a major planning exercise, involving some 2,000 scientists, engineers, economists, administrators and industrialists, has been undertaken in India. A Science and Technology Plan, covering some twenty-four sectors of the economy has been prepared by the National Commission on Science and Technology, as an integral part of the Fifth Development Plan (1974–78) of that country.

Decision-making functions in almost all countries are assumed by the government, and decisions are taken either personally by the Prime Minister, by an individual minister, or by a council of ministers. Each of these acts on advice provided by ministries, departments or other outside bodies. The central parliament (consisting of elected representatives of the community) bears responsibility for the implementation and subsequent evaluation of the government's decisions. This includes legislation proposed by the government.

All major decisions, therefore, are formally taken at ministerial level with the approval of parliament. Some countries have a single ministry broadly responsible for all areas of science. In the Federal Republic of Germany, for example, the Federal Ministry for Education and Science (known until October 1969 as the Federal Ministry for Scientific Research), is responsible for both fundamental and applied research, the latter including nuclear, space, data-processing and marine research and technology, and the promotion of advanced technology such as transport systems and bio-medical engineering. The federal ministry co-ordinates all the activities of the federal government in the field of science;

this includes both scientific research undertaken directly by the government, and its promotion of scientific activities undertaken by other bodies. A senior official from the Ministry for Education and Science is chairman of the Interdepartmental Science and Research Committee, on which sit representatives from all other ministries, and which acts as a co-ordinating body for the science plans of these ministries.

In other countries, various ministries or departments are responsible for particular areas of science and technology. In the United Kingdom, for example, the Department of Education and Science is responsible for work carried out in the universities and for the activities of the five British Research Councils, while the Department of Trade and Industry is responsible for the more applied aspects of R & D, including areas such as nuclear energy. In the United States, the Departments of Defense, Commerce and Agriculture, the Department for Health, Education and Welfare, and the Department of the Interior, each has its own budget for R & D, as do the independent agencies, including the Atomic Energy Commission, the National Science Foundation, and the National Aeronautics and Space Administration.

In yet other countries, such as Japan and India, while various departments or ministries are responsible for different areas of science and technology, a 'nodal' department also exists. The responsibility of this department—the Science and Technology Agency in the case of Japan and the Department of Science and Technology in India—usually have the following functions. First, to provide leadership in areas of science and technology which are interministerial in coverage. Second, to promote certain major national scientific or technological programmes. Third, to serve as the nodal point in government for issues and problems which are of a policy nature or which commonly affect scientific and technological organizations generally, for example technology policy or international scientific and technological co-operation. Four, to act

1. *National Science Policies in Europe*, Paris, Unesco, 1970 (Science Policy Studies and Documents, No. 17).

FIG. 75. Until the 1950s, the pluralistic system for the organization of science policy was the one most commonly adopted. Each government sector received funds for its R & D activities, and administered them independently. As this led to wasteful rivalry between the different sectors, and overlapping of their fields of competence, this system was universally abandoned. The most widely used alternatives are the co-ordinated and centralized systems, calling for a central body which advises or controls; the concerted-activities system is a combination of the two. The United States, the United Kingdom and France provide examples respectively of the co-ordinated, the centralized and the concerted-activities systems.

as the executive arm of whatever planning or co-ordinating machinery may exist at the official or ministerial levels in order to ensure that the total national effort in science and technology is knit together purposefully.

In almost all countries, various bodies are responsible for providing advice, either to individual ministers or to committees of ministers, or to the cabinet as a whole, on which decisions can be made. These bodies can be either completely autonomous as the Dutch Science Policy Council or the National Council of Science and Technology of Mexico, or part of the executive as the CNR (National Council for Research) in Italy or the National Commission on Science and Technology in India. In the United States, the President's Science Advisory Committee (PSAC) with its executive arm—the Office of Science and Technology (OST)—was until recently responsible

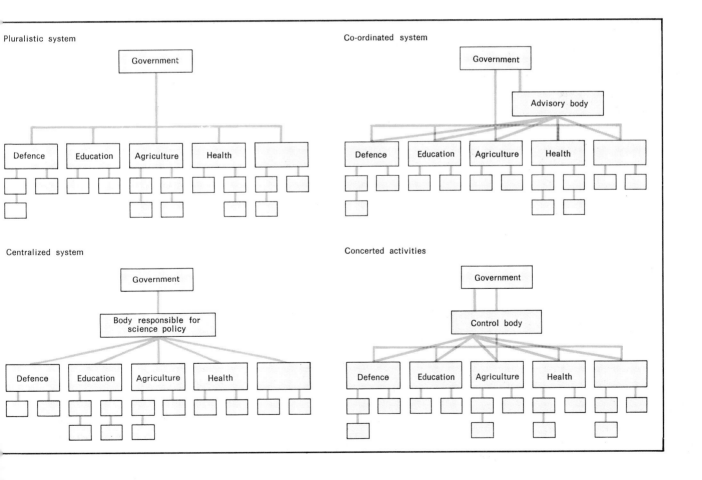

for directly assisting and advising the president on the major policy plans and programmes of the various departments and agencies, and the extent to which selected scientific and technological developments have an impact on national policies in other areas. Now, with the PSAC and the OST abolished, their responsibilities have been assigned to the Director of the National Science Foundation.

The disbanding of the major advisory and co-ordinating body for science policy in the United States might be looked at as a sign of a trend, to be found also in other Western countries such as France and the United Kingdom, indicating a lesser interest than in the past of the body politic in global planning of the research effort. Under the pressure of economic and social needs governments seem to favour sectorial actions to the detriment of an all-round strategy of scientific and technological development, made all the more evident by the accent put in recent years on applied research in areas of public interest such as development of energy resources, transportation, and housing.

Bodies with an important advisory role to government decision-makers include the CNR (National Council for Research) in Italy, the Science Advisory Council (whose chairman is the Prime Minister) in Sweden, the Science Council in the Federal Republic of Germany, the Science Council in Canada, and the consultative Committee for Scientific and Technical Research (CCRST) in France. Each of these tends to be made up of eminent academicians, scientists, technologists and industrialists, thus representing a wide range of views and interests. Although some, such as the CNR in Italy, may be responsible for carrying out government decisions, these advisory bodies have, in general, no executive powers within government. Thus in 1968 the United Kingdom Labour Government turned down a recommendation from the Council for Scientific 'Policy that the United Kingdom should take part in the building of a 300 GeV accelerator at CERN in Geneva (although the government later changed its mind after revised plans had been drawn up). The situation in the socialist countries is very different. First, the advisory function is divided between the State Committees for Science and Technology and the Academy of Sciences. The former advises more on technology policy issues, and problems concerning applied R & D, while the latter deals more with science policy questions, particularly policies and priorities for basic research. Secondly, and more importantly, both institutions combine advisory functions with significant executive powers and responsibilities for actually managing scientific and technological institutions.

Although science advisory bodies are meant to advise the political executive, such advice is usually tendered through the civil service structure. As a result, the instruments embodying executive measures needed to implement the advice provided by such bodies, for example cabinet papers, are usually prepared by groups of officials inside the ministry or ministries concerned. This often leads even an alert and competent political executive to obtaining scientific advice 'filtered' by generalists, especially from the viewpoint of what such generalists perceive to be the feasibility of the measures proposed.

Much as it would be agreeable to think that important decisions in the planning of science were taken on grounds of economic and social priorities, this is rarely the case. In most instances, they are little more than attempts to balance a range of conflicting pressures, including those (for example) from the industrial and military sectors, from the academic community and from those responsible for handing out government money, as well as a need to make as full use as possible of available resources. For example the British Government's White Paper on the reorganization of the British research councils,[1] by which decision-making on programme selection for applied research in medicine and agriculture will be largely removed from the research councils to the relevant government departments, emerged more from a desire to improve the efficiency of the research council system than from any calculated policy change.

There are some bodies, however, which try

1. *Framework for Government Research and Development*, London, HMSO, July 1972 (Cmnd. 5046).

to take a critical look at plans being put forward, and assess the extent to which they are or are not fulfilling a socially useful purpose. Usually these bodies operate at parliamentary level, and again their role is strictly an advisory one. The United States Congress, for example, has recently agreed to set up an Office for Technology Assessment. This acts as a type of 'think-tank' for Congress, advising legislators on the economic and social impact of proposed technological programmes, and suggesting areas where research and development might require strengthening or could be reduced. It is supervised by a Technical Assistance Board whose members include six senators, six congressmen, and a non-voting director.

An important congressional body with direct responsibilities in the planning of science is the Joint Committee on Atomic Energy, which was set up by the Atomic Energy Act of 1946 and which is composed of members of both houses. It is endowed with strong powers to keep a close eye on all American activities in the nuclear field, both for civil and defence purposes, thus exempting atomic energy from the protective cloak of executive privilege. A number of other committees of the House of Representatives hold hearings on government policy; these include the Appropriations Committee, which discusses the president's budget proposals, the Armed Forces Committee, which keeps watch on military R & D, and the Education and Labour Committee, which reviews the impact of technical progress on manpower. The Committee on Science and Astronautics of the United States House of Representatives, and particularly its Subcommittee on Science, Research and Development, have played an important role in recent years as a public forum for discussion of science policy criteria and as a source of information on the working of existing mechanisms.

In the United Kingdom, the House of Commons Select Committee on Science and Technology, which was first set up in 1966, is unique in that although it has no permanent, constitutional status, it undertakes on its own initiative surveys in depth, including public cross-examinations of important

individuals, of selected problems. Its reports and recommendations are then published, and topics covered include the British computer industry, problems of population, and the reorganization of the research councils. Again, however, its recommendations have no legal power and can easily be overlooked by those with responsibility for taking decisions. Many parliaments—including that of the United Kingdom—have select committees set up to discuss proposed legislation in areas covering science and technology.

One attempt to increase contact between parliamentarians and scientists is the Swedish Association of Members of Parliament and Research Workers (RIFO). This was set up in 1959, of whom over half were members of parliament and the rest scientists and technologists from various disciplines. The association meets a number of times each year to discuss questions on the planning of science. Hearings are sometimes arranged with members of the government, officials of the ministries, and other persons concerned.

The case of the People's Republic of China merits special attention, since, especially after the Cultural Revolution, it represents, in more than one way, a departure from the patterns of the organization of science at the national level thus far discussed.[1]

At that time (1966–69), Mao Tse-tung's criticism of the divorce from practice that characterized many of the research programmes of the 120 institutes of the Academy of Sciences led to a reassessment of their function. Many institutes were merged with industrial enterprises, or handed over to provincial governments to enable them to tackle specific localized problems. The Academy of Sciences still retains direct responsibility for about forty research institutes. The time devoted to university-level education is now reduced to two to three years. Students begin with a first-hand experience of production work in the factory or in the farm, then attend university courses for a short period and later return to productive activities to test the validity of their acquired knowledge.

1. M. Macioti, 'Hands of the Chinese', *New Scientist*, Vol. 50, 1971, p. 636–9.

The rural commune and the industrial factory in the People's Republic of China are not only production units, but also centres of training and research. At the government level, education and research nowadays are integrated in a single directive body—the Group for Education and Science—under the direct authority of the Council of Ministers. The general approach is much more flexible than it was before the Cultural Revolution. University research laboratories, for example, often carry out production tasks, in line with Mao's quotation that

besides meeting the needs of teaching and scientific research, all laboratories and affiliated workshops of engineering colleges which can undertake production tasks should do so to the best of their capability.

The so-called 'open door' approach, which seeks to link teaching, scientific research and production into a single system, makes it difficult to talk about R & D as it is referred to in the West. Neither can one refer to basic science since it is emphasized that the only research carried out is that which is relevant to practical social problems.

It is possible to distinguish, however, a number of features of Chinese science and technology. The first is an emphasis on 'worker-peasant' science, derived directly from the needs of agriculture and of small supporting industries. This gives Chinese science and technology its highly decentralized nature, since as far as possible workers are encouraged to develop their own scientific techniques and apply them to specific problems. The second feature is an emphasis on innovation. This can be seen at all levels, and provides an important stimulus for development. According to L. A. Orleans, in a report to the United States Congress,

in addition to the innovations in theory and policies, there has also been an explosion of substantive innovations in China's technology and science. Rather than concentrating all the efforts on applied research and development in the institutes and laboratories of the Chinese Academy of Sciences, both the professionals and the masses have been urged, each in his own way, to experiment and to innovate in order to improve ef-

ficiency and increase productivity on the farm and in the factory.

Despite radical differences with the organization of Western science, China is far from lagging behind in its scientific and technological development. The recent synthesis of the insulin molecule by Chinese scientists is perhaps one of their more spectacular successes. In fields of high technology, such as nuclear research, space research and their military applications, the People's Republic has caught up remarkably quickly with other countries despite a relatively late start. Indeed at the rate at which Chinese science was expanding at the beginning of the 1970s, some observers believed that the People's Republic was emerging to take over Japan's position as the third scientific and technical power of the world.

Goals, priorities and the allocation of resources

Having discussed how money is allocated to science at a national level, we now turn to the way in which scientific resources are fitted to specific goals and objectives. However much one can speak broadly of wide 'social and economic objectives', it is on the level of the individual research project that science is performed. It is therefore important to see how and why these projects are supported, and how attempts are made to co-ordinate a nation's research activity in particular fields.

Within the broad outline of the organization of science and technology as presented earlier (Chapters 26 and 27), resources are allocated in different countries in slightly different ways.

In the United States, each department and agency has its annual budget approved by Congress and distributes its R & D resources as it sees fit. Although more often than not research is contracted to outside bodies, some agencies, in particular the National Aeronautics and Space Administration, run their own laboratories and research institutions.

In many other countries, university expenditure is the responsibility of a single government department. Thus in the United Kingdom, the University Grants Committee

Fig. 76. Curves showing expenditure (expressed in percentage of gross national product) on R & D activities in the OECD countries. (From *The OECD Observer*, No. 76, July-August 1975.)

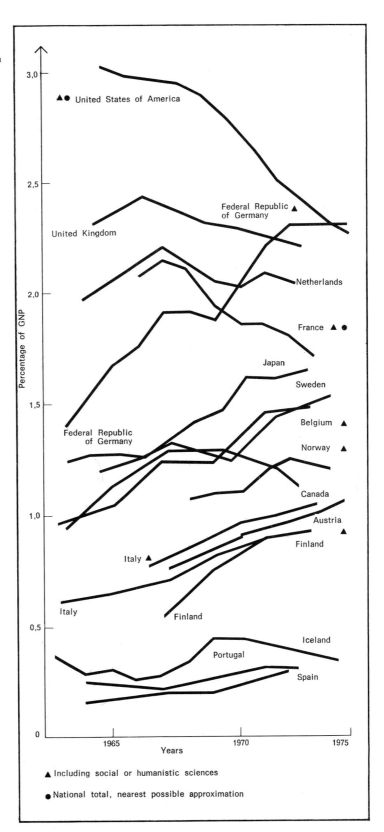

(UGC) (established in 1919 and now under the authority of the Department of Education and Science) receives an allocation from the government on a five-year basis. This is then distributed among the forty-two British universities, whose individual faculties and departments are free to choose their own research topics. The UGC thus acts as a buffer between the government and the universities, and as a formal guarantee of their autonomy. In France, higher education is divided between the *grandes écoles*, whose aim is to produce highly trained professional manpower, and the universities, where emphasis is more on the pursuit of knowledge than that of professional skills.

In the Federal Republic of Germany, most universities and technical universities are funded directly from the budgets of the individual *Länder*. The Conference of University Rectors (DRK) discusses matters concerning research, teaching and scientific training of common interest to all universities. It also serves as an information centre and a discussion forum for member universities, as well as providing a base for governmental discussion at the federal level. Only since an amendment to the constitution in 1969 has the federal government been able to co-operate with the *Länder* in educational planning.

In the U.S.S.R., and more generally in the socialist countries, the academy system, next to the ministerial system, is an important R & D performer. In the U.S.S.R., the Academy of Sciences controls some 240 research institutes and 'guides' the academies of other republics and their research institutes. The focus of the academy system had traditionally been the fundamental sciences.[1]

In a few countries, however, government responsibility for the performance of research is kept to a minimum. The Netherlands, for example, has a Central Organization for Applied Research (TNO), which is responsible for ensuring that applied scientific research 'is made to serve the public interest as effectively as possible'. It consists of a central organization and four specialized organizations, responsible for industry, food and nutrition, defence and health research, respectively. Most of TNO's budget comes from the government,

yet by law it is an autonomous body. It is therefore not subject to direct government control, although a number of government departments are represented on its board. The Netherlands Organization for the Advancement of Pure Research (ZWO) similarly enjoys independent status, while being entirely financed by the government. ZWO is run by scientists, and confines itself to organizing, co-ordinating and providing funds for fundamental research, although some foundations, financed by ZWO, have their own research institutes.

In market-economy countries, R & D activities in the private sector are formally planned by the individual enterprises (companies) themselves, with a minimum of government intervention. Government activities in this area are therefore confined to fiscal measures such as tax incentives for scientific and technological information, providing information and technical services, and offering R & D contracts. Often companies will come together to carry out joint research programmes, or to contract or support research elsewhere. Thus in the Federal Republic of Germany, the Stifterverband der deutschen Wissenschaft (Donor's Association for German Science) is a co-operative entity set up by industry and the professions, and to which over 5,200 federations, companies and individuals belong. Seventy per cent of the funds received as donations from industry are channelled to the German Research Association (DFG), and the rest to other important scientific institutions. In 1967, the Donor's Association had DM.30 million ($8 million) at its disposal, and as these were not earmarked for specific fields of research, the funds are intended to encourage 'flexible and imaginative' investment on the part of the recipients.

So far we have concentrated on the mechanisms that are available for funding and co-ordinating research. It is often possible, however, to distinguish two different types of research funding. The first is what has been called the *laissez-faire* approach, by which funds are awarded as required by the scientific

1. E. Zaleski *et al.*, *Science Policy in the U.S.S.R.*, Paris, OECD, 1969.

community. Research projects are judged as much on their scientific merit and the standing of the scientist who will be in charge as they are on the extent to which they fit into any overall plan for science. The second is a more oriented approach by which funds are made available to pursue a particular scientific goal (such as the search for a cure of cancer) or to stimulate a certain area of science. Both types of research are carried out in universities as well as in government agencies and private laboratories. The *laissez-faire* approach, for example, was the main philosophy behind the British research councils, whose position under the Department of Education and Science was intended, according to the original recommendations contained in the Haldane Committee report of 1918, to place

responsibility to Parliament in the hands of a Minister who is . . . immune from any suspicion of being biased by administrative conditions against the application of the results of research.[1]

In France, responsibility for financing and co-ordinating research outside the university sector rests with the Centre National de la Recherche Scientifique (CNRS). This was established (in its present form) in 1945, and provides France with an infrastructure of research institutes through which it has been able to establish and administer laboratories in newer fields of research which could not be placed within the French university structure. CNRS is a public agency under the authority of the Ministry of Education of France; it has been established 'to ensure the independence of research and facilitate the participation of scientists to its management'. At the time of its creation fundamental research was considered as completely independent from applied research; but since then things have changed, and the purpose, mission and methods of operation of the agency have been reviewed by a committee of experts in science policy.[2] The financing of CNRS falls within the responsibility of the economic planners, upon advice of the Délégation Générale à la Recherche Scientifique et Technique. The Italian Council for National Research (CNR) finances work in institutes, laboratories and centres on a similar basis, as do many of the organizations such

as the German Research Association (DFG) mentioned earlier. In the planned economy countries, responsibility for co-ordinating such non-directed research rests mainly with the Academies of Science, usually also in charge of carrying out research.

Various attempts have been made, however, to channel science and technology towards more specific goals.

In the last three decades, there have been many instances of scientific and technological efforts, both at the national and the international levels, being channelled towards specific goals. The most obvious of these have been government activities in the so-called high-technology areas of nuclear energy, space and defence research, as well as in major scientific programmes such as high energy physics and radio-astronomy. Most countries set up particular agencies or groups within existing ministries or departments to handle issues such as these as they arise. The exploitation of nuclear energy for peaceful purposes, for example, is the responsibility of the Atomic Energy Commission in the United States, the United Kingdom Atomic Energy Authority in the United Kingdom, the National Committee for Nuclear Energy in Italy, the Atomic Energy Commissariat in France, the Department of Atomic Energy in India, etc. Similarly, space comes under the National Aeronautics and Space Administration (NASA) in the United States, the National Centre for Space Studies (CNFS) in France, the Advisory Commission for Space Research (DKFW) in the Federal Republic of Germany and the Department of Space of the Government of India.

Sometimes government departments wish to stimulate research in a particular area without going to the extent of setting up a new statutory body. One example of the mechanisms that can be set up to support oriented research in this way are the *actions concertées* of the French Government for which special funds have been made available since 1960 (through the Scientific and Technical Research Fund).

1. *Report on the Machinery of Government*, London, HMSO, 1918 (Cmnd. 9230).
2. *Groupe d'Étude sur le Fonctionnement du Comité National*, Paris, CNRS, 1973.

These are selective and temporary contracts for particular programmes co-ordinated by the DGRST. Their function is to stimulate research in areas too new to receive sufficient support from traditional scientific organizations, to meet needs expressed by the economic sector, and to stimulate co-operation between scientists and laboratories in universities, those of the technical ministries and those in the private sector.

There were twenty *actions concertées* in 1968, and in 1972, twenty-five *actions concertées* and *actions complémentaires coordinées*. The latter have the same characteristics but a more flexible procedure, in that responsibility for each *action* lies in the hands of a group of experts nominated by the DGRST, rather than a committee nominated by ministerial resolution. Also within the CNRS and since several years a number of *actions thématiques programmées* (ATP) have been launched in order to stimulate research in certain areas through the device of contracts of limited duration; the selection of the 'themes' has been made in accordance with needs envisaged by the national plan. While political and administrative structures are quite different in France and the United States, one may find a remarkable convergence between the ATPs and the Research Applied to National Needs (RANN) programme recently started in the United States. This is administered by the National Science Foundation and, unlike most of NSF's activities, is concerned with problems such as pollution, fuel supplies, conservation and urban problems. The approach to these problems is supposed to be as basic as possible, emphasis, as with the French *actions concertées*, being placed on encouraging interdisciplinary and interinstitutional co-operation.

A third example of this type of policy can be found in the Department of Science and Technology set up in May 1971 by the Government of India; the department performs the function of supporting newer areas of science and technology and also interdisciplinary and inter-agency programmes. The R & D and production programme in the area of cryogenics, a magnetohydrodynamic power generation programme, based on coal gas (as opposed to natural gas), an ocean science and technology programme, and a national remote sensing programme are examples of what the DST is engaged in today. A similar 'lead-role' function is performed by the Science and Technology Agency of Japan.

Another example of the oriented approach to scientific programmes is the policy of the British Science Research Council to select a certain area, within a discipline or embracing a number of disciplines, for 'more favourable than average support' during a given period. This support is given, according to broad principles established by the SRC, 'on the basis of a review of their special potential for advancing basic science, or their economic or community value, or all three'. Such support enables a limited number of university departments, 'selected on the basis of their leadership, past achievement, present expertise, or other relevant factors (for example ability to collaborate with industry)', to concentrate effort on specific areas.

In the Federal Republic of Germany, various appeals by the Science Council led in 1968 to the initiation of a programme of 'special fields of research', with the aim of setting up a network of high quality inter-disciplinary centres. There were 148 such centres financed in 1972, almost twice the number of 77 in 1970.

Principal models for a policy

In the previous sections of this chapter, we have described some of the operational mechanisms concerned with the organization of scientific research. In this section we shall attempt to show how these mechanisms fit into an overall picture, and the extent to which this picture gives some indication—or not, as the case may be—of a somewhat elusive concept known as 'science policy'.

Ideally, of course, it would be nice if every country had a simple set of rules telling it which science should or should not be done at a particular time, and who should do it. We might then be justified in referring to this set of rules as a science policy, and indeed this is the way various attempts have been made to define such a concept. This type of conceptual precision, for example, can be found in the criteria that have been offered by the

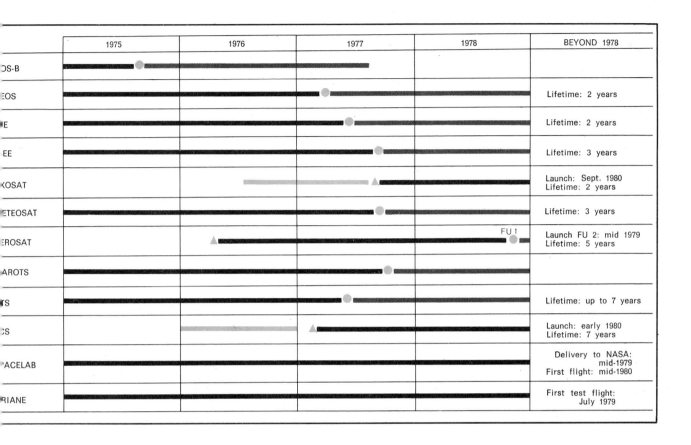

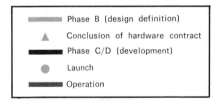

Phase B (design definition)
▲ Conclusion of hardware contract
Phase C/D (development)
● Launch
Operation

Fig. 77. The ten European countries which agreed to co-ordinate their space research (setting up first ELDO and ESRO and subsequently, in June 1975, the European Space Agency) have spent on their joint space programmes approximately the same sum as on their national space activities. The figure shows the projects being carried out by the European Space Agency, specifying the various phases. (*Source: La Recherche,* copyright P. Langereux.)

American physicist Alvin Weinberg as a means of making major scientific choices.[1] In two articles he outlined a set of criteria that might be used to analyse scientific projects competing for limited national resources. The main intrinsic criteria were whether the field was 'ripe' for exploration and whether the necessary manpower was available. Extrinsic criteria, on the other hand, were concerned with the relation of a given field to the rest of science (scientific merit), its possible commercial applications, and its military, prestige, educational and general social potential. Such criteria are now frequently used in making important decisions on scientific matters.

There are many who feel, however, that the unpredictable nature of science makes any attempt at formulating a 'policy' for science a dubious affair. For example, DeWitt Stetten,

1. A. Weinberg, 'Criteria for Scientific Choice', *Minerva*, Vol. 1, 1963, p. 159–71; 'Scientific Choice and Biomedical Science', *Minerva*, Vol. 4, 1965, p. 3–14.

director of the National Institute of General Medical Sciences in the United States, states that

If research is equated with a form of intellectual endeavour that cannot be planned, then the question 'How do you plan research?' translates into 'How do you plan what cannot be planned?' This is clearly a non-question.[1]

According to this view, which is widely held in the scientific community, science policy becomes equivalent to programming for the development of science; although objectives cannot reasonably be stated in strict terms, and hence, according to DeWitt Stetten, 're-search proper remains an unplanned exercise', it is possible to lay down the general criteria according to which science should develop and make plans accordingly.

Whatever view of science planning is taken, the basic problem remains the same. This is the problem of balancing the various pressures on how available resources—both financial and manpower—should be used, and the social, economic and cultural objectives that are expressed through these pressures.

Five reasons have been listed to indicate why such planning for science and technology is becoming increasingly important. These are: (a) the substantial growth of resources assigned to R & D in recent years and the increased concern on the part of governments that these resources be put to effective use; (b) the large number of qualified scientists and engineers now involved in R & D, the 1970 figure of 2.5 million being expected to double during the United Nations Second Development Decade; (c) the importance of science and technology in helping governments tackle problems over a wide range of fields, including the political, social, economic and military; (d) the problems created by the side effects of technological innovation, and (e) the duty of science and technology to forecast, forestall, and, if necessary, repair the harm done by the application of science and technology to development. According to Yvan de Hemptinne,[2]

to intervene 'scientifically', i.e. with foresight, in the destiny of nations is the ultimate goal of science policy of the nations and the *raisons d'être* of the institutions embodying it.

Within this broad context, it is possible to identify two different approaches to the problem of science planning. The first is generally referred to as the pluralistic approach, which divides responsibility both for planning and decision-making among a large number of bodies, and according to which an overall policy only exists as the aggregate of a number of individual policies. The second is the 'centralized' approach, by which attempts are made to design an overall national policy for science, and into which policies for the various areas of science are expected to fit.

Both systems of planning have their advantages and disadvantages. The 'pluralistic' approach, for example, offers the flexibility required to meet a range of operational needs, and to accommodate the unforeseen events that often occur in the development of science with minimum disruption to other sectors. According to a recent report published by OECD,[3] however, this approach has the disadvantage

that it places the long-term and short-term needs of the sector in direct competition with each other. As a result, the longer-term needs may suffer, especially in periods in which the total resources available to the sector are restricted or declining.

A second disadvantage, according to the report, is that goals tend to become static, and new opportunities, or goals based on new knowledge, tend to be overlooked.

The advantage of the centralized system, on the other hand, is that it can lead to a more efficient use of limited financial and manpower resources, and can thus increase the flexibility and adaptability of the overall system. The disadvantage, however, is that centralized planning can be less responsive to the operational needs of the various sectors, occasionally leading to rigid and conservative tendencies in sectoral policies.

1. D. Stetten, Jr, 'Research and Planning', *Science*, Vol. 177, 1972, p. 565.
2. Y. de Hemptinne, *Governmental Science Policy Planning Structures*, Paris, Unesco, 1972 (NS. Rou. 234).
3. *Science Growth and Society: A New Perspective. Report to the Secretary-General's* ad hoc *Group on New Concepts of Science Policy*, Paris, OECD, 1971.

What seems to happen in practice is that each country adopts a balance between these two approaches that fits in best with their own particular approach to economic and social planning. Thus even in the most heavily centralized economies, some of the advantages of decentralizing particular aspects of the planning process are being realized; similarly those with a *laissez-faire* economic tradition are beginning to realize the need for coherent planning, particularly in an era of 'big science', that can only be achieved through a certain degree of centralization. Most large science-based firms have adopted some balance of this nature as their research planning strategy. The OECD report points out that

such firms often have research arms attached to each of their operations or product divisions, in addition to a corporate laboratory for long-range research aimed at developing new products and operations and providing a general technical background for all the firms' activities and decisions.

A further important distinction to be made in the case of those countries which produce national plans for R & D, is the way in which these plans are prepared and submitted for approval to parliament. A recent Unesco survey of European science policies[1] has distinguished two different situations. In the first, R & D plans (for example five year and annual) are prepared and submitted for political decision making by the national body for general development planning as part of the national economic development plan. In this case, the plans are drawn up by a 'committee for science and technology' of the national planning body, in co-operation with the central science policy advisory body and other interested national organizations. In France, for example, responsibility for the nation's five-year plans rests with the Commissariat Général au Plan. The scientific and technological component of the plan is defined and elaborated by the commissariat's research commission, in close co-operation with the DGRST, before being submitted directly to parliament. Similar practices are followed in Spain, Hungary and Romania. In the second case, the national R & D plan is prepared by the central science policy advisory body, in co-

operation with all interested organizations, including the national economic planning body, and submitted directly to the political decision making entity (for example the cabinet) of the government and then to parliament.

However, there are countries, such as India, where a third variant obtains. The Science and Technology Plan is drawn up by the National Committee on Science and Technology, which is a supraministerial apex body, independent of the national planning organization. Although the preparation of science and technology plans and overseeing their implementation, are its prime function, the charter of the NCST also includes many responsibilities of the kind usually assigned to science policy advisory bodies. The Science and Technology Plan is operationally and financially integrated in the national economic development plan, through interaction between the NCST and the national economic planning body (the Planning Commission) followed by final approval by the cabinet. Where its science advisory functions are concerned, however, the NCST reports directly to the cabinet through the Minister of Science and Technology.

In countries such as the U.S.S.R. there is a highly structured plan, of which the programmes of the State Committee on Science and Technology, and those of other specialized State committees, such as the State Committee on Atomic Energy, the programmes of the Academy of Sciences, and of the various specialized academies (for example agriculture), are parts. Overall co-ordination and direction rest largely with the State Committee for Science and Technology, except for defence R & D. The Academy of Science plans primarily for fundamental research, the State Committee for Science and Technology concentrates on industrial R & D, including interministerial programmes aimed at developing specific technologies, while specialized bodies, like the State Committee for Atomic Energy, deal with their respective areas. In countries where no overall plan exists R & D plans are approved by the central science policy body itself, on behalf of the government. In Belgium, for example,

1. *National Science Policies in Europe*, op. cit.

the Ministerial Committee for Science Policy (CMPS) approves plans submitted directly by its own interministerial Commission (CIPS).

A further problem is that many countries depend heavily on foreign investment, which can itself have a large impact on the pattern of R & D, but lies outside the sovereignty of parliament. A study of central government investment in India between 1961 and 1966, for example, revealed that there had been a lag in the R & D concerned with the production of 'investment' goods, and that areas of relative lag in R & D were those characterized by heavy imports and foreign collaboration. Similarly, developing countries tend to find that they are paying more for the transfer of technology from the developed countries than they are for their own R & D; one corollary of this is that much of the research in the developing countries tends to be fundamental research carried out in the universities, with objectives that follow those of the international scientific community rather than the nation's economic and social needs.

The growing importance of science and technology in developing countries, together with an overall scarcity of resources for development, makes some form of planning for science as essential as it is in the developed countries. It would be wrong, however, to feel that this is likely to be achieved merely by the direct transfer of the mechanisms for science-policy making that have been evolved in the developed countries. Rather, each country must develop its own planning mechanisms in line with its own needs and in a way that will enable it to pursue its own priorities (see Part III, Chapter 27).

29 International dimensions of science

Motivations and problems

The rapid growth of science and technology in the twentieth century, and the increasing amount of trade and cultural contact between countries, have made the international aspects of science more important than ever. There are now over 300 international bodies concerned specifically with scientific and technological activities, and almost every large international intergovernmental organization from the United Nations to the Organization for African Unity has important responsibilities in these areas. For advanced countries, expenditure on international activities has accounted for an average of 5 to 15 per cent of the R & D budget in recent years.

In addition to these organizations and to the many exchange agreements that are made between scientists and research institutions in different countries, international conferences and congresses are held every year on scientific topics. There are also international centres where scientists from these countries can work together, such as the laboratories of the European Organization for Nuclear Research (CERN) in Geneva (in which twelve countries participate) or the Joint Institute for Nuclear Research at Dubna (Moscow), to which twelve socialist countries belong.

In the past, the international nature of science has had two main consequences. The first has been that centres of scientific activity have slowly moved from one part of the world to another, gravitating towards that area which provided the most fruitful social and economic environment. The earliest starts were made in Babylon, India, Egypt and Greece, then in Islam and mediaeval Christendom; but more recently it was conditions in western Europe in the sixteenth century, in particular an intellectual environment encouraging a neutral or even exploitative attitude towards nature, that led to the creation of modern science. Starting in Italy and southern Germany (one must place great weight on the origin of printing in Germany and the instrument crafts of Augsburg and Nuremberg), the centres of excellence moved north to France and England in the sixteenth and seventeenth centuries, spreading to Germany in the eighteenth and nineteenth centuries. In the twentieth century, the scientific lead has been taken over almost completely by Europe and the United States.

The second consequence of the international dimension of science has been the movement of scientists from one country to another. There are many stories of how, even when two countries have been at war, scientists from one country have been allowed to visit the other under special protection.

International scientific activities increased dramatically both before the First World War (a steady 8 per cent per year growth from the Karlsruhe Conference and until about 1910) and since the Second World War. Typical of the areas of recent growth are nuclear and space research. In Italy, for example, 69 per cent of its R & D activities carried out with other countries in 1971 was concerned with nuclear research, and 23 per cent concerned with space research; these activities account only for 16 and 4 per cent respectively of Italy's national R & D activities. Even in the Federal Republic of Germany, 59 per cent of its international R & D budget was spent on space research, and only 3 per cent of its national budget. One of the reasons for this is that the size of projects now undertaken, particularly in fields of applied science such as space and nuclear research, have made it necessary for smaller countries to pool their resources if they are to have any chance of competing with the larger ones in these fields.

A second reason for increased international activity has been the growing government involvement in science at all levels, and its importance as a tool for social and economic development at the national and international levels. This is particularly true in fields of industrial applied science, where the scale of projects, for example in the nuclear energy or aerospace fields, has made international co-operation an economic necessity. Co-operation in fundamental science also has its own political value; the agreements between the United States and the U.S.S.R. to carry out joint research programmes in space negotiated during a visit of President Nixon to Moscow are perhaps the best example of this. All aspects of science form an integral part of the external relations of most advanced countries, many of whom have scientific officers attached to their embassy staff in foreign countries. Together with the growth of science at the national and international levels, we have witnessed an increased involvement of ministries of foreign affairs in scientific activities. Often diplomats welcome the possibility of establishing agreements of scientific co-operation, but sometimes they look suspiciously at what appears to them to be an excessive zeal on the part of scientists who are inclined to have a simplistic tendency to equate their own goals with those of the nation as a whole. The relationships between scientists and diplomacy have been recently discussed in a report of the Science Council of Canada.[1]

International co-operation is not without its problems. Expensive collaborative projects can make substantial demands on available resources, and require adequate national co-ordination and activity if a country is to derive full benefit from international participation. In addition, although fundamental research and the results derived from it are seen by some as ends in themselves, the concept of applied research implies some form of social or economic pay-off; this has led in the past to considerable difficulties when countries have begun to ask for some form of fair return on the money they have invested in international projects. Although no less justified than in the case of national projects, the international nature of these projects can lead to political recrimination and other consequences.

Partly for this reason, and partly from a general disenchantment with international co-operative activities without clearly defined aims that seemed to grow spontaneously in the late sixties, many countries have now begun to cut back on their international expenditure on R & D. In the Federal Republic of Germany, for example, contributions to international governmental organizations concerned primarily with science and technology (IGSTOs) fell from 13.5 per cent of the country's publicly financed R & D budget in 1966 to 7.0 per cent in 1970; in Italy it fell from 24.2 to 12.8 per cent over this period, and the United Kingdom from 8.2 to 3.5 per cent between 1965 and 1970. In other countries, such as France, the fall has been less dramatic, while in some, particularly the smaller countries, international contributions as a proportion of the R & D budget continue to rise (see Table 53).

To a great extent, this decline was the result of a reduction in the budgets of two

1. *Canada, Science and International Affairs*, Ottawa, Science Council of Canada, April 1973 (Report No. 20).

major European organizations, the European Atomic Energy Community (EURATOM) and the European Launcher Development Organization (ELDO). These two, together with the European Space Research Organization (ESRO) and the European Organization for Nuclear Research (CERN), accounted for 95 per cent of the expenditure of the eighteen major IGSTOs in 1970.

The level of a country's involvement in international R & D activity is determined by the way its needs are expressed in each of these areas. For the developing countries, the main importance of international co-operation is that it provides them with a means of building up their own potential for science and technology; in the more advanced countries, international co-operation can reduce the cost of maintaining the 'critical mass' often needed to compete in areas of big science or high technology. The CERN laboratories in Geneva, whose large accelerators provide European nuclear physicists a chance to compete with their American and Soviet colleagues, provide a good example of this. In the field of applied science, a joint research project undertaken by scientists of the German Democratic Re-

public and Czechoslovakia to obtain food protein from petroleum is claimed, in an article on scientific and technical relations between socialist countries by Lloyd Jordan,[1] to have saved Czechoslovakia 6 million koruna (about U.S.$400,000) and the German Democratic Republic 3 million marks (this currency was not convertible) in wages, and scientific and technical co-operation is said to have earned 93 million koruna (about U.S.$6 million) for the Czechoslovak economy alone in 1966.

The large number of international organizations concerned with science and technology can be divided broadly into three categories.[2] The first is the operational organizations, characterized by co-operative research and study programmes and, in some cases, by a common budget and central laboratories for R & D activities. Most of these organizations are intergovernmental, and the category includes a number of United Nations bodies. It covers scientific (for example IAEA, CERN) and technological bodies (ELDO, or INTELSAT), as well as those concerned with essentially political, economic or cultural objectives (such as CMEA or OECD), or pursuing military aims (for example NATO). This category also includes international laboratories such as the International Centre for Theoretical Physics in Trieste and other non-governmental bodies, of which the most important is the International Council of Scientific Unions (ICSU).

The second category is organizations concerned with testing and standardization, such as the Bureau for Weights and Measures or the International Organization for Standardization. These are mainly non-governmental, as are most of those making up the third category, centres or bodies for professional contact and exchange of information. These are international organizations concerned with increasing contact among specialists, discussing common problems, or exchanging information. They range from the European

TABLE 53. Budgeted contributions to eighteen IGSTOs by countries of Western Europe, 1966 and 1970[1]

	Total in $million		As percentage of public-financed R & D	
	1966	1970	1966	1970
Italy	53.3	48.4	24.2	12.8
Belgium	16.9	20.4	10.5	7.8
Federal Republic of Germany	85.3	93.8	13.5	7.0
Norway	1.96	2.92	5.1	6.7
Denmark	2.77	4.36	—	5.5
Netherlands	13.2	13.6	6.8	4.4
Austria	1.53	1.83	—	4.2
Sweden	5.9	9.13	2.8	3.6
United Kingdom	68.5	48.6	5.0	3.5
France	76.0	66.4	4.8	3.4

1. These figures refer to expenditure in ELDO, ESRO, IAEA, EURATOM, DRAGON, HALDEN, EUROCHEMIC, ENEA, CERN, EMBC, OEEP, ESO, OECD, Unesco, WHO, WMO, IBWM, EUROCONTROL and NATO civil science programmes.

1. L. Jordan, 'Scientific and Technical Relations among Eastern European Communist Countries', Minerva, Vol. 8, 1970, p. 376–95.
2. M. Macioti, International Co-operation in Science and Technology, Paris, OECD, 1971.

Industrial Research Management Association (EIRMA) to the Pugwash Conference on Science and World Affairs, and their scope may be either world-wide or regional.

Other areas of co-operation which have become increasingly important are bilateral and multilateral agreements between two or more countries, and the activities of the large multinational corporations, which are able to distribute their R & D activities between a number of countries (see the end of this chapter).

In the light of past experience, various attempts have been made to lay down criteria by which projects for international co-operation between two or more countries might be assessed before being embarked upon. The Economic Commission for Europe (ECE) of the United Nations Economic and Social Council, for example, has suggested that any project should correspond to the motivation of the smaller countries which are at a relative disadvantage in fields of big science; that the participation of individual countries in a project, even one sponsored by an international organization to which they belong, should in principle be voluntary; that preference should be given to co-operative research where direct economic competition between participating countries is absent; that co-operative projects should concentrate on a few selected fields; that close attention should be paid to the feasibility of a project when it is in the preparatory stage, and that when the necessary data on inputs and outputs have been assembled, a cost-benefit analysis should be undertaken. Although this list is by no means exhaustive, it points to ways in which mistakes made in the past might be avoided.[1]

Another area in which the international aspect of science is important is in the movement of qualified personnel from one country or part of the world to another—the so-called 'brain drain'. Once a scientist has reached a particular level of experience, there is a natural tendency for him to move to an environment in which his skills will be most appreciated, both financially and otherwise. In 1967, for example, a survey carried out by the United States Council on International Educational and Cultural Affairs found that almost

25,000 professional, technical and other similar workers were admitted to the United States as immigrants. It is worth adding that most brain drain results not from 'pull' but from 'push' due to underutilization or political penalties in the country of origin. Also it is of some interest that there seems to have begun, in some fields, a reverse brain drain out of the United States) in recent years.

Although this flow has been diminishing (and even reversing, partly as a result of the tightening job situation for scientists in the United States, the brain drain still poses a severe problem to those developing countries trying to build up their own potential in medicine, science and technology; the total number of highly qualified personnel migrating from developing to developed countries has been estimated to be approaching 40,000 a year, and as such, is larger than the movement of technical assistance personnel from the developed to the developing countries.[2]

International non-governmental organizations

The first international scientific organization of a non-governmental type to be established was the Universal Society of Ophthalmology, which was founded in Paris in 1861. This was soon followed by the European Association of Geodesy (Berlin, 1864) and the International Meteorological Committee (Leipzig, 1872). Each of these was founded to meet a need that transcended national boundaries, and each was non-governmental in the sense that their members were individuals and bodies from a number of different countries, rather than the countries themselves.

Since the beginning of this century, the number of international non-governmental organizations concerned with science and technology (NGSTOs) has grown rapidly. Accord-

1. *Choice of Research Projects for International Co-operation in Science and Technology*, Geneva, Economic Commission for Europe, Science and Technology, 1971.
2. *Some Facts and Figures on the Migration of Talent and Skills*, Washington, D.C., United States Department of State, 1967.

ing to the *Yearbook of International Organizations*[1] there were 133 organizations concerned with technology and 184 concerned with science in 1972; in addition, a number of bodies in fields such as agriculture, transport, industry and health and medicine also have interests or carry out activities in the research field.

Most of these organizations fulfil the same role on an international level as professional societies and associations do on a national scale. They exist to serve the interests of their members by organizing international conferences and meetings, publishing journals, and generally facilitating contact between members of a particular branch of the scientific community. They also perform a function on the social and political level in that they contribute 'to the normative, economic, organizational and occupational integration of the international system'.

The most important of these international organizations is the International Council of Scientific Unions (ICSU). At the time of writing this comprises seventeen scientific unions, each of which is itself international, and covers almost every field of science, as follows: International Astronomical Union (IAU); International Union of Geodesy and Geophysics (IUGG); International Union of Pure and Applied Chemistry (IUPAC); International Union of Radio Science; International Union of Pure and Applied Physics (IUPAP); International Union of Biological Science (IUBS); International Union of Crystallography (IUC); International Geographical Union (IGU); International Union of Theoretical and Applied Mechanics (IUTAM); International Union of the History and Philosophy of Science (IUHPS); International Mathematical Union (IMU); International Union of Physiological Sciences (IUPS); International Union of Biochemistry (IUB); International Union of Geological Sciences (IUGS); International Union of Pure and Applied Biophysics (IUPAB); International Union of Nutritional Sciences (IUNS); International Union of Pharmacology (IUP). It also comprised the following committees: Scientific Committee on Oceanic Research (SCOR); Committee on Space Research (COSPAR); Scientific Committee on Water Research (COWAR); Committee on Science and Tech-

nology in Developing Countries (COSTED); Committee on Data for Science and Technology (CODATA); Committee on the Teaching of Science; Scientific Committee on Problems of the Environment (SCOPE); Scientific Committee on Antarctic Research (SCAR); Special Committee on Solar Terrestrial Physics.

ICSU was founded in 1919 at Brussels. It has two principal objectives: to facilitate and co-ordinate the activities of international unions in the field of the exact or natural sciences; and to act as a co-ordination centre for the national organizations which are the members of the council. In addition, the council tries to encourage international scientific activity and to promote the development of science and scientific research in individual countries.

ICSU has two categories of members: national members and scientific unions. Any territory having recognized independent scientific activity may be accepted as a national member, and in 1975, sixty-five countries belonged to ICSU, while ninety-three countries belonged to at least one of the scientific unions. The annual budget of ICSU is about $2.5 million for 1975 and of this $250,000 or 10 per cent provided by Unesco, and the rest by members' dues and voluntary contributions.

Services provided by ICSU include an abstracting board, concerned with the abstraction of scientific literature, and the Federation of Astronomical Geophysical Services. Six inter-union commissions are concerned with frequency allocations for radio astronomy and space science, radio meteorology, solar terrestrial physics, spectroscopy, geodynamics and studies of the moon.

One major activity organized by ICSU was the International Geophysical Year (IGY), which took place in 1957–58, and for which a special *ad hoc* committee mobilized during about eighteen months some sixty-five nations of both East and West, about twenty international organizations (both governmental and non-governmental) as well as over a hundred national centres. One of the highlights of

1. *Yearbook of International Organizations*, 14th ed., Brussels, Union of International Associations, 1972.

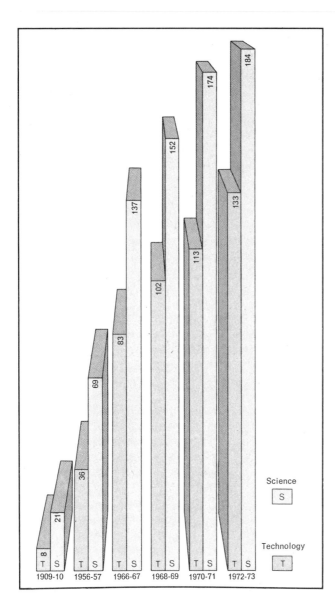

FIG. 78. Increase in the number of
international professional associations, in
science and technology, during this
century. (*Source*: *Yearbook
of International Organizations.*)

the IGY was the launching of the Soviet
Sputnik, and the success of the whole venture,
mainly involving international co-operation on
observation and the taking of measurements,
demonstrated the results that can be achieved
when many scientists from different countries
are co-ordinated in a single project. A similar

international programme of great interest to
the whole of mankind was the International
Years of the Quiet Sun (IYQS) during the
early 1960s and the International Biological
Programme (IBP).

Although ICSU is by far the largest, it
is by no means the only body concerned
with co-ordinating scientific activities at the
international level. Other similar bodies are
the International Council for Medical Sciences,
the World Federation of Engineering Organizations, the International Union for the Conservation of Nature, the International Astro-
nautical Federation, and the World Energy
Conference, each of which plays an important
role in co-ordinating activities in its own particular area.

Besides their purely functional role of providing a channel of communication between
scientists in a particular field and establishing
international standards, international non-
governmental organizations can have other
purposes, particularly that of establishing
contact between countries when political
barriers still remain at the official level. The
nature of science makes it particularly open to
this, and one study has shown that since it is
an area in which consensus appears to be quite
high, participation of professional associations
(by far the most numerous, and probably the
most influential type of NGO) from the United
States and the U.S.S.R. is higher than in NGOs
in which consensus is low, such as those dealing
with international relations, art or religion.[1]

Another area in which there has been
considerable non-governmental co-operation is
that of standards. The International Organization for Standardization (ISO), for example,
has as members the national standards bodies
of sixty countries, of which twenty-four are
European. ISO is responsible for the promotion of international standards as a basis
for developing co-operation in spheres of intellectual, scientific, technological and economic
activity.

Besides the international non-governmental
organizations mentioned above, there are also

1. L. Kriesberg, 'U.S. and U.S.S.R. Particpiation in
   International Non-governmental Organisations', *in*:
   L. Kriesberg (ed.), *Social Processes in International
   Relations*, New York, John Wiley, 1968.

international laboratories where scientists can come and work with colleagues from other parts of the world. In Europe, the most important of these are the Zoological Station of Naples, and the Jungfraujoch scientific station. The Naples Station was built in 1873 by a wealthy German naturalist, Anton Dohrn, near the sea in the city of Naples and is now an important centre for the study of Mediterranean marine biology. The centre is financed partly by the Italian Government and partly by renting out 'research tables' at $3,000 a year to scientists from sixteen countries, mainly in Western Europe.

The Jungfraujoch International Foundation for High Altitude Research (as it is properly called), founded in Berne in 1931, similarly offers scientists unique research facilities, but at an altitude of 3,457 metres. Studies are made of glaciers, the physiology of plants and animals at high altitude, cosmic radiation and the upper atmosphere.

Another international laboratory was recently established in Europe run by the European Molecular Biology Organization (EMBO). The organization was founded in Geneva in 1964 with the aims of promoting the 'development of molecular biology in Europe and the advanced training of young scientists' and to study the creation of a European molecular biology laboratory. At present it has over 200 scientists from fourteen countries as members, and in its early stages was partly financed by a grant of $600,000 from the Volkswagen Foundation. More lately, EMBO's activities have been financed by the participating governments. Until recently, EMBO's work has been limited to arranging scientific exchanges, meetings and educational courses, but in July 1972 it was agreed to go ahead with plans to build a laboratory at a site in Heidelberg (Federal Republic of Germany). There will eventually be two EMBO satellite laboratories in addition to Heidelberg, of which one at Hamburg will be using X-ray beams ten times stronger than any yet available to European biologists.

The International Centre for Theoretical Physics at Trieste should perhaps also be mentioned here. Strictly speaking, because the centre is jointly financed by Unesco and IAEA (each contributing about $225,000), and the Italian Government (contributing $350,000 a year), it is an intergovernmental organization. The centre attracts theoretical physicists from all parts of the world, and plays a particularly important role in bringing scientists from the developing countries into contact with their colleagues from the advanced countries to work on joint programmes. The centre reserves 50 per cent of its posts for scientists from the developing countries, provided that they devote the greater part of their activities to research in their own countries.

A new kind of international scientific organization, oriented entirely to the solution of large-scale social problems, is the International Institute of Applied Systems Analysis (IIASA), created and financed under the auspices of the academies of thirteen nations, both communist and non-communist. In 1973, IIASA began to examine how operational research and other science-based analytical processes can be brought into direct social service in dealing with massive processes such as public transport systems and energy supply, optimization of the supply and use of fresh water, and similar problem areas common to many societies regardless of politico-economic system. The institute is situated in a renovated former country residence of the Habsburg emperors, Schloss Laxenburg, south of Vienna.

Other very important international laboratories contributing directly to the improvement of living conditions in the Third World are the International Rice Research Institute in the Philippines and the International Centre for Maize and Wheat Improvement in Mexico, already mentioned.

Another important category of non-governmental organization includes the groups of scientists set up for mainly non-scientific reasons. Perhaps the best known of these is the Pugwash series of conferences, initiated following the publication of the 'Russell-Einstein manifesto' in July 1955 (see Part IV). A conference of scientists was subsequently held at the village of Pugwash in Nova Scotia in 1957, financed by a Cleveland industrialist, Cyrus Eaton; it is from the village that the conferences take their name. In the first meeting were discussed the potential hazards aris-

FIG. 79. The United Nations system.

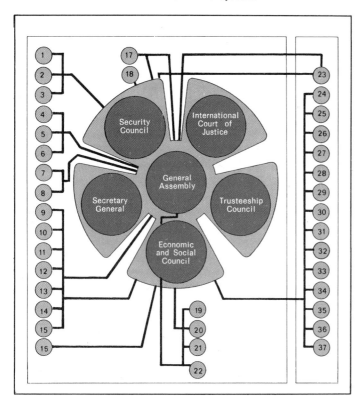

United Nations Organizations and Agencies

1 United Nations Truce Supervision Organization in Palestine (UNTSO)
2 United Nations Military Observer Group in India and Pakistan (UNMOGIP)
3 United Nations Peace-keeping Force in Cyprus (UNFICYP)
4 Main Committees
5 Standing and Procedural Committees
6 Other Subsidiary Organs of the General Assembly
7 United Nations Relief and Works Agency for Palestine Refugees in the Near East (UNRWA)
8 United Nations Conference on Trade and Development (UNCTAD)
9 Trade and Development Board
10 United Nations Development Programme (UNDP)
11 United Nations Capital Development Fund
12 United Nations Industrial Development Organization (UNIDO)
13 United Nations Institute for Training and Research (UNITAR)
14 United Nations Children's Fund (UNICEF)
15 United Nations High Commissioner for Refugees (UNHCR)
16 Joint United Nations - FAO World Food Programme
17 Disarmament Commission
18 Military Staff Committee
19 Regional Economic Commission
20 Functional Commissions
21 Sessional, Standing and **Ad Hoc** Committees
22 United Nations Environmental Programe (UNEP)
23 International Atomic Energy Agency (IAEA)
24 International Labour Organization (ILO)
25 Food and Agriculture Organization of the United Nations (FAO)
26 United Nations Educational, Scientific and Cultural Organization (UNESCO)
27 World Health Organization (WHO)
28 International Monetary Fund (IMF)
29 International Development Association (IDA)
30 International Bank for Reconstruction and Developmet - World Bank (IBRD)
31 International Finance Corporation (IFC)
32 International Civil Aviation Organization (ICAO)
33 Universal Postal Union (UPU)
34 International Telecommunication Union (ITU)
35 World Meteorological Organization (WMO)
36 Inter-Governmental Maritime Consultative Organization (IMCO)
37 General Agreement on Tariffs and Trade (GATT)

ing from the use of nuclear energy, the control of nuclear weapons, and the social responsibilities of scientists.

Two similar conferences were held in 1958. Following the second of these, held at Kitzbühel in Austria, the 'Vienna declaration' which has been called the 'tenet of the Pugwash movement', was published. This contained seven sections: the necessity to end wars; requirements for ending the arms race; what war would mean; hazards of bomb tests; science and international co-operation; technology in the service of peace; and the responsibilities of scientists. The declaration was later endorsed by several thousand scientists from all parts of the world, including many from the Soviet Union.

Early conferences concentrated mainly on the military uses of science, with topics such as arms control, chemical and biological warfare, and disarmament. The association was instrumental in helping to bring about the Nuclear Non-proliferation Treaty, which was signed in 1962, and in setting up the Stockholm International Peace Research Institute (SIPRI) as well as the International Centre of Insect Physiology and Ecology at Nairobi (ICIPE).

Pugwash conferences on 'Science and World Affairs' are now held once a year, organized by a small secretariat in London. There are national Pugwash groups in twenty countries. The 1972 conference held in Oxford (United Kingdom) was attended by over 300 scientists.

The World Federation of Scientific Workers is another association which seeks to unite scientists from all parts of the world in a discussion of common interests. WFSW, which brings together scientific associations of about thirty different countries, was set up following a meeting organized in London in 1946 by the British Association of Scientific Workers. Its aims are

to link together organisations or scientific workers whose activity is directed towards the safeguarding of scientific rights, the improvement of scientists' working conditions and status, and the fullest possible use of science for peaceful purposes.

WFSW carried out continuous activities concerned with the position and actions of scientists, international scientific collaboration and the social implications of science. Membership of its affiliated organizations now totals about 350,000 in twenty-seven countries, most of which are trade unions with corresponding organizations in another twenty countries. The federation has strong links with Eastern and Western trade unions, and provided a valuable channel of contact with scientists in these countries when relations between East and West were somewhat strained in the 1950s and early 1960s.

International governmental organizations

International governmental organizations can be divided into two categories, those the membership of which is independent of either geographical situation or political ideology, and those which are organized on a regional or a subregional basis, or which group together a number of countries sharing a common interest.

The main body in the first category is the United Nations, to which 132 countries at present belong. There are also fourteen international organizations (Specialized Agencies, subsidiary bodies and others) associated with the United Nations, such as Unesco and the World Health Organization. Nine of these have clearly defined responsibilities for science and technology. Countries which belong to one of these organizations need not necessarily belong to the United Nations, although in the majority of cases they do.

Besides the members of the United Nations family, there are also several smaller international bodies with world-wide interests. The budget of these organizations (bodies such as the International Institute of Refrigeration and the International Hydrographic Bureau) is seldom more than $150,000 a year. Each usually has a well-defined area of responsibility, working closely with the larger bodies that share some interest in the same field.

Although a number of countries revealed a growing disenchantment in the 1960s with the activities of international governmental bodies, most of this disillusionment was concentrated on regional agreements (such as

EURATOM or ELDO) rather than on the activities of global bodies (such as the members of the United Nations family). During the decade 1961–70 the total budget of all the members of the United Nations family increased from $142.9 million to $402.4 million, representing an average annual increase of 12.1 per cent.

The major function of the international governmental organizations is to deal with global issues and problems that require a high degree of international co-operation. In doing so, they fulfil a number of subsidiary, but equally important functions. They provide the developing countries, for example, with an opportunity to play a role in global activities that would be denied them in terms of purely economic or political strength; and by transferring aid and assistance to these countries, the organizations manage to redress some of the imbalance caused by present patterns of world trade.

These organizations also provide a forum for participation and debate on issues that lie outside the immediate political arena, and hence can sometimes avoid some of the problems inherent in the latter.

Political problems, however, remain. The success of the Stockholm Conference on the Human Environment was marred by the absence of the U.S.S.R. and a number of other socialist countries in protest at the refusal of the conference to include the German Democratic Republic as a participant. This was due to the opposition of a number of Western countries to the admission of that government to any member of the United Nations family, which then would have made the German Democratic Republic eligible to participate in the conference.

Other problems can arise from the very nature of these organizations. Since each country which belongs to an organization has a single vote, regardless of size or contribution to the organization's funds (usually based on the economic size of a country), frustration is often experienced by the larger countries when they find themselves unable to wield the power to which they are accustomed in other political circles. The governmental nature of the organizations means also

349

that they are unable, apart from exceptional circumstances, to criticize the activities of any member; hence they tend to find themselves working with the lowest common political denominators.

The United Nations itself has a number of committees and other organs concerned with science and technology, although usually as part of the function of bodies with wider responsibilities. The Economic and Social Council, for example, has five regional economic commissions, each of which takes an interest in certain aspects of science and technology. The Economic Commission for Europe provides one of the few forums in which those concerned with science and technology policy from both Western and Eastern Europe can meet to discuss common problems, and areas dealt with so far include the transfer of technology to the less-developed countries in Europe, technological forecasting, the dissemination of scientific and technological information, and environmental problems. Other bodies of the Economic and Social Council with an interest in science and technology include the Advisory Committee on the Application of Science and Technology to Development (ACAST), which has four regional and twelve *ad hoc* working groups, and published the 'World Plan of Action'; the Committee on Science and Technology for Development; the Population Commission; the Committee on Natural Resources; and the Office of Science and Technology.

Five other programmes or organizations within the United Nations which share an interest in science and technology are the United Nations Development Programme (UNDP), the United Nations Conference on Trade and Development (UNCTAD), the United Nations Institute for Training and Research (UNITAR), the United Nations Industrial Development Organization (UNIDO), the Food and Agriculture Organization and the United Nations Environmental Programme (UNEP).

UNCTAD is concerned with the transfer of technology to developing countries, and other factors that affect their economic and trading position, such as the concept of 'appropriate technology'. UNCTAD also provides technical advice on problems associated with international trade, such as transport and economic planning. The Santiago conference also recommended that the developed countries should

allocate at least 10 per cent of their research and development expenditure to programmes designed to solve problems of specific interest to developing countries generally, and as far as possible devote that expenditure to projects in developing countries.

This figure is considerably higher than the 5 per cent of non-military R & D suggested by the World Plan of Action.

The main activities of UNIDO in the scientific and technological sphere are concerned with technical training and information transfer. UNIDO also helps developing countries to create an adequate infrastructure in the field of technical information by training people to run information services in their own country, and is also actively involved, like UNCTAD, in developing various concepts of 'appropriate technology'.

The World Health Organization (WHO) was founded in 1948 with the broad aim of the attainment by 'all peoples of the highest possible level of health'. Its total expenditure of $75.8 million in 1970 was the highest of any member of the United Nations family.

The research activities of WHO are coordinated by its Advisory Committee on Medical Research, which was set up in 1960. Collaborative research projects are contracted out to established institutions and constitute by far the largest item in the research programme. WHO also supports its own field research teams, and makes grants to individual investigators working on subjects of interest to the organization. Other objectives of the research programme are: the provision of services for research, for example by setting up reference centres or issuing the reports of scientific groups; the training of research workers, as being one of the most effective ways of promoting research and increasing research potential in member countries; and the improvement of communication between scientists, for example through exchange programmes or promoting meetings, symposia and seminars.

Many of WHO's research programmes, such as those on malaria and smallpox control, are started at an international level and then handed over to institutions in the country involved. Its main area of research is in the communicable diseases, but other areas that are growing in importance are automated screening techniques applied to medical and mass-screening programmes, which will enable WHO to build up an epidemiological picture of the distribution of disease for both diagnostic and research purposes, the development of short-cuts in medical care, for example through cheap prosthetic devices such as optical and hearing aids, adapted to the production facilities that a particular country can provide; the study of the family as a medical unit, as well as general aspects of human reproduction, such as maternal and child health.

The United Nations Education, Scientific and Cultural Organization (Unesco) is perhaps the international governmental organization most deeply involved in science. Succeeding the League of Nations International Committee of Intellectual Co-operation in 1945, its aims are 'to contribute to peace and security by promoting collaboration among nations through education, science and culture'.

Responsibility for the natural sciences is divided between two departments: the first is concerned with science policy, scientific information, and with scientific and technological research and higher education; and the second with environmental sciences and research on natural resources.

The science policy section has carried out a number of surveys on the policies and potentials of member countries, and has done much important work on the standardization of statistics and techniques for their collection. Four regional conferences of ministers responsible for science and technology were held: in 1965 in Santiago (Chile) (CASTALA); in 1968 in New Delhi (India) (CASTASIA); in 1970 in Paris (MINESPOL); in 1974 in Dakar (Senegal) (CASTAFRICA).

In the environmental sciences, the International Oceanographic Commission was set up by the General Conference of Unesco in 1960, and the Co-ordinating Committee for the International Hydrological Decade

(1965–74) in 1965. In addition, the International Geological Correlation Programme (IGCP) has been launched jointly by Unesco and the International Union of Geological Sciences; IGCP was approved at the seventeenth session of the General Conference of Unesco held in 1972. The aim of this programme, which forms the core of Unesco's activities in geology, is to arrive at a better understanding of the geology of the earth and hence at a better understanding of mineral resources, through the regional, interregional and intercontinental correlation of geological formations and phenomena.

A similar project is the Man and Biosphere programme (MAB), which was set up at the 1970 General Conference of Unesco. This is an interdisciplinary programme of research undertaken by Member States whose general objective is to develop the basis within the natural and social sciences for the rational use and conservation of the resources of the biosphere and for the improvement of the global relationship between man and the environment.

Apart from WHO, the International Atomic Energy Agency (IAEA), which was established in 1956, is the only member of the United Nations family which is able to contract research directly to outside bodies. It aims

to accelerate and enlarge the contribution of atomic energy to peace, health and prosperity throughout the world to ensure, so far as it is able, that assistance provided by the agency, or at its request, or under its supervision or control, is not used in such a way as to further any military purpose.

IAEA employs about 350 professional staff, of whom almost half are scientists. It runs two laboratories, one of which (at Seibersdorf in Austria) provides analytical and calibration services for agriculture, chemistry, physics, hydrology, medical physics and electronics. The second laboratory, at Monaco, is concerned with studying the effects of radioactivity in the sea and is helping to investigate marine pollution problems in general.

In the programme adopted by the general conference of IAEA in 1963, it was agreed that scientific and technological developments in the various applications of isotopes and radiation sources warrant continued and increased efforts to obtain as soon as possible, and particularly in the developing countries, tangible results in medicine, agriculture, hydrology and industry.

Priority has been given to co-ordinated programmes of training, the exchange of scientists and experts, the provision of equipment, and the granting of research contracts, with these aims in view.

IAEA makes a special effort to ensure that a large amount of its support for research goes to institutions in developing countries, as this is one important way in which the scientific and technical infrastructure of these countries can be improved. Nearly three-quarters of research contract funds were awarded to institutes in developing countries in 1968, and the programme often serves as an important stimulus to nuclear research in general.

Interregional, regional and subregional organizations

International co-operation is seldom justified as an end in itself, but more often as a means of reaching a common objective. Since such objectives, particularly those of a social or economic nature, are often defined by geographic factors, interregional, regional and subregional co-operation can be more important to a particular country than co-operation on a global scale. The goals of regional organizations can range from fundamental research, such as the European Organization for Nuclear Research (CERN) or the Joint Institute for Nuclear Research, to direct politico-military purposes, such as the scientific activities of the North Atlantic Treaty Organization (NATO). Most tend, however, to concentrate on areas of applied research, such as nuclear energy or space research, with promises of both short- and long-term applications to social and economic problems.

Any country's participation in co-operative programmes is limited by its own objectives and resources; few can afford displays of altruism. In addition, such programmes are inevitably affected by technical and political

limitations, and by no means all branches of science and technology are suitable for co-operative projects, or can be dealt with by the setting up of intergovernmental organizations. Jean-Jacques Salomon[1] has suggested that governments will not undertake large-scale combined action except when prompted by one or more of the following motives: the research is to be devoted to an essentially extra-national subject, such as meteorology or oceanology; it requires expenditure which no country could meet from its own resources, particularly applying to nuclear and space research; the scientific activities in question are believed to contribute to some wider economic or political project for which the countries are pooling their efforts, or that participation in a particular form of co-operation will enhance the prestige of the country, nationally or internationally.

Europe is an ideal arena for such scientific and technological co-operation. The proximity of a number of States with common cultural background and interests together with, to a certain extent, shared capabilities has led to a number of fruitful co-operative projects. Many of these projects have been prompted by a desire to keep up with the technological giants of the United States and the U.S.S.R., and have been particularly valuable to the smaller European countries (such as Belgium, Norway or Sweden) since they present an alternative to the need to build up a 'critical mass' within expensive areas of research. In addition, such co-operation performs a useful role in terms of general social, economic and cultural integration.

Regional organizations dealing with science and technology are also found in many other parts of the world, such as the Pan-American Health Organization, the East Africa Medical Research Council, or the Asia-Pacific Forestry Commission.

A certain disillusionment with the activities of intergovernmental organizations has been referred to in previous sections. Most of this disillusionment stems from programmes of regional co-operation, and is reflected in the fact

1. J.-J. Salomon, 'Feasibility of Multilateral Co-operation', *Nature*, Vol. 218, 1968, p. 819–21.

that, for example, only two important or-ganizations were established in Europe be-tween 1966 and 1970, although eight had been set up between 1961 and 1965. According to a report produced by the Science Policy Re-search Unit at the University of Sussex[1] some of the factors that have led to this disillusion-ment have been: (a) that many of the pro-grammes have been considered to be too ex-pensive, with inadequate return on money invested; (b) a general mistrust in the ac-tivities of international organizations, and a feeling that too many already exist; (c) the duplication of research efforts that some-times arises; and (d) susceptibility to the financial crises of participating countries. The type of problems that can arise, for example, when countries co-operate in areas where com-mercial exploitation is not far away, is dem-onstrated by the experience of EURATOM, whose programmes suffered from conflict with the national programmes of its members.

Perhaps the most successful international co-operative ventures in Europe have been CERN and JINR. The European Organization for Nuclear Research (CERN) has constructed two large machines, a 600 MeV synchrocyclo-tron and a 28 GeV proton synchroton, at its site near Geneva. CERN is also one of the earliest attempts at co-operation, the conven-tion having been signed by twelve countries in Paris in 1953 after several years of negotiation within the Unesco framework and ratified a year later.

A body with similar aims to CERN is the Joint Institute for Nuclear Research (JINR) in Dubna. This has ten active member countries, mostly from Eastern Europe, but including the Democratic People's Republic of Korea, Mongolia and the Democratic Republic of Viet-Nam (China and Albania do not contrib-ute at present). JINR was set up in 1956; its aims are to ensure common theoretical and experimental research in nuclear physics by scientists of its member States, to secure the exchange of experience and achievement and to maintain contact with national and international scientific research in develop-ing nuclear physics and finding additional possibilities for the peaceful use of atomic energy.

Other regional groups concerned with fun-damental research include the European Or-ganization for Astronomical Research in the Southern Hemisphere (ESO), created by six countries of Western Europe to construct and use an astronomical observatory in Chile, where a large optical telescope is currently being developed; the International Council for the Exploration of the Sea (ICES), which has sixteen European countries and Canada as members, and encourages maritime research, playing a co-ordinating role among the pro-grammes of participating States; and the European and Mediterranean Plant Protection Organization (EPPO), which co-ordinates re-search activities and control measures in the area of noxious pests and diseases of cultivated plants and stored foodstuffs.

Within the Eastern European subsystems, the leading co-ordinator of the social, econ-omic, scientific, and technological efforts of member countries is the Council for Mutual Economic Assistance (CMEA). Established in 1949, its current members are the U.S.S.R., Poland, the German Democratic Republic, Romania, Hungary, Bulgaria, Cuba, Czecho-slovakia and Mongolia; Albania is a non-active member and Yugoslavia is an associate mem-ber. The aims of CMEA are to

promote the acceleration of economic and tech-nological progress in its member countries, and the raising of the level of industrialisation in the industrially less-developed countries.[2]

The pivot of CMEA's comprehensive pro-gramme is found to be in the harmonization of major trends for economic, scientific and technological policy of the CMEA member countries, co-ordination of long-range national economy plans and on this basis: expansion of co-operation in the field of production activity and scientific and technological development.

The CMEA member countries have decided to pool their efforts with an eye to further

1. *A Preliminary Examination of Inter-governmental Co-operation in Science and Technology affecting West-ern Europe (Project Perseus)*, Science Policy Re-search Unit, University of Sussex, 1971.
2. *Information about the Activities of the Council for Mutual Economic Assistance in the Field of Scientific and Technological Co-operation*, Moscow, 1972.

fostering and promoting scientific and technological co-operation by:

1. Arranging systematic mutual consultations on major problems of scientific and technological policy.
2. Elaborating scientific and technological forecasts and jointly planning the elaboration of selected scientific and technological problems making use of the most effective forms of co-operation.
3. Co-operating and co-ordinating scientific and technological research.
4. Organizing co-operation in the field of material and technical supply of scientific and technological research activities with apparatus, instruments and materials.

By the mid-1970s, about forty-five multilateral agreements for co-operation on eighteen key scientific and technical problems had been concluded through the efforts of the council. Such agreements include the organization of thirty-four co-ordinating centres, three collectives of scientists, three joint laboratories, and seven scientific and technical councils which will in turn co-ordinate activities within their purview of interest. So far, half a dozen international bodies (the International Technical and Scientific Information Centre for Agriculture and Forestry, the International Centre for Scientific and Technical Information, the International Institute for Standardization, the International Laboratory of Strong Magnetic Fields and Low Temperatures, and the International Collective of Scientists) have been set up.

Even though only one major international laboratory, the Joint Institute for Nuclear Research, exists outside the framework of CMEA, it should be noted that there are several joint research programmes in Eastern Europe which apparently do not call for joint institutions. The ES Computer Programme, the programme for the exploration and peaceful uses of outer space ('Intercosmos' satellites, 'vertical' rockets), and the co-operative exchanges of the Academies of Sciences are examples of this.

Although the Organization for Economic Co-operation and Development (OECD), which is based in Paris, is not strictly regional, it is usually included in this category as the organization's membership is confined primarily to market-economy countries. OECD was established in its present form in 1960 and by 1973 had as members the governments of 24 countries.[1] The convention under which OECD has been set up provides that it shall promote policies designed:

1. To achieve the highest sustainable economic growth and employment and a rising standard of living in member countries, while maintaining financial stability, and thus to contribute to the development of world economy.
2. To contribute to sound economic expansion in member as well as non-member countries in the process of economic development.
3. To contribute to the expansion of world trade on a multilateral, non-discriminatory basis in accordance with international obligations.

The Directorate of Scientific Affairs of OECD has played a very important role over the years in the development and diffusion of science policy studies; indeed, in the catalogue of OECD publications one may find some of the most significant contributions to the field, prepared by very competent experts. In the preparation of this Part III many of these studies have provided an invaluable source of information.

It is in space research that European countries, under the dominating shadow of American and Russian supremacy, have placed the largest stakes on international co-operation. The two major organizations in this field are the European Launcher Development Organization (ELDO) and the European Space Research Organization (ESRO).

ELDO was established in 1962 to 'develop and construct space vehicle launchers and their equipment suitable for practical applications and for supply to eventual users'. Its members are the governments of seven countries, Australia, Belgium, France,

1. The members of OECD are Australia, Austria, Belgium, Canada, Denmark, Finland, France, Federal Republic of Germany, Greece, Iceland, Ireland, Italy, Japan, Luxembourg, Netherlands, New Zealand, Norway, Portugal, Spain, Sweden, Switzerland, Turkey, United Kingdom and United States; Yugoslavia has observer status.

Federal Republic of Germany, Italy, Netherlands and United Kingdom; Denmark and Switzerland have observer status.

ESRO's convention came into force in 1964, and was signed by ten members. Its aims were to

provide for and to promote collaboration among European States in space research and technology exclusively for peaceful purposes, and to carry out a common programme of scientific research and related technological activities.

Over a period of about ten years, in spite of some remarkable scientific and technological achievements, because of financial difficulties and dissatisfaction in the use of national contribution the countries participating in ESRO and ELDO decided in December 1972 to merge the two organizations (together with the Centre Européen de Technologie Spatiale—CETS) into a single European Space Agency (ESA).

Political bodies likely to have an increasing impact on science and technology are European Communities (EC): the European Economic Community (EEC), better known as the Common Market, is perhaps the most important one. Its six original members—France, Federal Republic of Germany, Italy, Belgium, Netherlands and Luxembourg—were joined by the United Kingdom, Denmark and Ireland at the beginning of 1973. Initially little attention was paid to science and technology beyond the fields of nuclear policy (through EURATOM) and coal and steel (through ECSC). In 1965, however, a scientific and technical research policy group was set up to study the problems of co-ordinating policy in this area, and in July 1972 the EEC Commission in Brussels published proposals[1] for a joint science and technology policy; these proposals were reexamined and extended to the enlarged community in the last decisions of the EC Council (January 1974).

One of the more important bodies to have grown up around the European Economic Community (EEC) is the European Atomic Energy Community (EURATOM). This was established as part of the Treaty of Rome (signed on 25 March 1957), the first article of which stipulated that

it shall be the aim of the community to contribute to the raising of standards of living in member states and to the development of commercial exchanges with other countries by the creation of conditions necessary for the speedy establishment and growth of nuclear industries.

The original members of EURATOM were the signatories of the Treaty of Rome, namely, France, Federal Republic of Germany, Italy, Belgium, Luxembourg and Netherlands. Three more countries, the United Kingdom, Ireland and Denmark, were admitted to EEC and hence to EURATOM, at the beginning of 1973.

EURATOM works through two main channels. It contracts research to outside bodies, including research associations, and also runs its own Joint Nuclear Research Centre. This centre has included four main establishments, at Ispra in Italy, Mol in Belgium, Petten in Holland and Karlsruhe in Germany. It has employed about 2,000 scientists, engineers and supporting staff, and the focus of the centre's research has recently been shifting from the nuclear to more socially oriented fields.

In 1973 the first large co-operative project to emerge from the commission's working parties—a Medium-term Meteorological Forecasting Centre sited at Reading (United Kingdom)—was approved by seventeen Western European countries and Yugoslavia.

The Council of Europe in Strasbourg, which endeavours to co-ordinate the social and economic policies of seventeen Western European countries,[2] has a number of interests in science and technology. It has standing committees on economics, science and technology, agriculture and population. Conventions and agreements have already been signed by its members on admission to universities, diploma equivalences, patents, and social and medical assistance. Operational activities have included two 'Biostack' experiments flown to the moon by *Apollo 17* and *Apollo 18*. Another 'Biostack' will be included in *Apollo/Sojuz* mission.

In 1973, the European Economic Community, jointly with the Council for Europe,

1. *EEC Annual Reports (ACABQ)*, Brussels, 1971.
2. These are Austria, Belgium, Cyprus, Denmark, Federal Republic of Germany, France, Iceland, Ireland, Italy, Luxembourg, Malta, Netherlands, Norway, Sweden, Switzerland, Turkey and United Kingdom.

decided to establish the European Science Foundation, with the following objectives: (a) to advance co-operation in basic research; (b) to promote mobility of research workers; (c) to assist in the free flow of information; (d) to facilitate the harmonization of the basic research activities supported by member bodies. The foundation consists of a sort of federation of the national research councils of the participating European countries.

Reference should also be made to the North Atlantic Treaty Organization (NATO) which, although primarily a military organization designed to promote political stability in the North Atlantic area through military co-operation, also carried out a number of civilian activities in science and technology as part of its 'integrative' strategy. NATO has fifteen member countries[1] in Western Europe and North America, and has special committees on economic co-operation, armaments, scientific co-operation, information and cultural relations, and 'challenges to modern society'.

Outside Europe, there are few regional bodies dealing with applied research. On the American continent one of the most important of these is the Inter-American Institute of Agricultural Sciences, which was established in 1944 to

encourage and advance the development of agricultural sciences in the American republics through research, teaching and extension activities in the theory practice of agriculture and related arts and sciences.

In 1965 its budget was $2.3 million, and its research programme included both traditional agricultural methods and the application of atomic energy to agriculture.

Another important American organization is the Pan-American Health Organization (PAHO), which is based in Washington, D.C. Established in 1902 as the International Sanitary Bureau, it changed its name in 1958 and aims to 'promote and co-ordinate Western hemisphere efforts to combat disease, lengthen life and promote physical and mental health of the people'.

There are several regional organizations for scientific research in East Africa. The first is the East African Agriculture and Forestry Re-

search Organization at Arusha in the United Republic of Tanzania, which was established in 1948, has a staff of sixty, and whose members are the governments of Kenya, United Republic of Tanzania and Uganda. The purposes of the organization are to provide an international research service and act as a scientific advisory body to its member governments. Similarly the East African Medical Research Council undertakes for its members (Kenya, United Republic of Tanzania and Uganda) research into diseases of East Africa—mainly human diseases, but also animal diseases where relevant. The funds are received partly from the Central Legislative Assembly of the East African Community, and partly from the British Government; it runs six research establishments, covering malaria and vector-borne diseases, trypanosomiasis, viral research, leprosy, tuberculosis and general medical research. These two organizations, together with the East African Veterinary Research Organization and other smaller bodies, form the East African Common Services Organization.

Two other intergovernmental bodies, although primarily concerned with economic and political issues, include in this certain responsibilities for science and technology. The Organization for African Unity was founded in May 1963 to co-ordinate and harmonize the economic, educational, health, welfare, scientific and defence policies of its member countries. It has specialized commissions for co-operation in economic and social, educational and cultural, health, defence and scientific matters. The Organization of American States was established in 1890, although its charter and present name were adopted in 1951. It aims to strengthen the peace and security of the American continent through promoting co-operative action on the economic, social and cultural development of its twenty-three member states. Its total budget for 1971 was $40 million.

1. Belgium, Canada, Denmark, Federal Republic of Germany, France, Greece, Iceland, Italy, Luxembourg, Netherlands, Portugal, Spain, Turkey, United Kingdom, and United States of America.

Bilateral and other agreements

Bilateral agreements offer to governments a way round some of the problems encountered by intergovernmental organizations. They provide a higher degree of flexibility, and even if extended to three signatories, it is more likely that the interests of the partners coincide than when a number of countries are involved. Often they are claimed to be an efficient way of having certain programmes carried out. Bilateral agreements have been called by Professor Pierre Aigrain, a former French administrator (*délégué-général*) of industrial and scientific development, 'the third generation of co-operation', superseding the informal contacts existing between scientists before the Second World War and the growth of international organizations that followed that war.

Bilateral agreements (or trilateral agreements for that matter) can fulfil a number of functions. In Europe, where they play an important part in political and economic integration, many are based on joint projects, particularly in high technology fields. For the United States, which has little need for technological co-operation with other countries (some specialists claim, nevertheless, that the United States is a net importer of technology), bilateral agreements such as the proposals for joint United States/U.S.S.R. projects on space research have a predominatingly political role, the initiative often coming directly from the president or his advisers.

The exchange of research scientists is another form of such international co-operation, and often forms part of a joint research project. The Royal Society in England, the CNRS in France, the Consiglio Nazionale delle Ricerche in Italy, and the National Academies of Science in both the United States and in Eastern European countries, are just a few of the organizations which arrange such exchange programmes.

In addition, many countries welcome foreign students both from developed and developing countries. In 1969, more than 24,000 foreign students from 130 countries were studying in 300 higher and specialized secondary institutions in the U.S.S.R. In the United

Kingdom, 14,000 students from the developing countries were studying at universities in 1965–66, nearly 8 per cent of the total university population; almost half of these were studying science or technology, and half again were post-graduate students.

Further aid to developing countries is provided by the setting up of links between pairs of institutions, one in a developed country and one in a developing country. Both the Federal Republic of Germany and the United States have established a large number of such institutional bonds, the latter particularly in the Middle and Far East, while France still retains strong links with some of the French-speaking countries in and around Africa, such as Madagascar (see Table 54).

Turning to individual European countries, France is perhaps the Western European nation which has already achieved the most from bilateral agreements. An important series of agreements with the U.S.S.R. was initiated as far back as 1957, when a permanent commission charged with examining cultural relations between the two countries was first organized. Under a general agreement signed in 1966, it was agreed that the two countries would carry out joint programmes in science and technology, and a special committee, which became known as the 'Petite Commission' was set up to take responsibility for this. Since then, major research programmes have been undertaken on topics that include the exploration of the sea's resources, meteorology, rail transport, and predicting the future development of science and technology, as well as general programmes dealing with atomic energy, space, medicine, health, agriculture and fundamental research.

Within Western Europe, France has entered into agreements with the Federal Republic of Germany to build *inter alia* a high flux reactor at Grenoble (the Laye-Langevin reactor) and a telecommunication satellite (Symphonie); with the United Kingdom to develop several aircraft of which the Concorde is the most famous; and with Sweden through the Franco-Swedish Research Association (AFSR), there are many general agreements on scientific and technological co-operation with other countries.

357

The Federal Republic of Germany also has a well-developed network of agreements. It has a programme with Belgium, the Netherlands and Luxembourg on breeder reactors, and has signed an agreement with the Netherlands and the United Kingdom to develop the gas centrifuge process of uranium enrichment for nuclear reactor fuel; it has *ad hoc* agreements with the United States, such as the development of a research spacecraft (*Helios*) designed to fly within 23 million miles of the sun. In recent years, the Federal Republic of Germany has signed an increasing number of agreements with countries in Eastern Europe. It has a well-established system of exchange programmes, run by private bodies such as the Humboldt Foundation—many of whose exchanges are carried out with countries of Eastern Europe—or semi-official bodies such as the Max Planck Gesellschaft, the Deutsche Forschungsgemeinschaft and the Deutscher Akademischer Austausch Dienst.

Austria is an example of a relatively smaller European country which relies heavily on agreements with other countries for developing its scientific and technological potential. Austria has signed fifty-eight bilateral agreements with twenty-three countries, six of which deal specifically with co-operation in science and technology, four with cultural and scientific co-operation, eight in cultural relations and intellectual co-operation, and the rest with general areas such as documentation, education and the exchange of films. A general treaty has been signed with the French on science and technology, and one on general cultural matters with the U.S.S.R. As part of its aid to developing countries, Austria has, for example, constructed schools in Istanbul, Tehran, Kabul and Guatemala.

For a number of economic and political reasons, countries in Eastern Europe have a far stronger tradition of co-operation than their Western counterparts.

The scientific academies of the socialist countries have, in fact, played an important part in these programmes of co-operation, possibly more so than some of their equivalent bodies in the West.

In 1966, for example, thirty-four Bulgarian scientists from the Academy of Science in Sophia were working with colleagues in institutions of the Soviet Academy of Sciences on projects that had been included in a joint plan of the two academies. Similarly in 1967, scientists from the Academies of Sciences of the German Democratic Republic and the U.S.S.R. were said to be working jointly on eighty research problems, including the prediction of catalytic effects, magnetohydrodynamics of liquid metals and the properties of microporous absorbents.

The U.S.S.R. has agreements for bilateral co-operation with eighteen European countries, both in Eastern and Western Europe, of which the links with France are particularly strong. It also has similar agreements with forty-four countries outside Europe, including eighteen in Asia, twenty-two in Africa, and two in Latin America and the United States. In 1968, 280 Soviet R & D organizations were co-operating with 360 similar organizations in CMEA countries on joint working plans; in total, it has been estimated by Lloyd Jordan[1] that there are 700 research institutes both in the Soviet Union and in Eastern European socialist countries collaborating directly on research work, and a total of 1,037 institutes, project construction and design organizations of CMEA countries conducting joint work (see Table 55).

Most other Eastern European countries also have well-developed links with countries in Western Europe and other parts of the world. Romania, for example, had agreements for co-operation in science and technology with fifty-two countries in 1968, of which twenty-one were in Europe. It had established commissions for scientific and technical collaboration with thirteen socialist and non-socialist countries, and central institutions—such as the Ministry of Education, the State Committee for Nuclear Energy, and the Academy of Sciences—had arranged sixty-five direct agreements for scientific and technical co-operation with similar institutions in other countries.

Similarly Poland had signed agreements for general economic co-operation with Denmark, France, United Kingdom, Netherlands and Belgium and Luxembourg, building up a gen-

1. Jordan, op. cit.

TABLE 54. Number of links between higher education and other institutions in developed and developing countries

Developing countries[1]	United States	Belgium	Canada	France	Federal Republic of Germany	Nether-lands	United Kingdom
Argentina	3	–	—	2	11	–	2
Brazil							
Chile	4	–	—	—	6	–	—
Taiwan	9	–	—	—	1	–	—
Ethiopia	2	–	—	2	2	–	—
Ghana	2	–	4	—	1	–	3
India	24	1	2	2	15	1	5
Israel	4	–	—	5	5	–	2
Ivory Coast	—	–	—	14	—	1	—
Kenya	2	–	2	—	11	1	4
Madagascar	—	–	—	12	—	–	—
Mexico	5	–	—	—	5	–	—
Nigeria	11	–	1	—	9	1	10
Pakistan	10	7	2	—	8	–	1
Philippines	5	–	—	—	2	1	—
Thailand	9	–	—	—	3	–	1
Tunisia	3	1	—	2	2	–	—
Turkey	4	–	—	—	12	1	—
Venezuela	3	–	1	11	6	–	—
TOTAL	100	9	12	50	99	6	28

1. Selected developing countries are those with ten or more links with institutions in developed countries.
Source: Bilateral Institutional Links in Science and Technology, Paris, Unesco, 1969 (Unesco Science Policy Documents, No. 13)

eral framework for co-operation in applied R & D, although possibly its best international co-operation is with the United States. Agreements for the exchange of scientists had been established with the CNRS in France, the CNR in Italy, and the British Royal Society, as well as with Canada, Ghana, India and the United States (through the National Academy of Sciences). Poland and France have co-operated in the gas and power industries, and the Polish Institute of Nuclear Research co-operates with similar institutes in Austria, Belgium, Denmark, France, Italy, Norway and Sweden.

The dominant position of the United States in technology has given it little direct need for international collaboration at governmental level. The most important aspect of international co-operation is therefore the funds distributed by United States agencies to foreign countries to carry out specific programmes of R & D, which accounted for 0.5 per cent of the total R & D budget in 1970.

In 1966, for example, the National Aeronautics and Space Administration (NASA) had arranged for seventy-one countries to participate in its programmes or activities, many providing tracking facilities for rockets and satellites, while in 1967 the National Science Foundation spent $16.9 million on foreign or international programmes. Most of the foreign expenditure of the Atomic Energy Commission on R & D went to Japan for studies of the long-term effects of radiation on human beings.

Many of the bilateral relations established by developing countries take the form of direct links with institutions in the developed countries, as outlined at the beginning of this section, since the difference in levels of technology between developed and developing countries makes mutually beneficial collaboration difficult to arrange.

Many of these institutional ties are established with the aid of international organizations. The World Health Organization, for example, has helped to set up a scheme by which teaching staff from the medical faculty at the University of Edinburgh spend up to a year at the University of Baroda in India, in exchange for a number of Indian staff who go to work in Edinburgh. A similar arrangement has been established between The John Hopkins University School of Public Health in the United States and the National Institute of Health in the Philippines.

The International Atomic Energy Agency, as WHO, not only established its own affiliations with institutions in developing countries, but also helps to initiate ties between pairs of institutions. Typical arrangements are those that have been made between the University of Ankara and the Department of Physics and Mathematics at the Max Planck Gesellschaft in Mainz, and between the Nuclear Research Centre in Tunis and the IAEA's own laboratory at Seibersdorf in Austria.

A more elaborate scheme has been organized by Unesco in collaboration with Tehran Polytechnic in Iran. The polytechnic is made up of a number of institutes, and Unesco has helped arrange the setting up of viable bonds between these institutes and institutions in Europe.

Multinational companies, and the transfer of technology

According to the *Yearbook of International Organizations*, there were almost 8,000 firms with wholly owned subsidiaries in at least three other countries—the so-called 'multinational' firms—in 1970. More than half of these have their headquarters in either the United States or the United Kingdom and 680 (8.5 per cent) had ten or more foreign subsidiaries. The importance of these firms in terms of sheer size can be seen from the fact that the annual turnover of the largest, General Motors, in 1973 was, at $35,800 million, greater than the gross national product of either Switzerland or South Africa, the turnover of Royal-Dutch Shell in 1973 ($18,700 million) larger than the GNP of Denmark, and Nippon Steel's turnover larger than the GNP of Portugal and Ireland.

Many of these transnational firms conduct their R & D on an international basis. IBM,

TABLE 55. Number of research institutes of the Soviet Union and Eastern European communist countries in direct contact, 1957–69

Year	Research institutes		
	Eastern European communist countries	Soviet Union	Total
1957	195	201	396
1958	250	200	450
1959	n.a.	n.a.	600
1960	400	300	700
1969	n.a.	n.a.	700

Source: *Bilateral Institutional Links in Science and Technology*, Paris, Unesco, 1969 (Unesco Science Policy Documents, No. 13).

for example, has seven of its nineteen research laboratories situated in Europe; Shell has eight laboratories in the United Kingdom, three in the Netherlands, two in France, one in the Federal Republic of Germany, one in Japan, one in Canada, and eight in the United States (see Table 56). The R & D carried out in these laboratories can often make up an important part of a nation's R & D effort. This is particularly so in a highly competitive and concentrated field such as computers, or when multinational companies play a large role in a country's industrial sector, as happens in Canada or the Netherlands. In the developing countries, R & D performed by multinational firms can help raise a country's own scientific and technological and educational standards as well as providing a channel for the direct transfer of technology from the developed countries.

Professor Richard Robinson has defined the multinational firm as one in which, structurally and policy-wise, foreign operations are co-equal with domestic, and management is willing to allocate company resources without regard to national frontiers to achieve corporate objectives.

There are many reasons why such firms distribute their R & D activities among their subsidiaries in different countries. Some of these are: the manpower and other resource advantages that a foreign country might be able to offer (such as lower salaries for research scientists); a need to get round trade tariffs and other trade barriers; the need to protect patents, particularly for science-based chemical and pharmaceutical firms, where the only alternative might be to sell the patent to a potential marketing rival; and the fact that some companies were 'born' multinationally as the result of mergers between companies in different countries.

There are other reasons too. A survey of 100 United States firms carrying out research in Europe, made by the Stanford Research Institute, found that most of these companies spent less than 4 per cent of their total R & D budget there. For many, the main purpose of this was to gain access to the European scientific community and to monitor other R & D being carried out in Europe.

One United States firm, which has made a conscious effort to internationalize in research activities, is the computer manufacturer IBM. The computer industry is one in which the costs of research and development are high; IBM itself may have spent as much as $5,000 million on the development of its 360-series computer.[1] According to Louis Turner,[2] computer firms have now to spend between $25 million and $50 million a year on research in order to stay competitive in a world market; assuming this to represent 5 per

cent of turnover, it implies an annual turnover of $500 million to $1,000 million. In reality, IBM had a turnover of $9,532 million in 1972; its R & D costs in the same year were $676 million (of which $50 million went into research alone). The scale of enterprise required is one of the reasons that IBM retains such a firm hold on world markets (the United Kingdom and Japan are the only major market countries in the world in which IBM had less than 50 per cent share in the computer market in 1969).

IBM has now seven laboratories in Europe: a fundamental research laboratory in Zürich and development laboratories in the Federal Republic of Germany, Austria, the United Kingdom, France, Sweden and the Netherlands. Each country specializes in a certain set of components, and although IBM subsidiaries in different countries can subcontract research between themselves, most funds for research come direct from the American parent body in New York, which retains firm control over general research policy. In countries such as Sweden and Austria, IBM makes an important contribution to overall research effort in the computer sciences, opening up opportunities to scientists and engineers in the field.

Another firm with important international research facilities is the Dutch firm of Philips. Besides its main research laboratory at Eindhoven in the Netherlands, which employs about 2,200 people, it has five research laboratories in other European countries; one in the United Kingdom (600 employees), one in France (400 employees), two in the Federal Republic of Germany (about 400 employees each) and a smaller centre in Belgium with about 50 employees. Philips has national organizations in each of these countries, for which the research laboratories carry out directly a certain amount of work. Most research policy, however, is decided at an international level by a committee of research directors. Moreover, Philips has established a subsidiary in the United States.

TABLE 56. Total number of multinational firms with headquarters in a particular country

Country	Number
United States	2,468
United Kingdom	1,692
Federal Republic of Germany	954
France	538
Switzerland	447
Netherlands	268
Sweden	255
Belgium	235
Denmark	128
Italy	120
Norway	94
Austria	39
Luxembourg	18
Spain	15
Portugal	5

Source: Yearbook of International Organizations, 14th ed. Brussels, Union of International Organizations, 1972.

1. IBM will neither confirm nor infirm the figure. The order of magnitude of the sum appears to be correct.
2. L. Turner, The Multinationals, New York, N.Y., Hill & Wang, 1973.

A similar pattern is adopted by the British firm Imperial Chemical Industries Ltd (ICI). In addition to its twelve research units in the United Kingdom, the company has a 'European' research division, and other research units in the United States, Canada, Australia and India. Each of the main operating units of the company—i.e. divisions, subsidiaries and associates—has its proper R & D section, which is formally answerable to its own board, but which receives policy guidance from a central R & D organization. Each operating unit, however, is free to formulate its own R & D programme, and most of the initiative is taken at this level. The overseas units, in particular, pursue policies appropriate to their areas of operation and often quite different to those applicable in the United Kingdom.

Royal Dutch/Shell is the largest multinational company outside the United States, and the fourth largest in the world ($18,700 million of turnover in 1973). The main fields of R & D carried out by the company include exploration and production, oil products and processes, natural gas, marine research, chemical products and processes, metals, and general research. Shell Research Ltd, the company which owns and operates the eight British research laboratories, also has establishments in the Netherlands, France, Federal Republic of Germany, Japan, Canada and the United States. Research is carried out on a contractor-customer basis according to the demands of various companies in the Shell group; the exception to this is general research, which is of a more long-term and undirected nature, and in which the research company itself decides on programmes and the ways in which funds allocated to this area shall be spent. R & D performed outside the United States is co-ordinated by N. V. Shell Internationale Research Maatschappij, The Hague. In the United States, the Shell Oil Company in Houston carried out R & D in support of its own activities, although programmes are co-ordinated with those outside the United States.

Multinational companies are particularly important to smaller countries. Almost all the largest companies of Belgium, the Netherlands, Sweden and Switzerland, for example, sell more than 80 per cent of their production outside their home country. Such companies appear to have risen to their present size as a result either of exports or of their multinational structure. In Switzerland and the Netherlands, in particular, economic policies at home as well as the readiness of employees both to live and work abroad have encouraged the strategies of multinationals. Even in these countries, however, a greater portion of R & D activities has remained in the home countries than any other activity apart from central management. This underlines the importance of R & D to multinational companies, who tend to keep research as close to central management as possible, while moving production out to the countries which provide the main markets.

But this can lead to various problems. Some companies are tempted to remove the research activities of a newly acquired subsidiary to the parent company, leaving a country with a considerably reduced research potential in a particular field. Author Christopher Tugendhat quotes an IBM executive as admitting, for example, that 'people sometimes feel we are exploiting their country's brains and talents in the same way as a mining company exploits its natural resources'.[1]

Another problem is the question of conflict with national research policy. The Science Council of Canada, for example, has recently pointed out that a large amount of the R & D performed in Canada is done by companies with headquarters in other countries, particularly the United States.[2] The Science Council added that such research was not aimed at solving Canada's own needs and problems, but those of the parent company, which were often unrelated to Canada. Similarly authorities in the Netherlands have expressed concern at the concentration of multinational company research in their country. For in this way, they feel they are making an unfairly high contribution to the economies of other

1. C. Tugendhat, *The Multinationals*, London, Eyre & Spotiswoode, 1971.
2. *Innovation in a Cold Climate*, Ottawa, Science Council of Canada, October 1971 (Report No. 15).

nations, since most of the production and sales of the multinationals is outside the Netherlands.

In recent years the structure and function of multinational companies have been studied from many angles by economists, business executives, science policy experts and other specialists. At the Business School of Harvard University, for example, Professor Raymond Vernon has directed the Multinational Enterprise Project; during its first five years, the project was directed principally at an understanding of transnational enterprises whose parent firms were based in the United States. Early in 1972, however, this work was extended to cover the operations of multinational enterprises bases in Canada, Europe and Japan.

Multinational companies also play an important part in the transfer of technology from one country to another, particularly to the developing countries. The term 'transfer of technology' has been defined by Charles Cooper of the Science Policy Research Unit at the University of Sussex, in a report prepared for the United Nations Conference on Trade and Development (UNCTAD), as taking to cover

the transfer of those elements of technical knowledge which are normally required in setting up and operating new production facilities or in extending existing ones—and which are characteristically in very short supply (and often totally absent) in the developing countries.[1]

The importance of the transfer of technology to developing countries can be seen from the fact that many such countries pay more for patents, licences, technical know-how, and the like, than they spend on R & D. Sri Lanka, for example, paid 0.5 per cent of its gross national product on these kinds of technology transfer in 1970, but only 0.2 per cent on its own R & D; Nigeria spent 0.8 per cent on technology transfer in 1965, but only 0.5 per cent on R & D[1] (see Table 57).

The amount of money paid for technology transfer by the developing countries is also increasing rapidly. Estimates made from a survey carried out by the UNCTAD secretariat (TD/106, November 1971) revealed that payments for technology transfer were increasing

TABLE 57. Payments by developing countries for the transfer of technology and their relationship to gross domestic product and exports

Country	Year	Annual payments for transfer of technology		
		Total[1]	As proportion of gross domestic product	As proportion of exports
		$ million	%	%
Argentina	1969	127.7	0.72	7.9
Brazil	1966–68[2]	59.6	0.26	3.4
Colombia	1966	26.7	0.50	5.3
Mexico	1968	200.0	0.76	15.9
Nigeria	1965	33.8	0.78	4.2
Sri Lanka	1970	9.3	0.51	2.9
India	1969	49.0	0.12	2.7
Israel	1961–65[2]	3.9	0.17	1.2
Spain	1968	133.0	0.55	8.4
Turkey	1968	49.1	0.43	9.9

1. Includes payments for patents, know-how, trademarks, management and other technical services.
2. Annual average.

Source: C. Cooper and F. Sercovitch, *The Chains and Mechanisms for the Transfer of Technology from Developed to Developing Countries*, UNCTAD, 1971 (TD/B/AC 11/5).

in some countries at an annual rate as high as 36 per cent (Sri Lanka), 55 per cent (Nigeria) and even 65 per cent (Turkey). Extrapolating these and other results, the UNCTAD report estimated that at the end of the 1960s the developing countries were spending on average an annual sum equivalent to 5 per cent of their exports on technology transfer (a total of $1,500 a year), and that this figure was increasing at an average rate of 20 per cent a year. Excluding the major oil exporting countries, the figure was equivalent to two-fifths of the debt-servicing costs of the developing countries, and about 56 per cent of the flow of direct foreign investment to developing countries (including reinvested earnings).

There are two main mechanisms of technology transfer. The first is the direct mechanism, by which a government or firm buys

1. C. Cooper and F. Sercovitch, *The Channels and Mechanisms for the Transfer of Technology from Developed to Developing Countries*, UNCTAD, 1971 (TD/B/AC 11/5).

patents, licences, or technical knowledge in some other way from the owner (usually a commercial company) in another country. This is the way in which Japan rose to its present position of technological strength in the 1950s, even though this was only possible through a steady build-up of scientific competence since the Meiji Restoration (1867).

The second category of transfer mechanisms, the 'indirect' mechanisms that cover the wholly owned subsidiaries of multinational companies. These subsidiaries can be considered as recipients of technology which are completely controlled by the company which supplies them. In this case, the whole process of the transfer of technology is dominated by the parent company, and the only substantial negotiations which take place are those between the company and the government of a developing country.

Heavy reliance on the technical knowledge developed in the advanced countries is usually the only way a developing country can guarantee a sustained and rapid programme of industrialization. The growth of many of the less-advanced industrial countries, such as Ireland and Greece, as well as of developing countries, such as Brazil and South Africa, has been very largely due to foreign investment and the technology that this has brought with it.

Foreign investment and collaboration can, however, raise a number of problems. As Charles Cooper points out (see Chapter 30):

technology suppliers are concerned with the commercial advantages they can obtain through the transfer operation. There would be no problem about this if the pursuit of commercial advantage by the suppliers led to results which were in line with the economic and social requirements of the developing countries. There is, however, a good deal of evidence that this kind of coincidence is rare. Particularly in the case of indirect transfers, the developing countries face a contradiction: they need the technology and the capabilities which the supplying companies possess, but the terms on which they can get hold of it may be disadvantageous for economic and social development.

One way this can have an effect is through restrictive clauses on the commercialization of transferred technology. In Bolivia, for example, a study of 35 contracts for technology transfer found that 24 tied technical assistance to the use of patents, 22 tied additional know-how to the existing contract, 3 fixed the prices of final goods, 11 prohibited the production or sale of similar products, 19 required secrecy concerning know-how during the contract and 16 after the end of the contract, and 5 specified that any dispute or arbitration should be settled in the courts of the country of the licenser. Similarly in Chile, out of 175 contracts, 98 had clauses for quality control by the licenser, 45 controlled the volume of sales and 27 the volume of production.

This had led Junta del Acuerdo de Cartagena to suggest, in his report to UNCTAD, that often 'the only basic decision left to the licensee is whether or not to enter into an agreement for the purchase of technology. Technology', he continues, 'through the present process of its commercialisation, becomes thus a mechanism for controlling the recipient firms. Such control supersedes complements, or replaces that which results from ownership of the firm's capital'.[1]

In an attempt to combat this situation, the Commission of the Andean Pact established in December 1970 a series of policies aimed to regulate the mechanism by which foreign technology was acquired, both through legislation and institutional arrangements. These included a prohibition on the payment of royalties by a subsidiary to its parent company or to other affiliates, based on the principle that the effects of technological inputs should be reflected in the declared profitability of a foreign-owned subsidiary rather than transferred to another country's tax jurisdiction (Article 21). It also established a permanent system for the exchange of information among the five Andean countries about the terms and impact of the purchase of technology (Article 48).[1]

1. Junta del Acuerdo de Cartagena, *Policies relating to Technology of the Countries of the Andean Pact: Their Foundations*, UNCTAD, 1971 (TD/107).

30   Trends

Having presented some major traits of the
organization of the research system at the
national and international levels and in terms
of governmental and private industrial institutions, we will discuss briefly some trends
that can be perceived today in the science
policies of the industrially advanced countries
as well as in the developing world. These
comments will be preceded by some considerations on R & D's contribution to the economic
growth of nations, followed by a discussion on
the present need of internationalizing the
scientific effort.

Measuring scientific output

Governments and industries, as we have seen,
have been investing very large sums of money
in R & D, particularly in the course of the last
twenty-five years. The question has been
raised whether such investments were justified
by measurable improvement of the national
economies or in the productivity of industry.

The first difficulty is in deciding precisely
what one is trying to measure. The benefits of
scientific research can be described in many
different terms. The two most common of
these are the economic and social benefits that
flow from the application of science to practical problems, and the 'cultural' value of

science. There is also the value of creating a
pool of skilled manpower which can then be
turned to socially useful tasks, but this is
usually absorbed into an evaluation of social
and economic pay-offs.

It seems one of the unchallenged truths of
the scientific community that science is a
major driving force behind the economic
growth of advanced societies. The type of data
commonly quoted to support this point of view
is that which seeks to demonstrate the much
greater rate of growth of science-based industries when compared to that of the national
economy as a whole. It has been pointed out,
for example, that in the period 1945–65 the
average percentage growth of the economy of
the United States has been of 2.5 per cent
while certain industries with large involvement
in R & D had the following growth: Polaroid
13.4 per cent; 3M 14.9 per cent; IBM 17.5 per
cent; Xerox 22.5 per cent; Texas Instruments
28.9 per cent. Another version of this argument which is commonly heard is that the
gap between the times of a particular scientific discovery and its technological application
has shortened dramatically within the past
hundred years. An example of the way this is
said to be demonstrated is the fact that whereas
it took ninety-six years for polystyrene, first
developed in 1840, to find a practical application, by the mid-fifties, taking polypropylene

TABLE 58. Characteristics of the innovative process[1]

Innovation	Early recognition of need	Independent inventor	Technical entrepreneur	External invention	Government financing	Informal transfer of knowledge	Supporting inventions	Unplanned confluence of technology
Heart pacemaker	X	X	X	X	— X	X	X	—
Hybrid corn	—	—	X	X	X	X	X	—
Hybrid small grains	X	—	—	X	X	X	X	—
'Green revolution' wheat	X	—	X	—	X	X	X	X
Electrophotography	X	X	X	X	—[2]	X	X	X
Input-output economic analysis	X	X	X	X	X	X	X	X
Organophosphorus insecticides	X	—	X	X[3]	X	X	X	—
Oral contraceptives	X	—	X	X	—	X	X	X
Magnetic ferrites	X	—	X	—	X	X	X	X
Video tape recorder	X	—	X	X	—	—	X	X

1. Indicated as important (X) or unimportant (—) for each innovation.
2. But limited government funds were provided to a related development, giving indirect aid.
3. 'External invention' occurred only because the Second World War enabled American Cyanamid to market the innovation in advance of I. G. Farben.

Source: Interactions of Science and Technology in the Innovative Process: Some Case Studies, Columbus (Ohio), Battelle Institute, Columbus Laboratories, 1973.

and polycarbonates as two examples, this period had been reduced to five years.

Both these arguments, however, are open to dispute. On the first, it can be pointed out that the cases selected in Tables 58 and 59 are extreme examples of successful technological innovation, their success being in many ways just as much due to the state of the market at a particular time rather than any particular technical ingenuity. Other areas of the economy, such as the shipbuilding industry in Japan or the property market in many advanced countries, have achieved equally remarkable rates of growth on a very different technological base. The same case, although in a more sophisticated manner, can be made against the second argument. One British commentator, J. Langrish,[1] has raised three basic objections: that the special selection of examples is not a good way of testing theories, that the type of examples used are to a certain extent major innovations which may be quite untypical of the relationship between the science contained in thousands of publications and the small improvements being made all the time in industry, and that this approach centres around the difficulty of defining the scientific discovery on which an application is based. To illustrate this point, Langrish has constructed a diagram in which the trend goes in the opposite direction (see Fig. 80); as an extreme case, the basic reaction used in the highly important process for the conversion of ethylene to acetaldehyde by direct oxidation was discovered as early as 1894, yet lay buried in the literature until it was in effect rediscovered in the late 1950s.

This is not to question the importance of scientific and technological innovation in providing the basic tools by which economic growth is made possible. Rather it challenges the one-dimensional model which sees technological developments as flowing steadily from scientific discoveries, and thus implicitly places a high economic value on pure science. As Langrish has pointed out:

In the last 30 years . . . more has been spent on research and development in Britain than in any other country except the United States and the

1. J. Langrish, 'Does Industry Need Science?', *Science Journal*, December 1969, p. 81–4.

Soviet Union, yet the British economic growth rate has remained lower than that of Western Germany, Japan, France and Australia. It has therefore been assumed that Britain is not very good at using science, and that it has allowed other countries to make off with British scientific discoveries and exploit them elsewhere. The alternative possibility—that scientific discoveries do not really contribute to economic advance except in exceptional circumstances—has not been seriously considered.

Other studies have thrown further light on this problem. In particular, two studies carried out in the United States have demonstrated some of the complexities of the relationship between pure science and technology. The first of these, called Project Hindsight, was a study made by the Department of Defence into the development of weapons systems during the period 1945–63. The study found that of all the identifiable discrete contributions made to the development of twenty weapons over this period, 92 per cent came under the heading of technology, while the remaining 8 per cent were virtually all in the category of applied research, except for 0.4 per cent which fell in the category broadly referred to as pure research. The study concluded that 'it was not

able to demonstrate value for recent pure science', although it added that 'on the 50-year or more time-scale, undirected research has been of immense value'.[1]

The second study was carried out for the National Science Foundation in 1968 under the title TRACES (Technology in Retrospect and Critical Events in Science).[2] This examined the key scientific events that had led towards five major technological innovations.

Such key events were classified into three categories, as follows:
1. Non-mission-oriented research (NMOR); research carried on for the purpose of acquiring new knowledge, according to the conceptual structure of the subject or the interests of the scientist, without concern for a mission or application, even though the project within which such research was done may be funded with possible applications in mind.
2. Mission-oriented research (MOR); research carried on for the purpose of acquiring new knowledge expected to be useful in some application.
3. Development, the process of design, improvement, testing, and engineering, in the course of bringing an innovation to fruition.

The tabulation of the significant events, as related to the performer, gave the results shown in Table 60.

The conclusion was reached that, while nonmission or basic research provided the origins from which science and technology could advance towards the innovations which lay ahead, this seemed mainly to apply to work done between twenty and thirty years before the innovation.

Such study was further expanded very recently (1973) to a total of ten innovations:[3] the heart pacemaker; hybrid corn; hybrid small

TABLE 59. Duration of the innovative process for ten innovations

Innovation	Year of first conception	Year of first realization	Duration years
Heart pacemaker	1928	1960	32
Hybrid corn	1908	1933	25
Hybrid small grains	1937	1956	19
'Green revolution' wheat	1950	1966	16
Electrophotography	1937	1959	22
Input-output economic analysis	1936	1964	28
Organophosphorus insecticides	1934	1947	13
Oral contraceptives	1951	1960	9
Magnetic ferrites	1933	1955	22
Video tape recorder	1950	1956	6
Average duration			19.2

Source: Interactions of Science and Technology in the Innovative Process: Some Case Studies, Columbus (Ohio), Battelle Institute, Columbus Laboratories, 1973.

1. Project Hindsight. Final Report, Washington, D.C., Office of the Director of Defense Research and Engineering, 1969.
2. Technology in Retrospect and Critical Events in Science (TRACES), Vol. 1, Washington, D.C., 1968. Prepared by the Illinois Institute of Technology Research Institute for the National Science Foundation.
3. Interactions of Science and Technology in the Innovative Process: Some Case Studies, Columbus (Ohio), Battelle Institute, Columbus Laboratories, 1973.

TABLE 60. Significant events as related to the performer (percentages)

	University and college	Research institutes and government laboratories	Industry
NMOR	76	14	10
MOR	31	15	54
Development	7	10	83

grains; 'green revolution' wheat; electrophotography (copying by the application of electrostatics and photoconductivity); input-output economic analysis; organophosphorous insecticides; oral contraceptives; magnetic ferrites; video tape recorder. On the basis of the analysis of the case studies, a number of characteristics of the innovative process were found, as in Table 58. The duration of the innovative process for the ten innovations is given in Table 59.

Can innovation be managed? The answer given as the conclusion of the study is the following:

There has always been argument about the extent to which research and development can be managed. Whatever may be the merits of differing positions in this argument, we may confidently assert that, in the spectrum of science and technology, NMOR is the most difficult to manage, if it can be managed at all. Furthermore, as we have seen, significant NMOR events continue to occur up to the end of the innovative process; hence, we are forced to conclude that innovation cannot be completely controlled or programmed. Also, the actions of the technical entrepreneur, or the role of such motivational forces as recognition of need and recognition of technical opportunity, involve inventive or creative activities that do not lend themselves to detailed planning. Hence the high ranking of these factors in the analysis supports further the conclusion that innovation cannot be fully planned. We are therefore led to recommend that management, in trying to promote innovation, permit and encourage the opportunity to act upon ideas that fall outside the established or recognized pattern.

But if innovation cannot be fully controlled, we nevertheless can discern ways in which management can help it along. Our analysis reveals two such ways by demonstrating the importance of funding and of the confluence of technology. As to funding, it need not be munificent, at least in the early stages. It not only permits R & D to proceed, but probably also aids the innovative process by the confidence management generates in the R & D team through financial support. As to confluence of technology, it seems almost essential to innovation. Yet it too often occurs without planning, and one suspects that here is an opportunity for management, by promoting interdisciplinary R & D terms, to accelerate the innovative process.

The relationships between science and its applications, on the one hand, and economic development, on the other, are made complex by many intervening factors. We shall mention only a few. In the larger countries, a great fraction of R & D's investment goes to defence or prestige programmes. This obscures markedly the interrelations between R & D and economic growth in general, which is affected only marginally by those activities. The way economists measure economic growth excludes certain very important consequences of scientific research; thus the importance of the latter to a given country may appear to be less significant than it really is. This is particularly true when research and development are responsible for improvements in the quality of goods and services. Economic growth has hardly been affected by the discovery and diffusion, for example, of antibiotics which have had a major impact on human health. Further, in terms of the effects on productivity of an industry that spends relatively large amounts on research, this industry may contribute to the productivity of other industries. Consequently, the relationship between investment and productivity for that particular firm or group of companies would appear lower than it was to industrial productivity on the whole.

Innovations seem to be due, in general, to market opportunity much more than to new technical availability.[1] R & D follows economic development and is merely an essential overhead expense incurred by this de-

1. K. Pavitt, *The Conditions for Success in Technological Innovation*, Paris, OECD, 1971.

velopment. Another aspect of the problem that a number of economists have tried to comprehend is the contribution of technological change to economic growth. Again the situation is confused by the problem of definition—as pointed out by one economist writing in this area, 'the measurement of all economic activities involves an element of arbitrariness in designating border-lines'—and conclusions therefore lack a high degree of accuracy.

It does appear possible, however, to obtain certain results. Perhaps the most significant attempt at such measurement has been made by the American economist Edward Denison, who has calculated what economists call the 'residual factor' of technological change between 1950 and 1962 for eight Western European countries and the United States.[1]

The method used by Denison involves calculating the contribution of increases in both labour and capital inputs to growth in output. If the contribution to the growth in output of both of these input factors is deducted from the growth in total output, what is left over provides some indication of the contribution due to technological change. The exact interpretation is still controversial, and a number of criticisms can be raised, including the fact that what is left over includes a number of things, of which the factor labelled technological change may be one of the least impor-

tant. It also disregards those factors which cannot be quantified in monetary terms, and the 'residual factor' has in fact been labelled as mainly 'a measure of the economist's ignorance'.

The main conclusions of Denison's analysis are that the percentage of total output represented by the residual factor ranges from 41 per cent for the United States to 75 per cent for France, and represents on average about 50 per cent of the growth in total output of the countries considered (see Table 61). If it is appropriate to label this in some way as technological change, then it appears that the period under consideration saw very substantial contributions of technological change to economic growth of European countries during the 1950s and early 1960s. It is difficult to be precise about the exact size of this contribution, and this is a problem which a number of economists are now trying to deal with.

The problem of R & D's contributions to economic growth has been analysed also at the level of individual industries, in both agriculture and manufacturing.[2] Once more, a specialized author has pointed out how elusive is the attempt to pin down in quantitative

1. E. F. Denison, *Why Growth Rates Differ?*, Washington, D.C., The Brookings Institute, 1967.
2. E. Mansfield, 'R & D's Contribution to the Economic Growth of the Nation', *Research Management*, May 1972, p. 31–46.

TABLE 61. Contribution of inputs, and output per unit input, to economic growth, 1950–62 (percentages)

	Japan[1]	United States	Belgium	Denmark	France	Federal Republic of Germany	Netherlands	Norway	United Kingdom	Italy
Growth rate of real national income	10.1	3.32	3.20	3.51	4.92	7.26	4.73	3.45	2.29	5.96
Total factor input	4.03	1.95	1.17	1.55	1.24	2.78	1.91	1.04	1.11	1.66
Labour	1.31	1.12	0.76	0.59	0.45	1.37	0.87	0.15	0.60	0.96
Capital	2.72	0.83	0.41	0.96	0.79	1.41	1.04	0.89	0.51	0.70
Output per unit of input (residual)	6.1	1.37	2.03	1.96	3.68	4.48	2.82	2.41	1.18	4.30
Residual as percentage of total growth	60	41	63	56	75	62	60	70	52	72

1. Figures are for 1955–66.

Source: A. S. MacDonald, *Post-war Science Inputs and Outputs in Europe*, Geneva, United Nations Economic Commission for Europe, 1972 (Mimeo.).

terms the contributions of R & D to productivity—largely because of the complexity
of the problem under study.

Mention should be made here of attempts
that have been carried out in the Soviet Union
to evaluate the effectiveness of scientific and
technological research. In 1964, a document
was published entitled *Basic Code of Procedure
for Measuring the Economic Return on Research*

FIG. 80. The time elapsing between discovery and
practical application is not necessarily becoming any
shorter in our time: comparison between various
categories of innovations can lead, indeed, to
the opposite finding—that the R & D lead period is
growing longer.

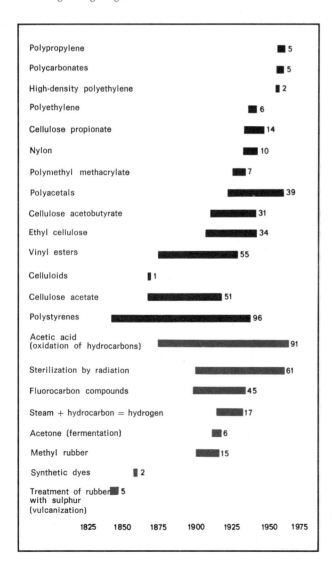

Work. The code was primarily intended as an
outline procedure to evaluate such developments as the creation of new technical processes, machines or other products, or the creation of new types of buildings or enterprises.

This procedure involves calculating the 'potential return' by comparing a new product or
process with a base product or process defined
as 'the highest level of technology already introduced, planned, or at the stage of completed
research, in the U.S.S.R. or abroad'. According
to a recent OECD report, since 1964, all proposed research projects included in the draft
plans of research institutes and design organizations are required to be supported by a
calculation based on the methods laid down in
the code.[1]

Undoubtedly the code has been of considerable significance in the planning of R & D
in the U.S.S.R. It is also interesting to note
that at the twenty-third Communist Party
Congress in 1966, Alexei Kosygin particularly
stressed the value of relationships based on
'economic accounting'—the term used to describe the system by which State enterprises
are expected to cover their outlays with the
payments they receive from the sale of their
products—between scientific institutes and
production enterprises. With Western science
beginning to feel the economic pinch, it seems
likely that market economy countries will soon
experience the gradual extension of such procedures beyond the laboratories of private
industry to which it has mainly been confined in
the past (see Chapter 28, 'Science Policies'). In
this context, it is to be expected that governmental agencies as well as industries will keep
close track on patents considered as 'output'
of scientific research (see Table 62).

We can conclude this section by quoting
the last paragraph of the recent article by
Mansfield already cited:

Yet, having taken pains to point out the limitations of the individual bits of evidence that have
been amassed, we must not lose sight of an
impressive fact: no matter which of the available
studies one looks at, the conclusions seem to point
in the same direction. In the case of those using
the judgmental approach, there is considerable

1. *Science policy in the U.S.S.R.*, Paris, OECD, 1969.

agreement that we (in the U.S.) may be under-investing in particular types of R & D in the civilian sector of the economy. In the case of the econometric studies, every study of which I am aware indicates that the rate of return from additional R & D in the civilian sector is very high.

Trying to measure the cultural value of science presents problems of a very different nature. These will be discussed in Part IV.

TABLE 62. Patents granted

	To nationals	To foreigners	Total
United States	47,100	17,300	64,400
U.S.S.R.[1]	30,600	1,800	32,400
France	8,500	17,800	26,300
Japan	21,400	9,500	30,800
Argentina	1,500	5,200	6,700
Brazil	500	2,200	2,700
Iran	50	600	650
Kenya	—	16	16

1. Includes inventors' certificates.

Source: Industrial Property, World Intellectual Property Organization (WIPO), 32, Chemin des Colombettes, 1211 Geneva 20, December 1972.

Changes in the research system

While well aware of the difficulties of forecasting the future, even over a short period, and the more so in an area like science where unexpected novelties may appear at any time, still it appears possible to identify certain trends in the existing research system, that may characterize the forthcoming evolution of scientific and technological research. In the attempt to pin-point such major trends we will have to limit our research to countries and regions where the government and the private sector are already investing large sums and where large scientific and technological communities are to be found, i.e. Europe, North America and Japan.

The study of recent documents by specialists in science policy, as well as the preoccupations expressed by politicians, advisers to governments and industry, and by scientists active in the laboratory, lead to the conclusion

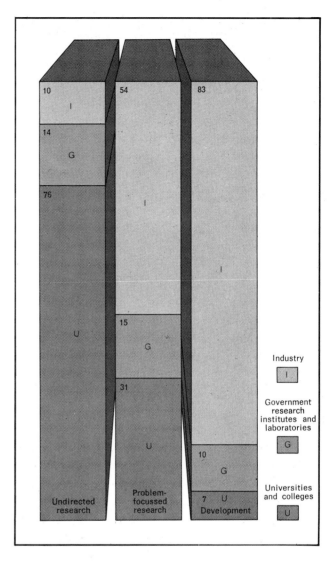

FIG. 81. Percentage distribution, by institutions, of the R & D activities which have led to five important technological innovations (magnetoferrites, videotape recorders, oral contraception pill, electron microscope, insulation matrix).

that the scene of science in the technologically advanced countries is likely to be influenced by three trends, which today are clearly perceptible but have not yet exerted any significant pressure on the research system; tomorrow they may become its dominating factors. Such trends can be discussed under three headings: integration, social relevance, and stability.

INTEGRATION

The presentation of the procedures followed in different countries for the planning and the management of the scientific effort, as discussed in previous chapters, may have led the reader to the conclusion that there have been many attempts to formulate a policy for science, but few instances in which major decisions at the governmental level were taken within the context of the general political, social and economic aims.

Science and technology policy has evolved through four stages. Up to 1945, there prevailed the attitude of *laissez innover*; from about 1945 to 1955, one could detect an aggregation of policies; from 1955 to 1968, there came a stage of co-ordination; and from that time to the present, there has been a tendency to integration.

In the first, *laissez inventer et innover*, stage, the private and public 'market' for discoveries, inventions and innovations determined what research, development and innovation work was doing. In the second, aggregate, stage, individual countries developed specific defence research policies, and policies for atomic energy, industrial research, and so on, but with very feeble attempts to co-ordinate them. Beginning in the 1950s, some countries began making concerted efforts to co-ordinate this aggregate of policies for science and technology by creating structures appropriate to the purpose within, or close to, the centres of political power. In the last stage, from the late 1960s to the present, the stimulation of the growth of research, development and innovation (and of their utilization) is being approached increasingly from the following point of view. This is the view that their growth is integrated with that of the national social system, not only as its needs are perceived now but with our eyes focused on future, long-range needs.

With the obvious exception of socialist countries, and possibly of France in some instances at least, governmental action in favour of S & T has consisted primarily in allocating rapidly increasing sums for research activities and in orienting some of them towards certain national objectives. The increase in financial resources was so rapid and uninterrupted that it was considered as an irresistible phenomenon beyond the control of political choices. Such a situation prevailed until about 1967, when, in some cases, the rate of investment growth tapered off. Questions began to be raised within the scientific and political communities on the criteria used by the decision makers. As we shall see shortly, science policy became a sort of independent variable within the general political process; now one notices an increasing and expanding awareness of the need for integrating scientific and technological development within the larger context of governmental decisions in politics, economics, education and social affairs.

One example of the theoretical discussion of such interactions and integration can be found in a recent study by Yvan de Hemptinne, where he examines a cybernetic model of the national R & D system.[1]

For our present purposes, the national R & D system will be defined as a set of organized scientific and technological resources and activities aiming at discovering, inventing, transferring, and promoting the application of new knowledge, with a view to achieving national objectives set by political authorities representative of the will of the people. The national R & D system is not the only subject of national science policy, but is perhaps its most important element. Science policy in the broad sense, as understood by countries in the forefront of modern progress, also covers such matters as training of researchers, technological innovation, international scientific co-operation, scientific monitoring of the human environment, etc.

A diagrammatic representation of this conception is given in Figure 82 which shows by way of example how various sectors of human activity link up with science policy.

In this study, we shall deliberately limit ourselves to the national R & D system (including related scientific and technical services), to the exclusion of aspects of national science policies at their point of contact with such other human activities as education, production, leisure, environment, etc., highly important as those aspects are.

1. Y. de Hemptinne, *Governmental Science Policy Planning Structures*, Paris, Unesco, 1972 (Unesco/NS/Rou. 234).

Similarly, for purposes of this analysis, the national R & D system will be treated as a cybernetic system, as shown in Figure 83. One of the chief characteristics of such systems is that they remain 'subservient' in spite of their autonomy, in the sense that they are always subordinate to something else. The government official stresses, and rightly so, the ultimate purpose of the system, in this case, national objectives, whereas the researcher rightly protects his freedom of action by stressing its autonomy, which as we know is the best guarantee of its efficiency. Ultimate purpose and autonomy are linked by means of signals, represented in the diagram by arrows. The national R & D system can thus fit in with its 'environment' and at the same time keep at arm's length the authority on which it depends.

Moreover, we know that every cybernetic action is the result of a synthesis of energy and information. Figure 83 therefore represents the flow of energy in the form of means (financial, human, material and other resources), while information is introduced into the system by the 'regulator' or comparator between the objectives set and the results obtained.

Finally, from the functional point of view three principal zones are to be distinguished in this cybernetic diagram:

Zone I corresponds to the planning and interministerial co-ordination of the system, i.e. the definition of its overall objectives and the mobilization and positioning of resources. It is at this 'strategic' level that the science policy of governments takes shape. This zone also includes the function of sectorial promotion of science policy, i.e. translation of general objectives into specific scientific tasks and projects.

Zone II corresponds to the actual performance of R & D, including management of research institutions, of R & D scientific services, and of both the processing and the 'packaging' of information produced by the system. It is at this 'tactical' level that the efficiency of the system is judged.

Zone III is the zone where scientific and technical information is used and applied in practice, and where one can see whether the system is ultimately justified and can gauge its effectiveness.

Similarly, the well-known 'Brooks report'—the report of a high-level panel appointed by the Secretary-General of OECD to advise the member governments, the vast majority of the advanced market economy countries[1]—includes the following statements:

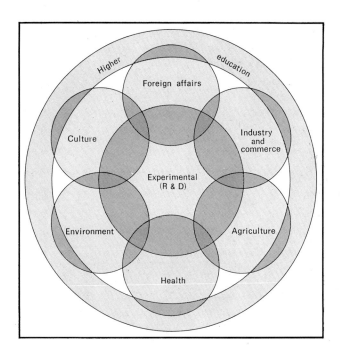

Fig. 82. Connexion between science policy and other sectors of the life of a country, according to Yvan de Hemptinne.

During the postwar era one can distinguish three phases of the overall political climate in which science has operated. The first was the longest and continued, with some fluctuations, from the end of the war to the beginning of the 1960s. It was characterised by high public faith in the efficacy of science, and high political prestige of scientists, chiefly in the countries that were victors in the war, and above all in the United Kingdom and the United States. During this period national security considerations and the evolution of the cold war dominated the formulation of national science policies. The science of physics held the centre of the stage in the development of postwar science as a whole, and the views and style of physicists held sway in the institutions of science and the councils of national science policy.

The second phase extended from about 1961 to 1967 and was characterised by the gradual emergence of economists and systems analysts as significant influences in science policy. Government attitudes towards science and technology emerged from the euphoric phase into an era in

1. *Science, Growth and Society: A New Perspective. Report to the Secretary General's* ad hoc *Group on New Concepts of Science Policy*, Paris, OECD, 1971.

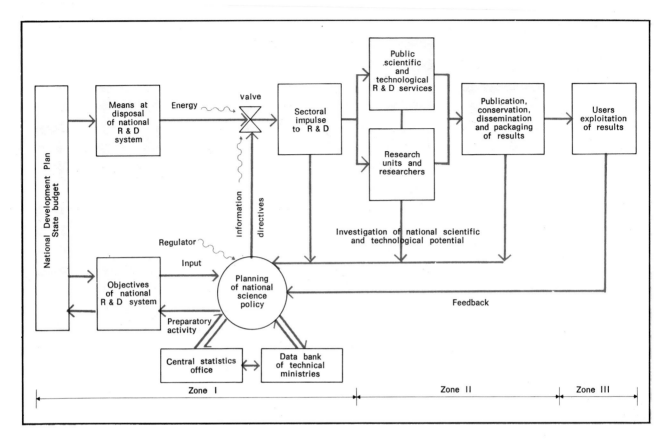

FIG. 83. Cybernetic model of national R & D system, represented as an organized system of resources and activities. (*Source*: Y. de Hemptinne, *Governmental Science Policy Planning Structures*, Paris, Unesco, 1972 (Unesco/NS/Ron. 234).)

which rational resource allocation and the role of science and technology in economic growth replaced the cold war and technological competition as central issues. Science began to be related to a much wider range of national problems, beginning with health and extending into the social and economic domains generally. Even the national security aspects of science were viewed partly in terms of their possible spill-over into economic and social development. This was the era of controversy over the existence of a technological gap, the process of transfer of science and technology to underdeveloped countries, research priorities in general, and the interpretation of academic studies of the economics and sociology of technological innovation.

But it was an era in which the attitude towards science and technology was fundamentally optimistic and hopeful. If the world was in trouble, it was because it had too little or the wrong kind of science or had not learned how to use it effectively. In this thrust R & D was regarded as a valuable investment in its own right, almost independent of its subject matter (provided it was sufficiently advanced and 'sophisticated') and independent of coupling to the operational problems faced by governments.

The third phase may be described as the period of disenchantment with science and technology. It began in the early 1960s in the United States; it made a more belated appearance in Europe, generally less acutely and in different forms. However, on neither side of the Atlantic did the recognitio of the leeway to be regained prevent questioning, first, of the major programmes theretofore assigned unquestioned priority and, then, of the goals to which the overall orientation of science and technology had been guided. However differently these problems may be perceived, the questioning of the aims of science policy finally merges into the questioning of the processes and ends of science itself.

Policy for science and technology is only one element of overall public policy; thus, it must be recognised that the search for the 'optimum', whether in relation to the aggregate amount allocated to R & D activities or to methods of management and execution, is more an art than a science, and it is unrealistic to expect otherwise. That is no reason, however, for not trying to link R & D effort more closely with a planning effort or, in other words, to introduce greater consistency between decisions affecting research activities and other policy decisions.

If science policy consists in getting the best out of the resources of science and in promoting technical innovation to attain national goals, it is essential that its links with other areas of government action directed towards the same ends be clearly understood. Study of the innovation process and a growing appreciation of the complex of technological, socio-economic and managerial factors have indeed revealed the limitations of the policies previously followed: first, the different elements of science policies were usually treated independently of each other; second, science policies themselves were often treated in relative isolation from other policy decisions.

The complete integration of the policies for science into the national planning of all components of governmental action will call for a major effort and great imagination on the part of the politicians and bureaucrats, with the close assistance of scientists and engineers. Necessarily it will be a gradual and slow process: in the long run, it should make the operation of all branches of government more rational, more scientific.

SOCIAL RELEVANCE

The growing concern of citizens and governments for the intricacy of the problems of industrial societies, for the consequences of runaway technologies, accompanied by the discontent of science as the panacea for all human problems (which we have briefly discussed in Part I and will elaborate in greater detail in Part IV) have convinced decision-makers that before embarking on enterprises with a large technological component, or with potential harmfulness to man, it will be necessary to evaluate their consequences at the level of society. Accordingly, in some countries special bodies have been created by governments and parliaments for technological assessment and steps are being taken to encourage research in areas which appear particularly relevant for the whole of society.

From the Brooks report we quote:[1]

In the last 30 years science policy priorities and choices have been determined largely by external challenges such as defence, national prestige, or competition in world markets rather than by other social needs such as the improvement of public services, the elimination of poverty, or the improvement of public health. These older priorities may be questioned on moral, political, social or scientific grounds; however, in terms of technological advances, it cannot be denied that many specific goals have been met. The success was due not only to the management methods thus developed but also to the fact that the targets were clear and straightforward, and the necessary fundamental knowledge was for the most part either in existence, or there was sufficient theoretical understanding to know where to look for it. If there is a lesson to be learned from past experience it is that clear goals, plus the requisite state of development of the underlying science, are perhaps the two main factors of technological success, and that, given these, a well-defined technological objective can usually be met, given enough money and an adequate supply of high-quality manpower.

The conditions for the success achieved in many space and defence projects are not yet evident in the social sphere. The goals are complex, unclear, and the subject of conflicting interests and a wide diversity of values and preferences among the people affected. The underlying basic knowledge is spotty, fragmented, and often not firmly established in a scientific consensus. Not only is it more difficult to achieve the ends proposed, but also, if not mainly, it is more difficult to determine what the ends themselves should be. In a democratic society, the problem of goal-setting gives rise to an arbitration between conflicts of interests in a context of imperfect information, ambiguity, negotiation, and divergent pressures. Competing value judgements multiply between budget considerations of economy and political considerations of needs and effectiveness.

To the extent that problems and goals have a non-technical component, the correct formulation of the problem so that it can even be addressed by the methods of science and technology becomes itself a demanding intellectual task. Systems analysis has been one approach to this task, but we do not possess a tested technique of systems analysis that is applicable in the presence of conflicting goals and values among different segments of the

1. *Science, Growth and Society: A New Perspective,* op. cit.

affected populations. Whether such an extended form of systems analysis can ever be developed is a matter of argument and speculation.

The setting of priorities in science and technology in the context of the social goals that are coming to the centre of the stage constitutes a formidable challenge to science policy. In many ways we will be forced to proceed along two parallel paths, and the setting of priorities between these paths will involve a delicate balancing of long-term and short-term considerations. One will be a short-term 'applied' approach based on current data and understanding, imperfect and incomplete as they are. This will involve high risk and may result in the pursuit of many wild geese. It will be expensive in operational and investment terms and may produce some spectacular failures with unpleasant political sequels. Some of the problems are too urgent to avoid this risk.

The fact that science and technology can no longer be considered as a sort of independent, and necessarily beneficial, variable of society—together with the realization that future scientific developments must be considered in a perspective different from that of the recent past—this is likely to influence profoundly the attitudes of citizens and governments towards science policy. In Part IV we will have an opportunity of discussing the various human implications of scientific advance.

STABILITY

The 'explosive growths' of which we are all aware—whether in the volume of scientific publications, in applications for university entry or in world population—are each of them a sign that the fundamental equilibria of earlier human life and society have been upset, by the improvement of medicine, by the removal of social barriers, or by the uninhibited spread of intellectual curiosity.

In these terms the philosopher S. E. Toulmin was discussing the problems of scientific growth in 1966 and was further raising the questions:[1] 'What are the chances that, here too (as in biological phenomena of growth), the current phases of rapid growth will come up against ceilings? Is there, indeed, any alternative?'

Already in 1963 Derek J. de Solla Price had discussed the characteristics of exponential curves, as related to scientific growth, in his classic book *Little Science, Big Science*,[2] and was hinting at the inevitable fact that the rate of growth must slow down, sooner or later, and reach a plateau.

In the real world things do not grow and grow until they reach infinity. Rather, exponential growth eventually reaches some limit, at which the process must slacken and stop before reaching absurdity. This more realistic function is also well known as the logistic curve, and it exists in several slightly different mathematical forms. Again, at this stage of ignorance of science in analysis, we are not particularly concerned with the detailed mathematics or precise formulation of measurements. For the first approximation (or, more accurately, the zeroth-order approximation) let it suffice to consider the general trend of the growth.

The logistic curve is limited by a floor—that is, by the base value of the index of growth, usually zero—and by a ceiling, which is the ultimate value of the growth beyond which it cannot go in its usual fashion (Fig. 84). In its typical pattern, growth starts exponentially and maintains this pace to a point almost halfway between floor and ceiling, where it has an inflection. After this, the pace of growth declines so that the curve continues toward the ceiling in a manner symmetrical with the way in which it climbed from the floor to the midpoint. This symmetry is an interesting property; rarely in nature does one find asymmetrical logistic curves that use up one more parameter to describe them. Nature appears to be parsimonious with her parameters of growth.

Because of the symmetry so often found in the logistic curves that describe the growth of organisms, natural and manmade, measuring science or measuring the number of fruit flies in a bottle, the width of the curve can be simply defined. Mathematically, of course, the curve extends to infinity in both directions along the time axis. For convenience we measure the width of the midregion cut off by the tangent at the point of inflection, a quantity corresponding to the distance between the quartiles on a standard curve of error or its integral. This midregion may be shown necessarily to extend on either side of the center for a distance equal to about three of the doubling periods of the exponential growth.

1. S. Toulmin, 'Is there a Limit to Scientific Growth?', *Science Journal*, August 1966, p. 80–5.
2. D. J. de Solla Price, *Little Science, Big Science*, New York, Columbia University Press, 1963. Reprinted by permission of the publisher.

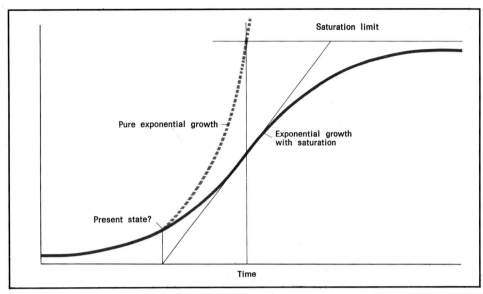

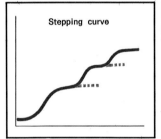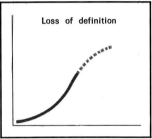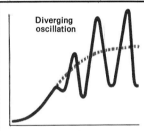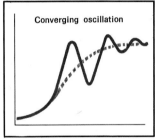

FIG. 84. General shape of the logistic curve and various ways in which organisms (both natural and artificial) following this curve may approach the saturation limit. Mathematically speaking, the curve extends to infinity, in both directions, along the horizontal (time) axis. Measurement of the size of the region cut by the tangent at the point of inflexion shows that this median region extends, on either side of the centre, for a distance equal to about three redoubling periods of the curve representing exponential growth. (From D. J. de Solla Price, *Little Science, Big Science*, New York, 1963.)

Now, with no stronger assumption than has been made about the previously regular exponential growth with a doubling period of 10 to 15 years, we may deduce, as we have, that the existence of a ceiling is plausible since we should otherwise reach absurd conditions at the end of another century. Given the existence of such a limit, we must conclude that our exponential growth is merely the beginning of a logistic curve in other guise. Moreover, it is seen that as soon as one enters the midregion near the inflection—that period of secession from accustomed conditions of exponential growth—then another 30 to 45 years will elapse before the exact midpoint between floor and ceiling is reached. An equal period thereafter, the curve will effectively have reached its limit. Thus, without reference to the present state of affairs or any estimate of just when and where the ceiling is to be imposed, it is apparent that over a period of one human generation science will suffer a loss of its traditional exponential growth and approach the critical point making its senile limit.

However, growths that have long been exponential seem not to relish the idea of being flattened. Before they reach a midpoint they begin to twist and turn, and, like impish spirits, change their shapes and definitions so as not to be exterminated against that terrible ceiling (Fig. 84). Or, in less anthropomorphic terms, the cybernetic phenomenon of hunting sets in and the curve

377

begins to oscillate wildly. The newly felt constriction produces restorative reaction, but the restored growth first wildly overshoots the mark and then plunges to greater depths than before. If the reaction is successful, its value usually seems to lie in so transforming what is being measured that it takes a new lease on life and rises with a new vigour until, at last, it must meet its doom.

So wrote D. J. de Solla Price in 1963. At the time, he expressed the view that the turn of the curve describing scientific production would probably occur about thirty years later than the first onset of difficulties at the end of the Second World War, i.e. approximately now. Looking at things as they are now in the developed countries, one can say that he was substantially right: there are clear signs in most of the industrialized countries that the rate of growth in public and private investments in scientific research is diminishing, and we can therefore expect that within a not too long time the magnitude of the scientific effort will reach a plateau.

It is possible, of course, to speak of saturation only at the national level: at the international level, growth might continue for decades before showing signs of deceleration. But we can identify clearly signs of 'over-developed' conditions in the United States (1967), Sweden (1968–69) and the U.S.S.R. (1972). We can then raise the question: How far are we from the time when a plateau will be reached? Will we reach it smoothly, through a progressive slowing down, through loss of definition, or divergent or convergent oscillation? Difficult as it is to foresee the future, it seems likely that, barring catastrophic events, the stage of stability could be reached through 'loss of definition'. If the trend towards integration, previously discussed, comes true, within a relatively short period of time many people now described as 'scientists' would shift their main interests away from work in the laboratory to become involved in social work, management and the like. Correspondingly, the sums earmarked for scientific research would diminish and reach a plateau, while other investments (today unpredictable) would become a major concern of governments and industry. Such 'smooth' change bound to the advent of a progressively more 'scientific' so-ciety is obviously optimistic, and will be discussed further in Part IV.

But one more consideration should be noted, if we extend our outlook to the rest of the world, in the case of the developing countries. There, as we shall argue in the next section, scientific and technological research is hardly noticeable in most cases. While it can be argued that those countries should not attempt to pattern their social and economic planning after the model of Europe and North America, there is indeed the possibility that in coming years we shall witness a major scientific development. If this should happen, exponential growth of the scientific effort would occur. On the gross assumption that in industrialized countries there is one graduate in science or technology for every 100 inhabitants, some 25 million persons should be trained in those fields in the Third World in order to reach a balance with respect to the rest of the world. This means that, at the global level, there would still be room for a tenfold expansion; this will require a major effort, extended over several generations. Obviously, the difficulty of the task would be even greater if those countries were to reach a stage in which 15 per cent of their graduates in science and technology were to perform research activities. It would be unduly optimistic to consider this expansion as a serious possibility in the course of a few decades. This leads us to the conclusion that the period of triumphal expansion of science, as we have lived it through the last thirty years, is over. A state of stability will have to be reached, probably within this century. Stability does not necessarily mean stagnation. The challenges of a society in equilibrium may be even greater than those of an expanding one, and the imagination and fantasy of the scientist may find even more pressing stimuli than at present.

Science for development

Early on in the First Development Decade (1960–70), the United Nations organized the Conference on the Application of Science and Technology for the benefit of the Less Developed Areas (Geneva, 1963). The 'revolution of raising expectations', which was occurring

TABLE 63. Policy-making bodies for science and technology in African countries[1]

	Ministry of science[2] or ministerial science policy committee	Science policy planning body-general	Multisectorial body for co-ordinating scientific research	Co-ordinating bodies for scientific research			
				Medical research	Agricultural research	Nuclear research	Industrial research
Algeria	—	—	—	—	—	—	—
Benin	—	—	—	—	yes	—	—
Botswana	—	—	—	—	—	—	—
Burundi	—	—	—	—	—	—	—
Cameroon	—	yes	yes	—	—	—	—
Central African Republic	—	—	yes	—	—	—	—
Chad	—	—	—	—	—	—	—
Congo	—	—	yes	—	—	—	—
Egypt	yes	...	yes	—	—	—	—
Equatorial Guinea	...	...	...	...	...	...	...
Ethiopia	—	—	—	—	yes	—	—
Gabon	—	—	—	—	—	—	—
Gambia	—	—	—	—	—	—	—
Ghana	—	yes[3]	yes[3]	—	—	—	—
Guinea	—	—	—	—	—	—	—
Ivory Coast	yes	—	—	—	yes	—	—
Kenya	—	—	—	—	—	—	—
Lesotho	—	—	—	—	—	—	—
Liberia	—	—	—	—	—	—	—
Libyan Arab Republic	—	—	—	—	—	—	—
Madagascar	—	—	yes	—	—	—	—
Malawi	—	—	—	—	yes	—	—
Mali	—	—	—	—	yes	—	—
Mauritania	—	—	—	—	—	—	—
Mauritius	—	—	—	—	—	—	—
Morocco	—	—	—	—	—	—	—
Niger	—	yes[3]	yes[3]	—	yes	—	—
Nigeria	—	yes	yes[3]	yes	yes	—	yes[3]
Rwanda	—	—	—	—	—	—	—
Senegal	—	yes[3]	yes[3]	—	—	—	—
Sierra Leone	—	—	—	—	—	—	—
Somalia	—	—	—	—	—	—	—
Sudan	—	yes	—	yes	yes	—	yes
Swaziland	—	—	—	—	—	—	—
Tanzania	—	yes[3]	yes[3]	—	—	—	—
Togo	—	—	—	—	—	—	—
Tunisia	—	—	—	—	—	—	—
Uganda	—	yes[3]	yes[3]	—	—	—	—
Upper Volta	—	—	—	—	yes	—	—
Zaire	—	yes[3]	yes[3]	—	yes	yes	—
Zambia	—	yes[3]	yes[3]	—	yes	yes	—

1. Table based on Vol. IV of the *World Directory of National Science Policy-making Bodies* published by Unesco (in preparation); ... indicates no information available.
2. Without other responsibilities.
3. The same body performs both functions.

379

in the less-developed countries, seemed to have identified in science and technology the panacea that was going to solve the problems of poverty, disease, backwardness, in a short time.

On the basis of the recommendations of the governmental conference, the United Nations established the high-level Advisory Committee on the Applications of Science and Technology and the Office of Science and Technology at their headquarters. In the following years the United Nations themselves, the Specialized Agencies, regional intergovernmental organizations and governments created a myriad of bodies, committees, councils, working groups, etc., which were supposed to come up with concrete suggestions on how to use modern scientific knowledge to accelerate economic and social development in the least-developed countries. One direct result of this was that the world was flooded by a deluge of good intentions and printed paper.

Ten years later the situation was best synthesized by Raoul Prébish: the Development Decade was the 'decade of frustration'. Some have even come to believe that the so-called applications of science and technology have worked against the efforts of development of the Third World. Thus Théo Lefèvre, Minister for scientific research of Belgium, wrote recently: 'Force nous est de constater que, sous l'effet de la science, l'écart ne fait qu'augmenter entre les pays industrialisés et ceux du tiers monde.' Something has gone wrong, but there is hardly a unanimity on the diagnosis of what has gone wrong. In the previous chapters we have often discussed the situation of the organization of science and technology in the least-developed countries, which was summarized (perhaps in an excessively pessimistic vein) by Stephen Dedijer as 'underdeveloped science in underdeveloped countries'. In reality, it is equally unjustified to consider all the least-developed countries as a homogeneous group in terms of both their socioeconomic conditions and their scientific and technological development. There are quite a few developing countries in which one finds a degree of sophistication in the organization and deployment of science and technology comparable to that of the smaller European nations,

for example, India, Egypt and Argentina. Nevertheless, conditions of general underdevelopment prevail, as if there were no correlation between scientific and economic progress. Tables 63-66 bring together a few data for comparison.

Innumerable books and articles have been published, and an untold series of meetings has been organized throughout the world, to analyse the problem, express faith in the thaumaturgic virtues of modern science and technology, and recommend to governments and private bodies the investment of larger shares of their revenues for a much more intensive and widespread use of science and technology. The gospel of science has been preached all over the developing world, and governments have been urged to set up special bodies for the development of science policy at the national level. These appeals have been met by a remarkable response (Table 63 to 66). In spite of a deluge of words and much 'scientific tourism' by European and North American 'experts', there was hardly any serious commitment of real effort to build up scientific and technological capabilities and direct them—deliberately—at development problems. Unfortunately, the social and economic conditions of the least developed countries have, with few exceptions, been worsening, when compared with those of North America, the U.S.S.R., Europe and Japan. The wealthy are becoming wealthier and the poor poorer; and the same trend seems to prevail within individual countries. Exceptions might be countries with great capacities for self-reliance (People's Republic of China) or massive doses of foreign investment (Republic of Korea).

The phrase 'the developing world' is no more than a convenient (or inconvenient) shorthand symbol, embracing countries as diverse ethnically, culturally, and in every way, as mankind itself. These diversities extend to economic development: to levels of past achievement, to rates of present growth, to future prospects.

Measures designed to aid the developing world as a whole thus face the problem of being effective for, and relevant to, Brazil and Bhutan, Singapore and Somalia, Turkey and United Republic of Tanzania. In particular, many development specialists argue that the very poorest countries, being without the bases of economic development, lack the capacity to absorb the sophisticated assistance essential for the more rapidly developing countries.

Against this background, in November 1971, the United Nations General Assembly approved a list of twenty-five least-developed countries. The criteria for selection were three-fold: *per capita* GNP not more than $100 yearly; literacy rates not more than 20 per cent (for the population aged 15 and over);

a manufacturing sector totalling no more than a tenth of GNP. Some of the twenty-five countries met all three criteria; others met two and were borderline in the third.

Of the twenty-five least-developed countries, sixteen are in Africa. Of these, fourteen are in an unbroken, dog-leg band that runs from the Atlantic across the Sahara to the Red Sea before turning south—Guinea, Mali, Upper Volta, Dahomey, Niger, Chad, Somalia, Uganda, Rwanda, United Republic of Tanzania, Burundi and Malawi. The Yemen Arab Republic, the only least-developed

TABLE 64. Policy-making bodies for science and technology in Latin American countries[1]

	Ministry of science[2] or ministerial science policy committee	Science policy planning body–general	Multi-sectorial body for co-ordinating scientific research	Co-ordinating bodies for scientific research			
				Medical research	Agri-cultural research	Nuclear research	Indus-trial research
Argentina	yes	yes	yes	—	yes	yes	yes
Barbados	—	—	—	—	—	—	—
Bolivia	—	—	—	—	—	yes	—
Brazil	—	—	yes[3]	—	yes	yes	yes
Chile	—	yes	yes	—	yes	yes	—
Colombia	—	yes[4]	yes[4]	yes	—	yes	yes
Costa Rica	—	—	—	—	—	yes	—
Cuba	—	yes[4]	yes[4]	—	—	—	yes[5]
Dominican Republic	—	—	—	—	—	—	—
El Salvador	—	—	—	—	yes	—	—
Ecuador	—	—	—	—	—	yes	—
Guatemala	—	—	—	—	—	yes	—
Guyana	—	—	—	—	yes	—	—
Haiti	—	—	yes	—	—	—	—
Honduras	—	yes	—	—	—	—	—
Jamaica	—	—	yes	—	—	—	—
Mexico	—	—	yes	—	yes	yes	yes
Nicaragua	—	—	—	—	—	—	—
Panama	—	—	yes[6]	—	—	—	—
Paraguay	—	—	yes[7]	—	—	—	—
Peru	—	yes	yes	yes	yes	yes	—
Trinidad and Tobago	—	—	yes	—	—	—	—
Uruguay	—	—	yes[3]	—	—	yes	—
Venezuela	—	yes[4]	yes[4]	—	yes	—	—

1. Table based on Vol. III of the *World Directory of National Science Policy-making Bodies* published by Unesco (Paris, 1968). The table includes Central America and the Caribbean Islands.
2. Without other responsibilities.
3. This body also performs general science planning functions at national level.
4. The same body performs both functions.
5. Research into sugar cane by-products.
6. Private organization.
7. Natural sciences.

TABLE 65. Policy-making bodies for science and technology in European and North American countries[1]

	Ministry of science[2] or ministerial science policy committee	Science policy planning body–general	Multi-sectorial body for co-ordinating scientific research	Co-ordinating bodies for scientific research						
				Medical research	Agricultural research	Nuclear research	Industrial research	Space research	Oceanographic research	Environmental research
Europe										
Albania	—	yes[3]	yes[3]	—	—	—	—	—	—	—
Austria	yes	yes	yes	—	—	yes	yes	—	—	—
Belgium	yes	yes	yes	yes	yes	yes	yes	—	—	yes
Bulgaria	yes	yes	yes	—	yes	yes	—	—	—	—
Byelorussian S.S.R.	—	yes	yes	see note 4 below						
Cyprus	—	—	yes	—	yes	—	—	—	—	—
Czechoslovakia	yes	yes	yes	—	—	—	—	—	—	—
Denmark	—	yes[3]	yes[3]	yes	yes	yes	yes	—	—	—
Federal Republic of Germany	yes	yes	yes	—	—	yes	—	yes	yes	—
Finland	—	yes[3]	yes[3]	yes	yes		yes	—	—	—
France	yes	yes[3]	yes[3]	yes	yes	yes	yes	yes	yes	yes
Greece	—	yes[3]	yes[3]	—	yes	yes	—	—	—	—
Hungary	yes	yes[3]	yes[3]	yes	yes	yes	yes	yes	—	—
Iceland	—	yes[3]	yes[3]	—	—	—	—	—	—	—
Ireland	—	yes[3]	yes[3]	yes	—	—	—	—	—	—
Italy	yes	yes[3]	yes[3]	yes	yes	yes	—	—	—	—
Luxembourg	—	yes[3]	yes[3]	—	—	—	—	—	—	—
Malta	—	—	—	—	yes	—	—	—	—	—
Monaco	—	—	yes	—	—	—	—	—	yes	—
Netherlands	yes	yes[3]	yes[3]	yes	yes	yes	yes	yes	—	yes
Norway	—	yes[3]	yes[3]	—	yes	—	yes	—	—	—
Poland	yes	yes[3]	yes[3]	yes	yes	yes	—	—	—	—
Portugal	—	yes[3]	yes[3]	—	yes	yes	yes	—	—	—
Romania	yes[3]	yes[3]	yes	yes	yes	yes	yes	—	—	—
Spain	yes	yes	yes	yes	—	yes	—	yes	—	—
Sweden	—	—	—	yes	yes	—	yes	—	—	yes
Switzerland	—	—	yes	—	—	yes	—	yes	—	—
Turkey	—	yes[3]	yes[3]	—	—	—	—	—	—	—
Ukrainian S.S.R.	—	yes	yes	see note 4 below						
U.S.S.R.	yes	yes	yes	see note 4 below						
United Kingdom	—	yes[3]	—	yes	yes	yes	yes	—	—	yes
Yugoslavia	—	yes[3]	yes[3]	—	yes	—	—	—	—	—
North America										
Canada	yes	yes[3]	yes[3]	yes	yes	yes	—	—	—	—
United States	—	yes	yes	yes	—	yes	yes	yes	yes	yes

1. Table based on the permanent Unesco survey of national science policy-making bodies in Member States.
2. Without other responsibilities.
3. The same body performs both functions.
4. Each Ministry or Department has (a) a scientific directorate; (b) a technological directorate; (c) a technological and scientific council; (d) a council for co-ordinating scientific activities.

TABLE 66. Policy-making bodies for science and technology in Asian countries[1]

	Ministry of science or ministerial science[2] policy committee	Science policy planning body–general	Multi-sectorial body for co-ordinating scientific research	Co-ordinating bodies for scientific research						
				Medical research	Agricultural research	Nuclear research	Industrial research	Space research	Oceanographic research	Environmental research
Afghanistan	—	—	—	—	—	—	—			
Australia	yes	yes	yes[3]	yes	yes	yes	yes[3]			
Bahrain	...	...	...	...	...	...	...			
Burma	—	yes[3]	yes[3]	yes	—	—	yes			
China	...	...	...	...	...	...	...			
Hong Kong	...	...	...	...	...	...	...			
India	yes	yes	yes[3]	yes	yes	yes	yes[3]	yes		yes
Indonesia	—	yes[3]	yes[3]	—	—	yes	—			
Iran	yes	yes[3]	yes[3]	yes	—	yes	—			
Iraq	—	yes[3]	yes[3]	—	—	—	—			
Israel	—	yes	yes	—	yes	yes	—			
Japan	yes	yes	yes	yes	yes	yes	yes			
Jordan	—	—	yes	—	yes	—	—			
Democratic Kampuchea	—	—	—	—	—	—	—			
Republic of Korea	yes	yes	—	—	—	—	yes			
Kuwait	—	—	—	—	—	—	—			
Laos	—	—	—	—	—	—	—			
Lebanon	—	yes[3]	yes[3]	—	—	—	—			
Malaysia	—	—	—	yes	yes	—	yes[4]			
Mongolia	—	yes[3]	yes[3]	—	—	—	—			
Nepal	—	—	—	—	—	—	—			
New Zealand	yes	yes	yes	yes	—	yes	—			
Pakistan	...	yes	yes[3]	yes	yes	yes	yes[3]			
Philippines	yes[5]	yes	yes	yes[6]	—	yes	—			
Qatar	—	—	—	—	—	—	—			
Saudi Arabia	—	—	—	—	—	—	—			
Singapore	yes	yes[3]	yes[3]	—	—	—	—			
Sri Lanka	—	—	—	yes	yes	—	yes			
Syrian Arab Republic	—	yes	yes	—	yes	—	—			
Thailand	—	yes	yes	—	—	yes	—			
Republic of Viet-Nam	—	—	yes	—	yes	yes	—			
Western Samoa	...	...	...	...	...	...	...			
Yemen	—	—	—	—	—	—	—			
People's Democratic Republic of Yemen	...	...	...	...	...	...	...			

1. Table based on Vol. II of the *World Directory of National Science Policy-making Bodies* (Paris, 1968) and Vol. IV (in preparation); . . . indicates no information available.
2. Without other responsibilities.
3. The same body performs both functions.
4. Rubber industry exclusively.
5. The National Science Board is responsible for co-ordinating medical, agricultural, nuclear, industrial and all other research activities.
6. Private organization.

country in the Middle East, is separated from this band only by the narrows of the Red Sea. Further south in Africa are landlocked Botswana and Lesotho—in fact, ten of the sixteen African countries are without coastlines.

Four of the Asian least-developed countries border on the Hindu Kush or Himalayan ranges—Afghanistan, Nepal, Buhutan and Sikkim. They are all landlocked, as is, away to their east, Laos. The remaining three are islands: Western Samoa in the Pacific, the Maldives scattered off India's south-west tip and the western hemisphere's only representative, Haiti.

As with the developing world as a whole, the least-developed display great diversity. For every Western Samoan, there are 250 Ethiopians. Six of the countries have population densities of under ten per square kilometre, while four are ten times as crowded. In area, the Maldive Islands would fit into the Sudan, 8,300 times with room to spare! As an expert group reported to UNCTAD, the problems 'vary a great deal from one country to another. It is therefore important to design an action programme for each of the least-developed countries'.

Naturally the least-developed have many shared characteristics lying behind the criteria that drew them into the same category in the first place. Most are peopled by peasants or nomads. Six families in seven live from the land, eating most of what they grow to keep themselves alive. The land is often marginal—desert, savannah or mountain—with limited and capricious water supplies. Livestock generally is of poor quality, stunted and toughened by the harsh environment. Where population grows, so does erosion, sweeping across land previously left under grass, and now needed for, and unsuitable for, crops.

The very conditions that make the peasant or nomad poor also make him unwilling to change. He feels he cannot risk it, for it is gambling with his family's survival. As a report by Ambassador Martin, Chairman of the Development Assistance Committee of OECD put it:

A revolution in village institutions must provide concrete help on the hard practical choices facing individual illiterate subsistence farmers: dare he try the new crop, when a distant market (which he does not fully understand) may offer less than adequate prices, and where even a good crop year may mean glut? Dare he desert the money lender for the local co-operative? Dare he disregard the lessons taught by father and ancestors to follow advice from a young extension worker?

In most parts of least-developed countries, transport is poor, with no railway and few local feeder roads. Scarce health services result in high infant mortality and malnutrition. Schools are few. Foreign investors keep their distance: banking facilities are weak, while the local market is usually small and with feeble purchasing power. With a few exceptions, mineral reserves are unknown (but may exist). As noted in the criteria, manufacturing is very limited: in 1969, less than 1 per cent of manufactures exported by developing countries came from the least-developed. Recent data are still more discouraging: the buying power of least-developed country exports in 1970–72 remained at best stagnant, and may have dropped. The exports themselves, mostly a narrow range of price-vulnerable primary products, have less and less interest for the rich market economies whose hard currencies the least-developed need.

The economic growth rates of the least-developed appear to be worsening. For all developing countries the GNP growth rate during the sixties was 5.1 per cent: for the least-developed it was 3.5 per cent. Surveys for 1970–72 show the least-developed with an average annual growth rate of 3 per cent, against nearly 6 per cent for all developing countries. Allowing for population increase, the least-developed *per capita* GNP growth for 1970–72 is near zero. In the crucial agricultural sector the news is also bad. During 1968–70, only five of the twenty-one least-developed countries for which the relevant data exist achieved agricultural growth rates equalling the 4 per cent Second Development Decade target. For most of them, food production is failing even to keep pace with population increase.

Aid flows to the least-developed have been well below those for the developing world as a whole. According to the Martin report, the least-developed received 18.5 per cent less official economic assistance *per capita* than other developing countries. The lowest *per capita* receipts of official aid for 1969–71 were Bhutan $0.18, Sudan $0.69, Haiti $1.03 and the Yemen Arab Republic $1.55, with Nepal, Afghanistan and Ethiopia only slightly higher. The Martin report contrasts these figures with Israel $29.21, Singapore $20 and Panama $14.70.

However, the Martin report continues:

The fact that the least-developed countries as a group are relatively less well treated . . . does not necessarily imply the need for an immediate and general increase in the volume of assistance. It can equally well be taken as a sign that there are difficulties in finding appropriate uses for it, at least within the conventional structure and forms of assistance. . . . An essential role for external assistance is therefore to create conditions for the effective use of assistance.

How can science and technology be applied to improve on the conditions of people whose styles of life are fundamentally different from those of the countries where scientific and engineering research have been developed? Criteria followed by governments and industries in the advanced countries simply cannot be applied in vast regions of the world. Indeed the volume *The Careless Technology*[1] offers a variety of examples of major mistakes incurred with the best of intentions. On the basis of the analysis of past failures and of the complexity of the problems at hand, novel policies are just beginning to be outlined. For example, the report of the eighth session of the United Nations Committee for Development Planning[2] appears as an important and radical document. It contains a number of ideas that are both fundamental to the concept of development planning, and to the use of science and technology for that purpose. The following list of eleven points is intended as a quick digest: they are arranged under the headings of Technology, Agriculture, Environment and Conclusion.

TECHNOLOGY

1. A contributing factor to the increasing gap between rich and poor in developing countries 'has been the labour-saving bias of certain kinds of technological change associated with development'.
2. Encouragement should be given to small- or medium-scale industries making light equipment using labour-intensive methods and local resources. Such industries are more efficient than the mass production of gadgets.
3. Science and technology should be used to produce economic growth by discovering labour-intensive techniques which reduce unemployment and save capital.
4. Strong and perceptive international assistance should be given for the development of alternative technologies.

AGRICULTURE

5. 'The objective of a prosperous rural sector needs to be at the centre of the strategy against mass poverty.'
6. Small-scale agriculture should be given particular interest because 'output per acre is usually higher on small holdings than on large farms'. It is even recommended that very small plots be given to families to feed themselves.
7. Agricultural products should be diversified (rather than increasingly specialized).
8. The policy of agricultural mechanization should be re-examined wherever it affects unemployment.

ENVIRONMENT

9. Development and environmental concerns are not competing but are complementary and mutually supporting.
10. Lack of development is itself a reflection of a poor nutritional and sanitary environment.

1. *The Careless Technology; Ecology and International Development*, edited by T. Farvar and J. P. Milton, Garden City, N.Y., The Natural History Press, 1972.
2. *Report of the Committee for Development Planning. 8th Session*, New York, United Nations, 1972 (E/5126).

CONCLUSION

11. 'It is in the common interest of mankind in the coming decades to formulate global development strategies and take more and more efficient measures for their implementation, in order to achieve a more balanced social and economic development throughout the world and a more harmonious relationship between man and nature.'

Such a document seems to indicate that we are approaching a turning-point in the philosophy of aid for development, both in economic terms and in the uses of scientific and technological research.

From the point of view of the economist, the new approach has been recently well-expressed by Samir Amin, director of the African Institute for Economic Development and Planning, in an article with the significant title 'Growth is not Development'.[1] He writes:

The Euro-American school of economists would have us believe that the prosperity of the world depends on the extension of Western institutions to the developing countries which find themselves on the periphery of the capitalist system. They overlook the fact that the evolution of human societies does not progress in orderly fashion through the extension of institutions from one nation to another or across cultural frontiers. It takes place in spurts of creative energy when a formerly dependent society breaks the bonds that fettered it. The centre has the power, but the future lies with the periphery. This is not the kind of view which the master civilizations like to hear, and when it is voiced, they tend to ignore it. More than a century ago, Alexis de Tocqueville warned that power was shifting away from Europe to America and Russia, but it took the bankers of Queen Victoria, of the Third French Republic and of William I of Germany another 50 years to find out how right he had been.

From the point of view of the scientist, it is worth quoting Dr C. M. Varsavsky, an Argentinian physicist:

The subject of aid to science and technology in developing countries is extremely complicated and deserves further study. Although many agencies have proposed, and even implemented, various policies, it is not at all clear that such policies are the best; they are in fact considered by many to be detrimental to the nations receiving the aid. . . .

Many of the aid programmes to scientists in underdeveloped countries are, in my opinion, aberrations that develop from the mistake of confusing Science with the Practice of Science. Many well-intentioned planners believe that since Science is universal, the same type of science can, and should, be practised everywhere. Few realize that although the law of gravity may be universal, it is very different for an American or a Russian physicist to work on gravity waves than it is for a Bolivian or Angolian one. And what makes it different is not just the immediate surroundings of the scientist (libraries, laboratories, and assistants) but—far more fundamental—it is the differences in cultural, social and economic background of the scientists as citizens of the different countries. . . .

The time has come for a reappraisal of the objectives of man. We have the experience of several thousands of years of civilization, a phenomenal technology, and the scientific method. With such tools, and a staff of both social and exact scientists, we should be able in a Happiness Research Institute to formulate objectives—and ways of reaching them—to make humanity happier.[1]

According to points of view of the type quoted, the major mistake made thus far is derived from the unwarranted assumption that there exists only one path to economic and social development of a country, that which was followed by the European and North American countries. In view of the great difference in their physical conditions and resources, in their cultural and traditional background, and in their local interests, the developing countries should adopt original and probably unprecedented patterns of development. For such purpose there is a need for building up an indigenous scientific and technological potential aimed at contributing directly to the self-realization of each country or region, consistent with the aspiration of the people.

A thorough assessment should be carried out of what the least-developed countries can

1. S. Amin, 'Growth is not Development', *Development Forum*, Vol. 1, No. 3, p. 1–3, Geneva, United Nations, 1973.
1. C. M. Varsavsky, 'Pugwash and Underdevelopment', *Pugwash Newsletter*, Vol. 7, 1970, p. 99–106.

realistically expect from science and technology; the study should be made not by the usual committee of experts, but by teams of scientists, engineers, economists and sociologists which should essentially belong to or be able to spend extensive periods in the countries or regions concerned, so as to reach a concrete 'feeling' of the needs. Imagination and inventiveness will be required. Probably, in the years to come, a new discipline will emerge: science for development, that would include those aspects of natural and allied social sciences that relate to the problems of least-developed countries.

The task ahead is that of bringing to the attention of scientists everywhere, and particularly of scientists in the developing countries, the existence of exciting and important problems, to make available research money, to publicize that availability and to provide the communications necessary for the establishment of a cohesive new discipline: science for development. There exists in the world an immense reservoir of talented scientists and engineers who are anxious to demonstrate to the world and to themselves that their efforts are important and useful. The mobilization of this largely untapped resource is likely to become the chief instrument for achieving a balanced human and economic development of countries of the Third World, through the realization of their human potentialities and increased productivity from their available natural resources.

Such a novel approach would require substantial changes also at the level of international institutions involved in the application of science and technology to development. In 1968 the Advisory Committee on the Applications of Science and Technology to Development (ACAST) called the attention of the United Nations to the need for streamlining methods, procedures and structures. Since then the situation has deteriorated further, through the establishment of an even larger number of bodies, committees, groups, panels, etc. At present the number of these bodies dealing with one or the other aspect of science and technology for development under direct United Nations responsibility, including the Regional Economic Commissions

but excluding the United Nations Specialized Agencies, exceeds 100! The cause of this excessive growth lies in the fact that new committees and working groups have been added to those already active to comply with new resolutions passed in a number of different United Nations bodies. These new bodies are often created without regard for the responsibilities and functions of previously existing ones, and compound the problem of bureaucratic proliferation.

As the most recent example of such a state of affairs, we can refer to the high-level governmental committee created by the Economic and Social Council of the United Nations, in which fifty-four countries participate, called the Committee on Science and Technology for Development (CSTD). The new body was established to plan and co-ordinate United Nations policies for bringing science and technology to bear on the problems of developing countries. It was created with the hope that it would provide a strong and effective voice in the development of United Nations science policies. Unfortunately, as reported in the scientific journal *Nature*:

. . . that hope dimmed a little last month when the committee's first meeting got bogged down in a political dispute, failed to tackle most of the items on its agenda and ended in uproar.

The state of confusion now reached and the novel approach to the problem, involving interrelations between science, technology and development, call for prompt measures to make the United Nations system more effective in this area, and also to co-ordinate its actions with those of other organizations with similar aims. Such measures cannot be identified or implemented unless a thorough study is made of the present situation. It would consist of the following parts:

1. Survey of United Nations and other bodies active in the promotion of science and technology in the least-developed countries. This part would consist of a detailed survey of the existing agencies, bodies, committees, panels, groups, etc., active in the area. The survey would include the whole United Nations system, intergovernmental agencies not belonging to the

United Nations system, national agencies of wealthier countries, private foundations and multinational corporations operating for the benefit of the least-developed countries. The survey would include data on terms of reference, date of establishment, reasons why the body was created, relationships to other bodies, etc.

2. Quantitative analysis of 1 in terms of financial investments and results. This part would present data on the costs of operation of the various bodies identified under 1 and will attempt to evaluate the concrete actions that have resulted from the activities of the said bodies.

3. Critical analysis of the present situation. On the basis of data presented under 1 and 2, a critical analysis would be carried out of the efficiency of current operations, attempting to identify the major pitfalls.

This part of the study should candidly appraise not only present inefficiencies in terms of proliferation of structures, but should also extend the evaluation to methods of operation of various bodies, relationships among them, organization of conferences, production of reports, ratios between staff on long-range contracts and short-range consultants, criteria for selection of consultants and experts, duplication and repetition of efforts, relations between work and time in headquarters and in the fields, amounts of paper used per structural and time unit, costs of translations, etc.

For this purpose studies *in loco* should be made not only at the seats of the various agencies, in consultation with their executive heads, but also in a representative sample of some twenty least-developed countries, in order to determine real levels of achievement as compared with original expectations, and also to identify difficulties in the national machineries devoted to the request, acceptance and utilization of foreign aid.

The study and the launching of the new discipline—science for development—are admittedly ambitious tasks. But the world's intellectual potential, financial means and good-will must be mobilized to catalyse the development of the people according to their aspirations, without well-meaning impositions on the part of the powerful.

Internationalizing the scientific effort

Science by its very nature is universal. Indeed, we have discussed (in Part I, Chapter 2) how the awareness of participating in a universal enterprise which disregards political frontiers even in time of war contributes to the intellectual excitement of the scientist. But such universality is an ideal, rather remote, alas, from the real world of the laboratory, of the university and of the research council. The Baconian notion that knowledge is power, power of a nature that could be used for the improvement of the human condition, has become the reality of the twentieth century. The scientist has become the symbol of the State, as Jean-Jacques Salomon has pointed out, for he needs the financial support of the government for his work and the government needs the scientist for its power and prestige.

As Salomon wrote recently:[1]

Science, however, has become associated with the state both because science needed the state and because the state could draw advantages from science: evolution is 'in many respects inevitable and in the very nature of things'.

Science has the initiative by virtue of the problems which it presents to the modern world, but it loses it in the solutions of these problems. The passage from means to ends involves assumptions and choices in which not only the scientificity of science, but also the spiritual community of which scientists boast, are of no help: if the monopoly of knowledge goes beyond the field of its technical applications, it becomes part of the common problem of individual commitment. The scientist does not feel that he was made to live either 'for' or 'from' politics, and yet he is condemned to live *in* politics: the modern Prometheus is bound by his vocation to the ambiguities of a role which neither his training nor his profession has equipped him to play.

For those scientists who have not given up the idea of seeing science as something greater and better than a technique, this situation does not

1. J.-J. Salomon, 'The *Internationale* of Science', *Science Studies*, Vol. 1, 1971, p. 23–42.

offer any more alternatives if they call the world to witness than it does if they seek to influence the organs of decision at a national level. They are obliged either to consider that the mission of science is to enlighten, if not to formulate the objectives which political authority should follow; or to leave to the political authority the problem of defining the aims which science should pursue. Illusion or desertion: in both cases the scientists remain linked—bound—to the system of decisions on which depend the means which will permit them to respond to their vocation. Rebels, while members of the 'establishment' or 'established', even if they contest, both are equally caught in the pitfalls of the political system. The extreme limit of this situation can only be either to turn one's back on worldly matters and to devote oneself exclusively to research or to sacrifice one's vocation on the altar of militant action.

The predicament of the modern scientist has been analysed in extreme terms by the political scientist Joseph Haberer,[1] who noted that

three significant transformations of modern science have occurred in the last century: first, the politicalization of science, then the shift from an international to a national orientation of the scientific enterprise, and finally, the professionalization of the community of science.

His criticism of the so-called internationalism of science offers arresting examples of its shallowness, which we will not go into now. But it is worth quoting a few of his paragraphs on the politicalization and professionalization of science:

Science as a multifarious human activity is not only a body of knowledge or theory, it is also a methodology, a praxis, a network of habits and roles through which this knowledge is acquired, tested, and transmitted. Further, science is a philosophy, an ideology, even a mythology—in any case, an outlook that contains considerable connotative and symbolic potency. Finally, science is an institution rooted in society and as such inevitably becomes politicalized. Because of its social nature, science is infused with politics. Politics is that sphere of human activity which deals with public problems arising primarily from the aspirations, conflicts, and dilemmas of social existence. While science is inherently political, it has only become politicalized in the 20th century. By 'politicalized' I mean that both in its internal affairs and in its relations to the rest of society science has become deeply immersed in political problems, issues and processes.

The question of the social responsibility of scientists was ignored in favour of the expediency I have termed prudential acquiescence. Before Hiroshima, scientists believed that there was a natural conjunction between their intellectual product and the betterment of humanity. Such a belief did not require them to consider the ambiguous consequences or moral choices entailed in the utilization of their work. For three centuries this optimism made it possible to evade, repress, and ignore the question of social responsibility in all but its narrowest forms.

Modern science has been singularly devoid of any serious concern with fundamental questions—for example, those involving the relations between ends and means. Its overriding instrumentalism has been expressed in its desire to control and dominate nature, almost as an end unto itself. Not an intrinsic love of knowledge, but a Faustian hubris characterized modern scientific temperament. . . .

These attitudes suggest that the power drive defines modern science and its practitioners far more accurately than does the belief that basic science is a disinterested search for knowledge and for the betterment of man's estate. It seems to me to be of the utmost significance that Bacon and Descartes, the institutional founders of modern science, placed the entire question of social responsibility into a limbo where it remained for the next 300 years. More than that, their advocacy of prudential acquiescence set the stance that modern science subsequently adopted in its relations with ruling powers. Their theories and their conduct posited retreat or an apparent acquiescence as the appropriate response to any serious confrontation with state or church.

We, the professional scientists, must admit that there is a great deal of truth in criticisms from students of the scientific enterprise. Indeed, individually and as a community, we have become to a greater or lesser degree corrupted by political power and national goals, often incompatible with our commitment to the universality of science. The time has come for an 'agonizing reappraisal' of our work, of our goals, of our attitudes not only towards our country, but to the world.

1. J. H. Haberer, 'Politicalization in Science', *Science*, Vol. 178, 1972, p. 713–24. Copyright by the American Association for the Advancement of Science.

I will discuss some of these issues in Part IV of the volume, but before closing this section on the organization of scientific research, I think it worth while to examine how we could attempt to bring about a reversal of the trend and devote our efforts to the internationalization of the scientific effort.

A few years ago I discussed the need of supporting scientific research at the international level.[1]

Considering the events of the last few years and the criticisms voiced against the prevailing nationalistic trends in research, I would advocate that all scientists, disregarding the kind of country they happen to be working in, should stop paying lip-service to the international idealism of science, and become active within their own national communities to urge their colleagues and the public authorities to transfer decisional power and funds to international agencies—which should become the only sponsors of basic and even technological research. As I will try to make clear in Part IV, the development of science in the long run is incompatible with the existence of sovereign nation-States. It could take a long time before world government becomes a reality, but I believe that it is our duty as professional scientists to refuse the 'prudential acquiescence', which so far has characterized our behaviour, and become aggressive missionaries for the internationalization of science.

While such a difficult and ambitious plan will require the contributions of a large number of devoted scientists, one could outline, as a starting point at least, the following steps: (a) mobilize the conscience of working scientists, by making them feel ashamed of their betrayal of the international ideal of science and their lack of a planetary conscience; (b) establish active and expanding groups, dedicated to this same purpose, within national communities that would establish close contact with similar groups in other countries; (c) outline together, for groups of countries, world regions, and eventually for the whole world, concrete actions which scientists could carry out to force their governments to relinquish financial and programming responsibilities and cede them to international bodies supporting science without the interference of the national bodies politic.

Proposals like this can sound preposterously utopistic and therefore not worthy of much attention. We should remember, however, that approximately at the time when modern science was born the abolition of autocratic absolute monarchies was considered equally utopistic. Thanks to the initiative of numerous intellectuals, among whom were many scientists, the concept of the democratic State was elaborated and later (even if imperfectly) realized. We, the encyclopaedists of the latter part of the twentieth century, must undertake a similar enterprise. Only by forcing the world of politics to accept the international ideals of science can we redeem our activities, and claim again that scientific endeavour is one of the noblest activities of man. Scientists of the world, unite: we must revamp the integrity of our mission.

1. A. A. Buzzati-Traverso, 'Scientific Research: The Case for International Support', *Science*, Vol. 148, 1965, p. 1440–4.

Part IV

The human implications of scientific advance

We live in the age of science. This is true in more than one way: not only has science opened for us unexpected horizons in our knowledge of the physical world, science itself has become a major social institution. And, through technology, science has changed the way of living of modern man.

Western civilization—in both capitalist and socialist countries—has been shaped by the myth that science and its applications would lead the human race inevitably to a state of well-being, happiness and freedom. The illuminist faith in reason first led us to believe that science, and only science, could resolve the human predicament. But now after 300 years and at the peak of science's glory, with doubt cast on such ideology, we wonder whether we have chosen the right path. We realize that

something has gone wrong. Science is in a state of crisis. The sense of alienation and dehumanization, felt by the masses toiling in the crucibles of high technology, is reinforced by the distress and even despair of scientists in the laboratory. Even thoughtful people in the developing countries begin to wonder whether the attempt to follow the pattern of economic development of the industrialized countries makes sense, or whether perhaps alternative models might be better suited to the needs of those regions.

Why? What has happened? Are these reactions justified? How relevant to the welfare of future generations will be further development of the scientific effort? Is there need for a change in our attitudes?

31

A period
of transition

The crystal clear pristine image of science that prevailed for well over two centuries began to tarnish when the scientist relinquished, or was forced to leave, his ivory tower. When the starry-eyed and aristocratic seeker of truth started to contribute to the solving of practical problems, he lost his innocence and some of his hallowed privileges. But he gained some advantages, such as that of becoming an ordinary citizen and thereby merging with society as a whole.

Next, a period of transition set in, a period which by now is well under way. At first there were few people, particularly among scientists, who noticed this change; most continued to think that scientific progress is an independent variable, always beneficial to society. Later, when it became clear that science as a force could become an instrument of disruption and death, people began to reason about the relationships between science and society.

Today, with science and technology affecting everyone directly and indirectly—when we cannot grasp the meaning of the world we live in without the presence of those components—the discourse must change and be expounded in terms of science in society. In other words, the hegemonic position of science and technology in human affairs must come to an end. Science and its applications should be considered on a par with other parameters in

the structure and function of society and, whenever necessary, conditioned by these. Only this can permit us the hope of discovering a new meaning to progress.

The bomb

The symptoms of transition are numerous and varied. Before the Second World War, writings on the human implications of scientific advance were rare; indeed, one can quote only one important volume, *The Social Function of Science*,[1] by J. D. Bernal, who, as a communist, was particularly sensitive to the problem. After the war, and particularly after 1960, the problem of the impact of science on society has been discussed from many angles, so that today we already have an imposing literature on the subject. In the bibliography of a recent book on the sociology of science[2] there are listed twenty-four volumes on the subject, since 1962, mostly by social scientists (only two by natural scientists). But this does not mean that the latter are insensitive to the problem: on the contrary, as we shall briefly

1. J. D. Bernal, *The Social Function of Science*, London, Routledge, 1939.
2. L. Sklair, *Organised Knowledge. A Sociological View of Science and Technology*, London, Hart-Davis, MacGibbon, 1973.

review now, their involvement in the debate is becoming wider and deeper as time goes by.

Already, toward the end of the war, scientists aware of forthcoming events began to worry about the consequences of the sudden release of nuclear energy; but it was only with the explosion of the bombs on Hiroshima and Nagasaki that the concern of much larger numbers of scientists for the fate of mankind was abruptly awakened all over the world. In many countries, after the war, groups of scientists were formed to discuss publicly the implications of nuclear weapons and to influence public opinion and governments, particularly in the United States and in the United Kingdom, and to ban the use of such deadly arms. Thus, in the United States the Federation of Atomic Scientists (now called the Federation of American Scientists) was created, and the Atomic Scientists' Association in the United Kingdom was established. *The Bulletin of Atomic Scientists*, edited by Eugene Rabinowitch, played a very significant role in stirring the conscience of the scientific community around the world and, at the international level, the World Federation of Scientific Workers catalysed the discussion of problems created by the advancement of science and technology. The initiative for organizing the first international conference of scientists was taken by Bertrand Russell. In a speech before the House of Lords on 28 November 1945 Russell had foretold the terrific destructive power of the H-bomb and the awesome danger it held for the human race. At that time he also expressed the view that an encounter of Soviet and Western scientists could serve as an effective channel to transform the arms race into a movement of genuine co-operation. In 1954, when his forecast came true, Russell decided that something should be done promptly; he drafted a manifesto, which was signed by Albert Einstein first, and then by other famous scientists. According to their declaration, war should be abolished and scientists should meet to evaluate the dangers and then take action to foster world-wide understanding among peoples.

On 7–10 July 1957, a conference took place at Pugwash, Nova Scotia (Canada), attended by twenty-two participants, of whom fifteen were physicists, two chemists, four biologists and one lawyer; seven were from the United States, three each from the Soviet Union and Japan, two each from the United Kingdom and Canada, and one each from Australia, Austria, China, France and Poland. Thus began the Pugwash movement, which can pride itself in having succeeded in lowering the political tension between West and East.

The multiplicity of crises

Nuclear war, as we will discuss below, was obviously the most important target on which to arouse the consciousness of the scientific community. As other threats of scientific and technological developments became discernible, however, the social concern of many scientists widened. To give an idea of the nature and magnitude of the problems, which professional scientists are pondering and discussing to the detriment of their specialized research we may consult a chart prepared in 1969 by a biophysicist, Professor John Platt.[1] Three years later, he came to the conclusion that in order to face the crises that the human race has to face now and in the future 'councils of urgent studies' should be established at the international, the national and local levels. The crises are more complex and arising more rapidly than existing social and political institutions can adequately handle them. 'New mechanisms and perhaps new institutions are needed to cope with these crises', both Platt and Cellarius have stated. 'We need solutions that will substantially decrease the threat of nuclear war, that will lead us to ways of life in much greater harmony with our environment, and that will allow all members of the human race to progress toward realizing the potential which our genetic, cultural and technological endowment makes possible'.[2]

The feeling of living in a period of crises of growing magnitude, resulting largely from technological developments, of urgency to act

1. J. Platt, 'What we must do', *Science*, Vol. 166, 1969, p. 1115–21. Copyright by the American Association for the Advancement of Science.
2. R. A. Cellarius and J. Platt, 'Councils of Urgent Studies', *Science*, Vol. 177, 1972, p. 670–6.

TABLE 67. Classification of problems and crises by estimated time and intensity (world)

Grade	Estimated crisis intensity (number affected $\times$ degree of effect)		Estimated time to crisis[1]		
			1 to 5 years	5 to 20 years	20 to 50 years
1	10^{10}	Total annihilation	Nuclear or RCBW escalation	Nuclear or RCBW escalation	✠ (Solved or dead)
2	10^{9}	Great destruction or change (physical, biological, or political)	(Too soon)	Famines Ecological balance Development failures Local wars Rich-poor gap	Economic structure and political theory Population and ecological balance Patterns of living Universal education Communications-integration Management of world Integrative philosophy
3	10^{8}	Widespread almost unbearable tension	Administrative management Need for participation Group and racial conflict Poverty-rising expectations Environmental degradation	Poverty Pollution Racial wars Political rigidity Strong dictatorships	?
4	10^{7}	Large-scale distress	Transportation Diseases Loss of old cultures	Housing Education Independence of big powers Communications gap	?
5	10^{6}	Tension producing responsive change	Regional organization Water supplies	?	?
6		Other problems—important, but adequately researched	Technical development design Intelligent monetary design		
7		Exaggerated dangers and hopes			Eugenics Melting of ice caps
8		Noncrisis problems being 'overstudied'	Man in space Most basic science		

1. If no major effort is made at anticipatory solution.

Source: J. Platt, 'What we must do', *Science*, Vol. 166, p. 1115–21, Table 1. Copyright by the American Association for the Advancement of Science.

and of the need to identify the responsibility of scientists, has spread rapidly within the international scientific community. Countless study groups have been established in many countries by scientists and sociologists to analyse the problems of war. Of special significance in this field are the activities of the Stockholm International Peace Research Institute (SIPRI), which publishes annually the much valued *SIPRI Yearbook of World Armaments and Disarmament* and also extensive studies on specific topics. Similarly, numerous meetings have been held and associations established to discuss the ethical implications of scientific progress; subjects such as abortion, organ transplantation, genetic engineering, cultivation of human embryos *in vitro*, have been widely debated in recent times. At an international meeting organized in Paris in 1972 by the Council for International Organizations of Medical Science (CIOMS), in co-operation with Unesco and WHO, it was recommended that:[1]

1. CIOMS and its parent organizations, Unesco and WHO, in conjunction with other national and international bodies concerned about the subject, should explore the possibilities of establishing an international non-governmental body to explore [sic] and study the moral and social issues raised by new and forthcoming developments in biology and medicine.
2. Such an organism would include, as a minimum, biological, medical and social scientists, humanists, religious leaders, science policy makers.
3. This body should be backed by the possibility of initiating and promoting research in the applications of biological and medical discoveries and their impact on society.

In the United Kingdom the Science and Society Council was established in 1973, prompted by a two-year study on the social obligations of the scientist, convened by a barrister, Paul Sieghart.[2]

Futures

Another sign of transition in attitudes is to be found in the novel concern for developments likely to occur in the future, not within the ordinary five- or seven-year plans of governments and economists, but over more extended periods. Aware of the accelerating momentum of their potential influence on human affairs, an increasing number of scientists, technologists, social scientists, entrepreneurial managers and other people concerned with the status of the world realize the immense power and responsibility they have in selecting among the multitude of possible futures those that might be better suited to the needs of men. 'This rising social consciousness', wrote Olaf Helmer, when he was Director of the Institute of the Future, 'among scientists based on a sense of urgency as well as a sense of their own strength, is propelling them in new directions that promise to relate their activities more closely to policy-making. The search for truth *per se* will thus be replaced or at least augmented, for better or worse, by a search for what is both morally right and attainable. The purists' motto of "science for society's sake" may find meaningful application.'[3]

The early futurologists, and by this I mean those that have been the pioneers in the field and were working already in the early 1960s, were substantially optimistic about forthcoming events. In their attempts to forecast the future, the simplest approach was obviously that of extrapolating current trends. Examples of these optimistic futures are Figure 85, published by Helmer in 1967, and the list of one hundred technical innovations very likely in the last third of the twentieth century, published in the classic *The Year 2000*, by Herman Kahn.[4]

But other scholars concerned about the future of the planet and the human race realized that this type of forecast was too simplistic: too little attention was paid to interactions and conflicts arising from technological progress. Their outlook is quite pessi-

1. S. Betsh (ed.), *Recent Progress in Biology and Medicine: Its Social and Ethical Implications*, Geneva, CIOMS, 1972.
2. P. Sieghart *et al.*, 'The Social Obligations of the Scientist', *Nature*, No. 239, September 1972, p. 15–18.
3. O. Helmer, 'Science', *Science Journal*, October 1967, p. 49–53.
4. H. Kahn, *The Year 2000. A Framework for Speculation about the Next 33 Years*, New York, Macmillan, 1967.

Fig. 85. This table of future technological achievements today appears completely out of date, not so much because the few years which have elapsed since it was drawn up (in 1967) have made it clear that some of the innovations then envisaged are unlikely to be possible but because of the change of attitude towards technology. Instead of asking only what we shall do, we now ask how we shall do it, for whom, with what resources and what consequences. The answer is not always either simple or convincing; the optimistic confidence of the past is now tinged with scepticism and perplexity. (*Source: Science Journal.*)

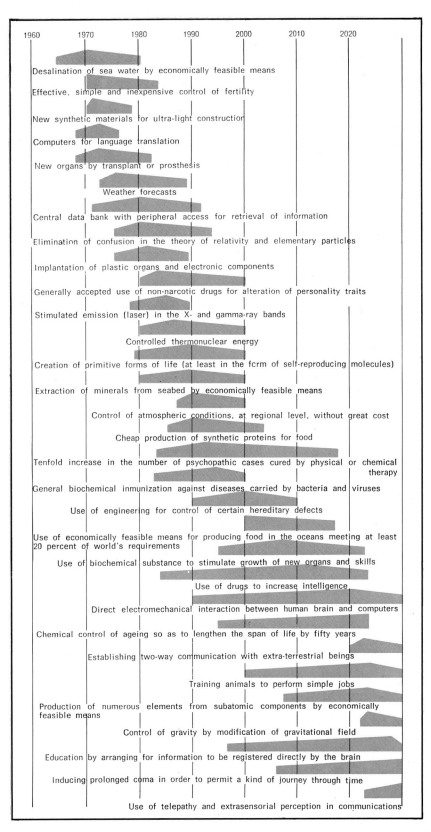

mistic because they think that more of what is happening today means disaster in the long run. Economic growth had become the synonym of progress during the last few decades, yet indefinite growth is impossible because the planet is limited in its resources. Table 68, prepared by the Hudson Institute in 1973, is an effective synthesis of these two opposing positions concerning the future.

Growth and its dangers

Of remarkable interest in this connexion is the action engaged by the Club of Rome to convince world leaders and the public at large that the course the world is on will lead inevitably to catastrophe unless its citizens change drastically their ways of living. The Club of Rome is an informal and, in more than one way, unusual association of some eighty-five individuals in more than thirty countries.

A little less than half its members are scientists and engineers; the others are business executives, industrialists, artists, economists, writers, thinkers and educators. The initial impulse for the creation of the club was a common concern regarding the deep crisis faced by humanity.

Men everywhere are perplexed by a range of elusive problems: deterioration of the environment, the crisis of institutions, bureaucratization, uncontrolled urban spread, insecurity of employment and loss of satisfaction in work, the alienation of youth, questioning of the values of society, violence and disregard of law and order, educational irrelevance, inflation and monetary disruptions in the face of material prosperity, the unbridged gap between rich and poor within and between nations, and many more like these. They appear to be world-wide symptoms of a general but as yet little understood malaise. It is this cluster of

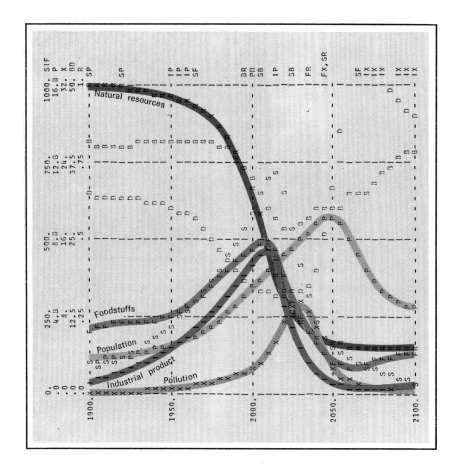

FIG. 86. The graph, taken from *The Limits to Growth*, shows the main variables in Jay W. Forrester's hypothesis, which was made with the assumption that no corrective intervention occurs. Jay W. Forrester was the originator of the first systemic model of the world. (*Source*: MIT.)

interwined problems which the Club of Rome terms the *problématique*. Their interactions have become so basic and are so critical that it is ever more difficult to isolate from the tangle of the *problématique* single, major issues and to deal with them separately. To attempt to do so only seems to increase the difficulties in other and often unexpected parts of the same mass.

After an inaugural period of intense discussion among the few original members of the club, it was agreed that concern with the world *problématique* was its central feature. A series of conversations was begun therefore with political and intellectual leaders in many parts of the world. Very soon it was realized, however, that policy and institutional change could result only from a more precise understanding of the nature and interactions of the main forces, whether material or ideological, that are influencing the evolution of modern

societies. It was felt that, since it was impossible to identify and analyse the global complexities at once, the club should limit its attention initially to the interactions between a few factors certainly significant to the problem at hand. It was then decided to attempt, by using any appropriate method which might exist or be developed, to quantify the scale and time dimensions of a limited number of them, such as the growth of population and of agricultural and industrial production. It was hoped that studies of this type might lead to the identification of values and institutional structures that need revision in order to achieve a better adaptation of man to his changing environment. After prolonged search for a suitable methodology, it was decided to invite the Systems Dynamics Group at the Massachusetts Institute of Technology to undertake the construction of a world dynamical model. An international team, led

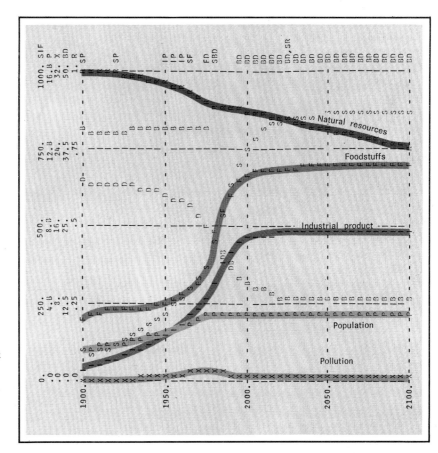

FIG. 87. The Director of the group of researchers who worked out the project of the Club of Rome is Dennis L. Meadows. The chart shows the evolution of the principal orders of magnitude, seen in their most favourable light, with judicious stabilization as of the year 1975. (Courtesy D. L. Meadows.)

by Professor Dennis Meadows, was formed to carry out the work. The findings of this first project were presented in the book *The Limits to Growth*,[1] which first appeared in March 1972 in the United States, and was later translated in some twenty languages.

The conclusions drawn in the book are, in substance, that if no rein is placed on the increasingly rapid growth of population and consumption, mankind will have reached and exceeded within a century the capacities of our terrestrial environment and the flowering of mankind will have been brought to a disastrous halt by forces beyond its control. In certain respects, this is no more than a question of dimension. On the one hand, the human family is represented by a group of countries where populations grow at a slow rate but use a rapidly increasing amount of raw materials and pollute the environment and by another group of countries with large populations growing tumultuously and where the expectations of a better life have spread rapidly. On the other hand, the planet Earth offers materially and theoretically limited resources; the latter are found only in finite quantities, whether in the form of deposits of raw materials which accumulated over lengthy geological eras—and which are therefore impoverished when man exploits them—or in the form of resources constantly being renewed through the conversion of solar energy.

The claim that there are material limits to man's growth on his planet thus appears obvious. The point under discussion, however, includes (a) whether or not we are actually approaching some of these limits with rapid strides, (b) what these limits are, and (c) what the consequences of not heeding them might be. According to the replies to these questions, the problems of population and development involve also, in one way or another, those of institutions and policies. And it is but a short jump from there before the values, social ways, and the very aims of human society are questioned.

The publication of Meadows' book had great impact throughout the world. Many of its aspects were criticized (see, for example, *Thinking about the Future: a Critique of the Limits to Growth*[2]): its methodology, the limitations in the parameters studied, the too simplistic approach of considering the planet as a whole without taking into account the great diversities existing in different regions of the world, and many others. But no one can deny that *The Limits to Growth* and the issues raised by the Club of Rome have opened an important debate on the validity of many beliefs of modern man, among which is the one that continuing material progress is unquestionably good. Many research projects are currently under way to study in depth in the *problematique* in general, and in the varied terms through which it manifests itself under differing ideological, political, geographic and economic conditions.

The most extreme criticism against economic growth, still being worshipped today as the panacea of all ills in the modern world, has been raised by a British sociologist, Colin Francome, who claims that the quality of life has lowered over the years 1961–70 in his country in spite of the fact that the British gross national product (GNP) has increased by 2.9 per cent per annum. The author has distinguished ten ways in which GNP can increase with no consequent rise in living standards:

1. Profits are often maximized by producing goods which have limited lifespan.
2. The increase in income could be due to the fact that goods and services which were free now have to be paid for.
3. In order to sustain a high level of production, it might be the case that people have to work harder or to do more tedious jobs.
4. Goods which people are able to buy may well be status goods.
5. Money is not rationally distributed throughout society, and there is evidence that the expansion of the economy in recent years has reflected an increase in money among the richer sections of the community.

1. D. Meadows *et al.*, *The Limits to Growth*, Washington, Potomac Press, 1972.
2. *Thinking about the Future: a Critique of the Limits to Growth*, London, Chatto & Windus, 1973.

TABLE 68. Two very basic and opposed positions (prepared by the Hudson Institute, 1973)

Neo-Malthusian beliefs and conclusions	The post-industrial perspective
1. We have a fairly good idea of what this world can provide mankind. Therefore, 'a fixed pie' is a good metaphor.	No one knows what the Earth holds or can produce. But 'an expanding bowl' has been and likely will be a good metaphor.
2. Man is rapidly depleting the Earth's food, energy, and mineral resources. Key resources will be expended in the next century.	If managed modestly well, resources will probably be available in plenty for everybody for the foreseeable future.
3. Exponential population and production growth are accelerating the exhaustion of resources.	The Earth can easily support populations and gross world products many times larger than today's. Population growth is now slowing down. Growth in GWP will probably also decline, but more because of the diminishing marginal utility of wealth than because of resource or pollution limits.
4. New discoveries of resources and new technology will delay the crisis, but not for long and their use often exacerbates the eventual crisis or even cataclysm.	New resources and technology often raise new problems and crises but they can still be used to improve efficiency and upgrade the quality of life to a permanently high plateau.
5. The technology and capital investment necessary to extract marginal resources will vastly increase pollution, probably to lethal levels. In any case we shall have to cope increasingly with diminishing marginal utilities.	New technology and capital investment are necessary to help prevent and clean up pollution. (But we must beware of far-fetched and unlikely but potentially catastrophic events due to misunderstood innovations or inappropriate growth.)
6. Dangerous gaps in incomes (both domestic and international) are widening rapidly. A world-wide class war or desperate political crisis is imminent. These possibilities are exacerbated by growth in the rich nations, particularly when it selfishly uses up the resources of the poor nations.	The next century will likely see the world-wide abolition of absolute poverty. Of course, some will continue to be much richer than others but the growing gaps will often accelerate the economic growth of the poor. However, arithmetical gaps will probably not decrease much until the middle or end of the twenty-first century.
7. The further industrialization of the Third World will be disastrous, and further growth of the developed world would be even worse. Therefore, the rich nations should halt their growth and share their current wealth with the poor. Further the poor should not be so willing to sell their increasingly valuable resources so cheaply.	Industrialization of the Third World and superindustrialization of the developed world will (and should) continue. It is foolish to imagine that the rich will voluntarily share significantly with the poor, especially if this sharing means a major deprivation for the rich, and it is probably nonsense to believe the poor will be strong enough to seize much wealth by force in the foreseeable future.
8. The rapidity of change, the complexity of the problems, and conflicting interests will make conflict management, resource management, pollution control, etc., all impossible. Some sort of slowing down and centralized world-wide decision making is imperative.	With some possible exceptions the level of management required is not remarkably high. Price and other market mechanisms can deal with most issues. Some low but practical degree of international co-operation can deal with most of the rest.
9. Unless revolutionary changes are made immediately the twenty-first century will see the greatest catastrophe since the Black Death. Billions will die of hunger, pollution, and/or war over shrinking resources. The total human misery will be immeasurable.	If present trends continue, the twenty-first century will see a humanistic postindustrial society in which the once-eternal economic problems of mankind will have been solved. Most misery will derive from the anxieties and ambiguities of wealth and luxury, not from physical suffering due to desperate scarcities.

6. The economic cost often does not take into account the cost to the community at large.
7. Apart from the problem of pollution, certain changes may benefit the individual but act against the interest of the community at large.
8. Advertising often increases needs: if these needs are then fulfilled there is no real improvement in living standards.
9. One of the limits of economics is that it is based solely on the present.
10. Growth in GNP may be facilitated by people neglecting their social relationships.

Well aware of the limitations of a Social Index, which attempts to evaluate the quality of life within a society, Colin Francome has compiled Table 69. 'All in all,' he concludes, 'while the difficulties in the Social Index are great, the problems are at least laid bare. The

TABLE 69. Social index[1]

Subject	1961	1966	1970
Number of children homeless and in care	4.1	5.1	5.7
Length of life from birth	M: 67.9 F: 73.8	M: 68.5 F: 74.7	M: 68.5 F: 74.7
Infant mortality rates	22.5	19.7	18.6[3]
Suicide rate			
Men	117	106	94
Women	115	113	94
Prison population	32.5	37.5	44.3
Murder and manslaughter rates	143	232	287
Crimes of violence	17.1	27.3	41
Unemployment rates	1.5	1.5	2.6
Working days lost			
Through stoppages	3.0	2.4	10.9
Through industrial injury	17	20	20
Number of hours worked	42.4	40.4	40.1
Proportion in poverty	370[4]	470	1,040
Percentage having holiday	30	31	34.5
Mental hospitals			
Beds occupied in	159	145	131
Day patients	351	1,111	1,813
Out patients	1,368	1,581	1,748
Air pollution due to cars and public transport	142.9	193.4	230.3
Air pollution due to smoke and SO_2	7.58	7.18	6.90
Chances of a child being placed in care because illegitimate	1.7	2.9	2.7
Maternal death rate	0.34	0.26	0.19
Percentage of children in classes of thirty or below	27.9	24.9	24.5
Increase in education	9.0	12.2	17.1

1. From 1961 to 1970 the GNP increased by 2.9 per cent per annum; the social index fell from 100 in 1961 to 83.33 in 1970.
2. (A) = England and Wales; (B) = United Kingdom; (C) = Great Britain. Population corrections based on United Kingdom figures
3. Figures for 1969.
4. Figures for 1960.
Source: C. Francome, *The Poverty of Growth*, Academics against Poverty, 25 Wilton Road, London SW1V 1JS.

Social Index is at least one step ahead of the gross national product because this just holds to a belief that increased wealth improves housing, helps people live longer and removes poverty. The Social Index tests whether this is really the case.'

One more clear sign of a new concern for the consequences of rampant technologies on the human environment is the vast ecological movement which has led to extensive action at the international and the national levels. Of special significance was the Conference on the Human Environment organized by the United Nations in Stockholm in June 1972, and the forerunner of the establishment of the United Nations Environment Programme (UNEP). Such a vast amount of literature has been produced in recent years on the subject of the 'environmental crisis' that it does not appear justified to discuss it here. But it is worth

Method of calculation[2]	Source	1961	1966	1970
Thousands index score based on number of children aged under 15: 1961=12.3 m; 1966=12.9; 1970=13.4 (A)	Social Trends, No. 2, p. 93	8	6.75	6.30
Official estimates (subject to revision) (C)	United Kingdom in Figures and Social Trends, No. 2, p. 98	10	10.11	10.11
Per 1,000 live births (B)	Social Trends, No. 2, p. 165 and Annual Abstract of Statistics 1969	6	6.86	7.26
Standard mortality rates 1968=100 (C)	Social Trends, No. 2, p. 99	4	4.40	4.93
Standard mortality rates 1968=100 (C)	Social Trends, No. 2, p. 99	4	4.07	4.90
Based on total population: 1961=52.82 m; 1966=54.65 m; 1970=55 34 m (B)	Social Trends, No. 2, p. 150	4	3.56	3.07
Conviction number based on population (A)	Annual Abstract of Statistics 1971, p. 78, 79	3	1.91	1.53
Thousands based on population (A)	Social Trends, No. 2, p. 144	4	2.53	1.71
Percentage of working force (C)	Social Trends, No. 2, p. 63	10	10.00	5.76
Millions based on working populations: 1961=24.77; 1966=25.58; 1970=25.04 (B)	Social Trends, No. 2, p. 63	4	5.70	1.08
Millions based on working population (C)	Social Trends, No. 2, p. 63	3	2.64	2.64
Number index assumes minimum of thirty hours (B)	Social Trends, No. 2, p. 64	2	2.38	2.41
Thousands related to population	Sir Keith Joseph. House of Commons, 10 November 1970	10	8.14	2.94
Percentage of population (C)	Social Trends, No. 2, p. 67	3	3.00	3.29
As in middle age one is more likely to be ill these figures are based on population aged 30–59 years: 1961=21.2; 1966=20.7; 1970=20.3 (C)	Social Trends, No. 2, p. 105	3 / 3 / 3	3.21 / 0.92 / 2.53	3.50 / 0.56 / 2.25
Thousands of passenger miles (C)	Social Trends, No. 2, p. 137 (2)	3	2.22	1.85
Million metric tons (B)	Social Trends, No. 2, p. 132	3	3.17	3.30
Thousands related to population under age of 15 (A)	Social Trends, No. 2, p. 93	2	1.28	1.37
Per 1,000 live births (A)	Annual Abstract of Statistics 1971, p. 36	2	2.62	3.58
Percentage of all children (A)	Social Trends, No. 2, p. 115	3	2.68	2.63
Percentage of pupils aged 17 expressed as percentage of the 13-year-old pupils four years earlier (B)	Social Trends, No. 2, p. 114	3	4.66	5.70
	Total	100	95.78	83.33

(Source: Annual Abstract of Statistics 1971, p. 7).

TABLE 70. Technical dilemmas and some social responses

Technical dilemma	Price response	'Fix-it' response
1. Pollution	Pollution inevitable and worth the benefit it brings	Solve pollution with pollution technology
2. Capital dependence	Technology will always cost money	Provide the capital; make technology cheaper
3. Exploitation of resources	Nothing lasts for ever	Use resources more cleverly
4. Liability to misuse	Inevitable, and worth it	Legislate against misuse
5. Incompatible with local cultures	Material advance is worth more than tradition	Make careful sociological studies before applying technology
6. Requires specialist technical élite	Undertake technical-training schemes	Improve scientific technical education at all levels
7. Dependent on centralization	So what?	No problem, given good management
8. Divorce from tradition	This is why technology is so powerful	Integrate tradition and technical know-how
9. Alienation	Workers are better fed and paid; what matters alienation?	More automation needed

Source: 'The Pressing Need for Alternative Technology', *Impact of Science on Society*, Vol. 23, No. 4, p. 257–72.

while to emphasize the fact that in this area, as in other aspects of the present predicament of man, crises arise because of hidden conflicts between the conduct of individuals and the collective outcome of the additive and sometimes mutually strengthening effects of the sum total of individual actions.

An additional symptom of the increased social involvement of scientists is their intellectual and constructive contribution to facilitate the cultural, social and economic development of the Third World. (The problem has already been discussed in Part III, Chapter 30.)

A similar trend is to be found in the attempt to define the rules that should be followed to obtain a more effective management and control of technology in the public interest, i.e. technology assessment (which has already

been discussed in Part I, Chapter 5) or what some call the 'social audit'.

The Marxist view

Somewhat at variance with this pessimistic outlook on the possibility of continuous economic expansion is the position of classic Marxist scientists and sociologists, as exemplified by the recent joint publication of the Academies of Science of the U.S.S.R. and Czechoslovakia with the title *Man—Science—Technology*.[1] The questions this work raises are posed in terms similar to those we have already reviewed:

1. *Man—Science—Technology. A Marxist Analysis of the Scientific-Technological Revolution*, Moscow-Prague, Academia Prague, 1973.

'Away-with-it' response	Alternative response	Radical political response
Inevitable result of technology; use less technology	Invent non-polluting technologies	Pollution is a symptom of capitalism, not of poor t˄chnology
Costs of technology are always greater than its benefits; use less	Invent labour-intensive technologies	Capital is a problem only in capitalist society
Use natural not exploitable resources	Invent technologies that use only renewable resources	Wrong problem: exploitation of man by man is the real issue
Misuse so common and so dangerous, better not to use technology at all	Invent technologies that cannot be misused	Misuse is a socio-political problem, not a technical one
Local cultures better off without technology	Design new technologies which are compatible	Local culture will be disrupted by revolutionary change in any case
People should live without what they do not understand	Invent and use technologies that are understandable and controllable by all	Provide equal chance for everyone to become a technical specialist
Decentralize by rejecting technology	Concentrate on decentralized technologies	Centralization an advantage in just social systems
Tradition matters more than technical gadgets	Evolve technologies from existing ones	Traditions stand in the way of true progress
Avoid alienation by avoiding technology	Decentralize; retain mass production only in exceptional cases	Alienation has social, not technical, causes

But where is this vehement and rapid progress leading us and in what will it ultimately result? Does it not harbour some grave dangers for man? What measures must be taken *now* to direct this progress entirely for the good and benefit of mankind so as to preclude any possible negative consequences it may have which may inflict, perhaps *even tomorrow*, some damage or even direct harm to man? These are some of the questions that are causing anxiety to people all over the world.

Then the theory follows:

If we were to try to outline the roads followed by the scientific and technological revolution under socialism, as distinct from its development under capitalism, we might imagine it as follows: from revolutionary change in science and technology, via changes in the structure of the forces of production and in the character of economic development, through changes in the content of human labour and the general level of qualifications, up to the mass formation of the all-around developed human personality—this is how the circle of social progress connected with the realization of the scientific and technological revolution is gradually closed up, becoming more and more complete. The net result of these processes under socialism finally reenters the process of development and application of science and technology as a new dynamic factor. The circle of interdependent changes caused by the scientific and technological revolution in socialist society become, in this way, a dynamic spiral of social progress. . .

Bourgeois ideology has proved itself incapable of scientifically analyzing the social implications and prospects of the scientific and technological revolution. Conceptions based on this ideology register the precipitous social transformations that are under way, but lack a general background theory of social evolution that could serve as a

405

scientific methodology in the study of the social processes and conflicts brought about by the scientific and technological revolution, and in forecasting of the future development of mankind.

The scientific and technological revolution, when in the service of socialist society, far from being used to the detriment of the masses of the working people, is fully oriented to serve the interest of the people, and to answer their needs (spiritual and material) to the maximum. Thus, the antagonism between scientific and technological progress and man, inherent to capitalism, is replaced in socialism by the harmonious unity of these opposites.

I thought it desirable to present this extensive quotation from the report mentioned even if it may appear to some, including me, a dogmatic assertion rather than a description of reality, but, of course, it is not the purpose of this book to enter into political debate.

Alternative technologies

The variety of reactions against the dehumanizing aspects of the techno-scientific complex that we have now briefly reviewed stem from the tacit or explicit assumption that some form of high technology society is here to stay. The recent awareness of the negative consequences of technological excesses is prompting a multitude of reactions aiming, so to say, at taming the beast in order that basic values can be defended and preserved. But the notion here is that the beast—science-based technology—is, within limits, or can be, beneficial to the human race.

At variance with such views still prevailing both in developed and emerging countries, a new trend must be mentioned. This aims at the development of novel 'appropriate' or 'intermediate' or 'alternative' or 'soft' technologies. The theory behind this movement, which already has an impressive number of adepts in the industrialized countries as well as many supporters in the developing ones, denies that the current uni-dimensional technocratic development prevailing in Europe and North America—the addiction to the way of thinking in terms of 'bigger', 'more', 'faster' and 'efficient'—is necessary and inevitable. Indeed, the failure of attempts to improve on

the social and economic conditions in technology, is ascribed by many thoughtful specialists to the fallacy that what has characterized the evolution of Western societies could and should be followed elsewhere.

To quote Robin Clarke, one of the most devoted and engaged advocates of the new trend:[1]

an alternative technology should be non-polluting, cheap and labour intensive, non-exploitive of natural resources, incapable of being misused, compatible with local cultures, understandable by all, functional in a non-centralized context, richly connected with existing forms of knowledge and non-alienating.

The alternative does not seek to jettison the scientific knowledge acquired over the past three centuries but instead to put it to use in a novel way. Space heating, in the primitive context, was achieved by an open wood fire. In the alternative context, it might still be achieved by burning timber—provided the over-all rate of use was lower than the natural rate of timber growth in the area concerned—but in a cheap and well-designed stove which optimizes useful heat output against the need for fuel.

This difference between primitive and alternative technology is important for it has in the past led to charges that the alternative is retrogressive, essentially primitive, and ignores the utility of modern scientific knowledge. This is not the case.

As we will discuss in the next chapter, there has been a variety of responses to the problems created by science and technology in the modern world. The optimists say that the advantages are greater than the disadvantages; the technocrats claim that more technology will cure the ills produced by technology; those completely disenchanted with science and technology will refuse them completely and preach the return to nature; and the young radicals in the West will concentrate their attack against the 'system', capitalism and imperialism and claim that only revolution may cure the ills of modern society. The stand of the advocates of alternative technology is different, and Robin Clarke has summarized the difference (in Table 70).

1. R. Clarke, 'The Pressing Need for Alternative Technology', *Impact of Science on Society*, Vol. 23, No. 4, 1973, p. 257–72.

Active followers of such ideas have estab-lished a multitude of communities, generally small, where they attempt to develop these new types of technologies and have organized information centres for disseminating ideas and methodologies. In the United Kingdom, for example, numerous groups are active in the country and the cities, studying and experimenting with new ways of living, farm-ing without using the products of chemical industry, using novel architecture that pro-vides dwellings heated by the Sun and does not use traditional sewage systems. In the United States the New Alchemy Institute (the motto of which is 'to restore the land, protect the seas, and inform the Earth's stew-ards') has developed a programme to pro-mote social and biological transformations in Massachusetts, California, New Mexico, and also abroad, in Costa Rica and Canada. Its director, Canadian scientist John Todd, is convinced that we should turn people into scientists as it might be too late to make humanitarians out of scientists! And similar activities are to be found in Sweden, the Netherlands, Federal Republic of Germany, France and Denmark.

For the conditions prevailing in the devel-oping countries, E. F. Schumacher, economist and philosopher, has coined the term 'inter-mediate technology'[1] to indicate the need for new scientific and technological approaches that would raise the economic status of in-digenous workers to a reasonable level: quite lower, however, than that of the industrialized countries, since the latter would call for major capital investments that are not available and, if available, would disrupt local traditional cultures. The desired technology would be in-termediate in that it would fill the gap be-tween primitive, indigenous tools that have remained unchanged since time immemorial, and the complex apparatus of modern, high-level, capital-intensive technology. Similar ways of thinking are often expressed in the developing countries. For example, Professor A. Rahman has recently examined critically the type of technologies that have been im-ported from the West into India; he has come to the conclusion that for his country, as well as for others of the Third World, there is an urgent need for alternative technologies, that would take advantage of modern knowledge, but would utilize it in forms better suited to local needs and ways of living than the sophisticated industrial products coming from Europe or North America.[2]

Such views foresee that the flight from the countryside to the city would be halted, a real development of the Third World could come about, the ecological crisis would find its best solution by reintegrating man within the natu-ral system, and science and technology would lose their present hegemonic position in the plurality of approaches to truth and of desir-able human activity. Undoubtedly this is an appealing vision, with an admittedly romantic flair. But might the present not be too late, at least for those countries that have generated the techno-scientific system?

1. E. Schumacher, *Small is Beautiful*, London, Blond & Bridge, 1973.
2. A. Rahman, 'Alternative Technology', *J. Scient. Ind. Res.*, Vol. 32, 1973, p. 97–100.

32 The end of an era

Science, as discussed in Part I, can be exhilarating, useful and dangerous; progress in a great number of scientific disciplines continues at a rapid pace, as presented in Part II; and the organization of scientific and technological research, as shown in Part III, has called for a major financial and managerial effort. Yet, an increasing number of scientists, as well as of laymen, particularly among the young are dissatisfied with the present state of affairs. They are doubtful that science can be the major means to salvation and wonder whether the course we are on may not lead to catastrophe. Science and technology have become, simultaneously, the greatest hope for human progress and one of the most serious threats facing contemporary man. This situation has arisen from the immense power of scientific technology, whether it be directed towards the elimination of human disease or the manufacture of weapons capable of eliminating the human race. In some cases the outcome of scientific applications have contradictory implications, as we have seen in Part I, Chapter 5. The mistrust in science is spreading at a time when the planet has to support almost 4,000 million people and can do so only through a sustained and rational use of technology. Yet the reckless use of technology geared primarily to exploitation is equally threatening. Only if science and technology can be brought under firm and humane control can catastrophe be avoided.

The perplexing difficulty of science today has been aptly summarized recently by Henryk Skolimowski as follows:[1]

The real bone of contention now is whether science is shaping the future of mankind in a desirable way, whether the unintended by-products of science do not place in jeopardy the whole enterprise of science; whether the world moulded by the machinery of scientific and technological progress is not deprived of what is essential to a humane, human and compassionate environment; whether we might not have to develop a set of altogether different principles for understanding nature (which we might call the New Science) so that we may interact with nature and with each other more harmoniously, thus less antagonistically, and thus in a way which is more conducive to our own survival. . . . Thus the dispute over the nature of science is nowadays the dispute over the value of our western Weltanschauung. An attack on science, if it is not a parochial scramble in the name of old religious ideals, is inevitably an attack on the whole rational-scientific civilization.

The era of unlimited faith in progress by means of science and technology, of which scientists of my age have witnessed the last exalting triumphs, has come to an end.

1. H. Skolimowski, 'Science in Crisis', *Cambridge Review*, 28 January 1972, p. 70–4.

The threats of science

A vast literature on the threats of science has appeared over the last fifteen years, accompanied by a smaller number of counter critiques. A bibliography compiled in 1970 by Peter Harper while he was at the University of Michigan already comprised some 300 items. It would be impossible to attempt a general summary of the debate, but it is worth while to review some of the criticisms that have been raised against science.

1. *Science is not evil but is too limited.* Many problems of the world and of man escape scientific analysis. In the study of man, for example, contemporary science—with emphasis on units, groups, populations rather than individuals, on detachment rather than on involvement—can be a serious handicap. There is a need for making science more vast and more flexible.[1]

2. *The analytical approach of science (reductionism) leads to a distorted picture of reality.* In fact, it is maintained, this approach cannot be used effectively when dealing with complex objects such as living things because the whole is more than the sum of its parts.[2] This can be dangerous because it can lead to 'mad rationality'. 'When the most abstract achievements of mathematics and physics satisfy so adequately the needs of IBM and the Atomic Energy Commission,' avers Herbert Marcuse,[3] 'it is time to ask whether such applicability is not inherent in the concept of science itself.'

3. *Science is not objective.* If accepted, this claim would undermine the very foundations of science itself. The major significance of this kind of attack has been clearly stated by Scheffler:[4]

The overall tendency of such criticism has been to call into question the very conception of scientific thought as a responsible enterprise of reasonable men. The extreme alternative that threatens is the view that theory is not controlled by data, but data are manufactured by theory; that rival hypotheses cannot be rationally evaluated, there being no neutral court of observational appeal, nor any shared stock of meanings; that scientific change is a product, not of evidential appraisal and logical judgment, but of intuition, persuasion and conversion; that reality does not constrain the thought of a scientist but is rather the projection of that thought.

Science, it is claimed, is not objective because it does not progress 'scientifically'. Ideas come first and data are then used to support the hypotheses. As the progression Kepler—Newton—Einstein shows, widely different versions of the truth are all allowable. Moreover, the nature of reality as revealed by science is questioned. What particle physicists now have to say about the nature of matter is clearly a long way from 'real' reality. They can talk only about mathematical abstractions formulated by an extreme form of reductionism (as above, point 2). Science does not reveal the 'truth' because it sees only what its techniques allow it to see. But what it does see is as valid as the nature of the truth revealed by other techniques of disciplined inquiry. Whether it is more valid is debatable.

Since the claim to objectivity represents the core and pride of science, it is worth while to quote a statement of one of its most articulate critics, Skolimowski:[5]

The ultimate argument in many discussions on the nature of science is that the objectivity of science must not be imperilled, that one of the greatest virtues of science is its objectivity, that the fruits of science are rooted in the idea of objectivity. What is the objectivity of science? Is it the cornerstone of all our cognitive pursuits on which knowledge and prosperity depends; or is it an abstract dogma which once played an important role in the growth of knowledge and in the growth of our civilization? Without obpuscating the

1. A. H. Maslow, *The Psychology of Science: a Reconnaissance*, New York, Harper & Row, 1966; M. Polanyi, *Personal Knowledge*, London, Routledge, 1958; T. Roszak, *The Making of a Counterculture*, London, Faber, 1968.
2. A. Koestler and J. R. Smythies (eds.), *Beyond Reductionism*, London, Hutchinson, 1969.
3. H. Marcuse, in: G. Holton (ed.), *Science and Culture. A Study of Cohesive and Disjunctive Forces*, Boston, Houghton Mifflin, 1965.
4. I. Scheffler, *Science and Subjectivity*, Bobbs-Merrill, 1967.
5. H. Skolimowski, *Science, Anti-science, the New Science*, Physics Colloquium, University of British Columbia, Vancouver, 25 February 1971.

point I shall declare that, in my opinion, the latter is the case, that objectivity is now a sterile dogma which keeps us at bay and is a hindrance. I can see the legions rising against this heresy.

4. *Science has become an ideology of its own—scientism.* While recognizing that science is not merely ideology, it is considered to be ideology as well. 'Ideology is an all-pervasive material force,' states Robert M. Young,[1] 'penetrating into our most intimate and subjective relationships as well as into our putatively disinterested inquiries in the biological and human sciences.' Ideologies are certainly not out of contact with reality, but they attempt to monopolize the whole of reality and interpret in terms which reflect the partial interests of limited groups.

5. *Science cannot be at the same time a dispassionate search for truth and serve as a paradigm of human values.* It is claimed that scientists are often at fault in confusing the two motives for their work. To satisfy their personal interest, they may claim it is in the cause of society when it is not or when its implications may in fact be malevolent. Similarly, they often try to justify their applied work on the grounds that the search for new knowledge must go on. The conflict between detachment and values appears particularly clear in the social sciences. Is a dispassionate social science possible at all?

6. *A distinction between science and technology is hardly feasible.* The Baconian prediction of 'the enlarging of the bounds of human empire, to the effecting of all things possible' has come true. The general law that whatever is possible is in fact applied means that scientific activity almost inevitably leads to technology. The distinction often heard in scientific circles that science is pure and clear of sin, while technology is the culprit, finds no support in today's world. Both science and technology are a part of the fabric of modern society and it would be artificial to draw a line between the two: both represent a continuum (see Part I, Chapter 3) and indeed many scientific advances have become possible thanks to tech-

nological progress. Since the time of Bacon, science has been considered the servant of man and therefore of society; the products of technology have made this concept real. Technology is the secular arm of science.

7. *All large-scale application of technology involve some internal contradiction.* These contradictions emerge not because the body politic or industrial tycoons make evil use of technologies, but rather as a result of 'the way technology is conceived, developed and applied by those very people who gave it birth: the scientists and technologists'.[2] As examples one may quote the fact that an increasing number of 'technological fixes' are often needed to keep within acceptable boundaries a variable or set of variables that have been disrupted by a related type of technology.

8. *Science is directly responsible for the majority of the unapproachable problems of modern society and also for the rate at which they worsen.*

In the last century we have increased our speeds of communications by a factor of 10^7; our speed of travel by 10^2; our speed of data handling by 10^6; our energy resources by 10^3; our power of weapons by 10^6; our ability to control disease by something like 10^2; and our rate of population growth to 10^3 times what it was a few thousand years ago.[3]

9. *Science is not neutral.*[4] If points 5 to 7 are accepted, then science cannot be neutral, and the scientist is responsible for what happens, for the use of his discoveries. His responsibility must necessarily be related to a specific social

1. R. M. Young, 'The Human Limits of Nature', in: J. Benthall (ed.), *The Limits of Human Nature*, London, Allen Lane Penguin, 1974.
2. L. Gallino, 'Rationality and Irrationality of Technology in Advanced Industrial Societies', *13th Pugwash Symposium: Social Aspects of Technological Change*, 1971, p. 11–27.
3. J. Platt, 'What we must do', *Science*, Vol. 166, 1969, p. 1115–21.
4. J. D. Bernal, *The Social Function of Science*, London, Routledge, 1939; M. Born, *Physics and Politics*, Edinburgh, Oliver & Boyd, 1962; A. D. Sakharov, *Progress, Co-existence and Intellectual Freedom*, London, Andre Deutsch, 1968; P. Ehrlich, 'Ecocatastrophe', *Ramparts*, September 1969.

context. He may be motivated in a very positive manner towards society, but his action will depend upon his conception of social reality and he may then choose the wrong course as judged from a different set of values.[1]

10. *Science controls people.* As a social institution and as the ideology of Western civilization, science influences powerfully even if surreptitiously, everybody's way of thinking. 'Our mental habits, our thinking and perceiving, have been predetermined by the network of knowledge within which we grew up. And this has been profoundly influenced by science.'[1]

11. *Science has not helped solve the problems of the developing countries.* Too often the forms of science and technology used in international aid have carried with them Western values and attitudes, totally disrupting the societies and cultures they have met, without replacing them with any viable alternative.[2]

The values of science and technology

Such critical stands concerning science (above) are only those that appear to be of great significance to our discussion; many more could be found in recent literature. Obviously, these viewpoints could easily be criticized in their turn, and, indeed, many supporters of science as an intellectual enterprise and as a social institution have counter-attacked. The fact remains, however, that critiques of sciences, technology, and of the scientists and engineers themselves have become notably more common during the last several years. Since science and technology, as we have seen, are admittedly the major parameter around which the modern world has evolved criticism of the kind cannot simply be dismissed as irrelevant to an understanding of the present world.

Throughout the history of ideas, over the centuries, the pendulum has been swinging back and forth between periods when thinkers believed that human reason unaided can attain objective truth and others when rationalism was put aside and greater values were attributed to intuition, inspiration, mysticism or

revelation. Stephen Toulmin[3] has pointed out that certain common topics are to be found in both the literature of the counter-cultures of today and its forerunners in earlier periods. The main parallels between current criticisms and those of the past, according to Toulmin, are five: anti-humanism, namely the alleged indifference and even callousness of scientists about humane issues; individualism, the superiority of the artist or the poet who expresses his subjective interpretation of the world as compared to the supposed conformist and collective activity of the scientist, who must be detached and refrain from personal passion; imagination, in that the romantic critic thinks that the scientist must conform to stereotyped and mechanical modes of argument; the issue of quality versus quantity, whereby the scientist—said to concern himself only with what is common to many different individual things, with general properties and with quantifiable or statistical magnitudes—becomes indifferent to the qualitative features in which every individual differs from his fellows; finally, the abstract character of scientific ideas and inquiries, whereby scientists begin by imposing certain arbitrary theoretical demands and standards on the variety of nature, and are then prepared to pay serious attention only to those aspects of nature they choose to accept as 'significant' by those standards. This time, however, the mounting tide of scepticism (if not outright refusal of science and reason, for the former is considered by many as the very embodiment of the latter) must be considered and met with greater attention and concern than before since the very foundations of so-called modern civilization might be at stake.

1. Skolimowski, 'Science in Crisis', op. cit.
2. B. Commoner, 'On the Meaning of Ecological Failures in International Development', in *The Careless Technology*, M. T. Farvar and J. P. Milton (eds.), p. xxi-xxix, Garden City, New York, Natural History Press, 1972; A. Rahman, 'Alternative Technology', *J. Scient. Ind. Res.*, Vol. 32, 1973.
3. S. Toulmin, 'The Historical Background to the Anti-science', in: *Civilisation and Science: in Conflict or Collaboration?*, p. 23–32, North Holland, Amsterdam, Elsevier—Excerpta Medica, 1972.

TABLE 71. Ten value systems for science and technology

	Science	Technology	Relationship between science and technology
1. Pure	Science most valuable cultural pursuit of all. It provides general theory exposing illogicality of all other political and social theory	Generally uninteresting. Value of science is science	Scientific understanding of life leads to ascetism and correct technical applications
2. Arch-scientific	Science extremely valuable cultural pursuit. It may lead to supra-political values and hence may solve social problems	Science is fount of technical progress; this is an important justification for its development	Politicians often fail to grasp scientific ideas and apply them badly. Scientists can help to sort out politicians
3. Liberal-scientific	Science is valuable cultural pursuit. It is compromised by the State and by undesirable technology	State exerts too powerful a role on the progress of science. Secrecy in science and subordination of university to politics is to be deplored	If technology misapplied, politics is to blame. Science cannot help in making political choices; it can only display 'facts' for public debate
4. Humanist	Science is valuable as it improves intellectual or material well-being. Man comes first, science second	Uneasy compromise: science for technological development resulting more important than science for intellectual satisfaction	Humanism alone can tell politicians how to apply technology for the best ends
5. Classic Marxist	Science among most important activities of State because it generates technical progress. There is no science for science's sake	Science's *raison d'être* is to provide momentum for technical and sociological progress. Science should be revolutionary	Coupled with Marxist analysis technology can solve all problems. Properly applied technology does not generate problems
6. Technocratic	Science generating technology is mainspring of progress. State and industries benefit from it. Science *per se* not interesting	Science's *raison d'être* is to provide momentum for progress. Science subordinate to state for improving conditions for all	Great value to technology and progress. 'No cost should be allowed to stand in the way of the latter', to quote Teller
7. Realist	Science and politics always inseparable. Science is good or bad as the state makes it	Scientists plead their cause by reference to science's applications but are inconsistent when they couple their plea for 'curiosity' research	Technology is inevitable result of science; responsibility for its effects shared by State and science
8. Doomsday	Science subordinate to its effects. Halt research in dangerous areas	Scientific progress leads to technical problems which threaten human extinction	Technology likely to extinguish man; would have been better if never invented
9. New Left	Science is a conservative art. It is one means by which the masses are kept under and humiliated	Science produces technology which is used by the establishment to preserve its power	Technology is the tool of the élite; it hampers social progress
10. Anti-science	Science is itself evil and its effects dehumanizing. It makes meaningless the 'great adventure of living'	As above	As above

TABLE 72

System	Man in relation to:	Impact and *cultural reactions* to impact
Ethical (religious)	God(s): perception of his own nature	Refutation of superstitious belief (Galileo; Darwin; Freud); *censorship*
		Discredit authoritarian doctrine; scepticism; *inquisition*
		Preponderance of the objective; short shrift to non-scientific forms of creativity and wisdom; *anti-intellectualism*
		Anomie; religious crisis; moral libertarianism; *counter-culture*
		Can man assume God-like responsibility for his own purposes? Whence values?
Biological	his body, pain and death	Public health, especially disappearance of infant mortality
		Population explosion and other demographic shifts; *ambivalence of benevolence*; unequal access to health facilities
		Moral dilemmas of resource allocation for preservation and prolongation of individual life
		Ethics of human experimentation; of sanctions on drug abuse; of abortion; of saving damaged foetuses
		Scientific deflation of racism
		Possibilities of autocratic manipulation of behaviour or of genotype with new biological tools
		Evolutionary consciousness and dilemmas of genetic intervention (*see*: Ethical)
Environmental (primary economic development) Ecology	nature: land	Mechanization of agriculture
		Potential alleviation of famine; population explosion
		Urbanization and alienation from land (*cf*. child labour)
		Wastage of wildlife, natural beauty
		Exploitation of commons (the multitude and the future)
		Environmental pollution and disease; *natural health movements*
Industrial (secondary economic development)	work, power	Industrial revolution
		Potential alleviation of toil (replacement of muscle-power with machines)
		Technological unemployment: industrial child labour: *Luddites; labour unionism*
		Rising expectation and imperfect realization of equalitarian opportunity; *communism; welfare politics*
Industrial capitalism (exhibited in United States, Europe, U.S.S.R.) Tertiary (ec. develop.)	systems of production	Unprecedented productivity
		Efficiency as autonomous (self-nourished) principle in the rational allocation of capital
		Assembly line; work alienation; mass consumption/advertising/culture; *counter-culture*
		Affluence; *anomie from disappearance of economic imperative (work ethic)*
		Leisure; disappearance of child labour and prolongation of economically dependent period beyond puberty; *generational reactions*
		Obsolescence of human skills; erosion of self-value upon displacement by machines
		Centralism; bureaucracy; market failures, externalities; *technology assessment consumer and environmental lobbies*
		Alienation of consumers from goods (unsafe products; cost of information); *government regulation*
		Index ethics (I.Q.) and economics (G.N.P.)—simplified models (social scientism) and suboptimal designs

System	Man in relation to:	Impact and *cultural reactions* to impact
Social and political Interstate	state and national culture	Acceleration of social change through redistribution of power; egalitarianism; social mobility and disappearance of hereditary castes and of extended family; *social confusion* (e.g. racial separatism *v.* integration); *future shock*
		Vulnerability of complex system to disruption by dissidents (hijacking; strikes)
		Rationalization of police powers; government power over citizens—including invasion of privacy
		Corporate wealth and unfair pressures on political process
		A new priesthood: 'the experts'; imputations of an unaccountable intellectual élite; relentless demands on citizens to keep up with evergrowing complexity; *participatory democracy and demands for simplification of political organization*
Interstate	national sovereignties	Ultimate weapons; risks of catastrophe (H-bombs; biological warfare); *international integration and law*
		Stable deterrence of global conflict
		Arms race; arms trade; military-industrial complex and domestic political consequences; *isolationism*
		Means, therefore temptations for conquest or insurgency

Source: J. Lederberg, 'The Freedoms and the Control of Science: Notes from the Ivory Tower', *Southern Californian Law Review*, Vol. 45, p. 596–614.

Points of view about the position and function of science and technology in society vary from one extreme to the opposite, from those who think that science is the most valuable and necessary human activity to those who believe that science is intrinsically evil and incompatible with human development and survival. In between, we find an immense variety of shades of positive or negative evaluations and attitudes. In fact, everyone who has been thinking about the problem, whether philosopher, practising scientist, historian, politician or leader of the rebellion of the young, has reached, consciously or unconsciously, a personal stand. It would be practically impossible to display systematically the whole range of opinions on the subject but we may use a simplified taxonomy as the one presented in Table 71.

Obviously, such classification has no pretence of completeness; it aims only at presenting in a few brief sentences a selection of attitudes or values towards science and technology that can be identified in recent literature, ordered according to ten ideological positions. As we will discuss in greater detail in Chapter 33 below, none of these views expressed synthetically in the table can be accepted as 'right' and the others 'wrong'; judgements are necessarily dependent and conditioned by political, economic, cultural and social priorities.

In a somewhat similar vein, Joshua Lederberg[1] has attempted a classification of concerns about the impact of science and technology on various systems (see Table 72). He points out that the classification does not pretend to be precise and that each category interacts with every other, 'nor does it explore the network of actions and reactions that would be the cultural history of western society.' A few cultural reactions are indicated by italics. The author adds:

1. J. Lederberg, 'The Freedoms and the Control of Science: Notes from the Ivory Tower', *Southern California Law Review*, Vol. 45, 1972, p. 596–614.

In further work, it would probably be useful to separate the perspectives of different historical epochs—that is, to suggest how this problem would have been viewed at different periods (and by different cultures). Then, and most appropriately, more attention could be focused on the special problems of the modern (and post modern?) era.

For those interested in further information on problems and attitudes referred to in the two tables, reference should be made to the list of works at the end of this chapter.

The interest of such tentative classifications lies primarily in the fact that they show, even if in a crude fashion, how science and technology are intimately related to all aspects of society and of the ideologies and values of man. To quote Kenneth Boulding:

There is a certain implicit assumption today that science is something above and beyond society, a kind of genie out of a bottle which promises or threatens to do all sort of good and bad things to us, but which belongs, as it were, to another order of creation. But this view of science as a genie outside of society, whether angelic or demonic, will not stand up to serious examination. Even though the rise of science might have something of the impact of a 'revelation' in sociological terms—that is, as a creation of evolutionary potential which is realized as the years go by—it is still a revelation which is very firmly embedded in human society and must be visualized as a phenomenon taking place, as far as we know, wholly within human society.

The myth of the Promethean function of science is over. Man is indeed an animal endowed with reason, but also with unreason. And, what is more significant for our discussion, reason and science cannot always and necessarily be identified. Seeking for scientific knowledge no longer represents an absolute good that should not be interfered with on any account.

Further reading

ARON, R. 1968. *Progress and disillusion: the dialectics of modern society.* London, Pall Mall Press.

BOULDING, K. E. 1968. *Beyond economics.* Ann Arbor, University of Michigan Press.

CHOMSKY, N. 1969. *American power and the Mandarins,* Cambridge, Mass., MIT Press.

COMMONER, B. 1966a. *Science and Survival.* New York, Viking Press.

——. 1966b. 'The integrity of science', *Science journal*, April 1966.

DUBOS, R. 1970. *Reason awake: science for man.* New York, Columbia University Press.

EHRLICH, P.; EHRLICH, A. 1970. *Population, resources, environment.* San Francisco, Freeman.

ELLUL, J. 1964. *The technological society.* New York, Knopf.

FERKISS, V. C. 1969. *Technological man. The myth and the reality.* New York, New American Library

FEUER, L. S. 1963. *The scientific intellectual. The psychological and sociological origins of modern science.* New York, Basic Books.

GALBRAITH, J. K. 1967. *The new industrial state.* Boston, Houghton Mifflin.

GOODMAN, P. 1969. 'Can Technology be Humane?' *New York review of books*, 20 November.

GREENBERG, D. S. 1969. *The politics of pure science.* New York, New American Library.

HARDIN, G. 1968. 'The tragedy of the commons', *Science*, Vol. 162, p. 1243.

——. 1969. *Population, evolution and birth control. A collage of controversial ideas.* San Francisco, Freeman.

KUHN, T. S. 1968. *The structure of scientific revolutions.* Chicago University Press.

MARCUSE, H. *One-dimensional man.* London, Routledge.

MISHAN, E. J. 1969. *The costs of economic growth.* London, Staple Press.

MULLER, H. J. 1970. *The children of Frankenstein.* Bloomington, Indiana University Press.

PECCEI, A. 1969. *The chasm ahead.* London, Collier MacMillan.

PLATT, J. 1966. *The step to man.* New York, Wiley.

POLANYI, M. 1961. *The logic of liberty.* Chicago University Press.

ROSZAK, T. 1972. *Where the wasteland ends.* New York, Doubleday.

TOFFLER, A. 1970. *Future shock.* New York, Random House.

ZIMAN, J. 1968. *Public knowledge. The social dimensions of science.* Cambridge University Press.

33

A new enlightenment

Perhaps modern man's major mistake was that of considering science as the quintessence of reason and of believing, accordingly, that scientific progress is synonymous with human progress.

In the famous Prague lecture 'Are the sciences really in crisis in spite of their unfailing successes?' Edmund Husserl stated, as early as 1935:[1]

As a starting point we shall take the drastic change in the general evaluation of science, which occurred at the turn of the century. It did not affect its scientific essence but rather what science—the sciences in general—means and may mean for human existence. In the latter half of the nineteenth century, the general *Weltanschauung* of modern man became exclusively dominated by positive science, man became fascinated by the ensuing 'prosperity', and this brought about an estrangement from those problems that are decisive for an authentic humanity.

As we have seen, since then this estrangement has progressed to the point that scientists themselves and societies at large have become aware of the predicament.

As we have already discussed (see Part I, Chapter 3), it is no longer possible to draw a line between science and technology: technology is science. For this reason we cannot any longer accept the view that science is a model of progress as expressed, for example, by Harvey Brooks in 1971:[1]

It [science] is the one area of human activity which can incontrovertibly be said to progress, not only despite, but because of the fact that the definition of progress, unlike that in other areas, is not anthropocentric. . . . Thus we cannot agree as to whether advancing technology constitutes progress, but in science knowledge and understanding do grow cumulatively independently of how they may be subsequently be used.

The relationships between science and society have become so terribly complex and interwined that statements of this kind become untenable.

At this stage, we should scrutinize the validity of assertions like those presented in the previous chapter, re-examine the role of the scientist in today's society, discuss the limits of the objectivity of science, analyse the foundations and the significance of ethical values, and attempt to outline a programme that may lead to a new meaning of progress. In other words, it appears desirable at this

1. E. Husserl, *Die Krisis der europäischen Wissenschaften und die transzendentale Phänomenologie. Eine einleitung in die Phänomenologische Philosophie*, The Hague, Martinus Nijhoff, 1954.
1. H. Brooks, 'Can Science survive the Modern Age?', *Science*, Vol. 1974, 1971, p. 21–30.

time to take as a model our predecessors of more than two centuries ago, those who both hypothesized the advent of the age of reason and influenced the development of the modern world. At that time, the architects of the Enlightenment regarded knowledge as a whole, rather than a collection of separate facts, and found in man the central vantage point from where the immense field surveyed in the *Encyclopédie* could be perceived and assessed. Spurred by the marvels of the exponential growth of science, men of the twentieth century have forgotten that irreplaceable perspective, and have plunged into the whirlpool of runaway science. Painful and difficult as it may turn out to be, a major intellectual effort is called for, to chart a major turn in our *Weltanschauung*, to outline the trends of a new enlightenment. A 'Science of man' is now urgently needed, as our forefathers foresaw, to use the words of Diderot:[1]

One consideration, above all, must not be lost sight of. It is this: if man or the thinking and contemplative being is banished from the surface of the earth, the pathetic yet sublime spectacle of nature becomes no more than a sad and mute scene. The universe is quiet, belonging to silence and the night. Everything changes into an enormous solitude where phenomena, unobserved, occur in obscurity, mutely. It is man's presence which makes the existence of beings interesting. And what can one better propose in the history of all creatures than to submit to this proposition? Why not introduce man into our study just as he is situated in the universe? Why not make of him the common focus? Is there in the infinity of space some point whence we could, more advantageously, draw great lines in order to connect all other points? What a lively and sweet reaction would there not be between all beings and man, between man and all beings?

The scientist in society

If we attempt to understand why the era of unlimited faith in progress by means of science and technology has come to an end we must concentrate our attention on the changes that have occurred in science as a social institution and more particularly in the behaviour of the scientist, in his motivations, in his ambitions, in his allegiances and in the role that he has been playing in modern societies. Science, as we have seen, can no longer be looked at only as a body of knowledge or a particular method to acquire it; it must be seen as a social activity as well. The mutual interactions between scientific research, technological developments, and social and economic change and even ideologies are so numerous and profound that we cannot hope to understand the total picture and the variety of attitudes towards science and technology that we have mentioned without scrutinizing: the ways in which scientists behave towards each other; the relationships that have been established between the scientific community, on the one hand, and political and economic power, on the other: the structures that have been established and have evolved at national and international level to cope with the increasing impact of scientific activities on society; and how new scientific knowledge is produced, disseminated and utilized.

We have already reviewed many aspects of these types of interactions in Part III of this volume; but for the question under discussion, it appears justified to concentrate on the transformations that have occurred during this century but especially in the last few decades in the interests, the activities and the outlook of the protagonist of the scientific enterprise: the professional scientist. We are bound to touch upon many themes that have been discussed in previous chapters; for this reason we will limit our discourse now only to those aspects appearing relevant to the issues at hand.

Over the years there has occurred a progressive professionalization, politicization and nationalization of science[2] and of scientists. The fact that our 'advanced' societies feed more and more on the results produced by scientific research has brought about increasingly close interactions between the scientists and political and economic power. Govern-

1. Diderot, 'Encyclopédie', in: *Encyclopédie ou Dictionnaire raisonné des Sciences des Arts et des Métiers*, Paris, 1751.
2. J. H. Haberer, 'Politicalization in Science', *Science*, Vol. 178, 1972, p. 713–24. Copyright 1972 by the American Association for the Advancement of Science.

ments and industry need massive production of new knowledge and technologies, while scientists have been relying at an increasing rate on the financial support and the social rewards of power. Jean-Jacques Salomon[1] has proposed the term *technonature* to indicate the area within which the interests and attitudes of scientists are inevitably linked with those of power and (at one and the same time) are responsible for the needs of power and also contribute to its objectives.

The transformation in the role of the scientist that has occurred in the last decades has brought about a fast growth of the scientific enterprise, or research system, as I have often indicated: the number of research workers has correspondingly increased at unprecedented rate, with the result that what was once considered as a dedication of one's life to the pursuit of truth has become a profession not substantially different from the other ones. And with the increasing demand for ever more sophisticated weapons by the national authorities, the scientists have become increasingly involved in the 'industrial-military complex' relinquishing as an obsolete myth the ideal of science as international endeavour.

Quite a number of active scientists of today, particularly of the older generations, do not think that this type of description applies to their own interests and aspirations or to those of their close pupils. They are still convinced, rather, that they personify the ideals of the dispassionate, detached, humble, pure search for truth. It cannot be denied that a limited number of such relic specimens may still exist. But the exception confirms the rule: the statements of two well-known physicists, one from the U.S.S.R. and the other from the United Kingdom, bear witness to our diagnosis. In 1966 Piotr Kapitza said:[2]

The year that Rutherford died (1938) there disappeared forever the happy days of free science work which gave us such delight in our youth. Science has lost her freedom. Science has become a productive force. She has become rich but she has become enslaved and part of her is veiled in secrecy. I do not know whether Rutherford would continue nowadays to joke and laugh as he used to do.

And John Ziman wrote in 1968:[3]

My impression is that the sort of scientist with which we are mainly concerned in this book—that is, more 'pure' scientist than a technologist—often feels no more than cupboard love for the organization for which he ostensibly works. . . . Of course he wants lots of money for his apparatus and may learn to become very cunning and selfish in special pleading for it, but the major purposes for which the great corporations exist—education, defence, profitable production, national prestige—may be of little moment to him. If a radar research laboratory, devoted to the development of military technology, happens to be the best place he can find for his study of compound semiconductors then he will be quite happy to have a niche there, feeling virtuous in the thought that 'wicked' defence spending is being used to support such 'good', ploughshareworthy activities as his own researches.

If science has lost its innocence, it is primarily because of the way scientists have behaved, particularly in recent times: it is for the scientists themselves that the age of innocence is past. A progressive convergence of the requirements of power and the rapid advance of highly competitive research fields has transformed the scientist as a component of modern societies. Believing that his action was guided by loyalty to Science with a capital S and to his country or ideology (another form of nationalism), the scientist of the last forty years or so has, on many occasions, been ready to serve clearly immoral ends. He is no longer neutral, nor can he hope to regain his pristine virtue under present dispensation. The quantitative change in the scale of scientific research that has occurred since the Second World War has brought about the qualitative change under discussion. At one time there used to be relatively few persons who had chosen scientific research as a way of devoting their lives to a noble intellectual cause, unmarred by contacts with the worlds

1. J.-J. Salomon, *Science et Politique*, Paris, Éd. Seuil, 1970.
2. Kapitza, P. L., 'Recollections of Lord Rutherford', *Nature*, Vol. 210, p. 782–3.
3. J. Ziman, *Public Knowledge. The Social Dimensions of Science*, 1968. Copyright by permission of Cambridge University Press.

of power and production; today science has become a profession for the thousands and the hundreds of thousands. And that profession is completely integrated in the process of industrial production, whether for war or for humane purposes.

In 1926 the French Julien Benda wrote his famous essay *La Trahison des Clercs*[1] in which he accused the intellectuals of his time—those that had the duty to fight for freedom, justice, peace and the dignity of man—of having betrayed their mission in that they were ready to preach nationalism, racism, class hatred and other kinds of dangerous irrationalism. Julien Benda wrote:

And, indeed, for more than two thousand years and until only recently, I note throughout history an uninterrupted succession of philosophers, theologians, writers, artists, scientists—whose action was in formal opposition to the realism of the multitude. With particular reference to political passion, these scholars expressed their opposition in two ways. Either, completely turned away from this passion, they set the example—as did a da Vinci, a Malebranche or a Goethe—of taking up activity purely disengaged from that of the spirit, and creating belief in the supreme value of this kind of existence; or else, properly in the role of moralists concentrating on the conflict in human selfishness, they preached—as did an Erasmus, a Kant or a Renan—in the name of humanity or justice, the adoption of an abstract principle, superior and directly contrary to the passion to which I have alluded... Thanks to them, one can say that (for two millenniums) humanity did wrong but praised right. This contradiction was the distinction of the human species and represented the gap through which civilization could slip along. But by the end of the 19th century, a capital change came about: *scholars began to play the political game.* Those who had served as a brake on popular realism became its stimulators.

At the present time, about one century later, we can affirm that we members of the scientific community (including myself) have witnessed and participated in the *trahison des scientifiques* for we have been ready to accept support for our research from whatever source, without realizing that in so doing our activity was becoming tainted, and with no countermeasures in sight.

The treason of the scientist is not very different from that of the intellectuals of the 1920s, in as much as we have accepted uncritically the trends of our time because they have made possible the rapid growth of our beloved research. We have not realized that our work had immense and threatening implications at the human level; we have not had the courage to refuse to partake in endeavours and enterprises that were endangering the survival of our species and, at the very least, the very values for which we had chosen to become scientists. We have also indulged in unethical practices of rivalry, competition and keeping secret the results or methodologies to make a discovery before our colleagues—practices that we have uncritically introduced into the research system borrowing them from industry and trade. Furthermore, as pointed out before, we have accepted to keep our research work secret for reasons of national defence or prestige.

The traditional attitude of the scientist to consider himself *au-dessus de la mêlée* is no longer justifiable. Precisely because science in our time has become a social institution, and a very significant one at that, the scientist must reconsider his position, his actions and his motivations within the social context. One thing is the role that one thinks he is playing, and another, often quite distinct, is his role as viewed by others. Yet, today, many scientists consider themselves as outsiders with respect to political life: scientific research is pure and neutral and, if evil use is made of its products, science and the research worker carry no responsibility. They still believe that their detached scientific attitude is the best antidote against the vagaries and ambiguities of the body politic and, for this reason, they find themselves in the best position to express objective views on whatever matter. But, as we have seen in the previous chapter, outside observers look at the scientist from a very different angle: they believe that the scientist is responsible for what is happening in the modern world and that he should accept such responsibility. There are clear signs that the

1. J. Benda, *La Trahison des Clercs*, London, Routledge & Kegan Paul Ltd, 1927.

traditional attitude is changing, however, and that an increasing number of scientific research workers are becoming aware of their new role and are concerned about the implications of such change.

But if we still believe—as I do—that reason is still the best foundation on which man makes sense of what happens around and inside him, if we still share the view of Peter Medawar that 'to deride the hope of progress is the ultimate fatuity, the last work in poverty of spirit and meanness of mind',[1] if we think that the habit of truth and the criticism of prejudices and myths that has characterized science throughout its development can only make man free, and that the objectivity of science still has a meaning, then we must face an agonizing reappraisal of our behaviour. We cannot afford any longer to be self-effacing and timid. Participating in well meaning activities—such as those of the various associations for the social responsibility of scientists, or for the study of war and peace—worthwhile as these may be, this is not enough. Scientists and their academies and societies should embark also in a critical examination of the recent past and attempt to identify when and where wrong decisions were taken, and when and where compromises with power have occurred. Scientists should be ready to give up their pet research projects for the sake of their conscience. Scientists, together with other scholars, should start a world-wide movement to rescue the scientific enterprise and attempt to identify a new meaning to progress.

The objectivity of science

As we saw in the last chapter, the very core and pride of science, its objectivity, is challenged. Statements of the type of those of Scheffler and Skolimowski are frequently found in recent writings of critics of the scientific enterprise. Are they justified? Should we take them seriously, or should we simply disregard them as expressions of mystical romanticism? What do those critics mean when they claim that 'data are manufactured by theory' or that 'objectivity is now a sterile dogma which keeps us at bay and is a hinderance?' Personally, I do not believe that

we should dismiss such assertions without comment, for they often come from the pen of the serious scholar.

Obviously, the subject of the objectivity of science would call for a systematic analysis and discussion of the vast literature on the subject. A treatment of this nature, however, would go much beyond the limited scope of this book and particularly of this chapter. I shall limit myself, therefore, with trying to identify the weak points of such criticisms and the limits within which the objectivity of science is still to be regarded as the necessary and, in a way, the unique foundation of organized and interpersonal knowledge. In other words, what follows is an attempt to explain why, in only relatively recent times, the objectivity of science has been questioned and to discuss some of the reasons why such critique should be rejected.

The nature of science and the way the scientist proceeds to discover his 'truth' has been the subject of scholarly discussions since the time of Francis Bacon; obviously, we need not attempt a presentation of the variety of interpretations proposed. Suffice it to say that until relatively recent times the prevailing schools of philosophers and historians of science, such as the 'Vienna Circle', the neopositivists, the inductivists and the hypothetico-deductivists like Karl Popper, considered science as a pure fruit of the intellect, an element of the world of ideas, utterly uncontaminated by human activities and therefore uninfluenced by the mores of the time.

Since approximately 1960, however, new trends of thought have appeared. At a symposium in Oxford on the history of science—the year was 1961—Thomas S. Kuhn presented a paper for which he chose the title 'The Function of Dogma in Scientific Research'. At about the same time his book, now famous, The Structure of Scientific Revolutions, appeared.[2] Kuhn considers the history of science as characterized by a succession of periods of 'normal science' and of 'revolutions', fol-

1. P. B. Medawar, The Hope of Progress, London, Methuen & Co., 1972.
2. T. S. Kuhn, The Structure of Scientific Revolutions, Chicago University Press, 1962.

lowed in their turn by 'normal science'. The revolution occurs when a new 'paradigm' is proposed and accepted to interpret a set of phenomena and events. Normal science is what happens when specialists of a certain discipline contribute to the advancement of knowledge within a general paradigm that has been formulated to account for what is known. But at a certain stage of scientific development, newly acquired knowledge may become incompatible with that paradigm and then a crisis occurs, which is resolved by a novel theory or paradigm that will replace the previous one.

Kuhn stresses the point that 'competing paradigms are incommensurable', that is, they are relatively incomparable. According to his interpretation of scientific advance, when the crisis comes the choice between the old and a new paradigm, while being rational and not emotional, could be influenced by factors such as values attributed by different scientists to 'accuracy, scope, simplicity, fruitfulness and the like'. If any of the neo-positivists had used the term 'paradigm' to indicate a major theoretical change (say, Einstein's general relativity as constrasted with Newton's universe) they would have thought that the change was inherent in the nature of science, that it was brought about by the necessity of thought, and that it was a response to logic and experiment alone. Kuhn's interpretation, instead, allows for external influences: 'Simplicity, scope, fruitfulness, and even accuracy', he writes, 'can be judged quite differently (which is not to say they may be judged arbitrarily) by different people.'

Thomas Kuhn's work was received with great interest and thoroughly discussed in scientific circles at a time when there occurred a flourishing of studies on the sociology of science, as we have seen in the previous chapter. Several scholars of the field refused the traditional 'internal' explanation of the development of science, and preferred to consider this not as an independent variable of society, but rather as an activity that, even if not determined by economy as the Marxists claim, can be affected by external social forces. The latter approach found recently a clear formulation in Leslie Sklair's statment:[1]

My criticism of some current thinking in the sociology of science, therefore, can be interpreted as part of a general strategy to undermine the view that the intrinsic nature of science is such as to require special explanations that set it apart from other social activities. Further, my criticism of certain philosophies of science may be interpreted as part of the same strategy to undermine the view that science is so special an activity (or rather that scientific knowledge is so special) that no sociological factors are very useful in explaining how it works. Science is part of the everyday world, it can be illuminated in a sociological fashion, and it requires no very special sociological factors to explain how it operates.

A third aspect that we should consider when studying the sources of current criticism to scientific objectivity, is to be found in a number of recent writings (and which we have reviewed in Part I, Chapter 2) on the act of creation of the scientist; this creativity was found in many cases to be similar to that of the poet, the painter or the sculptor.

It seems to me understandable, even though not justifiable, that a number of people —having read or heard about the trends of thought here briefly referred to, and in many cases people who never have had a personal experience of conducting to successful conclusion a significant piece of scientific research—these people might have jumped to some unwarranted conclusions, of which a small sample was presented in the preceding chapter. It would be easy to demonstrate the nonsense of statements such as: 'theory is not controlled by data but data are manufactured by theory'; 'rival hypothesis cannot be evaluated' (obviously a superficial and distorted presentation of Thomas Kuhn's views); or of the circular argument of Henryk Skolimowski. But rather than indulge in such exercise, it is perhaps more appropriate to stress that probably no scientist today would claim that the 'objectivity of science' means that there exists one and only one reality, the nature of which can be discovered uniquely through scientific procedures. Current attitudes on this matter are certainly more mod-

1. L. Sklair, *Organised Knowledge—A Sociological View of Science and Technology*, London, Hart-Davis, MacGibbon, 1973.

est: most research workers, I would think, realize that 'reality' is multifaceted and that we have different ways of approaching it. But the kind of reality that science reveals has one peculiar trait: it is the same and equally valid for any person that takes the trouble to go through the process of acquiring knowledge in that particular way. In other terms, science is interpersonal knowledge and it is in this quality that consists its objectivity. This is also a trait clearly not shared by other ways of approaching 'reality', for example, artistic experience.

In this connexion, I think that Professor John M. Ziman (a theoretical physicist) has given us a very valuable treatment of the problem in his book *Public Knowledge—An Essay concerning the Social Dimension of Science*.[1] He states:

science is not merely *published* knowledge or information. Anyone may make an observation, or conceive an hypothesis, and, if he has the financial means, get it printed and distributed for other persons to read. Scientific knowledge is more than this. Its facts and theories must survive a period of critical study and testing by other competent and disinterested individuals, and must have been found so persuasive that they are almost universally accepted. The objective of science is not just to acquire information nor to utter all non-contradictory notions; its goal is consensus of rational opinion over the widest possible field.

Concerning objectivity, Ziman states:

Objectivity and logical rationality, the supreme characteristics of the Scientific Attitude, are meaningless for the isolated individual; they imply a strong social context, and the sharing of experience and opinion. . . . The rationale of the 'scientific attitude' is not that there is a set of angelic qualities of mind possessed by individual scientists that guarantees the validity of their every thought—as if they were, so to speak, well tuned computing machines whose logical circuits precluded them from error—but that scientists learn to communicate with one another in such tones as to further the consensible end to which they are all striving, and eventually train themselves to construct their own internal dialogue in the same language. A private psychological censor takes over from the public policeman or parent and conforms our behaviour to social norms. But he does not keep whispering into our ear, 'Be honest, be truthful, be objective', in a chorus of pious aspiration; he says, 'Have you checked for instrumental errors? Is that series convergent? What is the present status of that old bit of theory?', and so on.

If we subscribe to this interpretation of scientific objectivity—that it is interpersonal knowledge and consensus of rational opinion over the widest possible field—we can easily dispose of another criticism of modern trends in scientific research, namely its analytical or reductionistic approach. Reductionism or its opposites—the variety of wholisms or holisms that have been repeatedly proposed over the years—are not articles of faith but just different procedures through which we may reach interpersonal knowledge. It does not make any sense to affirm that the analytical approach is inadequate to explain complex systems, since such an approach is the only one that allows the advancement of scientific knowledge, as its history clearly shows in every discipline. If it appears that the properties of a complex system are 'more' than the sum of the properties of its parts we have a clear indication that we do not know enough about the system to identify what that 'more' consists of. If we examine the extreme complexity of an ecosystem, it is certainly true that we could predict its conditions of equilibrium or its cycles just by adding the notions that we have about the biological characteristics of the species present and of their interactions with the physical surroundings. But if this is the case, it means simply that we have not adequately analysed the multiplicities of interactions, feedback mechanisms, etc., that relate the components of the ecosystem, over the period of time studied. In that case, we would not have reached an objective description of the system for the description would not meet a consensus of rational opinion. But once we had reached the necessary level of understanding to account for the very complexity of the system under study, we would use a holistic description that would stand the test of interpersonal knowledge.

1. Ziman, op. cit.

The criticisms referred to in the previous chapter, that reductionism is dangerous at the level of human affairs, appear justified in as much as their utterance takes the (now) obsolete stand that only science could offer adequate solutions to social problems. Such a stand is untenable because we know now that things are much more complex than our predecessors thought; the stand would not be objective because it would not meet with the consensus of rational opinion.

Scientific ethics and ethical science

After reasserting the value of the objectivity of science, we must face the serious problem of how to reconcile the intrinsic risks to human needs deriving from scientific progress, indeed to the survival of the species. Once more, the problem would call for an extensive treatment much beyond the scope of this volume. But for the sake of our argument it seems to me sufficient to limit our discussion to two main issues: (a) are traditional ethical principles compatible, can we construct a new scientific ethics? (b) what should we, the scientists, do to regain peace in our conscience?

The most radical reply to the first question was made recently by Jacques Monod, who thinks that none of the ethical foundations of human societies remains tenable in face of scientific inquiry and that only the definition of a new foundation for their value system could prevent their collapse. He writes:[1]

Science, in its development, has gradually attacked and dissolved to the core the very foundations of the various value systems which, from prehistoric times, had served as ethical framework for human societies. I believe that most social anthropologists would agree with the statement that the structure of virtually all ancient or primitive myths, as well as of more advanced religions, is essentially ontogenic. Primitive myths provide histories which almost invariably refer to one or several god-like heroes, whose sagas both account for the foundation of the group and imperatively govern its social structure and traditions, that is, its basic ethical system. The great religions may be regarded as generalizations, attempting to embrace the whole of Mankind in a similar, except much wider, interpretative ontogeny, whose recognized function is to provide a transcendent and therefore permanent and indisputable basis for a system of values. Most of the great philosophical systems, from Plato to Hegel and Marx may also be regarded as attempts to establish on an ontogenic basis an untouchable, irrefutable basis for a system of values, in turn serving as foundation of the social system.

Whether an ethical structure is justified by reference to a founder-hero, to a universal god, to an absolute idea, to the 'laws' of history, or to some 'natural' foundation for human rights, all such systems, from the concepts of Australian aborigenes, to the ideas of Rousseau or Marx, share one essential characteristic: namely that values and ethics are not a matter of human choice. Whether they are supposed to stand on divine, or 'natural' foundations, they are beyond the realm of human freedom. Reason or faith may serve to identify and recognize values, but not to define or alter them. To Man, Ethics and values do not belong; he belongs to them.

It is indeed easy to see the psychological purpose and the social function of raising these concepts on to so high a pedestal as to put them beyond human reach and make them untouchable. The psychological purpose is to abate and perhaps satiate the hunger for meaning; for the meaning of such absolute realities as life and death. The social function is one of stability: no social structure could survive whose very foundations could be questioned, denied or dismissed by any one at any time. The foundations of a value system therefore must appear not only unquestionable, but as it were, inaccessible.

In spite of all the efforts of priests, statesmen and philosophers, cultural codes are not untouchable. They have changed through prehistoric and historic times, at a rate that no biological, genetic evolution could have approached. In spite of all these changes however, one concept remained, up to recent times, invariant: namely that *some* immutable foundation of the value system did exist, and could be found or recognized.

It is this concept, the most essential within any ethical system, the cornerstone of any social structure, the sole (even though unreliable) substitute for genetic coding, that science now has destroyed, reduced to absurdity, relegated to the state of nonsensical wishful thinking. . . .

Science, as it emerged and developed, has shaped the modern world, given modern nations their technology and power. Yet these societies

1. J. Monod, 'On Values in the Age of Science', *The Place of Value in a World of Facts (Nobel Symposium 14)*, p. 19–27, Stockholm, Almqvist & Wiksell, 1970.

have failed to accept, hardly have they understood the most profound message of science. They still teach and preach some more or less modernized version of traditional systems of values, blatantly incompatible with what scientific culture they have. The western, liberal-capitalist countries still pay lipservice to a nauseating mixture of Judeo-christian religiosity, 'Natural' Human Rights, pedestrian utilitarianism and XIXth Century progressism. The Marxist countries still throw up a stupefying smoke-screen of nonsensical Historicism and Dialectical materialism.

They all lie and they know it. No intelligent and cultivated person, in any of these societies can really believe in the validity of these dogma. More sensitive, more impatient, the youth perceives the lies and revolts against them, forcefully revealing the intolerable contradictions within modern societies.

No society can survive without a moral code based on values understood, accepted, respected by the majority of its members. We have none of this kind anymore. Could modern societies indefinitely master and control the huge powers which they owe to science on the criterion of a vague humanism admixed with a sort of hopeful materialistic Hedonism? Could they, on this basis, resolve their intolerable tensions? Or shall they collapse under the strain?

I have quoted Monod rather at length for I think that his position is well justified and, indeed, it has already become the starting point of a vast debate. It must be pointed out that, once shown that traditional values are obsolete, new ones must be found and taught; Monod himself, however, has not yet outlined what the new ethics should consist of. We may turn to the study of the way the scientist proceeds or should proceed to see whether his practice contains the embryo of a new ethics. Anatol Rapoport pointed out in 1957 that there are ethical principles inherent in scientific practice:[1]

the conviction that there exists objective truth; that there exist rules of evidence for discovering it; that, on the basis of this objective truth, unanimity is possible and desirable; and that the unanimity must be achieved by independent arrivals at convictions—that is, by examination of evidence, not through coercion, personal argument and appeal to authority.

Rapoport claimed that not only science is related to ethics but that science is becoming a determinant of ethics inasmuch as the ethics of science must become the ethics of humanity. Interesting as this position is, unfortunately it falls short of presenting a new system of ethical principles derived on the basis of scientific principles. There are not clear reasons, moreover, why such a system would be necessarily valid. I am afraid that underlying this kind of proposal is the conviction that science and reason need be identified. The formulation of a new system of ethical principles is certainly one of the major tasks ahead for man: very urgent one too, for, as Monod states, the alternative would be the collapse of our civilization. But I believe that such formulation should be left not only to the scientists, much as their participation would be needed. The task is an essential part of the programme we shall attempt to outline in the last section of this book. But, in the meantime, what should we do?

A large number of replies by individual scientists or by various groups and associations for the social responsibility of scientists have been given, and it would be impossible to offer a complete and balanced review. One aspect of this extensive literature is striking: no serious attempt has been made to formulate a precise code of conduct; in very few instances have individual scientists refused to continue their activities when they felt that their research was morally objectionable; even more rare are the examples of scientists who refuse financial support from objectionable sources; and few are the cases where serious attempts have been made to include courses on the human consequences of our actions in the curricula of science and engineering studies. In a few words, it seems to me that the vast majority of the scientific community of the world does hardly pay attention to such issues, and a small minority play lip service to humane principles but act otherwise. As an example of the prevailing spirit I shall quote only a few sentences from a recent article, proposing the establishment of a 'science society council' for the United Kingdom:[2]

1. A. Rapoport, 'Scientific Approach to Ethics', *Science*, Vol. 125, 1957, p. 796–9.
2. P. Sieghart *et al.*, 'The Social Obligations of the Scientist', *Nature*, Vol. 239, September 1972, p. 15–18.

The first thing to say is that we came to no radical conclusions. Those which have been proposed fall into three broad groups. First, that there should be something in the nature of a Hippocratic Oath or code of ethics for all scientists, whereby they bind themselves not to take part in work which will have socially harmful consequences; second, given that we will have an increasingly science based civilization, that moral and political decisions about the social application of scientific work should be left increasingly to scientists themselves; and third, that there should be radical—if not revolutionary—reform of the whole social system. Of these we think the first impracticable, the second dangerous, and the third beyond our competence.

There then follows the reasons why they think so and the proposal:

We think that this situation could be much improved if there were to be brought into existence a body, organized by the scientific community itself and expressly charged with the task of informing the public in general, and the organs of government in particular, at the earliest possible time, of all scientific work likely to have important social consequences for good or ill.

I hope I am wrong, but I fear that the dangers of collapse of our world are too great and immediate to be met only by the mild measures suggested. As mentioned at the beginning of this chapter, I believe that a substantial group of scientists of the world should abandon for a while their laboratories and concentrate their efforts to elaborate a novel conception of their role in society, to enter into a profound and systematic discussion of the fundamentals of a new ethics and to identify urgent actions to be taken to dispel the most immediate risks. Among these the danger of nuclear war and of a further expansion of the arsenal of horrors stands paramount. The time is ripe for a serious discussion of the problem and for summoning the members of the scientific community to take a stand. As a starting point I wish to propose for discussion the following 'theorem', admittedly somewhat utopian: but I am afraid that time is running out and that we must adopt extreme positions.

1. Science is universal, in the sense that its discoveries are equally valid under the most diverse political dispensations; science knows no national boundaries.
2. Scientific discoveries lead, sooner or later, to applications of various types, currently indicated as technologies, the tenet of Francis Bacon that 'the enlarging of the bounds of the Human Empire, to the effecting of all things possible' has come true.
3. Even the most humane technologies may become dangerous when applied indiscriminately (for example, antibiotics and other medical advances led to the population explosion, as shown in Part I, Chapter 5); furthermore, technologies are often used for exploitation (see Part I, Chapter 5) and for destruction (see Part I, Chapter 4).
4. Careless technologies, exploitation and destruction are made possible because men (scientists included) are loyal to national States, to parochial political ideologies, or to professional clans.
5. Further expansion of science is incompatible with human survival because science is universal, while *de facto* scientific activities (and especially technological developments) are sponsored and carried out to satisfy national or private ambitions.
6. It follows that further development of science is incompatible with the existence of nation States.
7. Mankind must, accordingly, choose either to stop the development of scientific activities or to eliminate nation States.
8. Scientists, personifying the imperishable values of science as an essential part of the culture of modern man, must refuse support for their research activities, unless it comes from genuine international agencies, and become missionaries for the establishment of world government.

A new meaning to progress

'The ethic of the scientific revolutions', as Lewis S. Feuer pointed out,[1] was that of an optimistic, expansive view of human life. It was filled with the conviction that science would enhance human happiness. It had confi-

1. L. S. Feuer, *The Scientific Intellectual*, New York, Basic Books, 1963.

dence in the human estate and in the aims and possibilities of human knowledge. It proposed to alleviate drudgery, and to transform work from an eternal curse to a human joy. It aspired, in its reading of the book of nature, to abrogate the tired dictum of Ecclesiastes: 'Knowledge increaseth sorrow'. That ethic proved correct: superstitions were superseded by rational explanation; physical pain could be diminished and even abolished; working hours were shortened and labour became less strenuous and severe; death rates diminished and life spans were extended; living standards of Western societies improved and social securities offered a certain amount of serenity to the workers; slavery was abolished and the longing for social justice found satisfaction over vast regions; individual liberty of choice and action became greater for men and women; personal movement, sexual freedom and access to culture became easier and responsible parenthood could be planned; and the frontiers of human knowledge rapidly expanded. But, as we have seen, the extraordinary success obtained by science has made us lose a balanced perspective on life. Modern man has put an identity sign between technological advance and progress and thus his primary concern is no longer with the human race. If we accept the definition of progress as 'the end point, temporary or permanent, of any social action that leads from a less to a more satisfactory solution of the problem of man in society , we must admit that the institutionalization of science can hardly be identified with progress, particularly in recent times. The questions to be asked now is not 'if not reason, what?'—as some commentators of the present predicament have done—but rather 'is it reasonable to identify reason with science and scientific advance with progress?'

The answer is obviously no, for science represents only one approach to knowledge of the world, and scientific advance may contribute to man's progress; but, once more, scientific progress must be considered only as a component of a more complex total situation.

The idea of progress is peculiar to the modern world (as it has often been pointed out), the world that was the product of the scientific revolution and of the Enlightenment. In 1750

Anne Robert Jacques Turgot could state at the Sorbonne:

Manners are gradually softened, the human mind is enlightened, separate nations draw nearer to each other, commerce and policy connect at last every part of the globe, and the total mass of the human race, by alternating between calm and agitation, good and bad, marches always, however slowly, towards greater perfection.

And in 1793 Marie Jean Antoine Nicolas Caritat Marquis de Condorcet, in his 'Sketch for a Historical Picture of the Progress of the Human Mind', showed

that nature has assigned no limit to the perfecting of the human faculties, that the perfectibility of man is truly indefinite; that the progress of this perfectibility, henceforth independent of any power that might wish to arrest it, has no other limit than the duration of the globe on which nature has placed us.

Such faith in the perfectibility of the human condition is still inherent in the prevailing attitude of man, two hundred years later; but we begin to realize that science is not enough to guarantee the progress of man and that there lies a fallacy in the technocratic argument which confuses material and moral progress.

Within the scientific community one can notice now an increasing awareness that modern societies face a certain number of problems—such as nuclear war, population explosion and environmental decay—which represent major obstacles to the progress of the human race and do not find their solutions in science and technology.

Our options today are limited in number: we might abjure our faith in reason and follow the 'Old Gnosis,' which would give us visionary powers, primordial energies, sacramental awareness, adventures of spiritual regeneration, organic wholeness and similar sombre evasions;[2] we might continue to believe that science is the only source of salvation and that more science and technology can cure the ills

1. L. Sklair, *The Sociology of Progress*, London, Routledge and Kegan Paul, 1970.
2. T. Roszak, *Where the Wasteland ends—Politics and Transcendence in Postindustrial Society*, Garden City, N.Y., Doubleday, 1972.

of today's society, hiding to ourselves the ominous signs of impending catastrophes; or we might humbly recognize that the problems at hand are larger than our ability to understand, that we must widen the limits of our concern much beyond strictly scientific issues, and that, in co-operation with sociologists, philosophers and other thinkers, we must attempt to find a new meaning for progress. Obviously, I consider only the latter option as viable, for it reaffirms the hope of progress and opens wide horizons for new adventures of the intellect.

In his book on *The Sociology of Progress*,[1] Leslie Sklair has made a distinction between innovational and non-innovational progress: the former is 'progress by means of the production of new things, ideas and processes, with maximum impact on society' (through the institutionalization of invention and discovery); the latter is 'progress by means of the maintenance and diffusion of familiar things, ideas and processes, with minimal impact on society'; the term 'impact' being used 'in a special sense to signify the effect that the different types of progress have on social structures'. It may be that, after many decades of frantic innovative progress, reason would advise us to enter a period of non-innovative progress or of profound social rather than technological innovations. At this time, of course, none of us seems to know which course we should take. For this reason I think that the moment has come for launching a far-reaching programme of research on our own nature, on our authentic needs—probably quite different from the fictitious ones that have led us astray in recent times—and to identify the basic components of the human design. Historically, over the last three centuries, man has chosen to venture upon the untrodden tracks of science, entering first the ones that seemed easy to scout. Thus, with the support of mathematics, physics developed first, then chemistry, biology and the earth sciences. Man has remained isolated, as it were, from nature. Through separation of the observing subject from the studied object, and of facts from values, science has scored its triumphs but lost its own control, for scientists have become bureaucrats. 'Atomic

physics has been manipulated by the very blind and uncertain forces which command and dispute over our historical societies,' Edgar Morin comments,[2] 'Biology in its turn will be manipulated and even anthropology, when this will become a real science; it will be corrupted in even more dangerous ways.' *Ainsi c'est de façon dramatique, incertaine, aléatoire que se pose aujourd'hui le problème de la nature de l'homme, de l'unité de l'homme, de la nature de la société*, the French sociologist continues.

The vision of the advent of a *Scienza nuova*, of a science of man and for man, as the one so recently outlined by Edgar Morin, is probably the harbinger of a major turning in the thought and attitude of modern man. As the *philosophes* of the early eighteenth century took it upon themselves to criticize the conditions of society in preceding times and laid the intellectual foundations of the new civilization which transformed the whole world, similarly we must engage in a drastic examination and reorientation of our ways of thinking, of our priorities, of our scientific, social, economic, ethic and political structures—leaving aside the ones that our predecessors have elaborated and which are now obsolete. The *philosophes* were not philosophers in the traditional sense of the word but rather a group of assorted intellects, ranging in their prevailing interests from mathematics to politics and from biology to moral sociology, and recruited from the aristocracy or the humble classes. If the complexity of the problems of that time required the convergence of a wide range of intellectual skills, the hyper-complexity of today's world calls for the concentration of the endeavours of a vast variety of specializations coping with the whole gamut of the innumerable, intertwined problems of today. To my mind, interesting and worthwhile as they are, the current attempts to develop a 'critical science'[3] or to develop methodologies for technological assessment are much too limited in scope, for

1. Sklair, op. cit.
2. E. Morin, *Le Paradigme Perdu: la Nature Humaine*, Paris, Ed. du Seuil, 1973.
3. J. R. Ravetz, *Scientific Knowledge and its Social Problems*, Oxford, Clarendon Press, 1971.

they seem to aim at minimizing the untoward effects of the scientific enterprise as we know it today, rather than to examine it under a novel perspective, leading perhaps to a radical criticism of its structure and function.

There is, I think, a real and desperate need for a major daring advance in pure thought. Up to now, diagnoses on the predicament of modern man and his societies have been too sporadic and individualistic, leading too often to irrational despair. The time has come to take a fresh approach towards the establishment of an unprecedented community of scholars—a novel, as it were, outward-looking ivory tower. Both our confidence in reason and our courage will warrant its success.

List of illustrations

List of tables

Index of major international bodies

Index of laws, models, configurations (by name)

Subject index

Errata

Page 112, left-hand column, line 20: *for* smthods *read* methods

line 21: *for* destances *read* distances

line 22: *for* tiations *read* stations

Page 135, right-hand column, line 16: *for* rings', *read* rings,

line 17: *for* shape *read* shape,

line 25: *for* pustons *read* electrons

line 27: *for* interesting *read* intersecting

Page 158, right-hand column, lines 24 and 25: *for* Van der Waal *read* Van der Waals

Page 188, left-hand column, line 2: *for* polynucletide *read* polynucleotide

Page 217, right-hand column, line 20: *for* Infaction *read* Infection

Page 237, under 'Further Reading': *for* Brochet *read* Brachet

Page 261, right-hand column, line 25: *for* scyzophrenic *read* schizophrenic

Page 314, left-hand column, line 29: *for* cach *read* each

Page 323, right-hand column, line 14 from bottom: *for* 1836 *read* 1886